James Palmer

jamesrichardpalmer@gmail.com

814-777-3422

COMPOSITES FOR CONSTRUCTION

COMPOSITES FOR CONSTRUCTION:

Structural Design with FRP Materials

Lawrence C. Bank

JOHN WILEY & SONS, INC.

This book is printed on acid-free paper. ♾

Published by John Wiley & Sons, Inc., Hoboken, New Jersey
Published simultaneously in Canada

For general information on our other products and services or for technical support, please contact our Customer Care Department within the United States at (800) 762-2974, outside the United States at (317) 572-3993 or fax (317) 572-4002.

Wiley also publishes its books in a variety of electronic formats. Some content that appears in print may not be available in electronic books. For more information about Wiley products, visit our web site at www.wiley.com.

Library of Congress Cataloging-in-Publication Data:

Bank, Lawrence Colin, 1956–
Composites for construction : structural design with FRP materials / Lawrence C. Bank.
p. cm.
Includes index.
ISBN-13: 978-0471-68126-7
ISBN-10: 0-471-68126-1 (cloth)
1. Fiber reinforced plastics—Textbooks. 2. Polymeric composites—Textbooks. 3. Structural design—Textbooks. I. Title.
TA455.P55B36 2006
624.1′892—dc22

2005030950

10 9 8 7 6 5 4 3 2 1

CONTENTS

PREFACE

Fiber-reinforced polymer (FRP) composite materials have developed into economically and structurally viable construction materials for buildings and bridges over the last 20 years. FRP composite materials used in structural engineering typically consist of glass, carbon, or aramid fibers encased in a matrix of epoxy, polyester, vinylester, or phenolic thermosetting resins that have fiber concentrations greater than 30% by volume. They have been used in structural engineering in a variety of forms: from structural profiles to internal reinforcing bars for concrete members to strips and sheets for external strengthening of concrete and other structures. Depending on the form of the FRP product used in structural engineering, the FRP material is supplied either as a ready-to-use structural component such as a wide-flange profile or a reinforcing bar, or it is supplied in its constituent forms as dry fiber and liquid polymer resin and formed and cured in situ to create a structural component. These two forms should be familiar to structural engineers, as they have analogs in conventional structural materials such as steel beams or steel reinforcing bars which are supplied in ready-to-use form from a steel mill, or portland cement concrete, which is supplied in the form of cement, aggregate, and water constituents and is formed in situ to create a structural element.

The purpose of *Composites for Construction* is to provide structural engineering students, educators, and professionals with a code-based text that gives detailed design procedures for FRP composites for civil engineering structures. The emphasis of the book is on the design of structural members and structural systems that use FRP composites as one or all of the structural materials in the structure. The emphasis of the book is not on the design of the FRP composite materials themselves, and the book provides only a brief review of topics related to constituent materials, micro- and macromechanics of composite materials, and manufacturing methods of composite materials. It is important to emphasize that this book is self-contained and that no prior knowledge of FRP composite materials is required to use the book and to learn how to design with FRP composites in structural engineering. Fiber-reinforced composite materials have been used for many decades in the aerospace, automotive, and the industrial and recreational products industries. Many excellent textbooks and reference books are available that cover the manufacturing, mechanics, and design of fiber composite materials. Nor does the book cover topics of structural analysis or structural mechanics, except where needed to explain design procedures or design philosophy.

Composites for Construction is intended primarily for use as a college-level text in civil and structural engineering curricula. It is intended for senior-level (fourth-year) undergraduate students in civil engineering programs or first-year graduate students in structural engineering programs and for their instructors. Users of the book will find that its form is similar to that of traditional structural engineering design textbooks used to teach subjects such as steel design, reinforced concrete design, and wood design. The book is intended to be covered in a one-semester three-credit lecture-style course on FRP composites in structural engineering. Alternatively, the book can be used to supplement course material in courses on reinforced concrete design, steel design, or wood design since many of the topics covered have parallels to analytical and design methods in these subjects and are logical extensions of the methods used in these subjects. When used for a stand-alone course in a civil or structural engineering program, it is expected that students will have had at least one course in structural analysis, one course in reinforced concrete design, and one course in structural steel design (or design of wood structures). If students have the appropriate background, the book can also be used in architectural engineering or in construction engineering curricula.

Composites for Construction is divided into four parts. The first part provides an introduction to FRP applications, products, and properties and to the methods of obtaining the characteristic properties of FRP materials for use in structural design. The second part covers the design of concrete structural members reinforced with FRP reinforcing bars. The third part covers the design of FRP strengthening systems such as strips, sheets, and fabrics for upgrading the strength and ductility of reinforced concrete structural members. The fourth part covers the design of trusses and frames made entirely of FRP structural profiles produced by the pultrusion process. From a mechanics point of view, the type of FRP material examined in the three design parts of the book increases in complexity from a one-dimensional FRP reinforcing bar to a two-dimensional thin FRP strip or plate to a three-dimensional thin-walled FRP profile section. As the geometric complexity of the FRP component increases, so does its anisotropy and hence so does the number of properties that need to be considered in the structural design.

The format for each of the three design parts of *Composites for Construction* is similar. It starts with a discussion of the design basis and material properties used for the specific application of the FRP material in the design considered in that part. This is followed by sections related to design of specific types of structural members and structural systems (such as beams and columns or trusses and frames) that are unique to the different types of FRP materials examined in each part. In each chapter, examples of the design of typical members and structures are provided in a step-by-step, annotated format that enables the reader to follow the design processes. Each chapter concludes with a set of recommended homework problems. Each design part of the book also discusses important construction- and constructability-related aspects of FRP composites in structures. FRP materials are generally new to structural designers, and most designers have not yet inculcated an intuitive understanding of their behavior as they have for traditional materials such as concrete, steel, and wood. It is therefore important to understand how FRP materials are erected or installed or applied in the field and how the construction process can influence the design process.

As noted, *Composites for Construction* is code-based, which means that it provides design procedures in accordance with published structural engineering design codes, guides, and specifications. The discussion of FRP reinforcing bars and FRP strengthening systems follows a load and resistance factor design basis and presents design procedures for FRP materials used in combination with reinforced concrete according to the most recent editions of the design guides published by the American Concrete Institute: ACI 440.1R-06, *Guide for the Design and Construction of Structural Concrete Reinforced with FRP Bars,* and ACI 440.2R-02, *Guide for Design and Construction of Externally Bonded FRP Systems for Strengthening Concrete Structures.* The final part of the book provides design guidance for FRP profile sections according to a combination of recommendations of the ASCE *Structural Plastics Design Manual,* the *Eurocomp Design Code and Handbook,* the *AASHTO Specifications,* and manufacturer-published design guides. Procedures are provided for the structural designer on how to use this combination of codelike documents to design with FRP profiles.

When *Composites for Construction* is used as a textbook for a one-semester (15-week) course, the following allocation of time is recommended: Part One, three weeks; Part Two, four weeks; Part Three, four weeks; Part Four, four weeks. *Composites for Construction* is based on course notes developed by the author while teaching this course over a number of years at the University of Wisconsin–Madison and at Stanford University (while on sabbatical in 2003–2004). Based on the author's experience in teaching this course in recent years, a number of more advanced topics will need to be omitted, depending on the students' background, to be able to cover all the material in a one-semester course. This includes an in-depth treatment of micromechanics and lamination theory (in Chapter 3), FRP bars for bridge decks (in Chapter 5), load–deflection response in FRP-strengthened structures (in Chapter 9), FRP shear strengthening of columns (in Chapter 10), FRP confinement for ductility enhancement in columns (in Chapter 11), FRP profile beam-columns (Chapter 14), and combined FRP flexural–tension members (in Chapter 14). A midterm examination covering Parts One and Two and a final examination covering Parts Three and Four is recommended. In addition, individual or group design projects that run the length of the entire semester are recommended.

The imperial *inch–pound–second* system of units is used throughout the book, as these units are commonly used in U.S. design practice. SI (System International) and metric *cm–g–s* units are used occasionally where standard practice is to report material properties in these units (even in the United States). However, a complete SI version of the design examples and problems is not included at this time, in the interest of brevity.

Composites for Construction focuses on the mainstream application areas of FRP composites in structural engineering at the time of writing. Over the years there have been a number of other applications and uses of FRP composites in structural engineering. In many cases, code-based guidance has only recently been developed or is currently being developed for these applications. The applications include FRP tendons for internal or external prestressing of concrete; FRP stay cables for bridges or guy wires for towers; FRP grids, meshes, and gratings for reinforcing concrete; FRP stay-in-place forms for concrete beams, slabs, or columns; FRP strengthening of prestressed concrete structures; FRP strengthening of masonry structures; FRP strengthening of steel, aluminum, or timber structures; mechanically fastened FRP strengthening systems; FRP pretensioned sheets for strengthening; and FRP strengthening for blast loads on structures.

ACKNOWLEDGMENTS

Composites for Construction is based on notes I developed for a three-credit lecture course taught at the University of Wisconsin–Madison and at Stanford University, and on notes developed for short courses taught through the Department of Engineering Professional Development at the University of Wisconsin and at the University of Arizona. I am indebted to all the students who attended my courses at the University of Wisconsin and at Stanford University who provided both feedback and motivation to improve the early drafts of the book and helped to provide critical insights for improving and refining the examples and homework problems in the text.

I am thankful for the advice, encouragement, and support that I have received from many of my colleagues and friends in the FRP composites for construction community. I extend my special thanks to Toby Mottram, Antonio Nanni, Carol Shield, Joseph Hanus, Brahim Benmokrane, Kenneth Neale, Sami Rizkalla, Urs Meier, Charles Bakis, Jin-Guang Teng, Jian-Fei Chen, Jack Lesko, Thomas Keller, and Edward Nawy. The support of my colleagues in industry who were always forthcoming with technical data and information was also critical to the development of the book. I would like to thank Doug Gremel (Hughes Brothers), Glenn Barefoot (Strongwell), Rick Johansen (ET Techtonics), Dustin Troutman (Creative Pultrusions), Jay Thomas (Structural Preservation Systems), Ben Bogner (BP Chemical), Dave White (Sika Corporation), and Pete Milligan (Fyfe Company) for their willingness to provide information.

Composites for Construction would not have been possible without the support of my colleagues at the University of Wisconsin–Madison. In particular, I would like to thank Jeffrey Russell for his constant guidance and feedback, and Michael Plesha for his insightful reflections on book writing. Sincere thanks also go to the Structures and Geomechanics Group of the Civil and Environmental Engineering Department at Stanford University who hosted me as a UPS Visiting Professor at Stanford University while on sabbatical from the University of Wisconsin in 2003–2004. It was during this year that much of the manuscript was completed. In particular, I would like to thank Kincho Law of Stanford University for his support over the past 20 years. My thanks also go out to colleagues who have worked with me over the years, especially Timothy Kao, Russell Gentry, and Terry Gerhardt, and to Jim Harper at John Wiley & Sons for his understanding and guidance.

Finally, this book would never have been written without the constant, continuing, and unwavering support of my wife, Rebecca, daughter, Anna, and son, Jacob. It was their persistence, and their faith in me, that made this book a reality.

LARRY BANK

Madison, Wisconsin
January 2006

1 Introduction

1.1 OVERVIEW

Over the last decade there has been significant growth in the use of *fiber-reinforced polymer* (FRP) *composite materials* as construction materials in structural engineering, which for the purposes of this book is defined as the field of engineering that covers the analysis and design of primary, load-bearing structural members and systems in buildings and bridges by civil and structural engineers. Also known as *fiber-reinforced plastics,* or *advanced composite materials* (ACMs), these materials have proven themselves to be valuable for use in the construction of new buildings and bridges and for the upgrading of existing buildings and bridges. As these materials have transitioned from the research laboratory to implementation in actual structures, so have codes and specifications developed for them for use in civil engineering structural design in the last decade. Now, at the beginning of the twenty-first century, the structural engineering community is about to enter a stage in which structural design with FRP composites is poised to become as routine as structural design with classical structural materials such as masonry, wood, steel, and concrete.

In this chapter we review the historical development of FRP composite materials for use in structural engineering. The review and design chapters that follow focus on the use of FRP composite materials in three primary areas: (1) reinforcements for new concrete structural members, (2) strengthening for existing structural members, and (3) profiles for new structures. In what follows, code-based design procedures are provided for these three topics.

The historical review concentrates on actual engineering structures that have been designed by professional engineers and constructed using construction techniques and technologies used routinely in the architecture, engineering, and construction (AEC) industry. The review is intended to give real-world examples of the use of FRP composite materials to construct structures, or important elements of structures, that have traditionally been constructed from conventional construction materials. In addition to the foregoing three topics, other applications in large-scale civil infrastructure engineering, such as electric power transmission-line towers, luminaries supports, masonry strengthening, and timber strengthening, are described, even though code-based design procedures have not been developed at this time for these applications. This is to provide evidence of the feasibility of constructing safe and reliable structures that have a direct bearing on the construction of buildings and bridges and as an indication of future applications for which code guidance is currently being developed. The historical development of advanced composite materials and their widespread applications in the aerospace, automotive, industrial, and recreational products, although of great general technological interest, is not of specific relevance to buildings and bridges and is not included in this review.

Following the review of the building and bridge structures that have been constructed using FRP composite materials, a survey of the key mechanical and physical properties of representative FRP composite material products that are currently available is presented.

1.2 HISTORICAL BACKGROUND

FRP composites have been used on a limited basis in structural engineering for almost 50 years for both new construction and for repair and rehabilitation of existing structures. Structural and civil engineers have been affixing their professional stamps to designs for buildings and bridges for many years, even though these materials have not been recognized by official building codes, and no code-approved design procedures have existed until very recently. These forward-thinking engineers have tended to be people who have specialized expertise in the use of FRP composites in structural engineering or who were in-house structural engineers directly affiliated with a manufacturer of FRP components. In addition to being registered structural engineers involved in engineering practice, many of these engineers have had fundamental knowledge of the materials, manufacturing methods, and fabrication and installation methods for FRP composite in civil engineering structures. Many have worked on teams to develop new FRP components for civil engineering structures (Bank, 1993b). Since the mid-1990s, however, other structural engineers and architects have begun to design with FRP composites on a fairly routine basis. In general, these engineers and architects have not had specialized training or exposure to FRP composites as construction materials. These designs are completed with the aid of published design procedures or by proof testing, and often, with the aid of an in-house engineer from an FRP product manufacturer, who advises on the details of the design and provides sample specifications for the FRP material for the contract documents. Many structural engineers of this type have also affixed their professional seals to these designs over the past decade.

The historical review provides selected examples of where FRP composites have been used in building and bridges in the past half-century. More attention is paid to the applications from the 1990s, which were designed in a routine fashion by structural engineers, as opposed to those from before the 1990s, which were generally designed by engineers with a specialized knowledge of composites. The review describes applications that used FRP components and products available at the time of their manufacture. Many FRP products used in these projects are no longer produced or have been replaced by improved parts and products.

The state of the art of the early work, from 1980 to 1990, in the area of FRP composites for reinforcing and retrofitting of concrete structures in the United States, Japan, Canada, and Europe is detailed in collections of papers and reports edited by Iyer and Sen (1991) and Nanni (1993b). In 1993, a series of biannual international symposia devoted to FRP reinforcement of concrete structures was initiated.[1] At about this time, international research interest in the use of FRP in concrete increased dramatically. Collections of papers on the use of FRP profile sections in structures have been published by the American Society of Civil Engineers (ASCE) since the early 1980s by the now-disbanded Structural Plastics Research Council (SPRC) and in proceedings of the ASCE Materials Congresses.[2] In 1997, the American Society of Civil Engineers (ASCE) founded the *Journal of Composites for Construction,* which today is the main international archive for reporting on research and development in the field of FRP materials for the AEC industry.[3] In 2003, the International Institute for FRP in Construction (IIFC) was established in Hong Kong. To date, thousands of research studies and structural engineering projects using FRP materials have been reported worldwide. Reviews of developments in the field from 1990 to 2000 can be found in ACI (1996), Hollaway and Head (2001), Teng et al. (2001), Bakis et al. (2002), Hollaway (2003), Van Den Einde et al. (2003), and Täjlsten (2004).

1.3 FRP REINFORCEMENTS FOR NEW CONCRETE STRUCTURAL MEMBERS

FRP reinforcements for new concrete structural members have been of interest to structural engineers since the earliest days of the FRP composites industry. In 1954, in his seminal paper on the development of pultrusion technology, Brandt Goldsworthy speculated that "the chemical inertness of this material allows its use in . . . concrete reinforcing and all types of structural members that are subject to corrosive action in chemical plants or other areas where corrosive conditions exist" (Goldsworthy, 1954). Since then, engineers have endeavored to find ways to realize this dream in a number of ways, and at

[1] Known as FRPRCS: fiber-reinforced polymers in reinforced concrete structures. They have been held in Vancouver (1993), Ghent (1995), Sapporo (1997), Baltimore (1999), Cambridge (2001), Singapore (2003), Kansas City (2005), and Patras (2007).

[2] Sponsored by the ASCE Materials Engineering Division (now incorporated into the ASCE Construction Institute) and held in 1990 (Denver), 1992 (Atlanta), 1994 (San Diego), and 1999 (Cincinnati).

[3] Additional international symposia and journal series focused on FRP composites in structural engineering of note are listed at the end of this chapter.

present, FRP reinforcements for new concrete structural members can be divided into three primary areas: (1) FRP bars or grids for reinforced concrete (RC) members, (2) FRP tendons for prestressed concrete (PC) members, and (3) stay-in-place FRP formwork for reinforced concrete members.

1.3.1 FRP Bars or Grids for Reinforced Concrete Members

The use of FRP reinforcing bars and grids for concrete is a growing segment of the application of FRP composites in structural engineering for new construction. From 1950s to the 1970s, a small number of feasibility studies were conducted to investigate the use of small-diameter ($\sim\frac{1}{4}$-in. or 6-mm) glass FRP rods, with and without surface deformations, to reinforce or prestress concrete structural members (Nawy et al., 1971; Nawy and Neuwerth, 1977). In the early 1980s, glass helical–strand deformed reinforcing bars were produced for structural engineering applications by Vega Technologies, Inc. of Marshall, Arkansas (Pleimann, 1991). These bars were used to build magnetic resonance imaging facilities, due to their electromagnetic transparency. At the time, these FRP bars were cost-competitive with stainless steel bars, which were the only other alternative for this application. Many single-story cast-in-place wall and slab structures, which look identical to conventional reinforced concrete structures, were built. Designs were performed by registered structural engineers using the working stress design basis, and the buildings were constructed using conventional construction technology.

In the late 1980s, interest in the use of FRP rebars received a boost as attention focused on ways to mitigate corrosion in steel-reinforced concrete structures exposed to the elements, especially highway bridge decks. In the United States, International Grating, Inc. developed a sand-coated glass fiber–reinforced bar that was used experimentally in a number of bridge deck projects. This was followed in the 1990s by the development of deformed FRP bars by Marshall Composites, Inc. A number of companies experimented with FRP bars with helically wound spiral outer surfaces. These included Creative Pultrusions, Glasforms, Vega Technologies, International Grating, Hughes Brothers, and Pultrall.[4] In the late 1990s a number of these producers also experimented with carbon fiber FRP bars with deformed, helically wound, and sand-coated surfaces. Extensive research was conducted in the 1990s on the behavior of concrete beams and slabs reinforced with various types of FRP bars (Daniali, 1990; Faza and GangaRao, 1990, 1993; Nanni, 1993a; Benmokrane et al., 1996a,b; Michaluk et al., 1998). Studies were also conducted on the use of glass fiber pultruded FRP gratings in reinforced concrete slabs (Bank and Xi, 1993). ACI published its first design guide, ACI 440.1R-01, for FRP-reinforced concrete in 2001 (ACI, 2001). The guide was subse-

[4] Many of the companies mentioned no longer exist or have ceased production of these products. The information is provided for historical purposes.

quently revised in 2003 (ACI 440.1R-03), and the current version, ACI 440.1R-06, was published in 2006 (ACI, 2006).[5] Figure 1.1 shows samples of a number of commercially produced glass- and carbon-reinforced FRP bars for concrete reinforcing.

At the same time that glass fiber FRP bars were being developed for reinforcing concrete in corrosive environments in the United States, a parallel effort was under way in Japan. The Japanese effort focused, to a large extent, on carbon fiber reinforcement, due to concern for the degradation of glass fibers in alkaline environments. A two-dimensional grid product called NEFMAC (new fiber composite material for reinforced concrete) was developed successfully and commercialized by Shimizu Corporation (Fujisaki et al., 1987; Fukuyama, 1999). Grids with carbon fibers, glass fibers, and carbon–glass hybrid fibers were produced. The majority of applications of NEFMAC appear to have been in tunnels, where the noncorrosive properties and panel sizes of the grids were very effective when used with shotcreteing methods. Design guides for reinforced concrete design with FRP bars were published in Japan in 1995 and 1997 (BRI, 1995; JSCE, 1997). Efforts were also directed to developing polyvinyl alcohol (PVA) FRP bars, but these were never commercialized. Figure 1.2 shows carbon and glass NEFMAC grids developed in Japan in the 1980s.

Today, FRP reinforcing bars for concrete with both glass and carbon fibers are produced by a number of companies in North America, Asia, and Europe. The use of FRP bars has become mainstream and is no longer confined to demonstration projects. However, this is still a niche area in structural engineering; competition among manufacturers is fairly fierce and only the strong

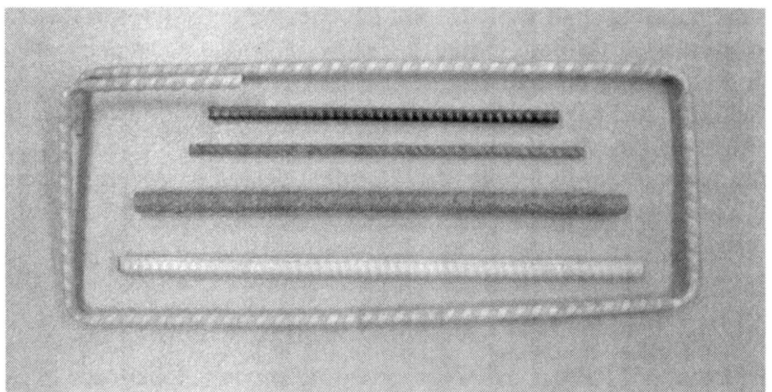

Figure 1.1 Glass- and carbon-reinforced FRP bars.

[5]This book provides design procedures in accordance with the 2006 version of ACI 440.1R-06 and should not be used with older versions. Substantial changes related to ASCE 7-02 load and resistance factors were introduced in 2006.

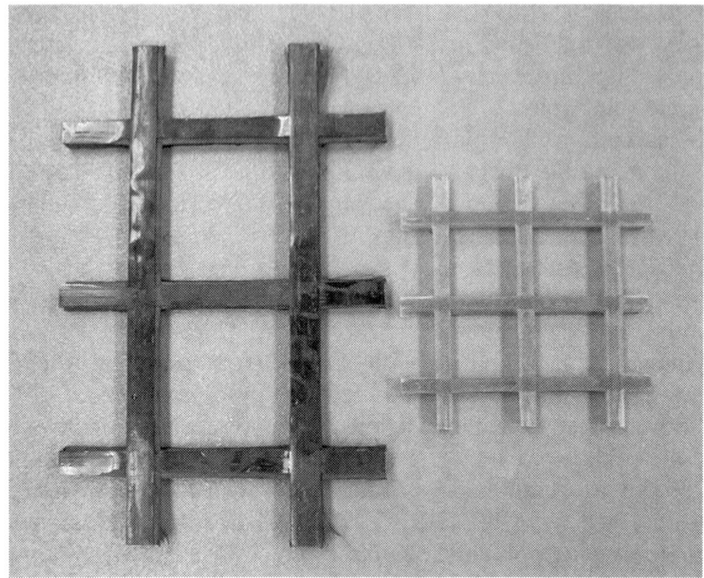

Figure 1.2 NEFMAC FRP grids.

companies are surviving. Applications have become routine for certain specialized environments. Most current applications appear to be in underground tunnels and in bridge decks. Figure 1.3 shows a large glass fiber FRP reinforcing cage being readied for placement in forms for a tunneling application. The FRP bars are used to create a "soft eye" for subsequent drilling through a vertical concrete shaft wall. These FRP bars are lap spliced to the steel bars in a conventional fashion. Figure 1.4 shows a bridge deck slab with a glass FRP top mat being poured.

In addition to FRP bars for reinforcing concrete, FRP dowel bars for concrete highway pavements and FRP ground anchors have also been commercialized successfully in recent years. Explicit design procedures according to the ACI are given in this book for the design of concrete structures reinforced with FRP bars.

1.3.2 FRP Tendons for Prestressed Concrete Members

Development of FRP tendons for prestressing concrete took place in Holland, Germany, and Japan in the early 1980s. This was motivated primarily by the desire to reduce corrosion in prestressed concrete elements. A rectangular aramid FRP strip called Arapree was developed in Holland by the Hollandsche Beton Groep (HBG) and Akzo, and a glass fiber tendon in an external polymer sheath called Polystal was developed by Strabag Bau AG and Bayer AG in Germany. These products were used in a number of bridge demon-

Figure 1.3 Reinforcing cage being readied for placement in forms for a tunneling application. (Courtesy of Hughes Brothers.)

Figure 1.4 Bridge deck slab with a glass FRP top mat being poured. (Courtesy of Hughes Brothers.)

stration projects in Europe, but their production has been discontinued. Arapree is now produced by Sireg in Italy. In the 1980s in Japan, a national project was undertaken to develop FRP reinforcements for concrete that focused primarily on tendons (Fukuyama, 1999). Many different aramid and carbon fiber products were developed and over 50 demonstration projects were completed. The Advanced Composite Cable Club (ACC Club) was founded to coordinate and commercialize FRP prestressing products in Japan. These included twisted-strand carbon tendons called carbon fiber composite cable (CFCC), known as Tokyo Rope, by Tokyo Seiko K.K.; aramid fiber rods and strips, called Arapree (under license from HBG), by Nippon Aramid Co.; aramid fiber rods, called Technora Rods, by Teijin Ltd.; aramid fiber tendons, called FiBRA, by Kobe Steel Cable; and carbon fiber tendons, called Leadline, by Mitsubishi Chemical Corporation. As conventional steel chucks and anchors could not be used, due to the low transverse strength of FRP tendons, all the manufacturers developed specialized prestressing anchors. This has proved to be the Achilles heel of the FRP prestressing industry. Both technical difficulties with the anchors and their high prices made them unattractive to the construction industry. Through the 1990s a number of demonstration projects were conducted in the United States and Canada using Japanese and European FRP prestressing products, but routine use of these FRP tendons has not occurred. These FRP rods have also been used as conventional reinforcing bars in a number of demonstration projects.

An FRP tendon and anchorage system has not been commercialized in the United States. The design of FRP pretressed concrete is discussed in the Japanese guide for FRP reinforcements (JSCE, 1997). A guide for the design of concrete structures with FRP prestressing tendons, ACI 440.4R-04, has been published (ACI, 2004b). FRP prestessing products produced in Japan in the 1980s and 1990s are shown in Fig. 1.5.

1.3.3 Stay-in-Place FRP Formwork for Reinforced Concrete Members

The use of FRP composites as stay-in-place (SIP) formwork has been explored for a number of years. FRP SIP formwork systems that act to reinforce the concrete after it has cured and systems used only as SIP forms have been developed. A stay-in-place bridge deck panel produced by Diversified Composites, Inc. in the United States has been used on two multispan highway bridges in Dayton, Ohio (1999) and Waupun, Wisconsin (2003). The deck form has a corrugated profile and is intended as a noncorroding substitute for stay-in-place metal and prestressed concrete deck forms. When used as a stay-in-place deck form, the FRP composite serves as the tensile reinforcement after the concrete has hardened. Figure 1.6 shows an FRP deck form being placed over concrete prestressed girders.

FRP tubular stay-in-place forms have also been used to manufacture beam and column members. They are also referred to as *concrete-filled FRP tubes.* A *carbon shell system* consisting of concrete-filled carbon FRP tubes has been

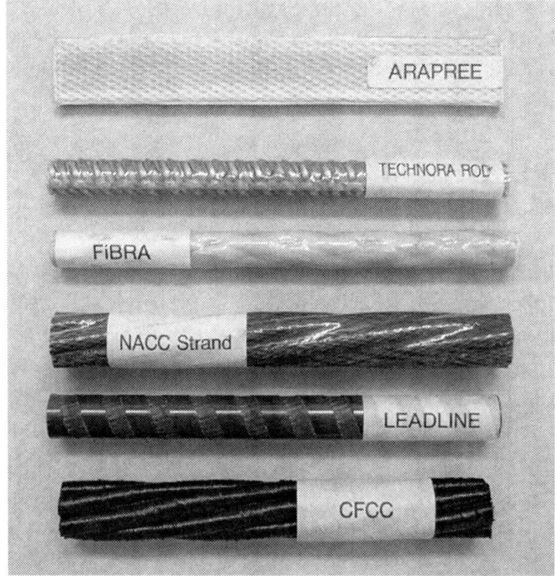

Figure 1.5 Japanese FRP prestressing tendons.

Figure 1.6 Placement of FRP stay-in-place forms for a bridge deck. (Courtesy of Adam Berg.)

used to produce girders for the superstructure of a highway bridge in California (Van Den Einde et al., 2003). Figure 1.7 shows FRP–concrete tubular piling manufactured by Lancaster Composites in the United States.

1.4 FRP STRENGTHENING OF EXISTING STRUCTURAL MEMBERS

FRP materials used to strengthen and repair load-bearing structural members are popular applications of FRP composites in structural engineering. Collectively, these applications are known as *retrofitting applications,* as they are used in existing structures and not in the construction of new structures. Retrofitting applications can be classified broadly into two types. One type is *strengthening,* where the original structure's strength or ductility (typically, its displacement capacity) is increased from the loads (or displacements) for which it was originally designed. This increase may be necessitated by the desire to make the structure compatible with existing building codes (particularly in the case of seismic retrofitting) or may be desired due to changes in use of the structure. FRP retrofitting to improve the performance (load carrying and ductility) of a structure when subjected to blast and impact loading has become of interest of late. The other type of FRP retrofitting can be classified as *repair.* In this case, the FRP composite is used to retrofit an existing and deteriorated structure to bring its load-carrying capacity or duc-

Figure 1.7 FRP piles. (Courtesy of Lancaster Composites.)

tility back to the loads or displacements for which it was designed (and hence is, in fact, a type of strengthening). Repair is necessitated when the original structure has deteriorated due to environmental effects, such as corrosion of steel reinforcing in concrete structures or when the original structure has been damaged in service or was not constructed according to the original design. For example, reinforcing bars may be omitted in a beam at the time of construction due to a design or construction error. Although these two types of applications are similar, there are important differences that are related primarily to evaluation of the existing structural capacity and the nature of the repair to be undertaken before FRP can be used. In many cases, a repair design will include strengthening to add a level of safety to the repaired structure and to account for uncertainty in the retrofit design.

FRP retrofitting has been used successfully on reinforced concrete structures, prestressed concrete structures, timber structures, and masonry and metal structures. At this time, code design guidance is only available for FRP retrofitting of reinforced concrete structures, particularly as applied to strengthening. Consequently, this historical review focuses primarily on the use of FRP retrofitting technologies for reinforced concrete and prestressed concrete structures.

Two primary methods are used to attach FRP composite materials to concrete structures (and to masonry, timber, and even metallic structures) for retrofitting purposes. One method employs premanufactured rigid[6] FRP strips [approximately 4 in. (100 mm) wide and $\frac{1}{16}$ in. (1.6 mm) thick] that are adhesively bonded to the surface of the structural member. The other method, known as *hand layup,* consists of in situ forming of the FRP composite on the surface of the structural member using flexible dry fiber fabrics or sheets of width approximately 6 to 60 in. (150 to 1500 mm) and liquid polymers. In recent years a new variant of the premanufactured strip method called *near surface mounting* (NSM) has been developed. In this method, a thin, narrow FRP strip [$\sim\frac{1}{8} \times \frac{3}{4}$ in. (3 by 18 mm)] or small-diameter round FRP bar [$\sim\frac{1}{4}$ in. (6 mm)] is inserted and then bonded adhesively into a machined groove at the surface of the concrete member.

FRP retrofitting has been used with bridge and building structures to strengthen static and quasistatic loads (such as increases in dead or live load in a bridge or building structure), and for dynamic loads (such as strengthening for improved seismic or blast response in a bridge or building structure). FRP composites have been used successfully for flexural strengthening of concrete beams and slabs, shear strengthening of concrete beams, and axial strengthening and ductility enhancement of concrete columns.

The use of FRP composites for retrofitting concrete structures appears to have evolved at approximately the same time, in the late 1980s, in Europe (particularly in Switzerland) and in Japan. Both of these initiatives were fol-

[6] *Rigid* is used here in the qualitative sense to describe a solid but nevertheless deformable material.

lowed very soon afterward by research and applications in the United States and Canada. Initially, research focused on flexural strengthening of structural members (concrete and timber). This was followed very closely by studies on confinement of concrete columns with FRP composite fabrics and sheets, known as *wraps,* to address a number of deficiencies in concrete columns, particularly in highway bridges, subjected to lateral loads due to earthquakes. Both of these FRP applications grew out of the experience gained with retrofitting of reinforced concrete using steel plates or steel jackets. The use of steel plates to strengthen reinforced concrete structural members was an accepted technology by the mid-1980s, particularly for bridge retrofitting (Eberline et al., 1988). The use of steel jackets to retrofit concrete columns developed into a routine practice in the United Sates following the Loma Prieta earthquake in 1989 (Chai et al., 1991). FRP shear strengthening of concrete beams has been studied since the early 1990s (soon after flexural strengthening) and numerous applications in concrete structures, particularly precast prestressed concrete T-beams, have been undertaken. Strengthening of concrete slabs for punching shear has been studied in recent years but is not yet an accepted technology.

Adhesively bonded precured carbon fiber-reinforced epoxy FRP strengthening strips for flexural strengthening of concrete beams was first studied and used in Switzerland in the late 1980s by Urs Meier and his colleagues at the Swiss Federal Laboratories for Material Testing and Research (EMPA). Activities at EMPA from the mid-1980s to the mid-1990s have been reviewed by Meier (1995). An FRP strengthening strip known as Carbodur was developed at EMPA and commercialized by Sika. The fire safety and fatigue life of the system have been studied at EMPA. These strips have been used in hundreds of building, bridge, and chimney retrofit projects in Europe (Taerwe and Matthys, 1999). An L-shaped precured strip has recently been developed by EMPA for shear strengthening of beams. Today, precured strips are produced by a number of manufacturers, including S&P Clever Reinforcement Company in Austria; Hughes Brothers, Strongwell, and FyfeCo in the United States; and Sika in Switzerland. Since the mid-1990s, following commercialization of the precured FRP strip, hundreds of research projects and thousands of FRP retrofits for static or nonseismic dynamic loads (e.g., vehicular loads, wind loads) have been carried out throughout the world. Figure 1.8 shows installation of a precured FRP strip.

Manufacturers of FRP products themselves have traditionally provided design guidance for the use of precured FRP strengthening strips. The first English-language edition of the Sika Carbodur design guide appeared in 1997[7] (Sika, 1997). Current editions have replaced earlier editions as more information has become available and new FRP strengthening products have been introduced. A design guide is published by the S&P Reinforcement Company

[7] Key design guides and specifications are listed at the end of this chapter and in the references.

Figure 1.8 Installation of a precured FRP strip. (Courtesy of Sika Corporation.)

that covers their products (S&P, 1998). More recently, professional organizations have developed general-purpose design guides for use of precured (and bonded) FRP strengthening products for concrete structures, which in time will probably replace manufacturers' guides (TR 55, 2004; FIB, 2001; ACI, 2002; CSA, 2002). Code guidance for NSM systems is currently being developed. Figure 1.9 shows the installation of NSM strips in a concrete slab.

Hand-layup FRP composite systems for retrofitting concrete structures were developed and commercialized in Japan in the late 1980s. This was soon followed by work in the United States. In Japan attention focused on the use of unidirectional carbon fiber sheets (known as *tow sheets*). The Tonen Corporation developed the Forca tow sheet, Mitsubishi Chemical developed the Replark sheet, and Toray developed the Torayca fabric system.

As opposed to the European focus, which was on strengthening for static load-carrying capacity, the Japanese effort was driven primarily by the need to retrofit building and bridge structures for earthquake-induced seismic loads. Consequently, much of the research was directed toward wrapping of columns to increase their lateral load-carrying capacity (Katsumata et al., 1988). In Japan during this time there was also work in the area of flexural and shear strengthening of concrete beams and slabs (Nanni, 1995). To date, Japanese FRP sheet products have been used in hundreds of bridge and building retrofit projects in Japan. A dramatic increase in FRP retrofitting projects in Japan was seen after the 1995 Hyogoken–Nanbu earthquake, which devastated the city of Kobe. Karbhari (1998) provided an extensive database of key projects completed in Japan at that time. In addition to manually applied hand-layup systems, the Ohbayashi Corporation pioneered the development of an automated carbon fiber winding machine for strengthening tall chimneys. A design

Figure 1.9 Installation of near-surface-mounted (NSM) strips. (Courtesy of Structural Group.)

guide was published by the Japan Society of Civil Engineers (JSCE) in 2001. An English edition of the Forca tow sheet technical manual was published in 1996 (Tonen, 1996), and an English edition of the Replark technical manual was published in 1999 (Replark, 1999).

Early work in the United States on FRP strengthening of concrete structures was undertaken for the purpose of seismic retrofitting of reinforced concrete columns. The Fyfe Company developed a glass fabric and epoxy hand-layup system for column retrofitting in the early 1990s that was tested at the University of California San Diego (UCSD) (Priestley et al., 1992). This work was followed by further development and testing of FRP column wrapping systems, with research being conducted primarily at UCSD (Priestley and Seible, 1995). Based on much of this earlier work, Acceptance Criteria 125 was published by the International Conference of Building Officials (ICBO) to provide guidance to contractors and designers who wished to use the method in the United States (AC 125, 1997).

Simultaneously, a number of researchers in the United States studied flexural strengthening of beams using hand-layup fiber sheets and fabrics of glass or carbon fibers (Saadamanesh, 1994). The Forca tow sheet carbon sheet strengthening system, called the MBrace System (MBrace, 1998), was commercialized in the United States by Master Builders, Inc. of Ohio together with Structural Preservation Systems of Maryland. The Fyfe Company developed and commercialized the Tyfo System (Tyfo, 1998). Installation of a fabric sheet is shown in Figure 1.10. Hexcel supplies fabric strengthening systems for Sika known as SikaHex. Nowadays numerous companies market a variety of fabric and sheet strengthening materials in the United States, including VSL, Edge Structural Composites, and Quakewrap. Figure 1.11 shows installation of an FRP wrap on a rectangular column.

Hand-layup fabrics and sheets have been used in thousands of projects in the United States and around the world. Since the 1994 Northridge earthquake in California, hundreds of highway columns have been retrofitted in California and neighboring states. Wrapping of circular columns with FRP is most effective; however, rectangular columns have also been retrofitted for both strength and ductility successfully using FRP wraps. FRP retrofitting of a highway column is shown in Fig. 1.12.

FRP strengthening sheets and fabrics have been used to strengthen precast T-beams for shear strength enhancement in numerous parking garage structures. Flexural strengthening projects for bridge beams and slabs with sheets

Figure 1.10 Installation of a carbon fiber fabric sheet. (Courtesy of Fyfe Company.)

Figure 1.11 Installation of FRP wrap. (Courtesy of Sika Corporation.)

and fabrics have been carried out in almost every state in the United States. In recent years FRP sheets and fabrics have been used to strengthen concrete shear walls for seismic retrofit and for blast strengthening (also know as *hardening*). Figure 1.13 shows carbon fabrics being installed on an interior concrete shear wall of a building for the purpose of seismic retrofitting.

Figure 1.12 FRP wrap being installed on a highway column. (Courtesy of Sika Corporation.)

Figure 1.13 FRP fabrics for building shear wall retrofit. (Courtesy of Racquel Hagen.)

Other strengthening systems, used in the United States primarily for column retrofits, have used prefabricated FRP shells and automated fiber winding. A prefabricated FRP composite shell system called the Snaptite system, which consists of flexible thin FRP shells that are bonded in multiple layers, was produced by the C.C. Myers company of California in the late 1990s (Xiao and Ma, 1997). Use of the Snaptite system is shown in Fig. 1.14. Large

Figure 1.14 Use of the Snaptite system. (Courtesy of Yan Xiao.)

cylindrical rigid shells, similar to steel jackets, have been produced by Hardcore Composites of Delaware and used in a number of experimental projects. In the late 1990s, XXSys of San Diego, California, developed a fiber winding system, similar to the Japanese Ohbayashi system, for bridge retrofits. None of these systems was ever used routinely in the United States for FRP retrofitting of concrete structures.

FRP sheets and fabrics have also been used extensively to retrofit concrete masonry wall systems as well as timber beams and steel and aluminum members. These uses are discussed in what follows as "other" applications of FRP composites in structural engineering since code-based design guidance is not yet available for these applications.

1.5 FRP PROFILES FOR NEW STRUCTURES

A cost-effective method of producing high-quality constant-cross-section FRP profile shapes, called *pultrusion,* was developed in the 1950s in the United States. Initially, small profiles were produced primarily for industrial applications, but the method was always envisioned as being used for developing FRP substitutes for conventional beams and columns in buildings and bridges. To quote Brandt Goldsworthy again: "The chemical inertness of this material allows its use in . . . all types of structural members that are subject to corrosive action in chemical plants or other areas where corrosive conditions exist" (Goldsworthy, 1954). By the late 1960s and early 1970s a number of pultrusion companies were producing "standard" I-shaped and tubular profiles. The Structural Plastics Research Council (SPRC) was established in 1971 by the American Society of Civil Engineers (ASCE), and a manual was developed for the design of *structural plastics* (McCormick, 1988). Called the *Structural Plastics Design Manual* (SPDM), it was originally published in 1979 as an FHWA report and subsequently by the ASCE in 1984 (ASCE, 1984). This guide was not restricted to pultruded profiles. In 1996, a European design guide for polymer composite structures was published (Eurocomp, 1996), and in 2002 the European Union published the first standard specification for pultruded profiles (CEN, 2002a).

The first large structures constructed from FRP profiles were single-story gable frames that were used in the emerging computer and electronics industry for Electromagnetic Interference (EMI) test laboratories. The electromagnetic transparency of the FRP profiles was a key benefit in these buildings which required no magnetic material above the foundation level. Custom pultruded profiles and building systems were developed and commercialized by Composites Technology, Inc. (CTI), founded by Andrew Green in Texas in the 1960s (Smallowitz, 1985). In 1985, CTI designed and constructed an innovative EMI building for Apple Computer. Similar structures were also constructed from FRP pultruded profiles produced by Morrison Molded Fiberglass Company (MMFG, now Strongwell) in Virginia for IBM and others in the 1980s. Figure 1.15 shows a FRP gable frame building during the installation of the FRP cladding. For much of this time, designing was done by an MMFG subsidiary called Glass-Steel. A design manual for the MMFG profile shape Extren was published in 1973. Creative Pultrusions began producing standard shapes in the late 1970s, called Pultex, and developed a design manual. Current editions of these manuals are published regularly and are available from these two companies.

The next major development in building systems for FRP profiles, which continues to be the largest market segment for large pultruded building components, was in the cooling tower industry. An FRP building system was developed in the 1980s by Composite Technology, Inc. for Ceramic Cooling Tower (CCT) and commercialized as the Unilite system (Green et al., 1994). The Unilite system consisted of a number of unique beam, column, and panel FRP pultruded components. Today, a number of pultrusion companies supply specialized parts for FRP cooling tower systems to a variety of cooling tower manufacturers as shown in Figure 1.16.

In addition to custom cooling tower structures, FRP profiles have been used in "stick-built" cooling towers since the late 1980s. These systems are typically constructed using tubular FRP sections 2 × 2 in. (50 × 50 mm) and 3 × 3 in. (75 × 75 mm) covered with an FRP or nonreinforced polymer cladding system. Designers typically use standard off-the-shelf pultruded profiles in these structures and design them according to applicable building codes. A stick-built cooling tower under construction is shown in Fig. 1.17.

To date, standard FRP profile shapes have not seen much use in multistory frame building structures for commercial or residential construction. One of the major difficulties with multistory frame structures using FRP profiles is the development of economical and effective means of connecting the individual members. Research has been conducted on the subject of FRP connections since the early 1990s, but no simple and effective connection system

Figure 1.15 FRP gable frame structure under construction. (Courtesy of Strongwell.)

Figure 1.16 FRP cooling tower. (Courtesy of SPX Cooling Technologies, Inc.)

Figure 1.17 Stick-built cooling tower under construction. (Courtesy of Strongwell.)

has yet been developed or commercialized for FRP pultruded profiles. Most current designs use steel-like connection details that are not optimized for FRP profiles. A prototype multistory framed building called the Eyecatcher, constructed by Fiberline Composites in Basel, Switzerland in 1999, is shown in Fig. 1.18. The building was constructed for the Swissbau Fair as a demonstration of the potential for FRP profile shapes (Keller, 1999).

In the field of bridge engineering, FRP profile shapes have seen increased application since the mid-1970s. Both the light weight of the FRP components and their noncorrosive properties serve to make them attractive as bridge decking panels and as superstructure members. Hundreds of 30 to 90 ft (9 to 27 m)-long short span pedestrian footbridges of the truss variety have been designed and constructed worldwide using small FRP profiles. Figure 1.19 shows a FRP bridge designed and constructed by ET Techtonics.

In 1992, a 131-m-long cable-stayed pedestrian bridge was constructed in Aberfeldy, Scotland, using the Advanced Composite Construction System (ACCS), a FRP plank system designed by Maunsell Structural Plastics (Burgoyne and Head, 1993; Cadei and Stratford, 2002). A fiber rope called Parafil was used for the cable stays. In 1997, a 40.3-m cable-stayed pedestrian bridge

Figure 1.18 Eyecatcher building. (Courtesy of Thomas Keller.)

Figure 1.19 Light-truss pedestrian bridge pultruded structure. (Photo and design by ET Techtonics.)

was constructed over a railway line using FRP profiles in Kolding, Denmark (Braestrup, 1999). Both of these structures are worth noting, due to their extensive use of FRP pultruded profiles that facilitated rapid and economical construction.

In the 1990s, a significant effort was undertaken by a number of FRP manufacturers to develop an FRP bridge deck system that could be used on conventional steel or concrete girders. Besides the potential for long-term durability of an FRP bridge deck, an added benefit exists when FRP decks are used to replace deteriorated reinforced concrete decks. Due to the significant decrease in dead weight of the structure, the live-load capacity for the re-decked structure can be increased, which may be beneficial, especially on bridges with load postings. FRP deck systems have been developed and commercialized by Creative Pultrusions, Martin Marietta Composites, Atlantic Research Corp., Hardcore Composites, and others (Bakis et al., 2002). As with FRP frame systems, the connections between the prefabricated FRP deck panels and those between the FRP deck and the superstructure have been sources of the greatest difficulty in realizing this technology. The high cost of glass FRP decks compared with conventional concrete decks does not appear to be able to offset savings in weight and construction productivity. Developing an approved bridge guardrail for FRP deck systems has also proved to be a challenge that has not been resolved satisfactorily at this time. In 2001, a glass–carbon FRP pultruded profile 36 in. high by 18 in. wide known as the *double-web beam* (DWB) was developed by Strongwell for use as a bridge girder. Figure 1.20 shows FRP girders on the Dickey Creek bridge in Virginia in 2001.

FRP rods have also been developed for use as guy wires for towers and for suspension cables for bridges. Pultruded glass FRP rods have been used since the mid-1970s in numerous applications in the United States as guy

Figure 1.20 FRP DWB girders on Dickey Creek bridge. (Courtesy of Strongwell.)

wires or bracing cables in antenna towers. The use of carbon FRP cables for bridges was first suggested in the early 1980s by Urs Meier of the Swiss Federal Laboratories for Materials Testing and Research (EMPA) in Dubendorf, Switzerland, as an alternative to steel cables for very long span suspension bridges, where the weight of the cable itself can be a limiting factor (Meier, 1986). Since then, EMPA and BBR Ltd. in Zurich, Ciba AG in Basle, and Stesalit AG in Zullwil, Switzerland, have worked together on developing carbon FRP cables for bridges. In 1997, two carbon FRP cables consisting of 241 5-mm (0.2-in.)-diameter carbon–epoxy FRP rods were used in the Storchen cable-stayed bridge in Winterthur, Switzerland. Figure 1.21 is a close-up of the multistrand carbon FRP cable used. FRP profiles have also been used as stays in the Kolding FRP pedestrian bridge constructed in Denmark in 1997, and carbon fiber cables were used in the Laroin footbridge constructed in France in 2002.

1.6 OTHER EMERGING APPLICATIONS OF INTEREST TO STRUCTURAL ENGINEERS

Strengthening of masonry structures with FRP strips, sheets, and fabrics is one of the largest emerging application areas of FRP composites in structural engineering. Unfortunately, no code-based design guidance is available for masonry strengthening at this time. Literature on the subject is reviewed in Nanni and Tumialan (2003). Most FRP strengthening of masonry has been on non-load-bearing in-fill walls and in historic masonry structures. Strengthening of both brick and hollow concrete masonry unit (CMU) structures is used to increase both the out-of-plane and in-plane capacity of masonry walls

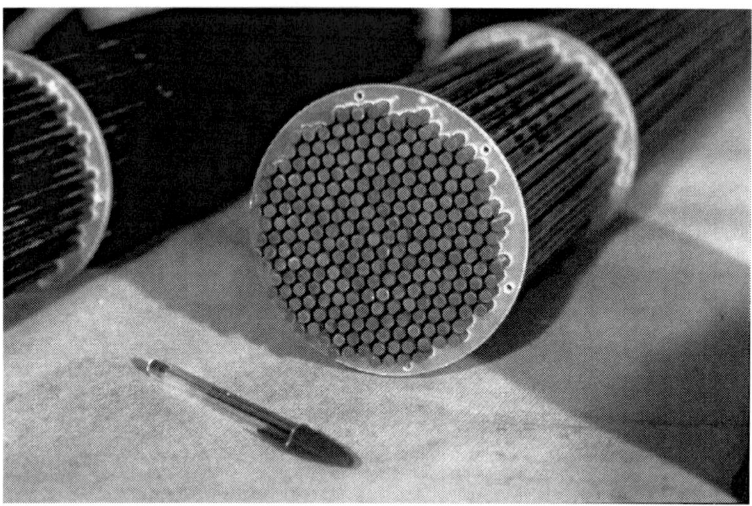

Figure 1.21 Carbon fiber cables used in the Storchen bridge in Switzerland. (Courtesy of Urs Meier.)

for both dynamic and static loads. Out-of-plane blast strengthening of masonry walls with FRPs is used to preserve the integrity of a masonry wall during a blast (Tan and Patoary, 2004). Anchorage of FRP composite material to the masonry itself and to the reinforced concrete frame in the case of an in-fill wall remains a challenge. Early experimental work on strengthening of masonry walls was conducted by Schwegler at EMPA (Schwegler, 1994) and by Priestley and Seible at UCSD (Priestley and Seible, 1995).

Strengthening of timber structures, primarily glulam (glue-laminated) beams, with FRP strips and manufacturing of glulam beams containing FRP layers to increase strength at the outer fibers has been used since the early 1990s. In the United States, early research and commercialization of the method was performed by Tingley (Tingley et al., 1997). Strengthening systems are designed on a case-by-case basis according to standard mechanics principles and bonded adhesively to the timber member. Mechanical fasteners such as lag bolts are often used also. Due to moisture in the wood, the choice of adhesive is critical, and differential thermal and moisture expansion between the FRP layers and the wood lamina can cause significant problems. At this time, code guidance is not available for FRP-strengthened timber structures.

Retrofitting of metallic structures has recently attracted some interest in the structural engineering community. In most cases the steel or aluminum structure is sufficiently stiff and strong. The FRP composite is generally used to increase fatigue resistance and to arrest cracks. In the United States, FRP composite wraps have been used successfully to repair fatigue damage on overhead aluminum sign support bridges (Pantelides et al., 2003). In this particular application, the FRP strengthening system is preferable to in situ welding, which is difficult and time consuming.

FRP composite products have been used for many years for luminaire poles and in electric power-line towers and components. Code guidance is provided by AASHTO (2001) for designing luminaire supports and sign bridge structures using FRP structural members. A recent manual published by ASCE addresses the use of FRP products in overhead utility line structures (ASCE, 2003). FRP products have also been used extensively in ladders and cooling towers, and standards and specifications for these specialized applications can be useful to the structural engineer (ANSI, 2000; CTI, 2003).

The use of mechanical fasteners instead of adhesive bonding as a method to attach FRP strips to concrete beams has been studied in recent years (Lamanna et al., 2001, 2004; Rizzo et al., 2005). Mechanical fastening has advantages that include very little surface preparation, rapid installation, lower costs, and immediate use of the strengthened structure. The method requires special FRP strips that have high bearing strength and longitudinal stiffness, and it cannot be used with conventional unidirectional carbon–epoxy FRP strips.

1.7 PROPERTIES OF FRP PRODUCTS FOR STRUCTURAL ENGINEERING DESIGN

To design a load-bearing structure, a structural engineer must have both qualitative and quantitative knowledge of the properties of the materials that are used in the design. Qualitative knowledge of the material properties is required to understand which of the physical and mechanical properties of the FRP composite material are important for structural design and why they are important. Quantitative knowledge of the material properties is required so that analysis can be performed to predict the behavior and capacity of structural members and structures that are wholly or partially made of FRP composite materials in order to size them appropriately to meet both strength and serviceability demands.

For conventional structural materials such as steel, concrete, and timber, existing design codes and specifications prescribe both specific material properties and standardized tests method to be used to obtain the properties for design. For example, in the case of reinforced concrete, standard cylinders are tested in accordance with ASTM C 39, Standard Test Method for Compressive Strength of Cylindrical Concrete Specimens. In addition, standard specifications published by the ASTM specify the properties of the constituent materials of the concrete (cements, aggregates, and admixtures) and of the steel reinforcing bars (ASTM, 2006). With a quantitative knowledge of the properties of the concrete and the steel reinforcing bars, a structural engineer can design a structural member to carry code-stipulated design loads safely and economically according to ACI 318, *Building Code Requirements for*

Structural Concrete (ACI, 2005). With a qualitative knowledge of the properties of concrete and steel reinforcing bars, a structural engineer can decide on such matters as whether or not to use certain admixtures in the concrete mix and whether or not to specify epoxy-coated reinforcing bars rather than uncoated bars.

Guide specifications and test methods for the use of FRP composites in structural engineering have been developed for the three design topics covered in this book: FRP reinforcing bars, FRP strengthening systems, and FRP profiles. These specifications are not as developed or as mature as those that have been developed for conventional materials over approximately the past 100 years. Nevertheless, suitable material specifications do exist and can be used for structural engineering design, as is demonstrated in the design chapters that follow. The perceived lack of standard material specifications and test methods for FRP composite materials in structural engineering, often given as a reason for not using FRP products, is no longer a convincing reason for not specifying FRP composites for use in structural engineering.

The primary theoretical and experimental methods used to predict and measure the characteristic values of the properties of FRP composite materials for use in structural engineering are provided in Chapter 3. Many excellent textbooks on the mechanics of laminated composite materials and experimental characterization of composite materials are available that cover this subject in great detail (Tsai and Hahn, 1980; Agarwal and Broutman, 1990; Daniel and Ishai, 1994; Adams et al., 2003). These books are focused primarily on advanced composites for aerospace applications.

Typically, the characteristic value of a material property is obtained from tests on a specified number of samples of the material. In this section we do not discuss how a specific property to be used in an analytical equation in a code-based design procedure to design a structural member, such as the specified compressive strength, is derived from the characteristic value of the property. The determination of a property that is used in a design procedure is described in later chapters, where the design of particular FRP structural members is discussed. The determination of the design property depends on the design basis used, such as the load and resistance factor design (LRFD) or the allowable stress design (ASD) basis. To determine the design value of a material property, a structural engineer needs to know how the characteristic value was obtained.

The characteristic properties of FRP composite materials for use in structural engineering design are usually obtained by experimental testing of FRP materials and products. Pertinent experimental test methods are reviewed in Chapter 3. The FRP composite properties may be determined experimentally on one of four levels: (1) fiber level (e.g., a single fiber in a ply of the flange of a wide-flange profile); (2) lamina level [e.g., one ply (or layer) of the flange of a wide-flange profile]; (3) laminate level (e.g., a flange of a wide-flange profile); and (4) full-section level (e.g., a wide-flange profile). The approximate ranges of the cross-sectional dimensions of the four levels, from smallest to largest, are: fiber level, less than 0.0004 in. (25 μm); lamina level, from

0.0004 to 0.04 in. (25 μm to 1 mm); laminate level, from 0.04 to 0.5 in. (1 mm to 1.3 cm); full-section level, greater than 0.5 in. (1.3 cm). In some cases, the FRP product in its entirety (e.g., an FRP strengthening strip) is at the lamina or laminate level, and the dimensional ranges given above may not be applicable. In some situations, properties at higher levels can be predicted from the experimentally determined properties at lower levels, using the theoretical models described in Chapter 3. For example, if the properties of a single ply of the FRP composite material are measured in a test (lamina level), theoretical models can be used to predict the properties on the laminate level and then on the full-section level. For structural engineering applications it is generally preferred to determine properties experimentally on the full-section or laminate level of the FRP product and not to rely on theoretical models. However, this is not always possible.

A commonly used method in engineering practice to obtain characteristic properties of materials is to use manufacturer-supplied specifications, often referred to as *spec sheets*. A structural engineer must know how to interpret the properties reported in manufacturer-published spec sheets for FRP products for use in structural engineering. At this time there is no uniformity in the manner in which manufacturers report their property data, and this can lead to confusion. A manufacturer will often report properties on a number of different levels for a single FRP product, and the structural engineer will need to develop the design properties from a mixed set of properties. In the design-oriented chapters that follow, specific design examples are provided in which manufacturer-published data are used.

Ranges of mechanical and physical properties of representative FRP products for use in structural engineering are provided below for the reader to get a feel for the order of magnitude of the values available. As noted in the historical review, the manufacturers of FRP products for use in structural engineering have tended to change over the years, and the products listed are representative of those in production in 2006.

It is important to note that the properties listed below, taken from manufacturer-published specification sheets, are not the maximum possible values for these FRP materials, nor do they necessarily correspond to the values from a specific test of a single FRP composite product. Neither are they design values or characteristic values. In the chapters on design procedures that follow, the calculation of design values from characteristic values obtained from specified numbers of tests is discussed in greater detail in the context of the design basis for a particular FRP product.

In Tables 1.1 to 1.5, typical properties of FRP composites of interest to the structural designer are listed. Since FRP composite products are anisotropic, properties are reported for the longitudinal direction, the transverse direction, and the shear planes separately, where available.[8]

[8]The reader should consult Chapter 3 for details on the mechanics of anisotropic composite materials and the effects of this anisotropy on the engineering properties.

TABLE 1.1 Properties of Typical Commercially Produced FRP Reinforcing Bars

	Glass-Reinf. Vinylester Bar[1,2] [0.5 in. (13 mm) diam.]	Glass-Reinf. Vinylester Bar[1] [1 in. (25 mm) diam.]	Carbon-Reinf. Vinylester[1] Bar [0.5 in. (13 mm) diam.]	Carbon-Reinf. Epoxy Bar[3] [0.5 in. (13 mm) diam.]
Fiber volume[a]	50–60	50–60	50–60	50–60
Fiber architecture	Unidirectional	Unidirectional	Unidirectional	Unidirectional
Tensile strength, longitudinal [ksi (MPa)]	90–100 (620–690)	80 (552)	300 (2070)	327 (2255)
Compressive strength, longitudinal [ksi (MPa)]	NR[b]	NR	NR	NR
Shear strength, out-of-plane [ksi (MPa)]	22–27 (152–186)	22 (152)	NR	NR
Bond strength [ksi (MPa)]	1.7 (12)	1.7 (12)	1.3 (9)	NR
Tensile modulus, longitudinal [Msi (GPa)]	5.9–6.1 (41–42)	5.9 (41)	18 (124)	21 (145)
Compressive modulus, longitudinal [Msi (GPa)]	NR	NR	NR	NR
CTE, longitudinal [10^{-6} °F^{-1} (10^{-6} °C^{-1})]	3.7–4.9 (6.7–8.8)	3.7 (6.7)	−4.0–0 (−7.2–0)	0.38 (0.7)
CTE, transverse [10^{-6} °F^{-1} (10^{-6} °C^{-1})]	12.2–18.7 (22.0–33.7)	18.7 (33.7)	41–58 (73.8–104.4)	NR
Barcol hardness	60	60	48–55	NR
24-hr water absorption (% max.)	NR	NR	NR	NR
Density [lb/in^3 (g/cm^3)]	0.072 (2.1)	0.072 (2.1)	NR	0.058 (1.6)

Source: [1] Aslan (Hughes Brothers); [2] V-rod (Pultrall); [3] Leadline (Mitsubishi).

[a] Estimated

[b] NR, not reported by manufacturers.

TABLE 1.2 Properties of Typical Commercially Produced FRP Strengthening Strips[a]

	Standard Modulus Carbon-Reinf. Epoxy Strip[1–3]	High-Modulus Carbon-Reinf. Epoxy Strip[1]	Glass-Reinf. Epoxy Strip[2]	Carbon-Reinf. Vinylester Strip[4]
Fiber volume[b]	65–70	65–70	65–70	60
Fiber architecture	Unidirectional	Unidirectional	Unidirectional	Unidirectional
Nominal thickness [in. (mm)]	0.047–0.075 (1.2–1.9)	0.047 (1.2)	0.055–0.075 (1.4–1.9)	0.079 (2.0)
Width [in. (mm)]	2–4 (50–100)	2–4 (50–100)	2–4 (50–100)	0.63 (16)
Tensile strength, longitudinal [ksi (MPa)]	390–406 (2690–2800)	188 (1290)	130 (900)	300 (2070)
Tensile strain (max.), longitudinal (%)	1.8	NR[c]	2.2	1.7
Tensile modulus, longitudinal [Msi (GPa)]	22.5–23.9 (155–165)	43.5 (300)	6.0 (41)	19.0 (131)
CTE longitudinal [10^{-6} °F^{-1} (10^{-6} °C^{-1})]	NR	NR	NR	−4.0–0.0 (−7.2–0)
CTE transverse [10^{-6} °F^{-1} (10^{-6} °C^{-1})]	NR	NR	NR	41–58 (73.8–104.4)
Barcol hardness	NR	NR	NR	48–55

Source: [1]Carbodur (Sika); [2]Tyfo (Fyfe); [3]Mbrace (Wabo/S&P); [4]Aslan (Hughes Brothers).

[a]All strips must be bonded with manufacturer-supplied compatible adhesives.

[b]Estimated.

[c]NR, not reported by manufacturers.

TABLE 1.3 Properties of Typical Commercially Produced FRP Sheet and Fabric Strengthening Materials[a]

	Standard Modulus Carbon Fiber Tow Sheet[1–3]	High-Modulus Carbon Fiber Tow Sheet[1,2]	Glass Fiber Unidirectional Fabric[1,2]	Carbon Fiber Multiaxial Fabric[2,4]
Thickness [in. (mm)]	0.0065–0.013 (0.165–0.330)	0.0065 (0.165)	0.014 (0.356)	NR[b]
Typ. width [in. (mm)]	24 (600)	24 (600)	48 (1200)	NR
Fiber architecture	Unidirectional	Unidirectional	Unidirectional	Various
Fiber tensile strength, longitudinal [ksi (MPa)]	550 (3790)	510 (3520)	220–470 (1520–3240)	508 (3500)
Fiber tensile strain (max.), longitudinal (%)	1.67–1.7	0.94	2.1–4.5	1.7
Fiber tensile modulus, longitudinal [Msi (GPa)]	33.0 (230)	54.0 (370)	10.5 (72)	33.0 (230)

Source: [1]MBrace (Wabo); [2]Tyfo (Fyfe); [3]Replark (Mitsubishi); [4]SikaWrap (Sika).

[a]All dry fiber sheet must be used with manufacturer-supplied compatible polymer resins.

[b]NR, not reported by manufacturers.

TABLE 1.4 Mechanical Properties of Typical Commercially Produced FRP Pultruded Materials

	Glass-Reinf. Vinylester Shapes (WF)[1,2] [0.25–0.5 in. (6–13 mm) thick]	Glass-Reinf. Vinylester Flat Sheet[1,2] [$\frac{3}{8}$–1 in. (9–25 mm) thick]	Glass-Reinf. Vinylester Rods[1,2] [0.25–1 in. (6–25 mm) diam.]	Carbon-Reinf. Epoxy Strips[3] [0.047 in. (1.2 mm) thick]
Fiber volume[a]	25–40	20–35	50–60	65
Fiber architecture	Roving and mat	Roving and mat	Roving only	Tow only
		Strength [*ksi* (*MPa*)]		
Tensile	30–46	20	100	406
Longitudinal	(207–317)	(138)	(690)	(2800)
Transverse	7–12	10	NR[b]	NR
	(48–83)	(69)		
Compressive	30–52	24	65	NR
Longitudinal	(207–359)	(165)	(448)	
Transverse	16–20	16	NR	NR
	(110–138)	(110)		
Shear	4.5–7	NR	NR	NR
In-plane	(31–48)			
Out-of-plane	3.9–4.5	6	8	NR
	(27–31)	(41)	(55)	
Flexural	30–49	30–35	100	NR
Longitudinal	(207–338)	(207–241)	(690)	
Transverse	10–19	15–18	NR	NR
	(69–131)	(103–124)		
Bearing	30–39	32	NR	NR
Longitudinal	(207–269)	(221)		
Transverse	26–34	32	NR	NR
	(179–234)	(221)		

TABLE 1.4 *(Continued)*

	Glass-Reinf. Vinylester Shapes (WF)[1,2] [0.25–0.5 in. (6–13 mm) thick]	Glass-Reinf. Vinylester Flat Sheet[1,2] [$\frac{3}{8}$–1 in. (9–25 mm) thick]	Glass-Reinf. Vinylester Rods[1,2] [0.25–1 in. (6–25 mm) diam.]	Carbon-Reinf. Epoxy Strips[3] [0.047 in. (1.2 mm) thick]
Modulus [*Msi* (*GPa*)]				
Tensile				
Longitudinal	2.6–4.1 (18–28)	1.8 (12)	6.0 (41)	23.9 (165)
Transverse	0.8–1.4 (6–10)	1.4 (10)	NR	NR
Compressive				
Longitudinal	2.6–3.8 (18–26)	1.8 (12)	NR	NR
Transverse	1.0–1.9 (7–13)	1.0 (7)	NR	NR
Shear, in-plane	0.43–0.50 (3.0–3.4)	NR	NR	NR
Flexural				
Longitudinal	1.6–2.0 (11–14)	2.0 (14)	6.0 (41)	NR
Transverse	0.8–1.7 (6–12)	1.1–1.4 (7.6–9.7)	NR	NR
Poisson ratio, longitudinal	0.33–0.35	0.32	NR	NR

Source: [1] Strongwell (2002); [2] Creative Pultrusions (2000); [3] Sika (2000).

[a] Estimated.

[b] NR, not reported by manufacturers.

TABLE 1.5 Physical and Electrical Properties of Typical Commercially Produced FRP Pultruded Materials

	Glass-Reinf. Vinylester Shapes (WF)[1,2] [0.25–0.5 in. (6–13 mm) thick]	Glass-Reinf. Vinylester Flat Sheet[1,2] [$\frac{3}{8}$–1 in. (9–25 mm) thick]	Glass-Reinf. Vinylester Rods[1,2] [0.25–1 in. (6–25 mm) diam.]	Carbon-Reinf. Epoxy Strips[3] [0.047 in. (1.2 mm) thick]
Fiber volume[a]	25–40	20–25	50–60	65
Fiber architecture	Roving and mat	Roving and mat	Roving only	Tow only
CTE, longitudinal [10^{-6} °F^{-1} (10^{-6} °C^{-1})	4.4 (7.9)	8.0 (14.4)	3.0 (5.4)	NR[b]
CTE, transverse [10^{-6} °F^{-1} (10^{-6} °C^{-1})	NR	NR	NR	NR
Barcol hardness	45	40	50–55	NR
24-hr water absorption (% max.)	0.6	0.6	0.25	NR
Dielectric strength, longitudinal [kV/in. (V/mm)]	35–40 (1380–1580)	35–40 (1380–1580)	60 (2360)	NR
Density [lb/in^3 (g/cm^3)]	0.060–0.070 (1.7–1.9)	0.060–0.070 (1.7–1.9)	0.072–0.076 (2.0–2.1)	NR

Source: [1]Strongwell (2002); [2]Creative Pultrusions (2000); [3]Sika (2000).

[a]Estimated.

[b]NR, not reported by manufacturers.

1.8 PUBLISHED DESIGN GUIDES, CODES, AND SPECIFICATIONS FOR FRP COMPOSITES IN STRUCTURAL ENGINEERING

In recent years a significant number of design guides, codes, and specifications have been published by technical organizations that give extensive guidance to structural engineers for design with FRP materials. Much of the seminal research conducted in the last 10–15 years on the subject of FRP composites for construction was published in a select number of conference proceedings and journals. The key publications are listed below.

1.8.1 FRP Reinforcing Bars and Tendons

ACI (2004), *Prestressing Concrete Structures with FRP Tendons,* ACI 440.4R-04, American Concrete Institute, Farmington Hills, MI.

ACI (2006), *Guide for the Design and Construction of Structural Concrete Reinforced with FRP Bars,* ACI 440.1R-06, American Concrete Institute, Farmington Hills, MI.

BRI (1995), *Guidelines for Structural Design of FRP Reinforced Concrete Building Structures,* Building Research Institute, Tsukuba, Japan. See also Design guidelines of FRP reinforced concrete building structures, *Journal of Composites for Construction,* Vol. 1, No. 3, pp. 90–115, 1997.

CSA (2000), *Canadian Highway Bridge Design Code,* CSA-06-00, Canadian Standards Association, Toronto, Ontario, Canada.

CSA (2002), *Design and Construction of Building Components with Fibre-Reinforced Polymers,* CSA-S806-02, Canadian Standards Association International, Toronto, Ontario, Canada.

JSCE (1997), *Recommendation for Design and Construction of Concrete Structures Using Continuous Fiber Reinforcing Materials,* Concrete Engineering Series 23, Japan Society of Civil Engineers, Tokyo.

1.8.2 FRP Strengthening Systems

AC 125 (1997), *Acceptance Criteria for Concrete and Reinforced and Unreinforced Masonry Strengthening Using Fiber-Reinforced Polymer (FFP) Composite Systems,* ICC Evaluation Service, Whittier, CA.

AC 187 (2001), *Acceptance Criteria for Inspection and Verification of Concrete and Reinforced and Unreinforced Masonry Strengthening Using Fiber-Reinforced Polymer (FFP) Composite Systems,* ICC Evaluation Service, Whittier, CA.

ACI (2002), *Guide for the Design and Construction of Externally Bonded FRP Systems for Strengthening Concrete Structures,* ACI 440.2R-02, American Concrete Institute, Farmington Hills, MI.

FIB (2001), *Externally Bonded FRP Reinforcement for RC Structures,* International Federation for Structural Concrete, Lausanne, Switzerland.

JSCE (2001), *Recommendation for Upgrading of Concrete Structures with Use of Continuous Fiber Sheets,* Concrete Engineering Series 41, Japan Society of Civil Engineers, Tokyo.

TR 55 (2004), *Design Guidance for Strengthening Concrete Structures Using Fibre Composite Materials,* The Concrete Society, Camberley, Surrey, England.

TR 57 (2003), *Strengthening Concrete Structures with Fibre Composite Materials: Acceptance, Inspection and Monitoring,* The Concrete Society, Camberley, Surrey, England.

1.8.3 FRP Pultruded Profiles

ASCE (1984), *Structural Plastics Design Manual,* ASCE Manuals and Reports on Engineering Practice 63, American Society of Civil Engineers, Reston, VA.

CEN (2002), *Reinforced Plastic Composites: Specifications for Pultruded Profiles,* Parts 1–3, EN 13706, Comité Européen de Normalisation, Brussels, Belgium.

Eurocomp (1996), *Structural Design of Polymer Composites: EUROCOMP Design Code and Handbook,* Clark, J. L. (ed.) E&FN Spon, London.

1.8.4 Manufacturers' Design Manuals

In addition to the association- or standards organization-developed design manuals listed above, manufacturers of FRP products for structural engineering applications also provide design guidance. In some cases, the association design guides have been developed from early versions of manufacturers' guides.

FRP Strengthening Systems

MBrace (1998), *MBrace Composite Strengthening System: Engineering Design Guidelines,* Master Builders, OH. Current edition at www.mbrace.com.

Replark (1999), *Replark System: Technical Manual,* Mitsubishi Chemical Corporation, Sumitomo Corporation of America, New York. Current edition at www.sumitomocorp.com.

Sika (1997), *Sika Carbodur: Engineering Guidelines for the Use of Sika Carbodur (CFRP) Laminates for Structural Strengthening of Concrete Structures,* Sika Corporation, Lyndhurst, NJ. Current edition at www.sikaconstruction.com/.

S&P (1998), *Clever Reinforcement Company,* Schere & Partners, Brunnen, Switzerland. Current edition at www.sp-reinforcement.ch.

Tonen (1996), *Forca Towsheet Technical Manual,* Rev. 5.0, Tonen Corporation, Tokyo.

Tyfo (1998), *Design Manual for the Tyfo Fibrwrap System,* Fyfe Co. LLC, San Diego, CA. Current edition at www.fyfeco.com.

FRP Pultruded Profiles

Bedford Reinforced Plastics Design Guide (2005), Bedford Plastics, Bedford, PA. Current edition at www.bedfordplastics.com.

Extren Design Manual (2002), Strongwell, Bristol, VA. Current edition at www.strongwell.com.

The Fiberline Design Manual (2003), Fiberline Composites, Kolding, Denmark. Current edition at www.fiberline.com.

The Pultex Pultrusion Global Design Manual (2004), Creative Pultrusions, Alum Bank, PA. Current edition at www.pultrude.com.

1.8.5 Key Conferences Series

FRPRCS* (*Fiber Reinforced Composites in Reinforced Concrete Structures*) *Series

1993 Vancouver, British Columbia, Canada, ACI
1995 Ghent, Belguim, RILEM
1997 Sapporo, Japan, JSCE
1999 Baltimore, Maryland, ACI
2001 Cambridge, UK, E&F Spon
2003 Singapore, New World Publishers
2005 Kansas City, Missouri, ACI

ACMBS* (*Advanced Composite Materials in Bridges and Structures*) *Series

1992 Sherbrooke, Quebec, Canada, CSCE
1996 Montreal, Quebec, Canada, CSCE
2000 Ottowa, Ontario, Canada, CSCE
2004 Calgary, Alberta, Canada, CSCE

ASCE Materials Congress Series

1990 Denver, Colorado, ASCE
1991 Las Vegas, Nevada, FRP Specialty Conference, ASCE
1992 Atlanta, Georgia, ASCE
1994 San Diego, California, ASCE
1999 Cincinnati, Ohio, ASCE

ICCI* (*International Conference for Composites for the Infrastructure*) *Series

1996 Tucson, Arizona
1998 Tucson, Arizona
2002 San Francisco, California

CICE (Composites in Civil Engineering) Series

2001 Hong Kong, China
2004 Adelaide, Australia
2006 Miami, Florida

ACIC (Advanced Polymer Composites for Structural Applications in Construction) Series

2002 Southhampton, UK
2004 Surrey, UK
2007 Bath, UK

1.8.6 Archival Journals

The major international English language journals that have published and continue to publish key technical papers on the topic of FRP composites in structural engineering are:

Composite Structures, Elsevier
Construction and Building Materials, Elsevier
Journal of Composites for Construction, ASCE
Structural Journal, ACI

PROBLEMS

1.1 Find information about the FRP composite structures and FRP-related organizations and publications listed below. For each item, provide a short written description of the product or application, a Web page link, and a picture of relevance.

Aberfeldy Bridge, Scotland
American Composites Manufacturers Association (ACMA)
Committee 440, American Concrete Institute (ACI)
Committee AC207, Transportation Research Board (TRB)
Composites Bridge Alliance Europe (COBRAE)
Composites Technology
EMPA, Switzerland
European Pultrusion Technology Association (EPTA)
Eyecatcher building, Switzerland
High Performance Composites

International Institute for FRP in Construction (IIFC)
ISIS Canada
Journal of Composites for Construction, ASCE
Kolding Bridge, Denmark
NetComposites Education
Pultrusion Industry Council (PIC)

1.2 Find information about the selected FRP manufacturers listed below and the products they produce for structural engineering applications. For each item, provide a short written description of the company and its product for use in structural engineering applications. Provide an image of the product and a URL.

BBR Carbon Fiber Stay Cables: FRP stay cables
Bedford Reinforced Plastics: pultruded products
Creative Pultrusions: pultruded products
Edge Structural Composites: strengthening systems
ET Techtonics: pultruded bridges
Exel Composites: pultruded products
Fibergrate Composite Structures: pultruded products
Fiberline Composites: pultruded products
Fyfe Co.: strengthening systems
Hughes Brothers: FRP reinforcements
Lancaster Composites: pilings
Marley Cooling Technologies: FRP cooling towers
Martin Marietta Composites: FRP bridge decks
Mitsubishi Composites: FRP tendons and strengthening systems
Pacific Composites: pultruded products
Powertrusion Poles: FRP poles
Pultrall: FRP reinforcements
Quakewrap: strengthening systems
Roechling-Haren: pultruded products
S&P–Clever Reinforcement Company: strengthening systems
Seasafe: FRP grating
Shakespeare: FRP poles
Sika: strengthening systems
Sireg: FRP tendons
Strongwell: pultruded products
Techfab: FRP reinforcements
Tokyo Rope: FRP tendons
Topglass Italy: pultruded products

VSL: strengthening systems

Wabo MBrace: strengthening systems

1.3 Visit your library (physically or virtually) and compile a list of all the journals and magazines that publish papers (exclusively or in part) and articles related to fiber-reinforced polymer composite materials. Try to determine what the focus of the publication is: for example, material characterization, theoretical mechanics, structural applications, mechanical engineering, aerospace engineering, polymer science, civil engineering.

2 Materials and Manufacturing

2.1 OVERVIEW

In much the same way that a structural engineer has a working knowledge of the composition of structural materials such as steel, concrete, and wood and how they are made into products for use in structures, a similar working knowledge of FRP composite materials is needed by the structural engineer. This includes having a qualitative knowledge of the constituent or raw materials and the processing methods used to produce the parts and how these affect the eventual mechanical and physical properties of the FRP part. The intent of this chapter is to provide the structural engineer with sufficient materials background to have a working knowledge of the FRP material that will be specified in the construction documents of a project.

Since, by and large, the same raw materials are used in a number of different manufacturing processes to manufacture FRP composite materials used in structural engineering, a general description of the characteristics and properties of these raw materials is first given in this chapter. Glass and carbon fibers are the principal synthetic fiber materials used to manufacture FRP products for use in structural engineering, and these fibers are discussed in detail. The polymer resins used most widely in FRP products for structural engineering applications are thermosetting epoxies, polyesters, and vinylesters, and these resins are discussed in detail. Less commonly used fiber and resin systems are discussed briefly, as they are not used widely in FRP products for structural engineering at this time.

This is followed by a discussion of the manufacturing methods used to produce FRP products or parts from the raw materials. The principal methods used to manufacture FRP composite products used in structural engineering are pultrusion and hand layup. These two processes are by far the most widely used processes for manufacturing FRP composites for structural engineering. Pultrusion is used to manufacture FRP profiles, FRP strengthening strips, and FRP reinforcing bars. Hand layup is used to fabricate formed-in-place FRP strengthening sheets, fabrics, and wraps. In addition, at present, detailed code-based design provisions have been developed only for FRP products for structural engineering made by pultrusion and hand layup. Less commonly used methods for FRP products for structural engineering, such as filament winding and closed molding, are discussed briefly.

2.2 RAW MATERIALS

To produce an FRP composite material, two primary raw material constituents are required, reinforcing fibers and a polymer resin matrix. In this section we review key properties and characteristics of the raw materials used to produce FRP products for structural engineering. We do not provide information on how the raw materials are produced; the reader is referred to the composite materials literature for coverage of this subject (e.g., Schwartz, 1997a,b). However, it is worth noting that all the raw materials are produced at high temperatures in industrialized processes that require highly specialized equipment and control. Analogous to the state of affairs that exists today, where a structural engineer does not typically have extensive knowledge of how portland cement is produced from limestone, a structural engineer is not expected to have extensive knowledge of how polymer resin is produced from crude oil or how glass fiber is produced from silica sand.

Except in very rare circumstances, the raw fiber and polymer constituents cannot be used in their as-produced forms to manufacture an FRP composite material. After the fiber filaments are produced, they are postprocessed in a number of secondary operations to produce fiber products such as strands, sheets, fabrics, and mats that can be used in a manufacturing process. Similarly, the raw polymer, which is generally referred to as the *base polymer* or *neat resin,* is often blended with other resins and mixed with a variety of additives and process aids to produce a resin system (or resin mix) for manufacturing. The fiber and resin systems are discussed later in the manufacturing sections of this chapter, as they are manufacturing method–dependent. Numerous companies manufacture and distribute both raw and postprocessed raw materials for use in production of FRP composites. The annual *Sourcebook* (Sourcebook, 2006) provides an extensive list of U.S.-based manufacturers and suppliers.

2.2.1 Reinforcing Fibers

The fiber phase of an FRP composite material consists of thousands of individual micrometer-diameter individual filaments. In the large majority of fiber forms used in FRP products for structural engineering, these fibers are indefinitely long and are called *continuous.* This is to differentiate them from short fibers of length 10 to 50 mm (~0.5 to 2 in.) that are used in the spray-up process for boat building and consumer products or in reinforced cementitious materials [known as glass-reinforced cements (GRCs) or fiber-reinforced cementitous (FRC) composites]. Continuous fibers are used at a relatively high volume percentage (from 20 to 60%) to reinforce the polymer resin: thus the term *fiber-reinforced polymer* (FRP). The mechanical properties of the fibers are typically orders of magnitude greater than those of the polymer resins that they reinforce; however, due to their filamentary nature they cannot be used as stand-alone construction materials and must be used in a synergistic fashion with polymer resins to realize their superior mechanical properties.

Glass Fibers Glass fibers are used in a multitude of FRP products for structural engineering, from FRP reinforcing bars for concrete, to FRP strengthening fabrics, to FRP structural profile shapes. Glass is an amorphous inorganic compound of primarily metallic oxides that is produced in fibrous form in a number of standard formulations or types. Silica dioxide (SiO_2) is the largest single compound in all glass formulations, constituting from 50 to 70% by weight of the glass. Different grades of glass fiber are identified by letter nomenclature. A borosilicate glass known as *E-glass* (electrical glass) because of its high electrical resistivity is used to produce the vast majority of glass fiber used in FRP products for structural engineering. *A-glass* (window glass) and *C-glass* (corrosion resistant, also know as *AR-glass* or alkali-resistant glass) are used to produce specialized products for use in structural engineering. *S-glass* (structural or high-strength glass) is used to produce the high-performance fibers used primarily in the aerospace industry.

The diameter of an individual glass fiber or filament ranges from approximately 3 to 24 μm (0.00118 to 0.00945 in.).[1] The 17-μm (0.0067-in.)-diameter fiber is most commonly used for FRP products for structural engineering. A glass fiber has a distinctive bright white color to the naked eye. Glass is usually considered to be an isotropic material. Approximate properties of commonly used grades of glass fibers are given in Table 2.1. Values presented in Table 2.1 are intended as a guide and should not be used in design calculations.

Glass fibers are produced at melt temperatures of about 1400°C (~2550°F). Individual filaments are produced with a surface coating called a *sizing* that serves to protect the filaments when they are formed into a bundle or a strand.

TABLE 2.1 Approximate Properties of Common Grades of Glass Fibers

Grade of Glass Fiber	Density [g/cm^3 (lb/in^3)]	Tensile Modulus [GPa (Msi)]	Tensile Strength [MPa (ksi)]	Max. Elongation (%)
E	2.57 (0.093)	72.5 (10.5)	3400 (493)	2.5
A	2.46 (0.089)	73 (10.6)	2760 (400)	2.5
C	2.46 (0.089)	74 (10.7)	2350 (340)	2.5
S	2.47 (0.089)	88 (12.8)	4600 (667)	3.0

[1] The properties of the fiber and resin constituent materials are typically given in metric units by manufacturers. However, in the United States, the properties of the fiber products (fabrics and rovings) and the FRP parts are typically given in imperial units.

The sizing also contains coupling agents, usually silanes, that are specially formulated to enhance bonding between the glass fiber and the particular polymer resin being used when making a glass-reinforced FRP composite material. Today, most commercially available glass fibers can be obtained with sizings that are compatible with the three major thermosetting resin systems used in structural engineering: epoxy, polyester, and vinylester. The commonly used term *fiberglass* is generally used to refer to the glass fiber-reinforced polymer composite material itself and not solely to the glass fiber constituent material. When referring to the fibrous reinforcement alone, the term *glass fiber* is preferred.

Glass fibers are particularly sensitive to moisture, especially in the presence of salts and elevated alkalinity, and need to be well protected by the resin system used in the FRP part. Glass fibers are also susceptible to creep rupture and lose strength under sustained stresses (Bank et al., 1995b). The endurance limit of glass fibers is generally lower than 60% of the ultimate strength. Glass fibers are excellent thermal and electrical insulators (hence, their extensive use in buildings and the electric power industry as insulation materials) and are the most inexpensive of the high-performance fibers.

Carbon Fibers Carbon fibers are used in structural engineering applications today in FRP strengthening sheets and fabrics, in FRP strengthening strips, and in FRP prestressing tendons. Carbon fiber is a solid semicrystalline organic material consisting on the atomic level of planar two-dimensional arrays of carbon atoms. The two-dimensional sheetlike array is usually known as the *graphitic form;* hence, the fibers are also known as graphite fibers (the three-dimensional array is well known as the *diamond form*). Carbon fiber is produced in *grades* known as *standard modulus, intermediate modulus, high strength,* and *ultrahigh modulus* (SM, IM, HS, UHM).[2]

Carbon fibers have diameters from about 5 to 10 μm (0.00197 to 0.00394 in.). Carbon fiber has a characteristic charcoal-black color. Due to their two-dimensional atomic structure, carbon fibers are considered to be transversely isotropic, having different properties in the longitudinal direction of the atomic array than in the transverse direction. The longitudinal axis of the fiber is parallel to the graphitic planes and gives the fiber its high longitudinal modulus and strength. Approximate properties of common grades of carbon fibers are given in Table 2.2.

Carbon fiber is produced at high temperatures [1200 to 2400°C (~2200 to 4300°F)] from three possible precursor materials: a natural cellulosic rayon textile fiber, a synthetic polyacrilonitrile (PAN) textile fiber, or pitch (coal tar). Pitch-based fibers, produced as a by-product of petroleum processing, are generally lower cost than PAN- and rayon-based fibers. As the temperature

[2] Carbon fibers are generally identified by their specific trade names. The properties of a carbon fiber produced by different manufacturers, even similar grades, can vary widely.

TABLE 2.2 Approximate Properties of Common Grades of Carbon Fibers

Grade of Carbon Fiber	Density [g/cm^3 (lb/in^3)]	Tensile Modulus [GPa (Msi)]	Tensile Strength [MPa (ksi)]	Max. Elongation (%)
Standard	1.7 (0.061)	250 (36.3)	3700 (537)	1.2
High strength	1.8 (0.065)	250 (36.3)	4800 (696)	1.4
High modulus	1.9 (0.068)	500 (72.5)	3000 (435)	0.5
Ultrahigh modulus	2.1 (0.076)	800 (116.0)	2400 (348)	0.2

of the heat treatment increases during production of the carbon fiber, the atomic structure develops more of the sheetlike planar graphitic array, giving the fiber higher and higher longitudinal modulus. For this reason, early carbon fibers were also known as *graphite fibers.* The term *carbon fiber* is used to describe all carbon fibers used in structural engineering applications. The term *graphite fiber* is still used in the aerospace industry; however, this term is slowly dying out. Similar to glass fibers, carbon fibers need to be sized to be compatible with a resin system. Historically, carbon fibers have been used primarily with epoxy resins, and suitable sizings for epoxy resin systems are readily available. Nowadays, carbon fibers are being used with vinylester and blended vinylester–polyester resins for FRP profiles and FRP strengthening strips. Sizing for carbon fibers for polyester and vinylester resins are not as common. Care must be taken when specifying a carbon fiber for use with a nonepoxy resin system to ensure that the fiber is properly sized for the resin system used.

Carbon fibers are very durable and perform very well in hot and moist environments and when subjected to fatigue loads. They do not absorb moisture. They have a negative or very low coefficient of thermal expansion in their longitudinal direction, giving them excellent dimensional stability. They are, however, thermally and electrically conductive. Care must be taken when they are used in contact with metallic materials, as a galvanic cell can develop due to the electropotential mismatch between the carbon fiber and most metallic materials. Some research has suggested that this can lead to degradation of the polymer resin in the FRP composite, especially in the presence of chlorides and to corrosion of the metallic material (Alias and Brown, 1992; Torres-Acosta, 2002).

Aramid Fibers Aramid fibers were used to produce first-generation FRP prestressing tendons in the 1980s in Europe and Japan; however, few manufacturers still produce aramid fiber FRP reinforcing bars or tendons. Aramid fabrics are occasionally used in FRP strengthening applications to wrap columns and as sparse-volume weft (fill) fibers in unidirectional glass or carbon fabrics for FRP strengthening. Aramid fibers consist of aromatic polyamide molecular chains. They were first developed, and patented, by DuPont in 1965 under the trade name Kevlar.

A combination of their relatively high price, difficulty in processing, high moisture absorption (up to 6% by weight), low melting temperatures [around 425°C (~800°F)], and relatively poor compressive properties have made them less attractive for FRP parts for structural engineering applications. Their advantages include extremely high tenacity and toughness, and consequently, they are used in many industrial products, either in bare fabric form or as reinforcements for FRP composites where energy absorption is required, such as in bulletproof vests (body armor), helmets, and automotive crash attenuators. They have a distinctive yellow color and are similar in cost to carbon fibers. Like carbon fibers, they have a negative coefficient of thermal expansion in the fiber longitudinal direction. They are the lightest of the high-performance fibers, having a density of around 1.4 g/cm^3 (0.051 lb/in^3). Depending on the type of aramid fiber, the fiber longitudinal tensile strength ranges from 3400 to 4100 MPa (~500 to 600 ksi), and its longitudinal tensile modulus ranges from 70 to 125 GPa (~10,000 to 18,000 ksi).

Other Fibers Other fibers that are now in the development phase for use in FRP products for structural engineering include thermoplastic ultrahigh-molecular-weight (UHMW) polyethylene fibers and polyvinyl alcohol (PVA) fibers. PVA fibers have been used in FRP bars and FRP strengthening sheets in Japan. UHMW short fibers are being used in the development of ductile fiber-reinforced cements (FRCs) but have not yet been used in FRP products for structural engineering. Inorganic basalt fibers, produced in Russia and the Ukraine, may see future applications in FRP products in structural engineering, due to their superior corrosion resistance and similar mechanical properties to glass fibers. Thin steel wires have been developed for use in FRP strengthening fabrics with either polymer or cementitious binders. Natural fibers such as hemp, sisal, and flax, as well as bamboo fibers, have been used in experimental applications to produce FRP composites, but no commercial FRP products are available that contain these fibers at this time. It is anticipated that FRP products in structural engineering that will be developed in the first half of the twenty-first century will probably use more of these natural fibers as sustainability and recyclability become more important drivers in the construction industry.

2.2.2 Polymer Resins

The term *polymer* is used to describe an array of extremely large molecules, called *macromolecules,* that consist of repeating units, or chains, in which the atoms are held together by covalent bonds. The term *polymer* is generally used to describe an organic material of this type; however, it can also be used to describe an inorganic material. The term *polymer resin,* or simply *resin,* is used in the composites industry to refer to the primary polymer ingredient in the nonfibrous part of the FRP material that binds the fibers together. This nonfibrous part is also known as the *matrix* or *binder.* When used in commercial and industrial products a polymer-based material is often known as a *plastic* and the acronym *FRP* is also often used to denote a fiber-reinforced plastic. The acronym *RP* is often used to connote a reinforced plastic, although this is mostly used to describe short-fiber-reinforced plastic products of lower strength and stiffness.

The differences between organic polymers depend on the functional groups present in the polymer chains and the extent of the interaction between these chains. Chains, which have a molecular backbone, may be linear or branched. (Seymour, 1987). Two primary groups of polymers exist today, thermosetting polymers and thermoplastic polymers. They are distinguished from one another by how the polymer chains are connected when the polymer is in its solid form. *Thermosetting polymers* are cross-linked, which means that their molecular chains are joined to form a continuous three-dimensional network by strong covalently bonded atoms. *Thermoplastic polymers* are not cross-linked, and their molecular chains are held together by weak van der Waals forces or by hydrogen bonds (Schwartz, 1997b). This affects their mechanical and physical properties. Due to cross-linking, a thermosetting polymer's structure is *set* when it solidifies or cures during the polymerization process, and it cannot be heated and softened and then re-formed into a different shape. On the other hand, a thermoplastic polymer does not set but remains plastic, and the molecular chains can "flow" when the solid polymer is heated such that it softens and can be reset into a different shape upon cooling.

Synthetic organic polymers are produced by polymerization techniques, either chain (or additional) polymerization or step (or condensation) polymerization. Most high-performance polymers are produced by condensation of difunctional reactants. The best known chain polymerization reaction is *free-radical polymerization,* in which an electron-deficient molecule (or free radical) is initially added to a monomer, forming a new, larger free radical. The chain reaction continues until the reactive constituents are expended.

The thermal response of polymers plays a large role in their processing, properties, and behavior. Pure crystalline solids such as metals undergo a phase change from solid to liquid at a transition temperature, called the *melting point,* T_m. This is the only thermal transition possible in a pure crystalline solid. Since polymers are semicrystalline solids that contain noncrystalline amorphous regions, other thermal transitions occur at lower temperatures than T_m. A thermal transition of particular interest to structural engineering, known as the *glass transition temperature,* T_g, occurs in the amorphous region of the polymer at a temperature below the melting temperature. At the T_g the physical (density, heat capacity) and mechanical (stiffness, damping) properties of the polymer undergo a change. When the temperature approaches T_g from below, the polymer changes from a rigid (known as *glassy*) to a viscous (known as *rubbery*) state, and vice versa, when the temperature approaches from above T_g.[3] The glass transition temperature may be referred to as the

[3]Note that the term *glass transition temperature* is not related to the glass fiber used to reinforce many FRPs. The term comes from the materials terminology that is used to distinguish between glassy or rubbery materials. A carbon fiber–reinforced polymer composite has a glass transition temperature.

heat distortion temperature or *heat deflection temperature* in polymer manufacturers' literature. Although not precisely the same property, these three transition temperatures are usually close in value (within 10°C) and are used interchangeably in the industry to describe roughly the same physical phenomenon. Both thermosetting and thermoplastic polymers have glass transition temperatures. A thermoplastic polymer liquefies at its melting temperature, whereas a thermosetting polymer begins to decompose at its melting temperature. Above the melting temperature a thermoplastic or thermosetting polymer will pyrolize in an oxygen-rich atmosphere.

In structural engineering, FRP composites must be used in their rigid states at operating temperatures below their glass transition temperatures. At temperatures higher than the glass transition temperature, the modulus of the resin, and hence the FRP composite, decreases. Since deflection criteria are used routinely in structural engineering design, an FRP part can become unserviceable at temperatures close to its glass transition temperature. In addition, the FRP part will be less durable and will have lower strength at temperatures above its glass transition temperature. On the other hand, elastomeric polymers like those used in asphalt binders are used above their glass transition temperatures in their viscous state. When these polymers are used below their glass transition temperatures, they become brittle and crack.

Polymer resins are good insulators and do not conduct heat or electricity provided that they have low void ratios. Water in the voids of a polymer composite can allow the composite to conduct electricity. For glass-reinforced polymer electric power parts, stringent limits are placed on the void ratio (usually less than 1%). Polymer resins are usually considered to be isotropic viscoeleastic materials. They creep under sustained stresses or loads and relax under constant strains or displacements. Most polymer resins are susceptible to degradation in ultraviolet light (White and Turnbull, 1994). Thermosetting polymer resins are generally not suitable for use at temperatures greater than 180°C (~350°F) and in fires if not protected in a fashion similar to steel structural members. Polymer materials have been shown to have acceptable fire ratings when used in appropriately designed protection systems and fire-retarding additives. Thermoplastic polymers have been developed for high temperatures up to 450°C (~800°F). Most liquid polymer resins have a shelf life between 6 and 12 months and should be stored at cool temperatures between 10 and 15°C (50 to 60°F).

Unsaturated Polyester Resin Polyester resin is widely used to make pultruded FRP profiles for use in structural engineering and is also used to make some FRP rebars. When greater corrosion resistance is desired in FRP parts, higher-priced vinylester resins are generally recommended, although the corrosion resistance of some polyester resins may be as good as that of vinylester resins. Polyester resins can also be used for FRP strengthening for structures. However, epoxy resins are preferred at this time for FRP strengthening applications because of their adhesive properties, low shrinkage, and environmental durability.

Unsaturated polyester resin is the most widely used resin system for producing industrial and commercial FRP composite material parts. It is referred to as an *unsaturated polymer* because the double-covalent bonds in its polymer chains are not saturated with hydrogen atoms. By dissolving the polymer in a reactive dilutent, typically a styrene monomer, an exothermic free-radical polymerization chain reaction takes place. This polymerization reaction occurs only in the presence of a catalyst, usually a peroxide. The reactive styrene is added at concentrations of 20 to 60 parts per hundred (pph) of the total resin mix by weight. Based on the type of acid monomer used in the production of the resin, three types of polyester resin, having increasingly better physical and mechanical properties, are produced: orthophthalic, isophthalic, and teraphthalic polyesters. The first unsaturated polyester resin was produced by Ellis and Rust in 1940 (Seymour, 1987).

Polyester resins are particularly versatile and can easily be filled and pigmented. They can be formulated in hundreds of different ways to tailor their properties to different manufacturing processes and end-use environments. In 2004, unsaturated polyester resins cost between \$0.60 and \$1.00 per pound (\$1.32 to \$2.2 per kilogram). They have a density between 1.15 and 1.25 g/cm^3 (0.042 and 0.046 lb/in^3). Depending on the polyester formulation and on the catalyst used, polyester resins can be cured at room temperatures or at elevated temperatures and can therefore have glass transition temperatures ranging from about 40 to 110°C (~105 to 230°F). Polyester resins usually are clear to greenish in color.

Epoxy Resins Epoxy resins are used in many FRP products for structural engineering applications. Most carbon fiber–reinforced precured FRP strips for structural strengthening are made with epoxy resins. In addition, epoxy resin adhesives are used to bond precured FRP strips to concrete (and other materials) in the FRP strengthening process. Epoxy resins are also used extensively in FRP strengthening applications, where the epoxy resin is applied to the dry fiber sheet or fabric in the field and then cured in situ, acting as both the matrix for the FRP composite and as the adhesive to attach the FRP composite to the substrate. When applied to dry fiber sheets or fabrics, the epoxy resins are often referred to *saturants.* Epoxy resins have also been used to manufacture FRP tendons for prestressing concrete and FRP stay cables for bridges. They are not used extensively to produce larger FRP profiles, due to their higher costs and the difficulty entailed in processing large pultruded FRP parts.

An epoxy resin contains one or more epoxide (or oxirane) groups that react with hydroxyl groups. Most common are the reaction products of bisphenol A and epichlorohydrin, called *bis A epoxies,* or those made from phenol or alkylated phenol and formaldehyde and called *novolacs.* The resins are cured (or hardened) with amines, acid anhydrides, (Lewis acids) by condensation polymerization and not, like polyesters, by free-radical chain polymerization. The epoxy resin and the curing agent (or hardener) are supplied in two parts and are mixed in specific proportions (usually about 2 to 3 parts to 1 part by weight) just prior to use to cause the curing reaction. The first epoxy resin was produced by Schlack in 1939 (Seymour, 1987).

Epoxy resins are particularly versatile and can be formulated in a range of properties to serve as matrix materials for FRP composites or to serve as adhesives. The epoxies used as the resins in FRP parts for structural engineering belong to the same family as the more familiar epoxies currently used in a variety of structural engineering applications, such as for concrete crack injection, as anchors for concrete, and for bonding precast concrete elements. Epoxy resins are known to have excellent corrosion resistance and to undergo significantly less shrinkage than polyester or vinylester resins when cured. Consequently, they are less prone to cracking under thermal loads. Epoxy resins have been developed for high-temperature applications of 180°C (350°F) and higher and have been the thermosetting resins of choice in the aerospace industry for the last 50 years. Epoxies based on bisphenol A resins cost about $1.10 per pound ($2.4 per kilogram); those based on the novolac resins cost about $2.00 per pound ($4.4 per kilogram) (2004 costs). The density of epoxy resin is about 1.05 g/cm^3 (0.038 lb/in^3). Epoxy resins can be cured at room temperature or at high temperature. In many aerospace applications, epoxy resin composites are postcured at elevated temperatures to raise their glass transition temperatures and to improve their physical and mechanical properties. The glass transition temperature of an epoxy is therefore highly formulation and cure temperature–dependent and can range from 40°C up to 300°C (~100 to 570°F). Epoxy resins usually are clear to yellowish or amber in color.

Vinylester Resins Developed in the last 20 years, vinylester resins have become attractive polymer resins for FRP products for structural engineering due to their good properties, especially their corrosion resistance and their ease of processing (Blankenship et al. 1989). Today, vinylester resins are used to make the majority of FRP rebars sold in the world and are also used widely in FRP pultruded profiles. Most manufacturers of pultruded profiles make profiles of identical shapes in both a polyester and a vinylester resin series. Vinylester resins have also been used to make FRP strengthening strips and FRP rods for near-surface-mounting applications. They are generally replacing polyester resins in FRP products in structural engineering, due to their superior environmental durability in alkaline environments.

A vinylester resin is a hybrid of an epoxy and an unsaturated polyester resin and is sometimes referred to as an *epoxy vinylester resin* or a *modified epoxy resin*. It is an unsaturated polymer that is produced from an epoxy and an acrylic ester monomer. When it is dissolved in styrene, it reacts with the styrene monomer in the same way as an unsaturated polyester does and cures by free-radical chain polymerization with a peroxide catalyst. Consequently, it tends to have many of the desirable physical properties of an epoxy resin and many of the desirable processing properties of a polyester resin. The two major groups of epoxies used to produce vinylester resins are bisphenol A and novolac epoxies (Starr, 2000).

Vinylester resins can be filled and pigmented. They have densities from 1.05 to 1.10 g/cm^3 (0.038 to 0.042 lb/in^3) and glass transition temperatures from 40 to 120°C (~100 to 250°F). They can be cured at room temperatures or at elevated temperatures. The cost of vinylester resins range from $1.20 to $1.60 per pound ($2.60 to $3.50 per kilogram) (2004 costs), making them more expensive than general-purpose unsaturated polyester resin. Vinylester resins have a color similar to that of polyester resins, ranging from clear to greenish.

Phenolic Resins Phenolic resins are the oldest and most widely used thermosetting resins; however, they have only recently been used for FRP products for structural engineering, due to the difficulty of reinforcing them and curing them by condensation polymerization. They were first developed by Leo Baekeland in the early 1900s and called Bakelite when filled with wood flour (Seymour, 1987). Until the 1980s they had to be cured at high temperatures from 150 to 300°C (~300 to 570°F). They are used extensively in the production of plywood and other engineered wood products. They are being introduced into FRP products for structural engineering because they have superior fire resistance, and they char and release water when burned. They can be filled and reinforced; however, they are difficult to pigment and have a characteristic brownish color. Their costs are similar to that of low-performance polyesters, about $0.60 per pound ($1.30 per kilogram) (2004 costs). Their density is around 1.50 to 2.0 g/cm^3 (0.054 to 0.072 lb/in^3). They have glass transition temperatures from 220 to 250°C (~430 to 480°F). At this time they are used in a limited number of FRP products, particularly in walkway gratings for offshore platforms and in FRP strengthening strips for timber structures.

Polyurethane Resins Thermosetting polyurethane resins have recently been introduced into the market as structural resins. They were first produced in the 1930s by Otto Bayer and consist of long-chain urethane molecules of isocyanate and hydroxyl-containing molecules (polyols). They have been used extensively in their thermoplastic formulation to produce insulation and structural polymer foam materials for decades. Only recently have they been produced in high-density forms that can be used in resin molding and pultrusion operations (Connolly et al., 2005). Polyurethane resins have high toughness and when used with glass fibers produce composites with high transverse tensile and impact strengths. Their cost is approximately the same as that of high-performance vinylester resins. Polyurethane resins do not require styrene to polymerize as do unsaturated polyester and vinylester resins.

Other Polymer Resins Thermoplastic resin systems such as polyethylene teraphthalate (PET, a saturated polyester), polypropylene, and nylon have been used in a very limited fashion to produce FRP parts for structural engineering. Thermoplastic composites based on polyether ether ketone (PEEK), polyphenylene sulfide (PPS), and polyimide (PI) thermoplastic resins, as well as many others, are being used extensively in the high-temperature aerospace composites market. The attractiveness of using thermoplastic resin systems in structural engineering is due to their ability to be heated, softened, and reformed, which may give the parts the potential to be joined by a local heating processes, akin to welding of metals. In addition, they are generally less expensive than thermosetting resins and are recyclable. However, they are difficult to process and generally have lower strength and stiffness than thermosets. They do, however, have higher elongations than thermosets (up to 20%), making them tougher and more ductile. A comparison of the properties of thermosetting resins for FRP products for structural engineering is given in Table 2.3.

FRP products produced for use in structural engineering can include significantly more ingredients than just the primary constituents: fibers and polymer resins. Fibers are produced with surface coatings called *sizings* and are supplied in many different strand and broadgood forms. Resins can contain fillers, catalysts, accelerators, hardeners, curing agents, pigments, ultraviolet stabilizers, fire retardants, mold release agents, and other additives. These have different functions, from causing the resin to polymerize to helping the processing to modifying the final properties of the FRP part. These many different supplementary constituents are manufacturing method–dependent and are discussed below.

2.3 MANUFACTURING METHODS

Two main manufacturing methods are used to produce FRP composite material products for use in structural engineering. The one method is an automated industrialized process, developed in the early 1950s, called *pultrusion,* in which the FRP products are produced in a factory and shipped to the construction site for fabrication and installation or erection. The other method is a manual method, known as *hand layup* or *wet layup,* in which the FRP product is manufactured in situ at the construction site at the time it is in-

TABLE 2.3 Approximate Properties of Thermosetting Polymer Resins

	Density [g/cm^3 (lb/in^3)]	Tensile Modulus [GPa (Msi)]	Tensile Strength [MPa (ksi)]	Max. Elongation (%)
Polyester	1.2 (0.043)	4.0 (0.58)	65 (9.4)	2.5
Epoxy	1.2 (0.043)	3.0 (0.44)	90 (13.1)	8.0
Vinylester	1.12 (0.041)	3.5 (0.51)	82 (11.9)	6.0
Phenolic	1.24 (0.045)	2.5 (0.36)	40 (5.8)	1.8
Polyurethane	varies	2.9 (0.42)	71 (10.3)	5.9

stalled. It is the original method used to produce fiber-reinforced polymer composites and dates back to the development of FRP materials in the 1940s. However, as described below, the hand-layup method as it is used in structural engineering is significantly different from that used in the rest of the composites industry.

The pultrusion process is used to manufacturer FRP reinforcing bars, FRP strengthening strips, and FRP profiles and is the most cost-competitive method for producing high-quality FRP parts for use in structural engineering. The hand-layup method is used to manufacture and install dry fiber strengthening sheets and fabrics and is also very cost-competitive, as it is particularly easy to use in the field. Other methods that have been used to produce specialized FRP products for use in structural engineering, such as filament winding and resin transfer molding, are discussed very briefly since code-based design guides for use of these products in structural engineering are either not available or are insufficiently developed at this time.

2.3.1 Pultrusion

Pultrusion is an automated and continuous process used to produce FRP parts from raw materials. Figure 2.1 shows a photograph of typical FRP pultruded parts used in structural engineering. A pultruded part can have an open cross section, such as a plate or a wide-flange profile; a single closed cross section, such as a hollow tube; or a multicellular cross section, such as panel with

Figure 2.1 Pultruded I-shaped beams. (Courtesy of Racquel Hagan.)

internal webs. The cross section does not have to have a constant thickness throughout. Although there is great flexibility in the shape, thickness variation, and size of the part cross section, the cross section must remain constant along its length. In addition, the part must be straight and cannot be cured into a curved shape. A pultruded part can be produced to any desired length, and if it is flexible enough, it can be coiled onto a spool for shipping (such as a thin FRP strip for strengthening or a small-diameter FRP rebar). Modifications to the pultrusion process have been developed for nonconstant cross sections or for producing curved parts; however, these are nonroutine variants of the pultrusion process. In other variants, a core material is used in a cellular part to fill the cavity in the part.

A pultrusion line or a pultrusion machine is used to produce the pultruded part. The first pultrusion machine, called the Glastruder, was developed by Brandt Goldsworthy in the early 1950s (Goldsworthy, 1954). In 1959, a U.S. patent for a pultrusion machine and the method for producing pultruded parts was awarded to Goldsworthy and Landgraf (1959). Pultrusion machines can now be purchased from a number of companies or can be built from scratch from available off-the-shelf materials and parts. Experienced pultrusion companies tend to develop and build their own pultrusion machines in-house. A schematic of a typical pultrusion line or machine is shown in Fig. 2.2.

To produce FRP parts for structural engineering, dry fibers impregnated with a low-viscosity[4] liquid thermosetting polymer resin are guided into a

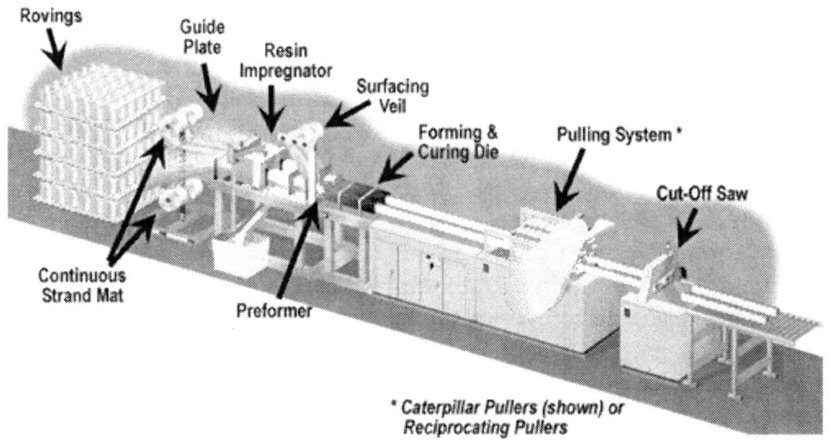

Figure 2.2 Pultrusion line. (Courtesy of Strongwell.)

[4]The viscosity of the polymer resin used in pultrusion depends on the part reinforcement and geometry and is a processing variable; it is usually between 500 and 3000 centipoise (cP), similar to that between motor oil and house paint.

heated chrome-plated steel die, where they are cured to form the desired FRP part. The FRP is cured as the material is pulled through the die by a pulling apparatus: hence, the name *pul*trusion. A number of variations of this basic process exist and are described by Meyer (1970). After exiting the die and extending past the pullers, the part is cut to length by a diamond blade cutoff saw. The rate of production of a pultruded part depends on the size of the part. Small $\frac{1}{4}$ in. (~6-mm)-round pultruded rods can be produced at rate of up to 60 in./min (150 mm/min), although larger complex multicellular parts can be produced at a rate of only a few inches per minute. As the surface area of the cross section increases, a greater amount of force is needed to overcome the frictional forces to pull the part through the die. Typical pultrusion machines have pulling capacities of 10,000, 20,000, and 40,000 lb (approximately 50, 100, and 200 kN).

The length of the heated steel die used to produce an FRP part is typically from 20 to 40 in. (500 to 1000 mm). The die, which is heated by electric resistance heaters or oil heat, is often heated in two or three separate regions along its length to different temperatures from 200 to 400°F (~90 to 180°C) in order to develop the best curing conditions for the type of resin system used in the part. The entrance to the die is typically cooled to prevent premature curing of the resin system. When parts are not required to have a smooth exterior surface, such as a deformed FRP reinforcing bar, a long die is not used. Rather, the impregnated fiber is pulled though a forming ring and is then cured by radiant heat in 6- to 8-ft (1.8- to 2.4-m)-long cylindrical ovens. The pulling apparatus and the cutoff saw are located downstream of the ovens. When thick parts are pultruded, radio-frequency (RF) heating is frequently used to preheat the resin before the part enters the heated die to assist the curing process.

The dry fiber is supplied in various forms on spools or rolls and is spliced "on-the-fly" to create a never-ending source of fiber. The dry fiber is usually impregnated or wet-out in a resin bath, which is continuously refilled and is usually open to the environment, which is ventilated. Most pultrusion manufacturers run their machines on a 24-hour continuous cycle and do not stop the process until the required lot of the FRP part is produced. Typically, only one operator is needed to oversee the line when it is running smoothly. Setup time for a pultrusion line can take from a few hours to a few days, depending on the complexity of the part, and typically involves two to three operators. Cleanup time for the line usually takes a few hours. Acetone is typically used to clean the parts of the pultrusion line that have been exposed to the polymer resin.

The raw materials that are used in the pultrusion process can be broadly viewed as breaking down into two main systems: the fiber system and the resin system. The *fiber system* contains all the dry reinforcements that are pulled into the resin system for wetting-out prior to entering the die. The *resin system* refers to the mix of ingredients that is used to saturate the fibers. The resin system is typically premixed in large batches [usually, 55-gallon (~200-L) drums] in a mixing room in a pultrusion plant before it is brought to the pultrusion line and pumped or poured into the resin bath.

Fiber System for Pultrusion The fiber system used in an FRP pultruded part can consist of different types and architectures of fiber materials. The raw fiber is processed and supplied either in strand form on a spool and known as *roving* or *tow,* or in broadgoods form on a roll and known as *mat, fabric, veil,* or *tissue.* The dry fiber is fed into the pultrusion die in a specific arrangement so as to locate the various different fibers and fiber forms in specific parts of the part cross section. High-density polyethylene (HDPE) or Teflon-coated plates with holes and slits or vinyl tubes, through which the individual fiber types pass guide the fibers through the resin bath and into the die. The guides also help to remove excess resin from the resin-saturated fibers before the fibers enter the die mouth. In FRP profiles, individual strands and mats or fabrics are usually laid out in symmetric and balanced alternating layers, giving the pultruded material a laminated or layered internal architecture.[5] Figure 2.3 shows a close-up of the corner of a pultruded tube with a $\frac{1}{4}$-in. (6.3-mm) wall thickness. The layup consists of mat layers on the exterior of the part and rovings in the interior. In FRP reinforcing bars and FRP strengthening strips, only fiber strands are used and the FRP part consists only of longitudinally aligned fibers and is referred to as *unidirectionally*

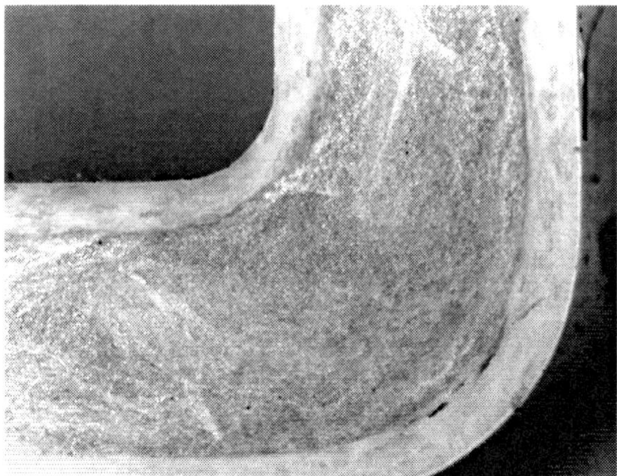

Figure 2.3 Pultruded tubular profile, showing roving and mat layers. (Courtesy of Saiphon Jacque.)

[5]The laminate must be symmetric and balanced, to prevent unwanted axial–shear coupling and bending–extension when it is loaded. See Chapter 3 for more detail on the mechanics of laminates.

reinforced. FRP reinforcing bars may also have helically wound fiber strands at the surface to create bar deformations for load transfer to the concrete.

Glass and carbon fiber are currently used to reinforce most FRP pultruded parts for structural engineering applications. A very small amount of aramid fiber is used in pultrusion. Glass fiber is used in pultruded profiles due to its low cost. Carbon fiber is used in FRP strengthening strips due to its high modulus. A small number of profiles and strips have been produced with a mixture of glass and carbon fibers to optimize the mechanical properties and costs of the part. These FRP pultruded parts are called *hybrids.* In addition to the reinforcing fibers, which give the FRP composite its strength and stiffness, nonreinforcing polyester and glass surfacing mats or veils are used at the surface of the part to aid in processing and to create a smooth surface finish.

Glass Fiber Rovings Individual continuous glass filaments are bundled, generally without a twist, into multifilament strands known as *rovings* that are used in the pultrusion process either as is or in fabrics produced from rovings. In the United States, roving quantity is traditionally measured in units of yield (yd/lb). Roving is produced in yields of 56, 62, 113, 225, 250, 450, 495, 650, and 675. Not all producers manufacture all yields. The number of filaments in an individual roving with a specific yield depends on the fiber diameter of the filament. The most common roving used in pultruded parts is a 113 yield roving, which has approximately 4000 filaments, usually having a diameter of 24 μm (93×10^{-3} in.) each. Figure 2.4 shows a spool of 113 yield glass fiber roving. In the metric system, roving quantity is measured in units of TEX (g/km). A 113 yield roving is equal to a 4390 TEX roving. Its cross-sectional fiber area is 0.00268 in^2 (1.729 mm^2). The following relationships (Barbero, 1999) relate these quantities:

$$\text{TEX(g/km)} = \frac{496{,}238}{\text{yield(yd/lb)}} \tag{2.1}$$

$$A_f(\text{in}^2) = \frac{1}{\rho_f(\text{lb/in}^3) \times \text{yield(yd/lb)} \times 36(\text{in./yd})} \tag{2.2}$$

Rovings are supplied on spools weighing approximately 50 lb (22 kg) each. Typically, tens to hundreds of separate rovings are pulled into a pultruded FRP part. Many large industrial companies manufacture glass fiber roving throughout the world. In 2004, the price of glass fiber roving was about \$0.70 per pound (\$1.5 per kilogram).

In the pultrusion process, the rovings are aligned along the direction of the pultruded part, which is known as the *machine direction* or *lengthwise* (LW) *direction.* In composite mechanics, this direction is known as the *longitudinal*

Figure 2.4 113 yield glass roving on a spool.

or *zero-degree* (0°) *direction* of the composite material. In structural design, this direction typically coincides with the longitudinal axis of the FRP bar, FRP strip, FRP beam, or FRP column. Consequently, the rovings provide the pultruded part with the majority of its axial and flexural strength and stiffness. For parts requiring only high longitudinal strength and stiffness, such as FRP reinforcing bars or thin FRP strengthening strips, high percentages of rovings on the order of 50 to 60% of the total volume of the FRP composite are used. However, such parts have low crosswise (CW) or transverse strength and stiffness. To develop transverse strength and stiffness in a pultruded part, fiber mats and fabrics are used in addition to the rovings. The polymer resin is typically relied on to provide the out-of-plane and transverse shear properties of the pultruded material. When producing tubular parts, glass fiber roving can also be wound in the circumferential direction around the tube to provide an outer layer of hoop reinforcement. Three-dimensional fiber reinforcement preforms for pultrusion that can give out-of-plane, or through-the-thickness, strength and stiffness are not used in regular parts, and a typical pultruded part has a platelike layered structure with alternating layers of roving and mats or fabrics through the plate thickness.

In the pultrusion operation, the roving spools are stacked on metal racks called *creels*. Creel racks should be metal and grounded to prevent electrostatic charge buildup during production. The roving is pulled off the individual spools and guided into the resin bath, where it is saturated with liquid resin (also known as *wet-out*). The roving is usually the first fiber reinforcement material to be wet-out. It is important to ensure that the roving bundles do not bunch together in the resin bath, as this can lead to rovings not being fully wet-out, which will lead to dry-fiber areas in the finished product, which is highly undesirable.

Glass Fiber Mats *Continuous filament mat* (CFM), also referred to in the United States as *continuous strand mat,* is the second most widely employed glass fiber product used in the pultrusion industry. CFM is used to provide crosswise (CW) or transverse strength and stiffness in platelike parts or portions of parts (e.g., the flange of a wide-flange profile). CFMs consist of random, swirled, indefinitely long continuous glass fiber filaments held together by a resin-soluble polymeric binder. They are different from copped strand mats (CSMs), which consist of short [1 to 2 in. (25 to 50 mm)] fibers held together in mat form by a resin-soluble binder[6] which are used mainly in sheet molding compounds. Because of the pulling forces exerted on the mats in the pultrusion processes, chopped stand mats are generally not suitable for pultrusion except when used together with a preformed combination fabric system (these fabrics are discussed below). Continuous filament mats also help to keep the individual rovings in position as they move through the die. Due to the random orientation of the fibers in the plane of the continuous mat, the mat, and hence the layer of the cured pultruded material that contains the mat, can be assumed to have equal properties in all directions (i.e., isotropic properties) in its plane.

CFMs are now produced by a number of companies, although for many years Owens-Corning Fiberglas (OCF) held a patent on the product and was the sole producer. E- and A-glass mats are used in pultrusion. For strength and stiffness, E-glass mats are used. A-glass mats are used primarily for better surface finish. A detailed discussion of the properties of various pultruded materials with E- and A-glass mats can be found in Smith et al. (1998) and Smith (2002). E-glass mats are typically available in weights of 1 to 3 oz/ft^2 (300 to 900 g/m^2) in $\frac{1}{4}$-oz/ft^2 (75-g/m^2) increments. A-glass mats are available in $\frac{1}{2}$- to $1\frac{1}{2}$-oz/ft^2 (150- to 450-g/m^2) weights in $\frac{1}{4}$-oz/ft^2 (75-g/m^2) increments. It is very important to note that in the United States, mat weights are given in ounces per square foot, while other broadgood materials, such as fabrics (discussed below) are given in ounces per square yard. Fiber mats are typically supplied on paperboard tubes 3 in. in diameter up to lengths of 84 in. (2.1 m). To use the mat product in pultrusion, the large roll is cut up into rolls of narrow widths, as required for the dimensions of the pultruded part being produced.

In the pultrusion operation, the CFM is typically wet-out after the rovings. A hose, funnel, or chute may be used to pour the liquid resin directly onto

[6]Be aware that in the industry the acronym CSM is often used for both continuous strand mat and chopped strand mat.

the mat surface as it is fed through guides. When the mat is light, it is often simply wet-out by the excess resin that is carried by the rovings into the forming guides. When the saturated mat and rovings enter the die, they are squeezed into the die opening. This squeezing produces an internal die pressure of around 80 psi (~550 kPa). The pressure tends to compress the mat layers. A rule of thumb used in the industry is that a 1-oz/ft^2 (300-g/m^2) E-glass CFM will occupy a thickness of 20 mils (0.50 mm) in the finished pultruded part and that a $\frac{1}{2}$-oz/ft^2 (150-g/m^2) A-glass mat will occupy a thickness of 16 mils (0.41 mm). Due to the structure of the swirled filaments in an E-glass CFM, the volume faction of fibers in a layer of pultruded material containing a E-glass mat is typically only in the range 20 to 25%. In 2004, the price of E-glass fiber continuous stand mat was about $1.20 per pound ($2.6 per kilogram). The increase in price over the cost of the roving material is due to the cost of the additional manufacturing process required to produce the mat from the filaments. Figure 2.5 shows an E-glass CFM.

Glass Fiber Fabrics Since unidirectional rovings give the pultruded composite reinforcement in its longitudinal direction, and the continuous stand mats give reinforcement in all in-plane directions equally (i.e., isotropically), the range of mechanical properties of the pultruded composite consisting of only rovings and mats is limited. To obtain a greater range of properties and to "tailor" the layup (or the fiber architecture) of the pultruded composite to yield specific structural properties, fabric reinforcements can be used in which fibers are oriented in specific directions and at specific volume percentages

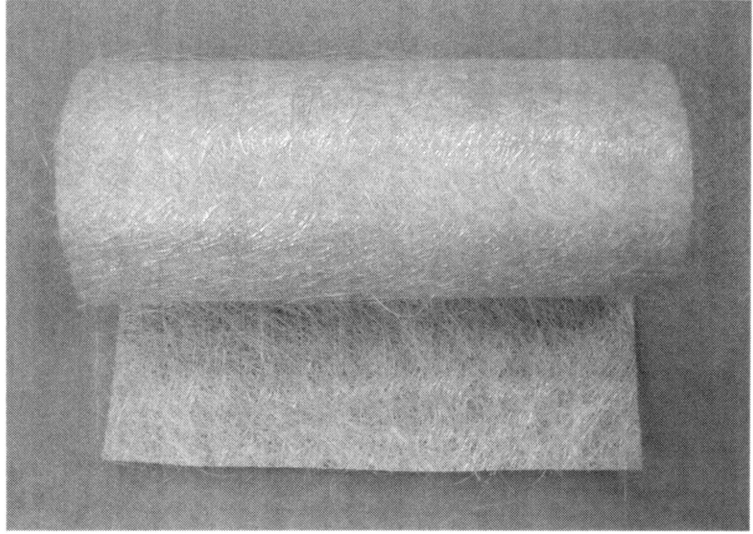

Figure 2.5 1.5-oz/ft^2 E-glass continuous filament mat.

to the pultrusion axis. This design approach, in which multiaxial plies (or layers) are used, is routine in the hand-layup technique; however, it is only in recent years that multiaxial fabrics have been used successfully in the pultrusion process. This is because pulling an off-axis ply (say, one with a 45° orientation to the pultrusion direction) is a nontrivial matter and special preformed fabrics are needed that can be pulled without causing distortion of the fiber orientations. Although now available, multiaxial fabrics are still used only in very special pultruded parts, as their costs can be considerably more than that of mats. They are not as easy to wet-out and pull as rovings and CFMs. It is, however, generally recognized in the composites industry that to optimize the mechanical properties of pultruded profiles for structural engineering, the next generation of profiles will need to use more multiaxial fabrics, possibly with hybrid fiber types, and have more sophisticated cross-sectional shapes. The recent double-web beam produced by Strongwell, shown in Fig. 2.6, is an example of such a engineered hybrid profile developed for bridge girders.

Glass fiber fabric materials for pultrusion are generally of two types. One type is a woven roving fabric; the other type is a stitched roving fabric. Woven roving is used routinely in hand-layup applications such as boat building and is supplied in weights between 6 and 48 oz/yd^2 (200 to 1600 g/m^2) and has fiber orientations of 0° and 90°. The percentage of 0° and 90° fibers [known as the *warp* and the *weft* (or *fill*) directions in the textile industry] depends

Figure 2.6 Double-web hybrid fiber pultruded beam. (Courtesy of Strongwell.)

on the weave pattern. Most woven fabrics made for use in pultrusion are of the plain or square pattern, with almost equal percentages of fibers in the two directions. To use a woven roving in a pultrusion process, it needs to be attached to a glass mat (usually, a chopped strand mat) to prevent it from distorting when pulled. Either powder bonding, stitching with a polyester or glass yarn, or needling are used to attach the woven fabric to the mat, which is then known as a *combination fabric.* Many different combinations of woven roving weights and mat weights are available. Commonly used types are 18-oz/yd^2 woven roving with a 1-oz/ft^2 mat (600-g/m^2 woven roving with a 300-g/m^2 mat). A close-up of a woven roving combination fabric is shown in Fig. 2.7.

The other type of fabric type that is used in pultrusion is a stitched fabric where the unidirectional layers of rovings in different directions are stitched together with or without a chopped mat. Popular types of stitched fabrics are biaxial (having equal percentages of 0° and 90° or +45° and −45° fiber orientations) and triaxial (having fibers in the 0°, +45°, and −45° fiber orientations). +45° and −45° fiber orientations are used to give a pultruded part high in-plane shear strength and stiffness properties. Unidirectional stitched fabrics in which the fibers in one direction are stitched to a mat can also be obtained. These are particularly useful when 90° fiber orientation is needed in a pultruded part to give it high transverse strength and stiffness. For unique applications, unbalanced stitched fabrics can be obtained. As noted previously,

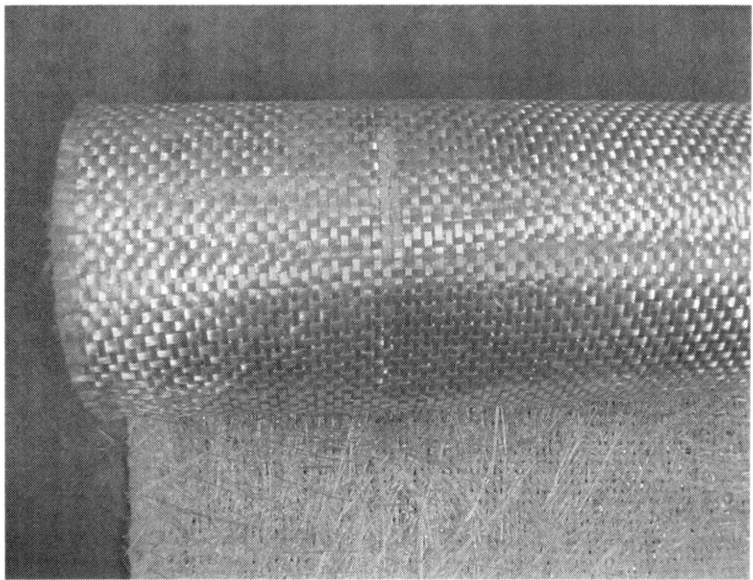

Figure 2.7 Woven glass roving combination fabric.

it is important to ensure that the resulting layup is both symmetric and balanced when using stitched and combination-stitched fabrics. A close-up of a stitched fabric is shown in Fig. 2.8.

The nomenclature used in the specialty fabric industry is often confusing and varies from manufacturer to manufacturer.[7] In one commonly used fabric nomenclature, the 18-oz/yd^2 woven roving with a 1-oz/ft^2 mat described previously is referred to as a *1810 woven-roving-mat fabric* or a *600/300 woven-roving-mat fabric* (in metric nomenclature). The first two digits give the weight of the bare fabric, in oz/yd^2, and the second two digits give the weight of the mat that is attached to the bare fabric, in oz/ft^2. In the metric system, both bare fabric and mat are given in g/m^2, and a slash is used to indicate the weight of the fabric and the mat, respectively. The digits are usually preceded by various letters to indicate the fiber type (glass, carbon, aramid), bare fabric type (woven or stitched), fiber orientations,[8] and whether or not a mat is used (M).

In addition to woven and stitched fabrics, braided and knitted fabrics are also produced for use in the pultrusion process. The braiding process is used

Figure 2.8 Stitched glass fiber fabric.

[7] The nomenclature used for specialized fabrics is highly producer dependent. The structural engineer is cautioned to be very familiar with the nomenclature used by the fabric producer before specifying a product for use in structural engineering.

[8] L, longitudinal (0); T, transverse (90); B, biaxial (0/90); X, ± 45 ply; TT, triaxial ($0/\pm 45$) Q, quadriaxial ($0/\pm 45/90$).

to produce tubular "sleeves" and narrow strips having biaxial and triaxial (in-plane) fiber orientations. A special mandrel and fiber-forming system is required to use a braided sleeve in the pultrusion process to make a tubular product. Figure 2.9 shows a braided sleeve.

Carbon Fiber Tows Carbon fiber strands called *tows* can be used in the pultrusion process and have been used since the 1970s to pultrude small specialized items such as archery arrows and solid rods. However, they have not generally been used to produce pultruded profiles for structural engineering, due to their high cost relative to glass fiber roving. In recent years, carbon fiber–reinforced epoxy pultruded FRP strengthening strips have increased the use of carbon fiber FRP products in structural engineering significantly. TEX is used to refer to the quantity of carbon fiber in a tow. Yield is not generally used for carbon fiber. A carbon fiber tow is identified by the number of individual carbon fiber filaments in the tow, which usually ranges from 1000 to 48,000 (known as 1K to 48K size tows); however, in recent years, high-yield tows up to 300,000 filaments (300K tow) have been produced. The relationships between tow size, TEX, cross-sectional area, and fiber density for a carbon fiber tow are (Barbero, 1999)

$$\text{TEX(g/km)} = \text{tow size(K)} \times A_f(\mu\text{m}^2) \times \rho_f(\text{g/cm}^3) \tag{2.3}$$

$$A_f(\text{cm}^2) = \frac{\text{TEX(g/km)}}{\rho_f(\text{g/cm}^3) \times 10^5} \tag{2.4}$$

Large standard modulus carbon tow sells for between \$7 and \$10 per pound (\$15 to \$22 per kilogram) ((2004), making it considerably more expensive than glass fiber roving. It has long been a stated goal of the carbon fiber producers to get carbon fiber down to \$5 per pound (\$11 per kilogram); however, this still seems unattainable, and cycles of feast and famine still plague the industry. Carbon fiber has significantly higher modulus and lower density than glass fiber, which can make it particularly advantageous to use in certain products for structural engineering (such as for FRP pultruded strengthening strips). Figure 2.10 shows a 12K carbon fiber tow on a spool.

When using carbon fiber in the pultrusion process the pultrusion line usually needs to be physically isolated from other lines in the plant. Carbon fiber is electrically conductive, and carbon fiber dust can cause short-circuiting in electrical boxes. In addition, when carbon fibers are being used in a plant where glass fiber pultruded parts requiring high electrical resistivity are produced (such as insulators for electric power lines), stray amounts of carbon fiber dust can contaminate the glass fiber part and cause significant quality control problems. Consequently, carbon fiber production facilities are often placed in separate sections or buildings in a pultrusion plant.

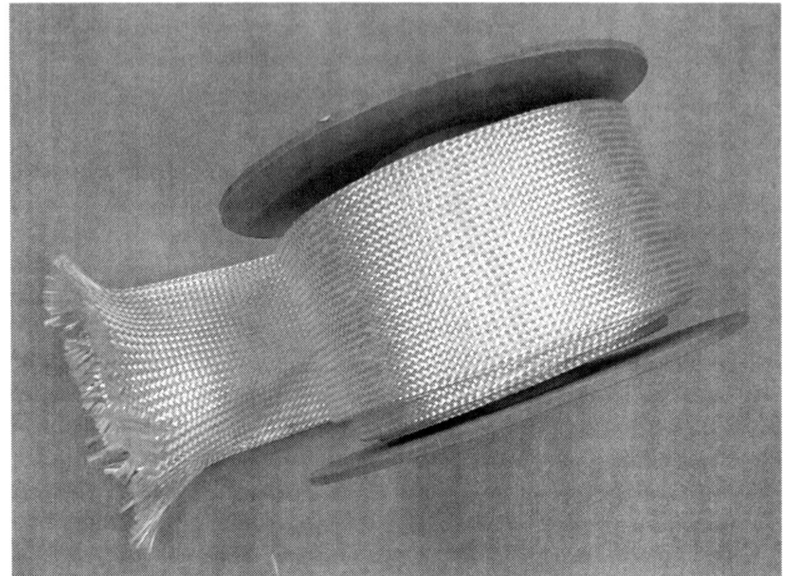

Figure 2.9 Glass fiber braided sleeve.

Figure 2.10 Carbon fiber tow on a spool.

Carbon Fiber Fabrics Carbon fiber fabrics are generally available in the same types of weaves and stitched goods as glass fiber fabrics. They range in weights from 6 to 32 oz/yd^2 (200 to 1050 g/m^2) and are typically produced in biaxial 0/90 and ± 45 bias-ply layups. They are produced by fewer manufacturers than glass fabrics and are typically produced as special orders from the major suppliers of specialty fabrics for unique pultrusion applications. At present, only a small number of pultruded FRP products are produced with carbon fiber fabrics for use in structural engineering. Carbon fiber fabrics are significantly more expensive than glass fiber fabrics. Continuous filament carbon fiber mat, like a glass CFM, can also be obtained from specialty manufacturers.

Hybrid Fabrics Most specialty fabric producers manufacture custom hybrid fabrics consisting of carbon, glass, and often also aramid fibers in different orientations and weights; however, these are not used in any regular FRP pultruded parts for structural engineering. They tend to be much higher priced and are used in one-off molding operations such as layup and resin transfer molding to produce more exotic composite material industrial products and sporting goods such as surfboards, skis, snowboards, and racquets.

Surfacing Veils An important characteristic of an FRP pultruded part is the quality of the surface. A smooth and regular surface without significant fiber fabric or roving pattern deformations on the surface is usually desired. In some pultruded parts, such as FRP rebars, this is not a concern, as surface deformation is desired. To achieve a good, smooth surface quality, often referred to as a *low-profile surface,* a very lightweight surfacing fabric, called a *veil* or *tissue,* is used as the outer layer of the pultruded part. Veils are typically made of nonwoven polyester filaments or C- and E-glass monofilament mats. Veils are much lighter than regular continuous filament mats and have nominal weights of 1 to 2 oz/yd^2 ($\sim$30 to 60 g/m^2). The surfacing veil tends to have a higher resin volume fraction than the other reinforcement layers, such as mats and rovings, and gives the part a resin-rich surface layer similar to that of a gel coat in a hand layup. The resin-rich surface also gives the part greater corrosion and ultraviolet resistance. Figure 2.11 shows a polyester veil.

Resin System for Pultrusion The three main thermosetting resins used in the pultrusion process are unsaturated polyesters, unsaturated vinylesters, and epoxies. To each of these base resins supplementary constituents are added to cause the polymerization reaction to occur, to modify the processing variables, and to tailor the properties of the final FRP pultruded part.[9] The ad-

[9]There is no single "resin mix recipe" for any of these resin systems; the mix depends on the FRP part to be pultruded and the pultrusion process variables.

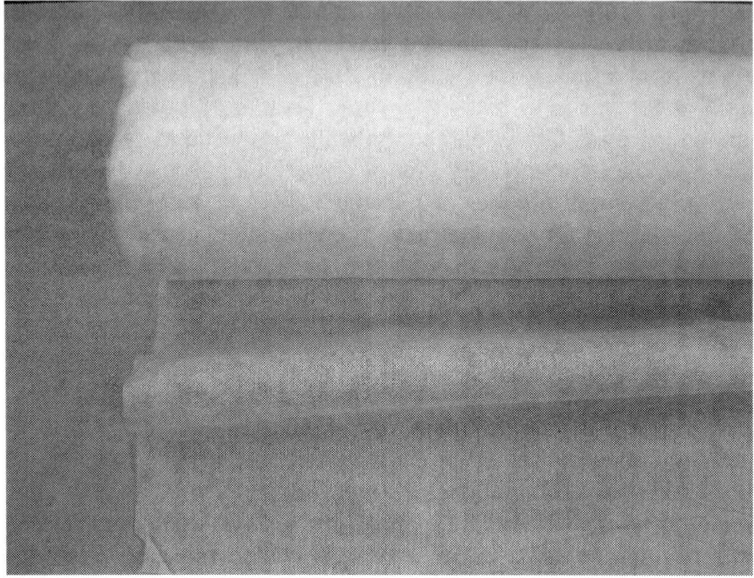

Figure 2.11 Polyester veil.

ditional constituents that are added can be broadly grouped into three main categories: polymerization agents, fillers, and additives. *Polymerization agents* are known as either *catalysts* or *curing agents* (or even *hardeners*), depending on the resin type being used. *Fillers* are sometimes called *extenders*. *Additives* are also known as *modifiers* or *process aids*. Examples of some sample pultrusion resin systems, with trade names and precise quantities of the various constituents used, can be found in Meyer (1985) and Starr (2000). Most pultruders regard their mixes as proprietary and therefore usually do not openly publicize the exact details of their resin systems. Resin systems are typically developed by a pultrusion company in collaboration with a resin manufacturer.

Polymerization Agents Unsaturated polyester and vinylester resins that react with a styrene monomer are catalyzed with organic peroxides. The peroxide is used to "kick-off" or initiate the curing reaction and is heat-activated by the die so that the resin does not begin to gel and cure in the resin bath.[10] Different catalysts are chosen to cause the polymerization processes to occur at a controlled time and lengthwise position in the pultrusion die. They are referred to as *kicker peroxides*, *medium-reactivity peroxides*, and *finishing*

[10]For room-temperature processing, a MEKP (methyl ethyl ketone peroxide) catalyst is generally used. An accelerator or promoter, usually a cobalt napthanate, is also required to initiate the reaction. In addition, a dimethyl analine (DMA) promoter is also required with a room-temperature-cured vinylester.

peroxides (Starr, 2000). Peroxides are added in quantities of between 0.25 and 1.5% by weight of the resin and are given as parts per hundred (pph) by weight.[11] It is also not uncommon for additional styrene monomer to be added to the *base resin* (also known as the *neat resin*) during the resin mixing operation. Ten to 15% by weight can be added to aid in processing (it decreases the resin viscosity) and to decrease costs. However, the quantity of styrene in the resin mix must be carefully controlled, as unreacted styrene in the cured part can lead to a part with poor environmental durability. Blended polyester and vinylester resins are also used in pultrusion to exploit beneficial processing and property characteristics of each resin type.

Epoxy resins used in pultrusion are polymerized by the addition of curing agents, typically of the amine type. The curing agent or hardener is usually referred to as *part B,* and the epoxy resin is referred to as *part A*. Curing agents are generally added at ratios of 25 to 50% by weight of the epoxy resin. Curing agent accelerators may also be used in small percentages (less than 1% by weight). Epoxy resins are significantly more difficult to pultrude than polyester or vinylester resins because of their lower shrinkage, higher viscosity, and longer gel times (Starr, 2000). Because of this, the pultrusion line is usually slower and requires more pulling force when an epoxy resin is used. Specially formulated epoxies that cure only at high temperatures are also required. Limited epoxy formulations are used in pultrusion. Epoxies are used where significant mechanical or physical property advantages can be obtained. At this time, no routinely produced standard FRP profile shapes are made with epoxy resin systems; however, most pultruded FRP strengthening strips are produced with epoxy resins.

Fillers Inorganic particulate fillers are used to fill or "extend" the base polymer resin used for pultrusion for three primary reasons: to improve processing dynamics, to reduce cost, and to alter cured part properties (Lackey and Vaughan, 2002). Inorganic fillers have particle sizes between 0.5 and 8 μm (0.0002 to 0.0032 in.) and can have spherical or platelike geometries. The three primary types of inorganic fillers used in pultrusion are kaolin clay (aluminum silicate), calcium carbonate, and alumina trihydrate (ATH). Typical polyester and vinylester pultruded FRP profiles shapes and FRP rebars have between 10 and 30% by weight of filler in the resin mix. Small pultruded parts having primarily unidirectional roving reinforcement usually have low filler percentages, from no filler to 5% by weight. Epoxy resin pultrusions can also be filled. Small parts such as FRP strengthening strips are generally not filled. The density of the inorganic fillers is between 2.4 and 2.6 g/cm^3 (0.0865 to 0.0937 lb/in^3), making it approximately equal to glass fiber and

[11] The percentage is given as parts per hundred of the base resin (i.e., 100 parts of base polymer resin) or parts per hundred of the total resin mix (i.e., 100 parts of the total resin mix, including all secondary constituents). In this case the base resin could be as low as 50 pph of the total mix.

about twice the density of the resin. Fillers are substantially less costly than either resins or fibers. In addition to serving as a filler, alumina trihydrate serves a dual role as a fire-retardant additive. Fillers usually decrease the key longitudinal mechanical properties and the corrosion resistance of an FRP pultruded part (Lackey et al., 1999). However, fillers can be used to improve properties, especially in the transverse direction and to modify the physical properties (Meyer, 1985) of an FRP pultruded part. Recently, pultruders have begun to experiment with nanosized particle fillers, and results show FRP parts with greater corrosion resistance.

Additives The third group of constituents that are added to the resin mix are those used to assist in the processing or to modify the properties of the cured FRP part (Lackey and Vaughan, 2002). Chemical release agents, typically metallic stearates, fatty acids, or waxes, are used to prevent the FRP part from sticking to the die interior. A foaming agent is often used to remove entrained air from the resin mix. Pigments, or colorants, are mixed in the resin to give the finished part different colors. Ultraviolet stabilizers are added to protect the resin in the cured part from the effects of sunlight. Additives that retard flame spread in the cured FRP part, such as antimony trioxide, may be added to FRP profiles to meet code-stipulated fire and flammability ratings.[12] In addition, thickening agents, toughening agents, and viscosity control agents may be used to modify the characteristics of the resin mix. These additives are generally added in quantities less than 1% by weight of the resin. The one additive that is often added in a significantly higher percentage is a low-profile or shrink additive. Low-profile additives are usually thermoplastic polymer materials and are added to the mix to prevent shrinkage cracking in the interior of thick parts and at the surface of thin parts. Silane coupling agents may be added to the resin mix to improve the bond between the fibers and the resin; however, these coupling agents are usually in the sizing on the fiber itself. It is important to note that all of these additives can influence both the physical and mechanical properties of the FRP part since they all affect the resin chemistry (Lackey and Vaughan, 2002). Even a seemingly innocuous additive such as a pigment can have a significant effect on properties, and changing the color of a FRP pultruded part is not necessarily a trivial matter.

2.3.2 Hand Layup

Hand layup is the term given to the manual method of constructing an FRP composite part by laying up, or rather, putting down, successive layers of fibers and impregnating them with a liquid polymer resin, which then cures to form a solid FRP composite element. The solid part takes the form and

[12] The use of heavy metals in plastics is being reevaluated due to possible health hazards. Lead-based pigments are no longer used in the industry.

shape of the mold or surface to which it is applied. The method is also known as *laminating* or *wet layup* or simply *layup* and is used to make laminates or panels of FRP composites. The hand-layup method is probably the oldest method of producing FRP parts and is used to make a variety of FRP products. The method is deceptively simple, and producing a high-quality FRP part using the method requires a significant degree of skill and good quality control.

For use in industrial products, laminates constructed using the hand-layup method are usually thin, less than $\frac{1}{8}$ in. (3.2 mm) in thickness. They can be much thicker in the marine industry, where FRP boat hulls can approach thicknesses of 6 in. (150 mm) or more. A foam or balsa core is often used to create a sandwich structure, with the FRP layers serving as the outer skins. Sandwich layups are used extensively in the boat-building and recreational products industries to produce surfboards of all types. The method is used with various degrees of sophistication, which all serve to improve the quality or the part, defined as a part with a high-volume fraction of fibers, precise fiber placement, and very few voids, the most significant of these being the use of higher temperatures and pressures to cure the layers after they have been impregnated with the resin, and the precise control of the amount of resin that is applied to the fibers. In its most sophisticated form it is widely used in the aircraft industry to manufacture airfoils and fuselage parts for aircraft. In this form the resin system is preimpregnated in precise amounts onto unidirectional fibers that are partially cured into sheetlike products called *prepregs*. The prepregs are then cut and shaped and placed in different orientations on a mold which is then inserted into a polymeric bag from which the air is extracted by vacuum (known as a *vacuum bag*) and cured at high temperature and pressure in an autoclave. Numerous manufacturers produce prepregs for the aerospace and recreational products industries.

Although the hand-layup method as it is used in structural engineering appears to be similar to that used in other industrial processes, it is different in one very important aspect. When an industrial product such as a sailboat hull or tennis racquet is produced, the product is made in, or on, a mold and it is removed from the mold after curing. In fact, the removal, or the release, from the molding surface is usually achieved by using a chemical or natural release agent to ensure that the FRP part does not stick to the mold or surface on which it was formed. In the structural engineering application, the situation is opposite. The FRP composite that is applied by hand layup onto the surface of an existing structural element needs to be firmly adhered to, or be firmly in contact with, the surface in order to perform its strengthening function. Herein lies the difficulty in using the hand-layup method in structural engineering. Not only is the method being used to produce the FRP strengthening element, but it is being used to create the interface between the FRP element and the existing structural element. This interface, or interfacial region, is in and of itself a vital part of the FRP strengthening system. (When premanufactured pultruded FRP strips are used for strengthening, it is only the adhesive interfaces that are created in the field, and the application of precured strips can be thought of as a form of the hand-layup method, even though the actual FRP composite itself is not produced by the hand-layup method.)

Because of this key difference between the industrial hand-layup method and the structural engineering hand-layup method, it is crucial to select the fiber and resin systems very carefully, such that both the adhesive function and the wetting-out function of the resin are present. Since the adhesive properties of the resin system will depend on the surface to which it is being bonded and the method of application, proven, well-tested combinations of fibers and resins and application methods should be used only in structural engineering applications. Although fiber fabrics and resins can be purchased almost anywhere, not any fiber and resin can necessarily be used to strengthen a structural element. At this time only code-based design guidance is available to structural engineers for fiber and resin systems that are used to strengthen reinforced concrete structural members. It is these FRP systems for structural strengthening, as they are known, that are discussed in more detail in what follows and for which design procedures are given in later chapters. In addition, specific guidance is given in published guide construction specifications for commercially available FRP strengthening systems for concrete as to how the FRP material must be applied in the field and what level of training and experience is required of the field contactor. Although there are many known applications of FRP strengthening systems to masonry and timber structures, a design guide does not exist at this time for strengthening these structures with FRP composites.

When used in structural engineering, the hand-layup method is typically used in its most elementary form, with the resin-impregnated fibers being cured at ambient temperatures without the use of externally applied pressure or high temperature. In this case, the FRP composite laminate is formed directly on the structural element to be strengthened, such as a beam or a column, and cured in place, in much the same way that reinforced concrete is cured in place. In a few rare cases, attempts have been made to use vacuum bags or elevated temperatures in the field. When dry fiber broadgoods and liquid resins are used to form the composite in the field on the substrate to be strengthened or repaired, hand layup is performed in one of two ways, depending on the type of dry fiber material that is used. When precured FRP strips are used, the FRP composite is produced by the pultrusion method described previously and affixed to the substrate using a hand-layup method.

In the case of a lightweight fiber tow sheet, the method is similar to that used to apply wallpaper. Following application of a primer sealant and filling of holes with putty, the surface is coated with a thin layer of the liquid resin system (also called the *saturant*) using a fabric roller, similar to the way in which paint is applied (in fact, the viscosity of the resin is typically like that of paint). An appropriate length of the fiber tow sheet is then cut and placed on the wet resin layer. Plastic serrated rollers are then used to depress the fiber sheet into the resin, causing the fibers to get wet-out by the resin and forcing excess resin and air out of the FRP composite. If insufficient resin was applied initially to wet-out the fibers, a second overcoat layer is applied. If additional layers of FRP tow sheet are required, additional resin is applied to the surface of the existing fiber layer and the procedure is repeated. Typically, no specific measures beyond rolling and forcing out the resin and air with hand rollers are used to consolidate the layup, and often no special measures are taken to ensure that a specific amount of resin is applied. Hence, the eventual volume fraction for the fibers in the cured composite and the thickness of the cured composite are not controlled directly as they would be in a closed molding operation. The installer decides whether or not the fiber is sufficiently wet-out. In the same way as a wallpaper installer has a "feel" for how much adhesive to apply and how to squeeze out and remove air bubbles, so an FRP strengthening system installer needs to develop a feel for, and experience with, the application process.

In the other case, where heavier woven fabrics are used, the method is similar; however, the fabric is usually wet-out prior to being lifted into place and applied to the structure. It is more difficult to wet-out thicker and heavier woven fabrics than to wet-out thin unidirectional tow sheets, and simply placing the fabric up on the previously wet surface will not yield sufficient liquid resin to wet-out the fabric. Many FRP structural strengthening systems suppliers have developed proprietary pieces of field equipment (typically called *saturators*) that consist of rollers and resin baths to wet-out the fabrics just prior to installation. After the fabric is saturated with the resin in the saturator, it is rerolled onto a plastic PVC pipe and then taken to the location where it is to be installed and unrolled again onto the concrete surface. After the saturated fabric is in place, it is rolled with hand rollers to expel surplus resin and air. Somewhat better control of the resin-to-fiber ratio can be achieved using a saturator than with a manual method; however, precise control of the amount of resin used is still difficult to achieve.[13]

The rate of installation in both methods depends very much on the *pot life* of the resin, which is the time that the mixed resin will stay in liquid form before beginning to gel and then cure. A typical epoxy resin used for structural hand layup will have a pot life of 1 to 4 hours between 60 and 80°F (15 and 27°C). At higher ambient temperatures pot life decreases, and at low temperatures some resins will not cure. Therefore, a surface area that can reasonably be covered in this amount of time by two installers is usually selected. If multiple layers of the FRP are to be used, all the layers must be cured simultaneously, and therefore all the layers must be applied when the resin is still wet.

[13]Some manufacturers recommend that a specific fiber-to-resin volume or weight ratio be used with their systems and provide instructions for weighing wet samples in the field to determine the appropriate resin quantity.

Fiber System for Hand Layup Two primary types of fiber systems are used when the hand-layup method is used for FRP strengthening: unidirectional tow sheets and uni- or multidirectional woven or stitched fabrics. Carbon and E-glass are the most commonly used fiber types; however, some manufacturers do supply aramid fiber fabrics and also hybrid fiber fabrics. AR-glass fiber fabrics can be obtained for corrosive environments.

Carbon Fiber Tow Sheets The term *tow sheet* is used to describe a wide, dry carbon fiber product in which individual carbon tows, usually 12K tows, are aligned parallel to each other and held in place by an open-weave glass fiber scrim cloth and epoxy-soluble adhesive. The scrim cloth is oriented at a $\pm 45°$ angle to the tow fiber as shown in Fig. 2.12. It is called a *sheet,* to differentiate it from a fabric, because it is very thin and is not woven or stitched. The tow sheet is supplied with a waxed paper backing to keep the thin layers separate and to aid in unrolling during installation. The thickness of a carbon fiber tow sheet is on the order of 0.005 to 0.015 in. (0.127 to 0.381 mm), and it weighs from 5 to 9 oz/yd^2 (150 to 400 g/m^2). The fibers in the tow sheet are straight and are called *stretched fibers* by some manufacturers. They do not have the undulating form that fibers in a woven fabric have due to the warp and weft weave. Unlike prepreg sheets, the fibers in tow sheets are not preimpregnated with a resin system and tow sheets should not be confused with prepregs, even though they are similar in appearance. Carbon fiber tow sheets are supplied in widths from 12 to 40 in. (300 to 1000 mm) and are available in standard and high-modulus carbon fibers. Carbon

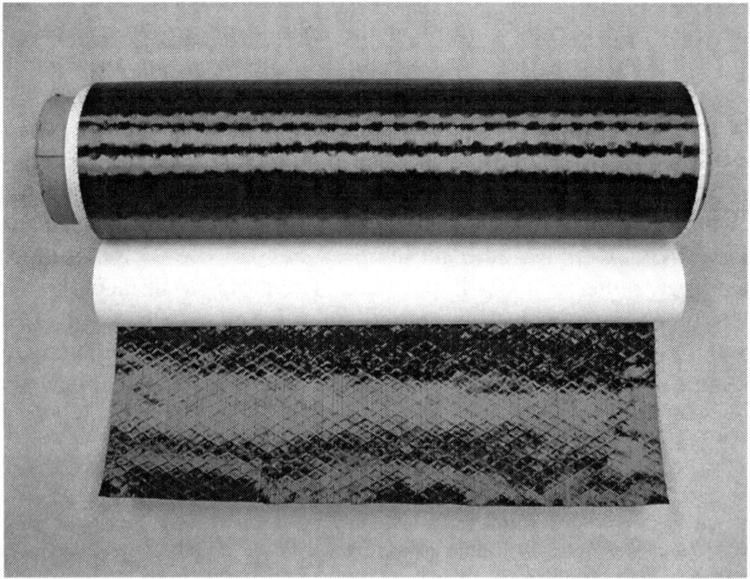

Figure 2.12 Carbon fiber tow sheet.

fiber tow sheets of this type were first produced in Japan in the 1980s by the Tonen Corporation (now the Nippon Steel Composites Company Ltd.) and were known as the Forca tow sheet (FTS).

The term *tow sheet* is the preferred term to use to describe a carbon fiber product that has continuous flat fibers held together by a very lightweight scrim cloth. However, this nomenclature is, unfortunately, not universal. Some manufacturers refer to conventional stitched fabrics (where the transverse fiber is a structural fiber and not a scrim fiber used solely to hold the primary longitudinal fibers in place) as tow sheets, and to tow sheets as fabrics. A number of manufacturers also refer to glass fiber tow sheets, which are usually woven with a very sparse and lightweight weft roving. For this reason, a structural engineer should always request to see a sample of the actual product being used in a structural design and listed in the project specifications and should not rely solely on verbal or written descriptions of the product.

It is important to note that the thickness measurement given for a carbon fiber tow sheet is a nominal measurement and is the average thickness of the fibers only and is not the thickness of the tow sheet after it has been impregnated with resin and cured. In addition, the longitudinal strength and stiffness properties generally reported by manufacturers for a tow sheet are the properties of the fiber only and are not the properties of the FRP composite material.

Carbon and Glass Fiber Fabrics The other large family of fiber products used for FRP strengthening applications are of the woven or stitched fabric type. These fabrics are similar to those used in pultrusion applications; however, they are typically supplied with a predominantly 0° fiber system (or a unidirectional fiber system), as shown in Fig. 2.13. Notice that the individual tows in the fabric product can be clearly identified, as opposed to the carbon tow sheet, where individual tows cannot be identified. The undulating pattern of the weave can also be seen in the fabric.

Bidirectional fabrics with fibers that are usually balanced in the 0° and 90° orientations are also available and can be used in cases in which bidirectional strengthening is desired, such as walls or two-way slabs. Both carbon and glass fabrics for structural strengthening are available and range in weights from 9 to 27 oz/yd^2 (300 to 900 g/m^2). Many of the unidirectional fabrics that are produced have a sparsely spaced (usually, at about 1 in. along the length) transverse weft fiber that holds the unidirectional carbon tows or glass rovings together. Fabrics are dry and are not supplied with a backing paper. Some unidirectional fabrics are supplied stitched to surfacing veils or mats similar to those used in pultruded products to hold the longitudinal fibers together. Fabrics are generally available in widths up to 84 in.

Resin System for Hand Layup Epoxy resins are used almost exclusively for structural hand layup for FRP strengthening applications. This is due largely to their superior adhesive properties and their low shrinkage when cured relative to polyester and vinylester resins that are used extensively in

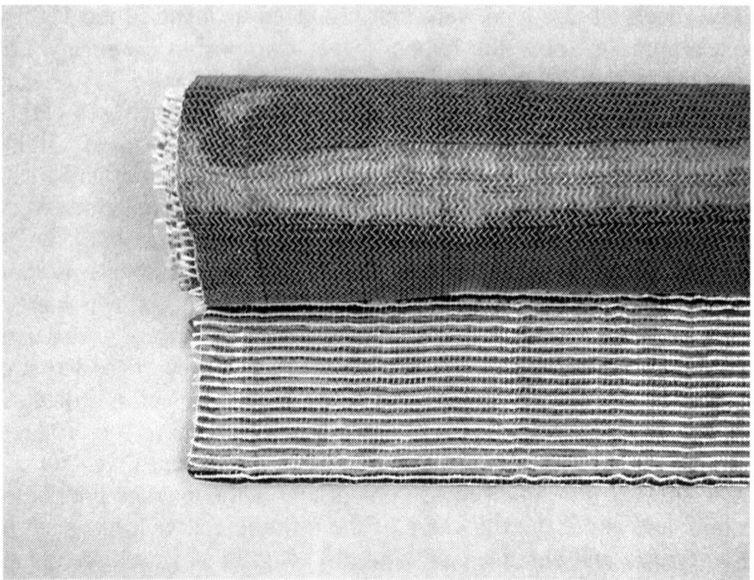

Figure 2.13 Carbon fiber fabric with an aramid weft fiber.

the industrial hand-layup method. Epoxy resins are reasonably easy to mix in the field, typically requiring the mixing of one large amount, the epoxy resin, together with another large amount, the curing agent. This is very different from polyester and vinylester resins, which require very small percentages (often less than 1%) of catalysts and processing aids to be added at the time of mixing to process the resin. In addition, the properties of the epoxy are not as highly dependent on the mix ratios as polyester and vinylester resins, which can be affected significantly by small amounts of additive or catalyst.

For FRP strengthening applications, the epoxy resin is usually sold in a packaged system together with the fiber sheet or fabric and additional surface primers and putties, which are also epoxy based. These packaged epoxies are typically formulated by FRP strengthening system manufacturers to be compatible with their fabrics and sheets and to be of the appropriate viscosity to be used in the field (either as a saturating resin or as an adhesive). These epoxies are typically unfilled systems that are clear to yellow or amber color. Many manufacturers pigment their brand-name epoxy resin formulations so that they are identifiable. Green, blue, and gray systems are common. They have a fairly low viscosity of around 500 to 1000 cP, similar to that of 10-weight motor oil. They are easily rolled with paint rollers. But due to their lower viscosity, they tend to run and drip if overapplied, making the installation process somewhat messy, especially in overhead applications.

It is important to note that where epoxy resins are used to saturate fiber sheets or to bond precured FRP strips to concrete substrates at room temperatures, the glass transition temperature of the cured FRP strengthening system will be in the range 50 to 60°C (120 to 140°F). This temperature is close to the upper limits of the regular operating temperature of a bridge or building structure on a hot day in some parts of the world. The glass transition temperature can be increased only by postcuring the layup at elevated temperatures, which is difficult to do in the field but has been tried by a number of researchers. Nevertheless, consideration should be given to postcuring FRP strengthening systems in the field with heaters if at all possible.

As noted previously, the condition of the surface to which the FRP strengthening system is applied in the hand-layup method is very important. When the FRP strengthening system is applied to a reinforced concrete member, for example, the surface of the concrete needs to be smoothed to a specific profile, sandblasted, and cleaned. If the concrete is damaged (e.g., due to corrosion), it must first be repaired. Protrusions must be ground down. The surface must be dried and a primer must be applied to seal the concrete. Thereafter, a putty is used to fill holes and cracks. After the FRP layers have been applied, a protective coating is usually applied as the last coat. An architectural finish can also be applied over the strengthening system if desired. These details must be specified by the structural engineer in the project specifications and should not be left to the FRP system installer as a "performance" item. Guide specifications and construction checklists published by the National Cooperative Highway Research Program in the United States should be consulted and followed (NCHRP, 2004).

Epoxy resin systems used as adhesives to bond precured pultruded FRP strips have different formulations from those used in the dry fiber and liquid resin hand layup method. These epoxy systems, which are referred to as *adhesive pastes,* are typically filled with inorganic fillers (silica sand, silica fume, clays) and have a much higher viscosity, in the range of 100,000 cP, and a consistency similar to that of peanut butter. They are applied with a spatula or trowel much like drywall compound (vinyl spackle).

2.3.3 Other Manufacturing Processes

A number of other processing methods are used in a limited way to produce FRP composite material parts for structural engineering applications. These processes are typically industrialized processes that produce single units (one-off) rather than being continuous production processes like that of pultrusion (Schwartz, 1997b). Tubular products used as stay-in-place column or beam forms, or as FRP piles, are manufactured using a filament winding process from either glass or carbon fibers with epoxy, vinylester, and polyester resins. In filament winding a resin-saturated fiber roving or tow is wound continuously around a cylindrical mandrel at a variety of wind angles. The part is cured using heaters and is then removed from the mandrel, which may be collapsible. Filament winding is used to make FRP pressure tanks and FRP poles. A planar variant of filament winding is used to manufacture FRP grids for concrete reinforcement.

FRP bridge deck panels, FRP fender piles, and rigid FRP jackets for column strengthening have been produced using different variations of open and closed molding, including resin transfer molding (RTM), resin infusion molding (RIM), or vacuum-assisted resin transfer molding (VARTM). In this process the dry fiber forms, usually fabrics, are arranged on molds or on core materials, and the entire fiber form is then saturated with resin and cured. If the molded is closed, the resin is transferred under pressure and the resulting parts have two finished sides. If the mold is single sided, it is usually enclosed in a vacuum bag and the resin is infused using vacuum pressure. The resin is sucked into the vacuum bag, flows through the fiber perform, and excess resin then exits the bag through special ports. Large bridge deck panels have been produced using this method (Bakis et al., 2002).

PROBLEMS

2.1 Find information about the FRP composite material constituents and products listed below. For each item, provide a short written description of the product, a Web page link, and a picture of the product.

A&P Technology: braided fabrics
Akzo: peroxides
Aldrich: dimethyl analine
AOC: polyester resins
Ashland: hetron resins
Dow: momentum resins
DuPont: aramid fibers
Engelhard: kaolin clay fillers
Entec Composite Machines: pultrusion equipment
Frekote: mold releases
Georgia Pacific: phenolic resins
Hexcel: carbon epoxy prepregs
Hollinee: A-glass continuous strand mats
Huber Engineered Materials: fillers
Magnolia Plastics, Inc.: epoxy resins
Nexus: veils
Owens-Corning Fiberglas: E-glass continuous filament mat
PPG: 113 yield glass roving
Reichhold: vinylester resins
Resolution Specialty Materials: epoxy resins
Saint Gobain Fibers: stitched fabrics
Sudaglass: basalt fibers
Toho: carbon fibers

Toray: carbon fibers
Vectorply Corporation: multiaxial fabrics
Zoltek: carbon fibers

SUGGESTED ACTIVITIES

2.1 Visit a pultrusion company. They usually welcome tours of students. Contact the American Composites Manufacturers Association (ACMA), the PIC (Pultrusion Industry Council), or the EPTA (European Pultrusion Technology Association) for information on nearby pultrusion companies.

2.2 Do a hand layup using standard boat cloth and polyester resin with MEKP catalyst. Obtain materials from a marine supply store (West Marine) or Fiberglast or other local source.

2.3 Obtain a sample of a pultruded profile. Cut, polish, and examine the cross section with an optical microscope to see the layered structure of mats and rovings. Drill and cut the specimen with a diamond blade circular saw. Understand the need for ventilation and vacuum cleaning composite dust. Get a feel for the products and how to work with them. Bond a strain gauge to an FRP part and measure strains.

2.4 Mix up some epoxy resins. Change the part ratios and study the effect. Use a thermometer to measure the exothermic temperature. Mix similar quantities in a tall, narrow cylinder and a low, flat pan and observe the differences.

2.5 Get some carbon sheet and fabric and try to cut it in a straight line! Experiment with various cutting tools. Try to cut an aramid fiber fabric with regular scissors. Record your observations.

2.6 Cut up a composite tennis (or squash or racquetball) racquet and look inside. Write a short paper on how composite material racquets are made. What kinds of fibers and resins are used?

2.7 Purchase a quick-setting two-component epoxy from your local hardware store (<$5). Mix up in a plastic container (e.g., the bottom section of a disposable 0.5-L water bottle). Hold the container in your hand and feel the temperature rise. If a thermocouple is available, insert it in the epoxy while the epoxy is curing and measure the exothermic reaction temperature. Observe the change of state (liquid to solid) as the temperature changes.

3 Properties of FRP Composites

3.1 OVERVIEW

To design a structure with an FRP composite, a structural engineer must have knowledge of the physical, and especially, the mechanical properties of the FRP material being used in the design. These properties are typically provided by the manufacturer of the FRP material product. However, at this time there is no standard format for providing mechanical and physical property data. Often, incomplete property data are provided, and in many cases it is not clear if the property data reported are actual test data or have been reduced by a safety factor for design purposes. Therefore, it is important for a structural engineer to understand how properties of FRP composites are determined so that independent checks or tests can, if necessary, be conducted to verify the FRP material properties reported.

Essentially, there are two ways to determine the properties of an FRP composite for use in design. The first is via theoretical calculations, and the second is via experimental measurements. Since an FRP composite is an inhomogeneous material, it can be characterized theoretically and experimentally on a number of different levels (or scales), as discussed in Chapter 1. Structural engineers should understand what these different characterization levels are when specifying material tests or when conducting numerical modeling, such as finite element analysis. As noted in Chapter 1, an FRP composite material can be characterized on one of four levels: (1) the fiber level, (2) the lamina level, (3) the laminate level, and (4) the full-section level. Below, methods are described for determining the properties at each of these levels by both theoretical and experimental methods.

3.2 THEORETICAL DETERMINATION OF PROPERTIES

3.2.1 Fiber Level

Many properties of FRP composite can be predicted from the properties of the fiber and polymer resin system (known as the *matrix* in mechanics terminology[1]). This fiber-level analysis is known as *micromechanics*. Both phys-

[1] The term *matrix* is used to designate the entire resin system, including the polymer, the fillers, the additives, and the processing aids.

ical and mechanical properties can be predicted using micromechanics models. Different levels of sophistication exist in micromechanics models that have been developed over the years (Bank, 1993a).

For FRP composites in structural engineering, micromechanics is used for three primary reasons: (1) to relate mass or weight fractions of the constituents to their volume fractions, (2) to get estimates of the stiffness properties and selected strength properties of unidirectional composites, and (3) to approximate the longitudinal strength and longitudinal modulus of a fiber sheet material by ignoring the mechanical contribution of the matrix material in the longitudinal direction. For structural engineering calculations, only simple micromechanics models, based on the rule of mixtures, are commonly used for these purposes.

To use micromechanics models to predict the properties of an FRP composite, the physical and mechanical properties of the fiber and matrix constituents need to be known. These are usually obtained by experimental methods. In simple models it is assumed that the matrix is a homogeneous material that is isotropic and linear elastic. At the first approximation the fibers are also usually assumed to be homogeneous, isotropic, and linear elastic.[2]

Assuming that there are no voids in the FRP composite, the relationship between the weight (or mass) fractions, w, and the volume fractions, v, of the fiber (subscript f) and matrix (subscript m) constituents of an FRP composite (subscript c) are

$$v_f = w_f \frac{\rho_c}{\rho_f} \tag{3.1a}$$

$$v_m = w_m \frac{\rho_c}{\rho_m} \tag{3.1b}$$

$$\rho_c = v_f \rho_f + v_m \rho_m = \frac{1}{w_f/\rho_f + w_m/\rho_m} \tag{3.1c}$$

$$v_f + v_m = 1 \tag{3.1d}$$

$$w_f + w_m = 1 \tag{3.1e}$$

$$v_f = \frac{w_f(v_f \rho_f + v_m \rho_m)}{\rho_f} \tag{3.1f}$$

[2] Glass fibers are, in fact, isotropic; however, carbon fibers are anisotropic. If available, the anisotropic properties of the carbon fiber can be used in the micromechanics equations (see Tsai and Hahn, 1980).

$$v_f = \frac{V_f}{V_c} \qquad w_f = \frac{W_f}{W_c} \tag{3.1g}$$

$$v_m = \frac{V_m}{V_c} \qquad w_m = \frac{W_m}{W_c} \tag{3.1h}$$

where V_i is the volume of constituent i, W_i the weight (or mass) of constituent i, and ρ_i the weight density (or mass density) of constituent i. The mass and volume fraction relationships above are important in the production of FRP composites, since quantities of material are usually measured by weight when mixed or when priced.

The four independent engineering properties of an orthotropic (or transversely, isotropic[3]) unidirectional FRP composite (E_x, E_y, E_s, ν_x) with known isotropic fiber properties (E_f, ν_f, G_f), isotropic matrix properties (E_m, ν_m, G_m), and fiber and matrix volume fractions (v_f, v_m) can be estimated at a first approximation by the following rule-of-mixtures equations:

$$E_x = v_f E_f + v_m E_m \tag{3.2a}$$

$$\frac{1}{E_y} = \frac{v_f}{E_f} + \frac{v_m}{E_m} \tag{3.2b}$$

$$\frac{1}{E_s} = \frac{v_f}{G_f} + \frac{v_m}{G_m} \tag{3.2c}$$

$$\nu_x = v_f \nu_f + v_m \nu_m \tag{3.2d}$$

$$\nu_y E_x = \nu_x E_y \tag{3.2e}$$

where E_x is the longitudinal modulus in the fiber direction, E_y the transverse modulus perpendicular to the fiber direction, E_s the in-plane shear modulus, ν_x the longitudinal (or major) Poisson ratio, and ν_y the transverse (or minor) Poisson ratio.[4] The longitudinal tensile strength of a unidirectional composite, X_t, which is dominated by the properties of the fiber, can be estimated at a first approximation by

$$X_t = v_f \sigma_{f,t}^{\text{ult}} + v_m \left(E_m \frac{\sigma_{f,t}^{\text{ult}}}{E_f} \right) \tag{3.3}$$

[3] The transversely isotropic composite is the three-dimensional version of the two-dimensional unidirectional composite. It has five independent constants, the four noted above plus an out-of-plane Poisson ratio (see Bank, 1993a).

[4] The stiffness properties of a unidirectional composite are discussed further below.

where $\sigma_{f,t}^{\mathrm{ult}}$ is the ultimate tensile strength (UTS) of the fiber and it is assumed that in the longitudinal direction the fiber fails before the matrix. The second term in the equation is usually small, and therefore the longitudinal tensile strength of a unidirectional composite can be approximated conservatively by

$$X_t = v_f \sigma_{f,t}^{\mathrm{ult}} \tag{3.4}$$

When the fibers are continuous, as is typically the case, the fiber volume fraction is often replaced by the fiber area fraction, a_f. This leads to the equation often used with FRP strengthening systems for the modulus (E_{FRP}) and the strength (σ_{FRP}) of the strengthening system when the contribution of the polymer matrix is ignored, as

$$E_{\mathrm{FRP}} = a_f E_f \tag{3.5}$$

$$\sigma_{\mathrm{FRP}} = a_f \sigma_f^{\mathrm{ult}} \tag{3.6}$$

The transverse tensile strength, Y_t, the longitudinal compressive strength, X_c, the transverse compressive strength, Y_c, and the in-plane shear strength, S, are all matrix-dominated properties and are significantly lower than the longitudinal tensile strength. Equations to estimate their values can be found in Agarwal and Broutman (1990); however, these are regarded only as order-of-magnitude estimates. Reliable theoretical equations for prediction of the properties in the matrix-dominated failure modes as a function of the properties of the fiber and the matrix are not available.

The expansion and transport properties of an FRP composite can also be predicted at a first approximation from the constituent properties of the fiber and the matrix material. In a unidirectional FRP composite, these properties are anisotropic, and values need to be determined for both the longitudinal and transverse directions of the composite. The coefficients of thermal expansion, α_x and α_y, and the coefficients of moisture expansion, β_x and β_y, are used to determine the longitudinal and transverse strains in an FRP composite when subjected to thermal or hygroscopic loads. The transport properties, which also have both longitudinal and transverse values, are diffusivity, D_x and D_y; the thermal conductivity, k_x and k_y; the electrical conductivity, e_x and e_y; and the magnetic permeability, m_x and m_y. Both the hygroscopic expansion coefficients and the transport coefficients are usually predicted at a first approximation using the rule of mixtures for the longitudinal and transverse directions, respectively, and the fiber and matrix constituents are usually assumed to have isotropic hygroscopic and transport properties (Bank, 1993a).

Other important physical properties of an FRP composite that have a significant influence on the long-term durability of the composite, such as the glass transition temperature, the fatigue life, creep compliance, ignition and combustion temperatures, impact resistance, and damping ratios, are not generally predicted from micromechanics (see Bank, 1993a). The glass transition temperature, T_g, of the unidirectional FRP composite is often assumed to be equal to the glass transition temperature of the matrix material.

Analysis Example 3.1[5] An E-glass/vinylester unidirectional rod is to be used as an FRP reinforcing bar. The rod has an 80% fiber weight fraction. Assuming that there are no voids in the composite, determine the fiber volume fraction and the stiffness properties of the composite rod. You are given the following properties for the fiber and the matrix:

E-glass fiber: $\rho_f = 9.4 \times 10^{-2}$ lb/in^3, $E_f = 10.5 \times 10^6$ psi, $\nu_f = 0.20$

Vinylester matrix: $\rho_m = 4.57 \times 10^{-2}$ lb/in^3, $E_m = 500{,}000$ psi, $\nu_m = 0.38$

SOLUTION

$$G_f = \frac{E_f}{2(1 + \nu_f)} = 4.375 \times 10^6 \text{ psi}$$

$$G_m = \frac{E_m}{2(1 + \nu_m)} = 181{,}159 \text{ psi}$$

$$\rho_c = v_f\rho_f + v_m\rho_m = \frac{1}{w_f/\rho_f + w_m/\rho_m} = \frac{1}{0.8/0.094 + 0.2/0.0457}$$
$$= 0.0776 \text{ lb/in}^3$$

$$v_f = w_f\frac{\rho_c}{\rho_f} = 0.8\left(\frac{0.0776}{0.094}\right) = 0.66$$

$$E_x = v_fE_f + v_mE_m = 0.66(10.5 \times 10^6) + 0.34(0.5 \times 10^6) = 7.1 \times 10^6 \text{ psi}$$

$$\frac{1}{E_y} = \frac{v_f}{E_f} + \frac{v_m}{E_m} = \frac{0.66}{10.5 \times 10^6} + \frac{0.34}{0.5 \times 10^6}$$
$$= 7.42 \times 10^{-7} \Rightarrow E_y = 1.35 \times 10^6 \text{ psi}$$

[5] In analysis examples in this book, results are often provided to three or even four significant figures to enable the reader to follow the calculations. However, such precision is not advised for practical engineering calculations since the properties of the constituents are known only to one or two significant figures.

$$\frac{1}{E_s} = \frac{v_f}{G_f} + \frac{v_m}{G_m} = \frac{0.66}{4.375 \times 10^6} + \frac{0.34}{0.181 \times 10^6}$$

$$= 2.029 \times 10^{-6} \Rightarrow E_s = 492{,}778 \text{ psi}$$

$$\nu_x = v_f \nu_f + v_m \nu_m = 0.66(0.20) + 0.34(0.38) = 0.26$$

$$\nu_y = \frac{E_y}{E_x} \nu_x = \frac{1.35}{7.1}(0.26) = 0.049$$

3.2.2 Lamina Level

A single ply (or *lamina*) of a planar FRP composite material that contains all of its fibers aligned in one direction is called a *unidirectional ply.* The unidirectional ply plays a key role in characterizing the behavior of FRP composites in structural engineering for the following reasons: (1) Many FRP products used in structural engineering, such as FRP rebars, FRP sheets, and FRP fabrics, are used in the unidirectional form as the FRP end product; (2) the properties of FRP composites are often obtained experimentally by testing unidirectional FRP materials; and (3) the unidirectional ply is the basic building block for calculating the properties of a multidirectional FRP laminate, often used to represent the walls of a pultruded profile or the structure of a hand-laid-up composite.

It is assumed that the unidirectional ply is a two-dimensional plate and that its thickness is very small compared to its in-plane dimensions. The constitutive relations (or the stress–strain relations) for a unidirectional ply define the relationships between the in-plane stresses (σ_x, σ_y, σ_s) and the in-plane strains (ε_x, ε_y, ε_s) in a local or ply x,y,s coordinate system.[6] The global or structural in-plane coordinate system is defined as a 1,2,6 coordinate system, and the stresses and strains in the global system are σ_1, σ_2, σ_6 and ε_1, ε_2, ε_6, respectively.[7] The ply is assumed to be linear elastic and orthotropic with the local x-axis aligned parallel to the fiber direction and the local y-axis perpendicular to the fiber direction. Tensile and compressive moduli are assumed to be equal. As shown in Fig. 3.1, the angle, θ, measured counterclockwise from the global 1-direction to the local x-direction, defines the in-plane relationship between the orthogonal coordinate systems.

[6] This chapter follows the notation, terminology, and sign convention of Tsai and Hahn (1980). In many other texts, the local coordinate system is identified as the (*L,T,LT*) system (longitudinal, transverse, and shear orientations). The global coordinate system is often identified as the (1,2,12) or (11,22,12) system.

[7] This form is known as the *contracted notation.* The shear direction is the s or the 6 direction. The shear strain in this notation is the engineering shear strain and is equal to twice the tensorial shear strain (i.e., $\varepsilon_s = \gamma_{xy} = \varepsilon_{xy} + \varepsilon_{yx}$ and $\varepsilon_6 = \gamma_{12} = \varepsilon_{12} + \varepsilon_{21}$).

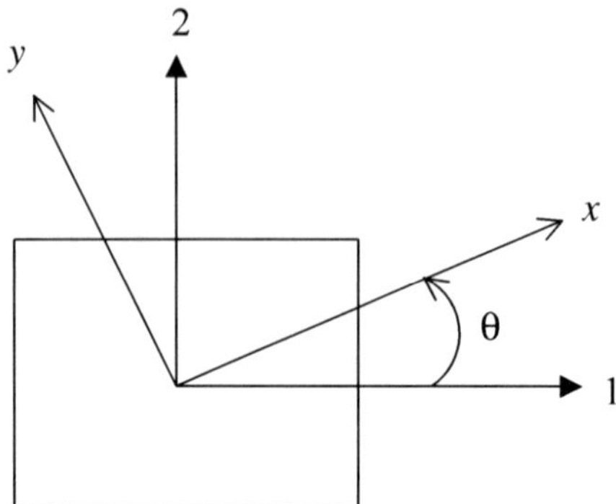

Figure 3.1 In-plane coordinate systems for a unidirectional FRP lamina.

The unidirectional ply is orthotropic in the *x,y,s* coordinate system but anisotropic in the 1,2,6 coordinate system.[8] The stress–strain relations in the *x,y,s* coordinate system, also called the *on-axis orientation,* in terms of the five engineering constants (E_x, E_y, ν_x, ν_y, and E_s) of the unidirectional ply are expressed most conveniently in compliance form as

$$\begin{Bmatrix} \varepsilon_s \\ \varepsilon_y \\ \varepsilon_s \end{Bmatrix} = \begin{bmatrix} \dfrac{1}{E_x} & \dfrac{-\nu_y}{E_y} & 0 \\ \dfrac{-\nu_x}{E_x} & \dfrac{1}{E_y} & 0 \\ 0 & 0 & \dfrac{1}{E_s} \end{bmatrix} \begin{Bmatrix} \sigma_x \\ \sigma_y \\ \sigma_s \end{Bmatrix} \tag{3.7}$$

or in indicial form as

$$\varepsilon_i = S_{ij}\sigma_j \qquad (i,j = x,y,s) \tag{3.8}$$

where the longitudinal modulus, E_x, the transverse modulus, E_y, the major Poisson ratio, ν_x, the minor Poisson ratio, ν_y, and the in-plane shear modulus, E_s,[9] are defined as

[8] The unidirectional ply is often referred to in the literature as *specially orthotropic,* since it is orthotropic in the global system only when θ is zero.

[9] The shear modulus is often denoted by the letter G and defined as $G = \tau/\gamma$.

$$E_x = \frac{\sigma_x}{\varepsilon_x} = \frac{1}{S_{xx}} \quad (\sigma_x \neq 0,\ \sigma_y,\ \sigma_s = 0) \tag{3.9a}$$

$$\nu_x = -\frac{\varepsilon_y}{\varepsilon_x} = -\frac{S_{yx}}{S_{xx}} \quad (\sigma_x \neq 0,\ \sigma_y,\ \sigma_s = 0) \tag{3.9b}$$

$$E_y = \frac{\sigma_y}{\varepsilon_y} = \frac{1}{S_{yy}} \quad (\sigma_y \neq 0,\ \sigma_x,\ \sigma_s = 0) \tag{3.9c}$$

$$\nu_y = -\frac{\varepsilon_x}{\varepsilon_y} = \frac{S_{xy}}{S_{yy}} \quad (\sigma_y \neq 0,\ \sigma_x,\ \sigma_s, = 0) \tag{3.9d}$$

$$E_s = \frac{\sigma_s}{\varepsilon_s} = \frac{1}{S_{ss}} \quad (\sigma_s \neq 0,\ \sigma_x,\ \sigma_y = 0) \tag{3.9e}$$

Since the compliance matrix, S_{ij}, is symmetric, only four of the engineering constants are independent. The minor Poisson ratio is found using the symmetry relations

$$\nu_y = \frac{E_y}{E_x}\nu_x \tag{3.10}$$

It is important to note that the shear modulus is an independent quantity for an orthotropic ply and cannot be determined from the isotropic relation:

$$G = \frac{E}{2(1+\nu)} \tag{3.11}$$

For an isotropic material plate (or ply), $E_x = E_y = E$, $\nu_x = \nu_y = \nu$, and $E_s = G$, of which only two of the three constants are independent. In addition, these elastic constants are independent of orientation.

For laminate calculations the stiffness form of the stress strain relations is preferred and is given as

$$\begin{Bmatrix} \sigma_x \\ \sigma_y \\ \sigma_s \end{Bmatrix} = \begin{bmatrix} \dfrac{E_x}{1-\nu_x\nu_y} & \dfrac{\nu_y E_x}{1-\nu_x\nu_y} & 0 \\ \dfrac{\nu_x E_y}{1-\nu_x\nu_y} & \dfrac{E_y}{1-\nu_x\nu_y} & 0 \\ 0 & 0 & E_s \end{bmatrix} \begin{Bmatrix} \varepsilon_x \\ \varepsilon_y \\ \varepsilon_s \end{Bmatrix} \tag{3.12}$$

or in indicial notation as

$$\sigma_i = Q_{ij}\varepsilon_j \qquad (i,j = x,y,s) \tag{3.13}$$

where the Q_{ij} are the ply on-axis stiffness coefficients.

When the unidirectional ply is rotated through a counterclockwise angle, θ, into an off-axis orientation as shown in Fig. 3.1, the stresses and strains in the local x,y,s ply coordinate system can be obtained from the stresses and strains in the global 1,2,6 coordinate system using the following transformation equations:

$$\begin{Bmatrix} \sigma_x \\ \sigma_y \\ \sigma_s \end{Bmatrix} = \begin{bmatrix} \cos^2\theta & \sin^2\theta & 2\cos\theta\sin\theta \\ \sin^2\theta & \cos^2\theta & -2\cos\theta\sin\theta \\ -\cos\theta\sin\theta & \cos\theta\sin\theta & \cos^2\theta\,\sin^2\theta \end{bmatrix} \begin{Bmatrix} \sigma_1 \\ \sigma_2 \\ \sigma_6 \end{Bmatrix} \tag{3.14a}$$

$$\begin{Bmatrix} \varepsilon_x \\ \varepsilon_y \\ \varepsilon_s \end{Bmatrix} = \begin{bmatrix} \cos^2\theta & \sin^2\theta & \cos\theta\sin\theta \\ \sin^2\theta & \cos^2\theta & -\cos\theta\sin\theta \\ -2\cos\theta\sin\theta & 2\cos\theta\sin\theta & \cos^2\theta\,\sin^2\theta \end{bmatrix} \begin{Bmatrix} \varepsilon_1 \\ \varepsilon_2 \\ \varepsilon_6 \end{Bmatrix} \tag{3.14b}$$

In addition, the anisotropic off-axis stiffness coefficients of the unidirectional ply can be obtained in terms on the orthotropic on-axis stiffness coefficients from

$$\begin{Bmatrix} Q_{11} \\ Q_{22} \\ Q_{12} \\ Q_{66} \\ Q_{16} \\ Q_{26} \end{Bmatrix} = \begin{bmatrix} \cos^4\theta & \sin^4\theta & 2\cos^2\theta\sin^2\theta & 4\cos^2\theta\sin^2\theta \\ \sin^4\theta & \cos^4\theta & 2\cos^2\theta\sin^2\theta & 4\cos^2\theta\sin^2\theta \\ \cos^2\theta\sin^2\theta & \cos^2\theta\sin^2\theta & \cos^4\theta + \sin^4\theta & -4\cos^2\theta\sin^2\theta \\ \cos^2\theta\sin^2\theta & \cos^2\theta\sin^2\theta & -2\cos^2\theta\sin^2\theta & (\cos^2\theta - \sin^2\theta)^2 \\ \cos^3\theta\sin\theta & -\cos\theta\sin^3\theta & \cos\theta\sin^3\theta - \cos^3\theta\sin\theta & 2(\cos\theta\sin^3\theta - \cos^3\theta\sin\theta) \\ \cos\theta\sin^3\theta & -\cos^3\theta\sin\theta & \cos^3\theta\sin\theta - \cos\theta\sin^3\theta & 2(\cos^3\theta\sin\theta - \cos\theta\sin^3\theta) \end{bmatrix} \begin{Bmatrix} Q_{xx} \\ Q_{yy} \\ Q_{xy} \\ Q_{ss} \end{Bmatrix} \tag{3.15}$$

The six independent off-axis compliance coefficients can be obtained directly from the on-axis compliance coefficients or by inversion of the 3 × 3 stiffness matrix as follows:

$$S_{ij} = Q_{ij}^{-1} \qquad (i,j = 1,2,6) \tag{3.16}$$

The compliance matrix for the off-axis ply can be written in terms of nine off-axis "effective" engineering constants (of which six are independent) as follows:

$$\begin{Bmatrix} \varepsilon_1 \\ \varepsilon_2 \\ \varepsilon_6 \end{Bmatrix} = \begin{bmatrix} \frac{1}{E_1} & -\frac{\nu_{12}}{E_2} & \frac{\nu_{16}}{E_6} \\ \frac{\nu_{21}}{E_1} & \frac{1}{E_2} & \frac{\nu_{26}}{E_6} \\ \frac{\nu_{61}}{E_1} & \frac{\nu_{62}}{E_2} & \frac{1}{E_6} \end{bmatrix} \begin{Bmatrix} \sigma_1 \\ \sigma_2 \\ \sigma_6 \end{Bmatrix} \tag{3.17}$$

or

$$\varepsilon_i = S_{ij}\sigma_j \qquad (i,j = 1,2,6) \tag{3.18}$$

The off-axis engineering constants are defined as

$$E_1 = \frac{\sigma_1}{\varepsilon_1} = \frac{1}{S_{11}} \quad (\sigma_1 \neq 0,\ \sigma_2,\ \sigma_6 = 0)$$

longitudinal modulus (3.19a)

$$\nu_{21} = -\frac{\varepsilon_2}{\varepsilon_1} = -\frac{S_{21}}{S_{11}} \quad (\sigma_1 \neq 0,\ \sigma_2,\ \sigma_6 = 0)$$

major Poisson ratio (3.19b)

$$\nu_{61} = \frac{\varepsilon_6}{\varepsilon_1} = \frac{S_{61}}{S_{11}} \quad (\sigma_1 \neq 0,\ \sigma_2,\ \sigma_6 = 0)$$

shear–longitudinal coupling ratio (3.19c)

$$E_2 = \frac{\sigma_2}{\varepsilon_2} = \frac{1}{S_{22}} \quad (\sigma_2 \neq 0,\ \sigma_1,\ \sigma_6 = 0)$$

transverse modulus (3.19d)

$$\nu_{12} = -\frac{\varepsilon_1}{\varepsilon_2} = -\frac{S_{12}}{S_{22}} \quad (\sigma_2 \neq 0,\ \sigma_1,\ \sigma_6 = 0)$$

minor Poisson ratio (3.19e)

$$\nu_{62} = \frac{\varepsilon_6}{\varepsilon_2} = \frac{S_{62}}{S_{22}} \quad (\sigma_2 \neq 0,\ \sigma_1,\ \sigma_6 = 0)$$

shear–transverse coupling ratio (3.19f)

$$E_6 = \frac{\sigma_6}{\varepsilon_6} = \frac{1}{S_{66}} \quad (\sigma_6 \neq 0,\ \sigma_1,\ \sigma_2 = 0)$$

in-plane shear modulus (3.19g)

$$\nu_{16} = \frac{\varepsilon_1}{\varepsilon_6} = \frac{S_{16}}{S_{66}} \quad (\sigma_6 \neq 0,\ \sigma_1,\ \sigma_2 = 0)$$

longitudinal–shear coupling ratio (3.19*h*)

$$\nu_{26} = \frac{\varepsilon_2}{\varepsilon_6} = \frac{S_{26}}{S_{66}} \quad (\sigma_6 \neq 0,\ \sigma_1,\ \sigma_2 = 0)$$

transverse–shear coupling ratio (3.19*i*)

Due to symmetry of the compliance matrix, only six of these constants are independent. Usually, E_1, E_2, E_6, ν_{21}, ν_{61}, and ν_{62} are determined experimentally from uniaxial and pure shear tests, and ν_{12}, ν_{16}, ν_{26} are calculated using the symmetry relations. Note that $S_{ij} = S_{ji}$ but that $\nu_{ij} \neq \nu_{ji}$.

The strength of a unidirectional ply is typically described by five independent ultimate stresses (called *strengths*) and five independent ultimate strains (called *rupture strains*). The five strengths of the unidirectional ply are the longitudinal tensile strength, X_t; the longitudinal compressive strength, X_c; the transverse tensile strength, Y_t; the transverse compressive strength, Y_c; and the in-plane shear strength, S. The failure strains are the longitudinal tensile rupture strain, ε_t^x; the longitudinal compressive rupture stain, ε_c^x; the transverse tensile rupture strain, ε_t^y; the transverse compressive rupture strain, ε_c^y; and the in-plane shear rupture strain, ε^s. The strengths are typically obtained from uniaxial and pure shear tests.

Both strengths and rupture strains are used to define the ultimate capacity of a unidirectional ply even though it is assumed for the purpose of stress analysis that the unidirectional ply is linear elastic (which would mean that the rupture strain could be calculated directly from the strengths and the moduli). However, in reality, the unidirectional ply is not linear elastic, especially in the transverse and shear directions at moderate to high stress levels (see Adams et al., 2003). Therefore, whereas the linear elastic assumption is reasonable for prediction of the behavior of the unidirectional ply at service load stresses, it is not appropriate when the unidirectional ply is subjected to loads that approach its ultimate stresses and strains.

When the unidirectional ply is subjected to a uniaxial state of stress or strain, failure is determined by comparing the applied stress or strain to the corresponding strength or rupture strain. In the case of FRP composites for structural engineering, this is often the case for unidirectional products such as FRP rebars and strips and sheets. However, if a multiaxial state of stress and strain exists in the unidirectional ply due to a complex set of applied loads, or if the ply is loaded in its off-axis orientation, a multiaxial failure criterion needs to be employed to predict the combination of loads under which the unidirectional ply will fail. The most popular failure criteria used today are the maximum stress failure criterion, the maximum strain failure criterion, the Tsai–Wu quadratic failure criterion, and the Tsai–Hill (or maximum work) failure criterion (Tsai and Hahn, 1980; Agarwal and Broutman,

1990; Daniel and Ishai, 1994; Adams et al., 2003). Of these criteria, the first two do not account for the interactive effect of multiaxial stresses and strains, whereas the latter two do account for these effects.

3.2.3 Laminate Level

A multidirectional laminate is constructed from a number of unidirectional plies that are stacked at various on- and off-axis angles relative to the global 1,2,6 axes. A stacking sequence is used to define the layup. The stacking sequence shows the individual plies of the laminate, from the bottom ply to the top ply, in an ordered sequence of angles of the plies, such as [0/90/+45/−45]. It is generally assumed that all plies have equal thickness. If they do not, or they are made of different materials, the stacking sequence notation is modified to indicate this. Depending on the stacking sequence, a multidirectional laminate may be orthotropic or anisotropic in its plane or out of its plane and may have coupling between in-plane and out-of plane responses. In-plane response is known as *extensional response* and out-of-plane response is known as *flexural response*.

Classical lamination theory (CLT) is generally used as a first approximation to describe the constitutive relations of a thin laminate. According to CLT, the in-plane response is described by three extensional stress resultants, N_1, N_2, N_6, and the out-of-plane response is described by three flexural stress resultants, M_1, M_2, M_6.[10] Corresponding to these stress resultants are six generalized strain resultants: the midplane strains, ε_1^0, ε_2^0, ε_6^0 and the plate curvatures, κ_1, κ_2, κ_6. The general constitutive relation for a multidirectional laminate is given as

$$\begin{Bmatrix} N_1 \\ N_2 \\ N_6 \\ M_1 \\ M_2 \\ M_6 \end{Bmatrix} = \begin{bmatrix} A_{11} & A_{12} & A_{16} & B_{11} & B_{12} & B_{16} \\ A_{21} & A_{22} & A_{26} & B_{21} & B_{22} & B_{26} \\ A_{61} & A_{62} & A_{66} & B_{61} & B_{62} & B_{66} \\ B_{11} & B_{12} & B_{16} & D_{11} & D_{12} & D_{16} \\ B_{21} & B_{22} & B_{26} & D_{21} & D_{22} & D_{26} \\ B_{61} & B_{62} & B_{66} & D_{61} & D_{62} & D_{66} \end{bmatrix} \begin{Bmatrix} \varepsilon_1^0 \\ \varepsilon_2^0 \\ \varepsilon_6^0 \\ \kappa_1 \\ \kappa_2 \\ \kappa_6 \end{Bmatrix} \tag{3.20}$$

The stiffness matrix above is often referred to as the *[A–B–D] matrix*. The A_{ij} terms define the extensional response, the D_{ij} terms the flexural response, and the B_{ij} terms the extension-bending coupled response.[11] Note that the entire [A–B–D] matrix is symmetric and that the [A], [B], and [D] submatrices themselves are symmetric so that the [A–B–D] matrix has at most 18 inde-

[10] Note that the in-plane resultants N have units of *force/length*, whereas the moments M have units of *force* since the resultants are defined per unit width of the laminate.

[11] The A terms have units of *force/length*, the B terms units of *force*, and the D terms units of *force × length*.

pendent terms. The [A–B–D] matrix can be obtained by summation of the stiffness coefficients of the individual on- and off-axis plies as follows:

$$A_{ij} = \sum_{k=1}^{n} Q_{ij}^{k}(z_k - z_{k-1}) = \sum_{k=1}^{n} Q_{ij}^{k} h_k \tag{3.21}$$

$$B_{ij} = \frac{1}{2}\sum_{k=1}^{n} Q_{ij}^{k}(z_k^2 - z_{k-1}^2) = \sum_{k=1}^{n} Q_{ij}^{k}(-\bar{z}_k h_k) \tag{3.22}$$

$$D_{ij} = \frac{1}{3}\sum_{k=1}^{n} Q_{ij}^{k}(z_k^3 - z_{k-1}^3) = \sum_{k=1}^{n} Q_{ij}^{k}\left(h_k \bar{z}_k^2 + \frac{h_k^3}{12}\right) \tag{3.23}$$

where z_k is the distance from the midplane to the top of ply k, $\bar{z}_k$ the distance to the middle of ply k, h_k the thickness of ply k, and n the total number of plies in the laminate, as shown in Fig. 3.2. Note that the coordinate z is positive for a ply above the midplane and negative for a ply below the midplane.

A very common class of laminates is the class in which the layup is symmetric about the midplane. This implies that for every material type, orientation, and thickness of ply below the midplane there exists an identical (a mirror image) ply above the midplane at the identical distance from the midplane. In this case, the B submatrix is identically zero. Only in very special circumstances are unsymmetric laminates used. Their use in structural engineering should be avoided unless their unique properties are beneficial. Unsymmetric laminates have extensional-bending coupling that can affect the response of thin laminates, especially those used for strengthening applications.[12]

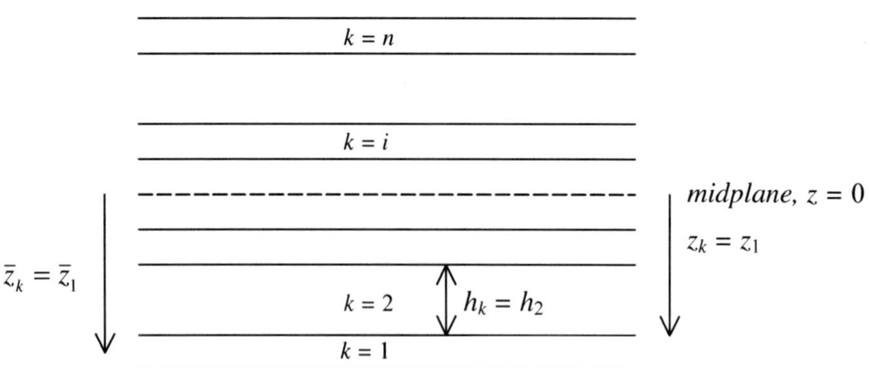

Figure 3.2 Laminate notation.

[12] Extensional-bending coupling can cause undesirable debonding of FRP strengthening strips.

In a symmetric laminate the extensional and bending responses are uncoupled (although they can have bending–twisting coupling) and the individual $[A]$ and $[D]$ submatrices can be inverted separately[13] to give extensional and bending compliance matrices, known as the $[a]$ and the $[d]$ matrices, as follows:

$$a_{ij} = A_{ij}^{-1} \quad (i,j = 1,2,6) \tag{3.24}$$

$$d_{ij} = D_{ij}^{-1} \quad (i,j = 1,2,6) \tag{3.25}$$

Effective in-plane engineering constants can then be defined for the symmetric laminate of thickness, h, by defining an average in-plane stress in the laminate, $\overline{\sigma}$, as follows:

$$\sigma_i = \frac{N_i}{h} \quad (i = 1,2,6) \tag{3.26}$$

The compliance relationship then becomes

$$\varepsilon_i^0 = ha_{ij}\overline{\sigma}_j \qquad (i,j = 1,2,6) \tag{3.27}$$

or

$$\begin{Bmatrix} \varepsilon_1^0 \\ \varepsilon_2^0 \\ \varepsilon_6^0 \end{Bmatrix} = \begin{bmatrix} \dfrac{1}{E_1^0} & -\dfrac{\nu_{12}^0}{E_2^0} & \dfrac{\nu_{16}^0}{E_6^0} \\ -\dfrac{\nu_{21}^0}{E_1^0} & \dfrac{1}{E_2^0} & \dfrac{\nu_{26}^0}{E_6^0} \\ \dfrac{\nu_{61}^0}{E_1^0} & \dfrac{\nu_{62}^0}{E_2^0} & \dfrac{1}{E_6^0} \end{bmatrix} \begin{Bmatrix} \overline{\sigma}_1 \\ \overline{\sigma}_2 \\ \overline{\sigma}_6 \end{Bmatrix} \tag{3.28}$$

The nine effective in-plane (also known as *extensional*) engineering constants are defined analogously to the engineering constants for the off-axis unidirectional ply [see equations (3.19*a*) to (3.19*i*)]; for example,

$$E_1^0 = \frac{\overline{\sigma}_1}{\varepsilon_1^0} = \frac{1}{a_{11}h} \quad (\overline{\sigma}_1 \neq 0, \overline{\sigma}_2 = \overline{\sigma}_6 = 0)$$

longitudinal in-plane modulus (3.29*a*)

[13] Note that in the general case, when $B_{ij} \neq 0$ the full 6×6 stiffness matrix must be inverted to obtain the compliance matrix.

$$E_6^0 = \frac{\overline{\sigma}_6}{\varepsilon_6^0} = \frac{1}{a_{66}h} \quad (\overline{\sigma}_6 \neq 0, \overline{\sigma}_1 = \overline{\sigma}_2 = 0)$$

shear in-plane modulus (3.29*b*)

$$\nu_{21}^0 = -\frac{\varepsilon_2^0}{\varepsilon_1^0} = -\frac{a_{21}}{a_{11}} \quad (\overline{\sigma}_1 \neq 0, \overline{\sigma}_2 = \overline{\sigma}_6 = 0)$$

Major in-plane Poisson ratio (3.29*c*)

and analogously for E_2^0, ν_{12}^0, ν_{16}^0, ν_{61}^0, ν_{26}^0, ν_{62}^0.

In a similar fashion, effective engineering flexural properties can be defined in terms of the effective stress at the outer surface, defined as[14]

$$\sigma_i^f = \frac{6M_i}{h^2} \quad (i = 1,2,6) \tag{3.30}$$

and the actual strain at the outer surface, defined as

$$\varepsilon_i^f = \frac{\kappa_i h}{2} \quad (i = 1,2,6) \tag{3.31}$$

The effective flexural stress–strain relation is then defined as

$$\varepsilon_i^f = \frac{h^3}{12} d_{ij}\sigma_j^f \quad (i,j = 1,2,6) \tag{3.32}$$

which in terms of the effective flexural engineering constants is written as

$$\begin{Bmatrix} \varepsilon_1^f \\ \varepsilon_2^f \\ \varepsilon_6^f \end{Bmatrix} = \begin{bmatrix} \dfrac{1}{E_1^f} & -\dfrac{\nu_{12}^f}{E_2^f} & \dfrac{\nu_{16}^f}{E_6^f} \\ -\dfrac{\nu_{21}^f}{E_1^f} & \dfrac{1}{E_2^f} & \dfrac{\nu_{26}^f}{E_6^f} \\ \dfrac{\nu_{61}^f}{E_1^f} & \dfrac{\nu_{62}^f}{E_2^f} & \dfrac{1}{E_6^f} \end{bmatrix} \begin{Bmatrix} \sigma_1^f \\ \sigma_2^f \\ \sigma_6^f \end{Bmatrix} \tag{3.33}$$

where, for example,

[14] This terminology follows Tsai (1988). In plate theory, a positive bending moment is generally defined as "tension on the top," which is opposite to that used in conventional beam theory.

$$E_1^f = \frac{\sigma_1^f}{\varepsilon_1^f} = \frac{12}{d_{11}h^3} \quad (\sigma_1^f \neq 0,\ \sigma_2^f,\ \sigma_6^f = 0)$$

longitudinal flexural modulus (3.34*a*)

$$E_6^f = \frac{\sigma_6^f}{\varepsilon_6^f} = \frac{12}{d_{66}h^3} \quad (\sigma_6^f \neq 0,\ \sigma_1^f,\ \sigma_2^f = 0)$$

twisting (shear–flexural) modulus (3.34*b*)

$$\nu_{21}^f = -\frac{\varepsilon_2^f}{\varepsilon_1^f} = -\frac{d_{21}}{d_{11}} \quad (\sigma_1^f \neq 0,\ \sigma_2^f,\ \sigma_6^f = 0)$$

longitudinal flexural Poisson ratio (3.34*c*)

and analogously for the remaining six effective flexural engineering constants, E_2^f, ν_{12}^f, ν_{16}^f, ν_{61}^f, ν_{26}^f, ν_{62}^f, following the same logic as equations (3.19*a*) to (3.19*i*) for the off-axis engineering constants defined previously.

In structural engineering, every effort is made to use laminates that in addition to being symmetric, have in-plane orthotropic properties so that no coupling exists between in-plane extension and in-plane shear (i.e., $A_{16} = A_{26} = A_{61} = A_{62} = 0$). Only specific types of symmetric laminates have no extension–shear coupling. These are: (1) cross-ply laminates, in which the laminate consists of only 0° and 90° plies (e.g., a 0/0/90/90/0/0 laminate); (2) balanced laminates, in which for every ply at a positive off-axis orientation there exists a ply at a negative off-axis orientation (e.g., a −30/+30/+30/−30 laminate); and (3) quasi-isotropic laminates, in which the ply angles are separated by an angle of 360° divided by the total number of plies in the laminate, n, where n is an even number equal to or greater than 6 (e.g., a −60/60/0/0/60/−60 laminate). Note that for a quasi-isotropic laminate the in-plane properties are not direction dependent and that the isotropic relationship between the engineering constants holds (hence the name *quasi-isotropic*[15]),

$$E_1^0 = \frac{E_6^0}{2(1 + \nu_{21}^0)} = \frac{E_6^0}{2(1 + \nu_{12}^0)} \tag{3.35}$$

Of the three type of laminates described above, only the symmetric cross-ply laminate is also orthotropic in its flexural properties (i.e., $D_{16} = D_{26} = D_{61} = D_{62} = 0$). A number of unique types of laminates exist that have flexural orthotropy but not in-plane orthotropy (an example of this is the unsymmetric

[15] The laminate is known as *quasi-isotropic* since it is not isotropic in the thickness direction nor isotropic with respect to its flexural properties.

angle-ply laminate, $+\theta/-\theta/+\theta/-\theta$, commonly produced by the filament winding process).

In structural engineering applications, laminates are typically thin relative to their two in-plane dimensions, and loaded in-plane, and therefore most attention is paid to their in-plane properties. In rare situations such as in thin-shell structures, where laminates are loaded out-of-plane, their flexural properties may be used. Consequently, in design procedures for FRP composites for structural engineering the in-plane effective engineering properties of the laminate are used in calculations involving sheets, strips, and thin-walled sections.

The strengths (and ultimate strains) of an FRP composite laminate are difficult to predict, and a variety of mechanics models are used to estimate when a laminate will fail due to the loads applied. The general procedure is to determine the stresses and strains in each ply of the laminate as the laminate is loaded in an incremental fashion. At each load increment the stress state in the ply is checked with regard to one of the failure criteria for the unidirectional ply described previously. When the first ply fails (i.e., satisfies its failure criterion), the condition is known as *first ply failure* (FPF). However, a multidirectional laminate can often continue to carry additional loads beyond its FPF load. A nonlinear analysis must then be performed using reduced stiffnesses for the laminate's surviving plies (see Agarwal and Broutman, 1990). Often, the first ply failure condition is used as a conservative criterion to define laminate failure for design purposes. Computer programs are available to perform simple nonlinear calculations[16] to determine the FPF and ultimate failure loads of two-dimensional laminates according to commonly used failure criteria. They are useful for obtaining an order-of-magnitude estimate of the ultimate strengths of a laminate when loaded by a specific set of applied loads.

For structural engineering design the strength and rupture strain properties of the laminate are usually obtained by experimental testing in a fashion similar to that used to determine the strength properties for an on-axis unidirectional ply. Since, as noted previously, most laminates used in structural engineering are orthotropic, in-plane testing along the orthotropy axes is an effective means of obtaining both the strength and stiffness properties of laminates for design purposes.

Analysis Example 3.2 Given is a three-layer E-glass/polyester pultruded composite material with a thickness of 0.125 in. The composite consists of two outer layers of continuous filament mat[17] (CFM) with a thickness of 0.020 in. each and an internal layer of unidirectional rovings with a thickness of 0.085 in. The continuous filament mat has a 23% glass fiber volume fraction and the following quasi-isotropic mechanical properties:

[16] CompositePro, Peak Innovations, Littleton, Colorado, for example.

[17] Also known as continuous strand mat (see Chapter 2).

$$E_x = 0.91 \times 10^6 \text{ psi}$$

$$E_y = 0.91 \times 10^6 \text{ psi}$$

$$E_s = 0.34 \times 10^6 \text{ psi}$$

$$\nu_x = 0.35$$

The unidirectional roving layer has a 51% fiber volume fraction and the following orthotropic mechanical properties:

$$E_x = 5.66 \times 10^6 \text{ psi}$$

$$E_y = 1.50 \times 10^6 \text{ psi}$$

$$E_s = 0.48 \times 10^6 \text{ psi}$$

$$\nu_x = 0.28$$

It is assumed that the tensile and compressive moduli are equal for both layers.

For the FRP composite material given, determine (a) the in-plane stiffness coefficients, Q_{ij}, for the continuous filament mat layer; (b) the in-plane stiffness coefficients, Q_{ij}, for the unidirectional roving layer; (c) the $[A]$ and $[D]$ matrices (indicate the units for the terms in the matrices); and (d) the effective in-plane and flexural engineering properties for the layup.

SOLUTION (a) Determine Q_{ij} for the CFM layers:

$$\nu_y = \nu_x \frac{E_y}{E_x} = 0.35$$

$$m = (1 - \nu_x \nu_y)^{-1} = 1.140$$

$$Q_{xx} = mE_x = 1.140(0.91 \times 10^6) = 1.037 \times 10^6 \text{ psi}$$

$$Q_{yy} = Q_{xx} = 1.037 \times 10^6 \text{ psi}$$

$$Q_{xy} = Q_{yx} = m\nu_x E_y = 3.630 \times 10^5 \text{ psi}$$

$$Q_{ss} = E_s = 0.34 \times 10^6 \text{ psi}$$

(b) Determine Q_{ij} for the unidirectional (UD) layer:

$$\nu_y = \nu_x \frac{E_y}{E_x} = 0.074$$

$$m = (1 - \nu_x \nu_y)^{-1} = 1.021$$

$$Q_{xx} = mE_x = 1.021(5.66 \times 10^6) = 5.780 \times 10^6 \text{ psi}$$

$$Q_{yy} = mE_y = 1.532 \times 10^6 \text{ psi}$$

$$Q_{xy} = Q_{yx} = m\nu_x E_y = 4.2789 \times 10^5 \text{ psi}$$

$$Q_{ss} = E_s = 0.48 \times 10^6 \text{ psi}$$

(c) To determine the $[A–B–D]$ matrix, determine z values assuming that the middle UD layer is divided into two equal-thickness layers about the midplane:

$$z_0 = -0.0625 \text{ in.} \qquad z_3 = 0.0425 \text{ in.}$$

$$z_1 = -0.0425 \text{ in.} \qquad z_4 = 0.0625 \text{ in.}$$

$$z_2 = 0.0$$

$$A_{ij} = \sum_{k=1}^{n} Q_{ij}^k (z_k - z_{k-1})$$

$$\begin{aligned} A_{11} &= 2\{(1.037 \times 10^6)[-0.0425 - (-0.0625)] \\ &\quad + (5.780 \times 10^6)[0.0 - (-0.0425)]\} \\ &= 2[(20{,}740) + (245{,}650)] = 532{,}780 \text{ lb/in.} \end{aligned}$$

$$A_{22} = 171{,}000 \text{ lb/in.} \qquad A_{66} = 54{,}400 \text{ lb/in.}$$

$$A_{12} = 50{,}985 \text{ lb/in.} \qquad A_{16} = A_{26} = 0$$

$[a] = [A]^{-1}$ (can invert upper 2×2 matrix and A_{66} separately, due to orthotropy)

$$|A| = A_{11}A_{22} - A_{12}A_{21} = 8.888 \times 10^{10} \text{ (lb/in.)}^2$$

$$a_{11} = \frac{A_{22}}{|A|} = \frac{1.717 \times 10^5}{8.888 \times 10^{10}} = 1.932 \times 10^{-6} \text{ in./lb}$$

$$a_{22} = \frac{A_{11}}{|A|} = \frac{5.328 \times 10^5}{8.888 \times 10^{10}} = 5.995 \times 10^{-6} \text{ in./lb}$$

$$a_{21} = \frac{-A_{21}}{|A|} = \frac{-0.510 \times 10^5}{8.888 \times 10^{10}} = -5.738 \times 10^{-7} \text{ in./lb}$$

$$a_{66} = \frac{1}{A_{66}} = \frac{1}{0.544 \times 10^5} = 1.838 \times 10^{-5} \text{ in./lb}$$

$$a_{16} = a_{26} = 0$$

$B_{ij} = 0$ since the laminate is symmetric.

$$D_{ij} = \tfrac{1}{3} \sum_{k=1}^{n} Q_{ij}^k (z_k^3 - z_{k-1}^3)$$

$$\begin{aligned} D_{11} &= \tfrac{2}{3}\{(1.037 \times 10^6)[-0.0425^3 - (-0.0625^3)] \\ &\quad + (5.780 \times 10^6)[0.0 - (-0.0425^3)]\} \\ &= 2[(173.594 + (443.731)] = 411.550 \text{ lb-in.} \end{aligned}$$

$$D_{22} = 194.137 \text{ lb-in.} \qquad D_{66} = 62.510 \text{ lb-in.}$$

$$D_{21} = 62.467 \text{ lb-in.} \qquad D_{16} = D_{26} = 0$$

$$[d] = [D]^{-1}$$

$$|D| = D_{11}D_{22} - \mathrm{D}_{12}D_{21} = 75{,}985.3 \text{ (lb-in.)}^2$$

$$d_{11} = \frac{D_{22}}{|D|} = \frac{194.1}{75{,}985} = 2.554 \times 10^{-3} \text{ 1/lb-in.}$$

$$d_{22} = \frac{D_{11}}{|D|} = 5.417 \times 10^{-3} \text{ 1/lb-in.}$$

$$d_{21} = \frac{-D_{21}}{|D|} = \frac{-62.5}{75{,}985} = -8.225 \times 10^{-4} \text{ 1/lb-in.}$$

$$d_{66} = \frac{1}{D_{66}} = \frac{1}{62.5} = 1.60 \times 10^{-2} \ 1/\text{lb-in.}$$

$$a_{16} = a_{26} = 0$$

(d) Effective in-plane engineering properties:

$$E_1^0 = \frac{1}{a_{11}h} = \frac{1}{(1.932 \times 10^{-6})(0.125)} = 4.141 \times 10^6 \text{ psi}$$

$$E_2^0 = \frac{1}{a_{22}h} = \frac{1}{(5.995 \times 10^{-6})(0.125)} = 1.334 \times 10^6 \text{ psi}$$

$$E_6^0 = \frac{1}{a_{66}h} = \frac{1}{(1.838 \times 10^{-5})(0.125)} = 0.435 \times 10^6 \text{ psi}$$

$$\nu_{21}^0 = \frac{-a_{21}}{a_{11}} = \frac{-(-5.738 \times 10^{-7})}{1.932 \times 10^{-6}} = 0.297$$

$$\nu_{12}^0 = \frac{-a_{21}}{a_{22}} = \frac{-(-5.738 \times 10^{-7})}{5.995 \times 10^{-6}} = 0.096$$

Symmetry check: $E_1^0 \nu_{12}^0 = E_2^0 \nu_{21}^0 = 3.97 \times 10^5$ psi

Effective flexural engineering properties:

$$E_1^f = \frac{12}{d_{11}h^3} = \frac{12}{(2.554 \times 10^{-3})(0.125)^3} = 2.406 \times 10^6 \text{ psi}$$

$$E_2^f = \frac{12}{d_{22}h^3} = \frac{12}{(5.417 \times 10^{-3})(0.125)^3} = 1.134 \times 10^6 \text{ psi}$$

$$E_6^f = \frac{12}{d_{66}h^3} = \frac{12}{(1.838 \times 10^{-5})(0.125)^3} = 0.384 \times 10^6 \text{ psi}$$

$$\nu_{21}^f = \frac{-d_{21}}{d_{11}} = \frac{-(-8.225 \times 10^{-4})}{2.554 \times 10^{-3}} = 0.032$$

$$\nu_{12}^f = \frac{-d_{21}}{d_{22}} = \frac{-(-8.225 \times 10^{-4})}{5.417 \times 10^{-3}} = 0.015$$

Symmetry check: $E_1^f \nu_{12}^f = E_2^f \nu_{21}^f = 3.6 \times 10^4$ psi

Analysis Example 3.3 Given is a three-layer carbon–epoxy composite material with a total thickness of 0.05 in. The composite consists of two outer layers (top and bottom) of +10° off-axis unidirectional composite with a thickness of 0.010 in. each and an internal layer of 0° unidirectional composite with a thickness of 0.030 in. Each unidirectional carbon–epoxy layer has a 60% fiber volume fraction and the following orthotropic elastic constants:

$$E_x = 20.05 \times 10^6 \text{ psi} \qquad E_s = 0.74 \times 10^6 \text{ psi}$$

$$E_y = 1.46 \times 10^6 \text{ psi} \qquad \nu_x = 0.26$$

It is assumed that the tensile and compressive moduli are equal for both layers.

For the given FRP composite material, determine (a) the in-plane stiffness coefficients, Q_{ij}, for the on-axis layer; (b) the in-plane stiffness coefficients, Q_{ij}, for the off-axis layer; (c) the [*A*] and [*D*] matrices (indicate the units for the terms in the matrices); and (d) the effective engineering properties for the layup.

SOLUTION (a) Determine Q_{ij} for the 0° (on–axis) layer:

$$\nu_y = \nu_x \frac{E_y}{E_x} = 0.0189$$

$$m = (1 - \nu_x \nu_y)^{-1} = 1.0049$$

$$Q_{xx} = mE_x = 1.0049(20.05 \times 10^6) = 20.15 \times 10^6 \text{ psi}$$

$$Q_{yy} = mE_y = 1.467 \times 10^6 \text{ psi}$$

$$Q_{xy} = Q_{yx} = m\nu_y E_x = 3.815 \times 10^5 \text{ psi}$$

$$Q_{ss} = E_s = 0.74 \times 10^6 \text{ psi}$$

(b) Determine Q_{ij} for the +10° (off-axis) layer:

$$\begin{Bmatrix} Q_{11} \\ Q_{22} \\ Q_{12} \\ Q_{66} \\ Q_{16} \\ Q_{26} \end{Bmatrix} = \begin{bmatrix} \cos^4\theta & \sin^4\theta & 2\cos^2\theta\sin^2\theta & 4\cos^2\theta\sin^2\theta \\ \sin^4\theta & \cos^4\theta & 2\cos^2\theta\sin^2\theta & 4\cos^2\theta\sin^2\theta \\ \cos^2\theta\sin^2\theta & \cos^2\theta\sin^2\theta & \cos^4\theta + \sin^4\theta & -4\cos^2\theta\sin^2\theta \\ \cos^2\theta\sin^2\theta & \cos^2\theta\sin^2\theta & -2\cos^2\theta\sin^2\theta & (\cos^2\theta - \sin^2\theta)^2 \\ \cos^3\theta\sin\theta & -\cos\theta\sin^3\theta & \cos\theta\sin^3\theta - \cos^3\theta\sin\theta & 2(\cos\theta\sin^3\theta - \cos^3\theta\sin\theta \\ \cos\theta\sin^3\theta & -\cos^3\theta\sin\theta & \cos^3\theta\sin\theta - \cos\theta\sin^3\theta & 2(\cos^3\theta\sin\theta - \cos\theta\sin^3\theta) \end{bmatrix} \begin{Bmatrix} Q_{xx} \\ Q_{yy} \\ Q_{xy} \\ Q_{ss} \end{Bmatrix}$$

$$Q_{11} = 1.906 \times 10^7 \text{ psi} \qquad Q_{66} = 1.263 \times 10^6 \text{ psi}$$

$$Q_{22} = 1.507 \times 10^6 \text{ psi} \qquad Q_{16} = 3.035 \times 10^6 \text{ psi}$$

$$Q_{12} = 9.048 \times 10^5 \text{ psi} \qquad Q_{26} = 1.597 \times 10^5 \text{ psi}$$

(c) Determine the $[A–B–D]$ matrix by determining the z values:

$$z_0 = -0.025 \text{ in.} \qquad z_3 = 0.015 \text{ in.}$$

$$z_1 = -0.015 \text{ in.} \qquad z_4 = 0.025 \text{ in.}$$

$$z_2 = 0.0$$

$$A_{ij} = \sum_{k=1}^{n} Q_{ij}^k h_k$$

$$A_{11} = 2(1.906 \times 10^7)(0.01) + (2.015 \times 10^7)(0.03) = 9.875 \times 10^5 \text{ lb/in.}$$

$$A_{22} = 7.415 \times 10^4 \text{ lb/in.}$$

$$A_{12} = 2.954 \times 10^4 \text{ lb/in.}$$

$$A_{66} = 4.747 \times 10^4 \text{ lb/in.}$$

$$A_{16} = 6.070 \times 10^4 \text{ lb/in.}$$

$$A_{26} = 3.194 \times 10^3 \text{ lb/in.}$$

Invert $[A]$ to obtain $[a]$ (full 3×3 matrix inversion required):

$$A = \begin{pmatrix} 9.857 \times 10^5 & 2.954 \times 10^4 & 6.07 \times 10^4 \\ 2.954 \times 10^4 & 7.415 \times 10^4 & 3.194 \times 10^3 \\ 6.07 \times 10^4 & 3.194 \times 10^3 & 4.746 \times 10^4 \end{pmatrix} \quad \text{lb/in.}$$

$$a = A^{-1}$$

$$a = \begin{pmatrix} 1.112 \times 10^{-6} & -3.828 \times 10^{-7} & -1.396 \times 10^{-6} \\ -3.828 \times 10^{-7} & 1.366 \times 10^{-5} & -4.295 \times 10^{-7} \\ -1.396 \times 10^{-6} & -4.295 \times 10^{-7} & 2.289 \times 10^{-5} \end{pmatrix} \quad \text{in./lb}$$

$B_{ij} = 0$ since the laminate is symmetric.

$$D_{ij} = \sum_{k=1}^{n} Q_{ij}^k \left(h_k \bar{t}_k^2 + \frac{h_k^3}{12} \right)$$

$$D_{11} = 2(1.906 \times 10^7)\left[0.01(0.02)^2 + \frac{(0.01)^3}{12} \right] + (2.015 \times 10^7)\left(\frac{0.03^3}{12} \right)$$

$$= 200.994 \text{ lb-in.}$$

$$D_{22} = 15.608 \text{ lb-in.}$$

$$D_{12} = 8.248 \text{ lb-in.}$$

$$D_{66} = 11.98 \text{ lb-in.}$$

$$D_{16} = 24.786 \text{ lb-in.}$$

$$D_{26} = 1.304 \text{ lb-in.}$$

Invert to obtain $[d]$:

$$D = \begin{pmatrix} 200.994 & 8.248 & 24.786 \\ 8.248 & 15.608 & 1.304 \\ 24.786 & 1.304 & 11.98 \end{pmatrix} \quad \text{in.-lb}$$

$$d = D^{-1}$$

$$d = \begin{pmatrix} 6.77 \times 10^{-3} & -2.429 \times 10^{-3} & -0.014 \\ -2.429 \times 10^{-3} & 0.066 & -2.107 \times 10^{-3} \\ -0.014 & -2.107 \times 10^{-3} & 0.112 \end{pmatrix} \quad (\text{in.-lb})^{-1}$$

(d) Effective in-plane engineering properties:

$$E_1^0 = \frac{1}{a_{11}h} = \frac{1}{(1.112 \times 10^{-6})(0.05)} = 17.99 \times 10^6 \text{ psi}$$

$$E_2^0 = \frac{1}{a_{22}h} = \frac{1}{(1.366 \times 10^{-5})(0.05)} = 1.464 \times 10^6 \text{ psi}$$

$$E_6^0 = \frac{1}{a_{66}h} = \frac{1}{(2.289 \times 10^{-5})(0.05)} = 0.874 \times 10^6 \text{ psi}$$

$$\nu_{21}^0 = \frac{-a_{21}}{a_{11}} = \frac{-(-3.828 \times 10^{-7})}{1.112 \times 10^{-6}} = 0.344$$

$$\nu_{12}^0 = \frac{-a_{21}}{a_{22}} = \frac{-(-3.828 \times 10^{-7})}{1.366 \times 10^{-5}} = 0.028$$

$$\nu_{16}^0 = \frac{a_{16}}{a_{66}} = \frac{-1.396 \times 10^{-6}}{2.289 \times 10^{-5}} = -0.061$$

$$\nu_{26}^0 = \frac{a_{26}}{a_{66}} = \frac{-4.295 \times 10^{-7}}{2.289 \times 10^{-5}} = -0.019$$

$$\nu_{61}^0 = \frac{a_{61}}{a_{11}} = \frac{-1.396 \times 10^{-6}}{1.112 \times 10^{-6}} = -1.255$$

$$\nu_{62}^0 = \frac{a_{62}}{a_{22}} = \frac{-4.295 \times 10^{-7}}{1.366 \times 10^{-5}} = -0.031$$

Symmetry checks:

$$E_1^0\nu_{12}^0 = E_2^0\nu_{21}^0 = 5.042 \times 10^5 \text{ psi}$$

$$E_1^0\nu_{16}^0 = E_6^0\nu_{61}^0 = -1.097 \times 10^6 \text{ psi}$$

$$E_2^0\nu_{26}^0 = E_6^0\nu_{62}^0 = -2.748 \times 10^4 \text{ psi}$$

Effective flexural engineering properties:

$$E_1^f = \frac{12}{d_{11}h^3} = \frac{12}{(6.77 \times 10^{-3})(0.05)^3} = 14.2 \times 10^6 \text{ psi}$$

$$E_2^f = \frac{12}{d_{22}h^3} = \frac{12}{(6.60 \times 10^{-2})(0.05)^3} = 1.454 \times 10^6 \text{ psi}$$

$$E_6^f = \frac{12}{d_{66}h^3} = \frac{12}{(11.2 \times 10^{-2})(0.05)^3} = 8.57 \times 10^5 \text{ psi}$$

$$\nu_{21}^{f} = \frac{-d_{21}}{d_{11}} = \frac{-(-2.429 \times 10^{-3})}{6.77 \times 10^{-3}} = 0.359$$

$$\nu_{12}^{f} = \frac{-d_{21}}{d_{22}} = \frac{-(-2.429 \times 10^{-3})}{6.60 \times 10^{-2}} = 0.037$$

$$\nu_{16}^{f} = \frac{d_{16}}{d_{66}} = \frac{-1.4 \times 10^{-2}}{11.2 \times 10^{-2}} = -0.125$$

$$\nu_{26}^{f} = \frac{d_{26}}{d_{66}} = \frac{-2.107 \times 10^{-3}}{11.2 \times 10^{-2}} = -0.188$$

$$\nu_{61}^{f} = \frac{d_{61}}{d_{11}} = \frac{-1.4 \times 10^{-2}}{6.77 \times 10^{-3}} = -2.068$$

$$\nu_{62}^{f} = \frac{d_{62}}{d_{22}} = \frac{-2.107 \times 10^{-3}}{6.60 \times 10^{-2}} = -0.322$$

Symmetry checks: $E_1^f \nu_{12}^f = E_2^f \nu_{21}^f = 5.3 \times 10^5$ psi

$E_1^f \nu_{16}^f = E_6^f \nu_{61}^f = -1.78 \times 10^6$ psi

$E_2^f \nu_{26}^f = E_6^f \nu_{62}^f = -2.73 \times 10^5$ psi

3.2.4 Full-Section Level

When the FRP composite is used in the field it is usually referred to as a *component* or a *part,* such as an FRP strengthening wrap or an FRP pultruded beam. In this case, theoretical methods can be used to determine the material and section properties of the entire FRP part. In most structural engineering design, one-dimensional theories such as bar and beam theories are used, which require determination of the effective *composite*[18] section properties of the FRP part.[19]

In the case of an FRP reinforcing bar or an FRP strengthening strip or wrap, the cross-sectional area, A, and the longitudinal mechanical properties, E_L and σ_L, are the key properties used in design. The full-section properties in the longitudinal direction are then typically obtained by the rule of mixtures to yield a composite axial stiffness, $E_L A$, and a composite axial capacity, $\sigma_L A$.

[18] Where full-section mechanics are concerned, the term *composite* is used to describe a part made of multiple materials in the mechanics of materials sense (e.g., composite construction of steel and concrete). This can be a source of confusion in the FRP composite terminology.

[19] More details on this concept are provided in the design chapters.

In these applications, the bending and shear stiffness of the FRP part are not usually considered in the design.

In the case of a symmetric FRP profile used as a beam or used as a column, the geometric properties of primary interest are the cross-sectional area, A; the second moment of area, I; and the shear area, A_s. When the profile consists of FRP laminates (or panels) having different in-plane properties, the effective properties are obtained by conventional one-dimensional section analysis[20] to give a composite flexural stiffness, E_LI; axial stiffness, E_LA; and shear stiffness, $G_{LT}A_s$ (Bank, 1987). The bending moment capacity, axial capacity, and transverse shear capacity are then determined for the full section based on stress analysis according to a first ply failure or ultimate failure criterion. If twisting is included, a composite torsional stiffness, $G_{LT}J$, is defined. If warping of the cross section is included, a warping stiffness, E_LC_w, is defined. If the panels of the profile are anisotropic in-plane, a one-dimensional theory may not be appropriate (Cofie and Bank, 1995). A simplified one-dimensional theory for bending and twisting coupling in FRP beams is presented in Bank (1990).

3.3 EXPERIMENTAL DETERMINATION OF PROPERTIES

Many standards organizations around the world publish standard test methods for FRP composite materials; among these are the American Society for Testing and Materials (ASTM) and the International Organization for Standardization (ISO) (ASTM, 2006; ISO, 2006). Literally thousands of standardized test methods are available for testing fiber, resin, and additives and for testing FRP composites on the fiber, lamina, and laminate levels. However, not many standards are available for testing FRP composites on the full-section level in a form that is required for structural engineering applications. In recent years a number of organizations have developed test method standards directly related to the use of FRP composites in structural engineering. These include the American Concrete Institute (ACI), the European Committee for Standardization (CEN), the Japan Society for Civil Engineers (JSCE), and the Canadian Standards Association (CSA). In this chapter we review the tests available for FRP composites for the fiber, lamina, laminate, and full-section levels. More discussion and detail is provided for full-section test methods that have been developed specifically for the use for FRP composites in structural engineering.

At this time there is no general agreement between materials suppliers, part manufacturers, structural engineers, and other stakeholders on precisely which test methods should be used. However, there is general agreement as to which physical and mechanical properties are important in design. The

[20] Such as the transformed section method for flexural stiffness, EI.

choice of the specific test method is often left to the constituent material supplier, the FRP composite material producer, or the FRP part manufacturer. Test results are reported according to the requirements of the test method chosen and can therefore be subject to interpretation as to their relevance for structural engineering design.

In an engineering project, a structural engineer is usually required to develop a material specification for the FRP composites used in the project for inclusion in contract documents. In this specification, the structural engineer is required to specify the test methods to be used and the way in which the FRP material properties are to be reported. In addition, the designer may specify specific values of certain properties that must be achieved by the FRP material in a particular test. Limits for FRP materials can be found in model specifications (e.g., Bank et al., 2003), in specifications for pultruded profiles published by the European Committee on Standardization (CEN, 2002a), and in the specifications for reinforced plastic ladders[21] published by the American National Standards Institute (ANSI, 2000). Determination of design values for specific design bases for FRP reinforcing bars, FRP strengthening systems, and FRP profiles is discussed in the design chapters that follow.

It is important to note that although full-section test methods have been developed for testing FRP composite products for use in structural engineering on a full scale, this does not mean to imply that tests on the fiber, lamina, and laminate levels are not needed or required. Full-section test methods have been developed to characterize FRP products for use in structural engineering where appropriate coupon-type tests are not felt to be applicable, due to a specimen's size or the data reduction methods used. An FRP material specification for FRP composite products for use in structural engineering can include tests of the FRP material on the lamina and laminate levels as appropriate, and a specification for the material constituents on the fiber level.

3.3.1 Fiber Level

For structural engineering applications of FRP composite materials, very little testing is performed or specified by the end user (i.e., the FRP part manufacturer or structural engineer) on the fiber level. Most specifications for FRP composites for use in structural engineering concentrate on specifying properties of the FRP composite after it has been manufactured or fabricated (i.e., on the lamina, laminate, or full-section level). Manufacturers of raw fiber, polymer resins, and additives typically conduct standard tests on the constituents and report them in product specification sheets.

[21] Although dealing with ladders, this specification is concerned primarily with pultruded profiles and full-scale performance testing of FRP products and is therefore pertinent to specifying pultruded profiles in structural engineering. The commercial success of FRP ladders over the last 20 years can be attributed largely to the development of this rigorous material and performance specification for pultruded products.

Many standard test methods for obtaining mechanical, physical, and chemical properties of polymer resins (also known as *plastics* for this purpose) in their liquid and hardened states are available from ASTM and other standards organizations.[22] These have been developed primarily by the plastics industry over the past 50 years. It should be noted that the properties of the polymer are typically obtained from tests on cast resin samples that have been post-cured at elevated temperatures.[23] Tests for the secondary constituents of the polymer matrix (see Chapter 2) are reported in a similar fashion according to ASTM or other industry-specific test methods.

To obtain the properties of reinforcing fibers used in FRP composites, testing can be conducted on single fibers taken from rovings or tows using a single-fiber test method such as ASTM C 1557 or D 3379.[24] However, single-fiber testing is difficult, and the data obtained from tests on single fibers are not necessarily representative of their properties when in the FRP composite itself. Consequently, fiber manufacturers usually report the mechanical properties of their fibers when impregnated with a commonly used resin and tested as an FRP composite. The impregnated fiber test used by glass fiber manufacturers is ASTM D 2343, and the test used by carbon fiber manufacturers is ASTM D 4018. Fiber properties are then calculated from the FRP composite test data using the rule-of-mixtures approximations given previously.

3.3.2 Lamina Level

By far the majority of tests conducted on FRP composites for structural engineering applications are conducted on coupons cut from as-fabricated FRP composite parts. When these tests are conducted on coupons cut from an FRP composite containing only unidirectional fibers (such as FRP strengthening strips and sheets or FRP reinforcing bars), the testing conducted is of a unidirectional ply (i.e., on the lamina level). When the coupon is cut from an FRP composite that contains multidirectional plies or mats (such as an FRP strengthening fabric or an FRP profile), testing is on a multidirectional plate (i.e., on the laminate level.) For FRP composites for structural engineering applications, every effort should be made to test coupons cut from an as-fabricated in situ part, and to test it at the as-fabricated full thickness of the part. When multiple plies of the same unidirectional composite are used to create full-part thickness in the field (as in the case of a multilayer FRP strengthening system), it is permissible to test one unidirectional ply and to calculate the properties of the composite based on the ply values.[25] When

[22] Many standard ASTM test methods for the properties of plastics are also used to obtain the properties of reinforced plastics (i.e., FRP composites), and these methods are discussed in Sections 3.3.2 and 3.3.3.

[23] This is also typically done when the polymer is used in the hand-layup method for FRP strengthening, even when the resin is cured at ambient temperature in the field.

[24] A list of the titles of all the ASTM test methods cited in this chapter is given in Section 3.4.

[25] ACI 440.2R-02 allows this approach.

multidirectional plies are used in FRP strengthening fabric, the as-fabricated full-thickness part should be tested to obtain the design properties.

When determining properties of FRP composites on either the lamina or laminate level, the same ASTM tests are used since the test is conducted on the macroscopic level and assumes, for the purposes of conducting the test, that the coupon is made of a homogenous material. Tests commonly used for an FRP lamina or laminate are given in Table 3.1. The physical tests noted should be conducted on samples taken from the FRP composite and not on samples taken from the polymer matrix alone. This is because the fiber can have an influence on many of the physical properties that are often assumed to be a function of the resin alone, such as the hardness, glass transition temperature, and flash ignition temperature. Another useful guide to testing of composite materials that can be consulted in addition to Table 3.1 is ASTM D 4762. This guide also provides test methods for additional properties of interest to structural engineers, such as creep, fatigue, and fracture properties of FRP composites. Additional guidance on conducting tests on composite material unidirectional and multidirectional laminates may be found in Adams et al. (2003).

The distinction between testing of the unimpregnated single-fiber level and the impregnated roving or tow on the lamina level is of particular importance for the structural engineer in the design of FRP strengthening systems. ACI 440.2R-02 (2002) allows two different methods for designing FRP strengthening systems. One method uses the properties of the FRP composite, which are calculated using the measured gross area of the FRP composite; the other method uses the properties of the fibers, which are calculated using the manufacturer-supplied area of the fibers in a dry sheet or fabric. However, according to ACI 440.3R-04 (2004) Test Method L.2,[26] when the area of the fiber method is used, the fiber properties are not obtained from single-fiber tests. For either design method both the longitudinal strength and stiffness of either the FRP composite or the fibers are obtained from a test on the FRP composite at the ply level. Consequently, the methods lead to identical designs.

3.3.3 Laminate Level

As noted above, the test methods used for a multidirectional laminate are technically the same as those used for a unidirectional ply, as shown in Table 3.1. However, when used for multidirectional composites, two key differences should be noted. First, as can be seen in Table 3.1, only longitudinal mechanical tests are conducted on the unidirectional ply, whereas both longitudinal and transverse tests are required for the multidirectional laminate. Second, the exact requirements of the ASTM tests cannot always be satisfied, due to the thickness of the FRP parts used in structural engineering and to

[26]Titles of ACI 440.3R-04 test methods are listed in Section 3.4.

TABLE 3.1 Recommended Test Methods for FRP Composites at the Lamina and Laminate Levels

Ply or Laminate Property	ASTM Test Method(s)	Test Required
Mechanical properties		
Strength properties		
Longitutinal tensile strength	D 3039, D 5083, D 638, D 3916	Unidirectional ply and multidirectional laminate
Longitudinal compressive strength	D 3410, D 695	Unidirectional ply and multidirectional laminate
Longitudinal bearing strength	D 953, D 5961	Unidirectional ply and multidirectional laminate
Longitudinal short beam shear strength	D 2344, D 4475	Unidirectional ply and multidirectional laminate
In-plane shear strength	D 5379, D3846	Unidirectional ply and multidirectional laminate
Impact resistance	D 256	Unidirectional ply and multidirectional laminate
Transverse tensile strength	D 3039, D 5083, D 638	Multidirectional laminate only
Transverse compressive strength	D 3410, D 695	Multidirectional laminate only
Transverse short beam shear strength	D 2344	Multidirectional laminate only
Transverse bearing strength	D 953, D 5961	Multidirectional laminate only
Stiffness properties		
Longitudinal tensile modulus	D 3039, D 5083, D 638, D 3916	Unidirectional ply and multidirectional laminate
Longitudinal compressive modulus	D 3410, D 695	Unidirectional ply and multidirectional laminate
Major (longitudinal) Poisson ratio	D 3039, D 5083, D 638	Unidirectional ply and multidirectional laminate
In-plane shear modulus	D 5379	Multidirectional laminate only
Transverse tensile modulus	D 3039, D 5083, D 638	Multidirectional laminate only
Transverse compressive modulus	D 3410, D 695	Multidirectional laminate only

Physical properties		
Fiber volume fraction	D 3171, D 2584	Unidirectional ply and multidirectional laminate
Density	D 792	
Barcol hardness	D 2583	
Glass transition temperature	E 1356, E 1640, D 648, E 2092	
Water absorbed when substantially saturated	D 570	
Longitudinal coefficient of thermal expansion	E 831, D 696	
Transverse coefficient of thermal expansion	E 831, D 696	
Dielectric strength	D 149	
Flash ignition temperature	D 1929	
Flammability and smoke generation	E 84, D 635, E 662	

Source: Adapted from Bank et al. (2003).

the fact that they may have fabric and mat layers. ASTM D 6856 provides recommendations on the modifications that need to be made to the standard test methods when fabric composites (which also tend to be thicker than traditional unidirectional ply layup composites) are used.

3.3.4 Full-Section Level

FRP composites are used in structural engineering applications in the form of FRP parts or components, such as FRP reinforcing bars, FRP strips or wraps, and FRP profile sections. When designing, a structural engineer performs iterative analytical calculations in terms of member stress resultants, such as the bending moment (M), twisting moment (T), shear force (V), and axial force (F), which are obtained from the applied loads to determine the member deformations and capacities (also known as *strengths*) using the geometric and material properties of the members. The member deformations are obtained using the member's rigidities, the bending rigidity (EI), the torsional rigidity (GJ), the axial rigidity (EA), and the transverse shear rigidity (GA_s[27]), which are obtained from the geometric dimensions of the member cross section and the material properties. The member load-carrying capacities are obtained from the material strengths and geometric properties of the section (e.g., $M = \sigma_b S$, $P = \sigma_a A$).

In homogeneous isotropic materials such as steel, the material strengths and stiffnesses are not location dependent in the section, and the load-carrying capacity is determined relatively easily. However, in FRP members, the stiffness and strength properties are location dependent within the cross section on the fiber, lamina, and laminate levels due to the inhomogeneity and anisotropy of the material. This tends to complicate the design process, as the calculation of full-section rigidities and load-carrying capacities are more difficult to write in the form of simple stress-resultant equations. However, as described in the design sections of this book, numerous simplifying assumptions can be made in the analysis of FRP parts for structural engineering and the design process is consequently also simplified significantly. One common assumption made is that the properties of the FRP part can be considered to be homogeneous on a full-section level for certain types of calculations. In this case, the full-section member properties needed for structural engineering design for an as-delivered or as-produced part are then obtained from tests on the entire cross section (and often the entire length) of the part. These tests, known as full-section tests, attempt to determine effective properties for design calculations based on the as-intended performance of the entire FRP part when it is in service.

FRP parts used in structural engineering are typically large [rebar and profile lengths greater than 10 ft (3 m)], with surface areas of strips and wraps

[27] Where A_s is the effective or shear area for transverse shear deformation, often given as kA, where k is known as the *Timoshenko shear coefficient*.

greater than 50 ft^2 (5 m^2), and may be relatively thick [>0.25 in. (6 mm)]. Therefore, testing them at their full section (or full size) often requires specialized fixtures and procedures that are different from those used to test FRP materials in small coupon form at the lamina and laminate levels. For the three different categories of FRP parts used in structural engineering, different full-section tests have been developed by a number of organizations.

Full-Section Tests for FRP Reinforcing Bars Full-section tests for FRP reinforcing bars that have been developed by the ACI, CSA, and JSCE are listed at the end of the chapter. These methods are used to test FRP reinforcing bars to develop design properties for quality control and assurance purposes as well as for research purposes. Since FRP reinforcing bars have a variety of cross-sectional shapes and surface irregularities that are required to enhance bond properties, a test method has been developed to determine the nominal cross-sectional area and nominal diameter of an irregularly shaped bar (ACI B.1; CSA Annex A and JSCE E-131). Since FRP bars have significantly lower compressive strength in the transverse direction than tensile strength in the longitudinal direction, they have a tendency to crush in the grips when tested in tension. Due to their irregular surface profile, standard grips used for smooth FRP rods recommended by ASTM D 3916 cannot be used to test FRP reinforcing bars. Therefore, special methods are used to anchor FRP bars in a tensile testing machine when conducting tension tests (ACI Appendix A and CSA Annex B). The provisions detail a number of methods to embed the FRP bars ends in a steel sleeve to enable testing to failure at full-section loads (which can approach 200 kips for larger-diameter FRP bars). Figure 3.3 shows FRP rebars with steel end sleeves attached.

Since FRP reinforcing bars have different moduli and surface textures from those of conventional steel reinforcing bars, provisions are given for determining either the development length (CSA Annex D) or the pull-out bond strength (ACI B.3; CSA Annex H, and JSCE E-539) for FRP bars. These test methods can be used to investigate the bond behavior of FRP bars in concrete and to compare their bond behavior to that of conventional bars under different loading and environmental conditions. Since the strength of the bent

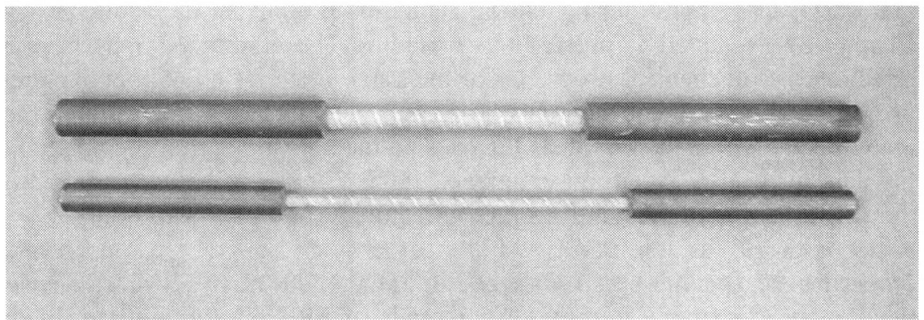

Figure 3.3 FRP reinforcing bar with steel sleeves for tensile testing.

portion of the FRP reinforcing bar can be significantly less than that of the same straight bar due to the particular FRP manufacturing process used, tests for bent FRP bars for use as stirrups in beams are provided (ACI B.5 and B.12 and CSA Annex E). The values obtained in the bond strength and bent-bar strength tests provide important data for design of concrete structures with FRP reinforcing bars. In some cases, these test data have been incorporated in design equations listed in the ACI 440.1R-06 design provisions. They are also used by manufacturers to provide properties for their bars that can be used in design directly by the structural engineer.

Since an FRP reinforcing bar is anisotropic, its shear strength cannot be calculated from its tensile strength. The transverse shear strength of an FRP reinforcing bar is needed to determine the dowel strength of FRP reinforcing bars. Test methods for determining the shear strength in direct shear (as opposed to in bending) are provided in ACI B.4, CSA Annex, and JSCE E-540.

Since FRP reinforcing bars are nonmetallic, they do not corrode in aqueous environments, due to electrochemical effects. However, they are susceptible to chemical and physical degradation of their polymer resins, additives, and fibers. Glass fiber–reinforced composites are particularly susceptible to degradation in alkaline environments. Test methods for durability (ACI B.6, CSA Annex O, and JSCE E-538) are used to determine the residual tensile strength of a bar (with appropriate testing anchorages) following accelerated conditioning in prescribed alkaline solutions that are intended to simulate the concrete pore-water alkalinity. To date, these methods have not been used to develop design procedures but are used only as a guide to the long-term durability of FRP bars in concrete. In a similar fashion, test methods are provided for creep rupture (ACI B.8, CSA Annex J, and JSCE E-533) and fatigue (ACI B.7, CSA Annex L, and JSCE E-534).

All of the tests noted above are conducted on lengths of FRP reinforcing bars using the full cross-sectional area of the bar with all of its surface irregularities. It is important to note that the properties of FRP reinforcing bars are bar size (usually given in terms of the nominal diameter)–dependent and that tests must be conducted for each discrete bar size. This is due to the anisotropic nature of the FRP materials used and the fact that resins and additives used for thicker bars (0.5 in. and larger in diameter) are usually different from those used for smaller bars. Thicker bars tend to develop thermal shrinkage cracks during curing, and low-profile shrink additives (see Chapter 2) are added to prevent this cracking. These additives may have an effect on the mechanical properties of the bars. Since different manufacturers use different constituent materials, processes, and surface deformations, each manufacturer generally conducts the tests on the bars they produce and reports the data for use by structural engineers. JSCE-E-131 can be used to identify the constituent materials and geometric features of an FRP reinforcing bar in a standardized fashion. JSCE-E-131 also provides guaranteed mechanical properties for specific bars referenced in JSCE (2001). An alternative specification for identifying constituent materials of FRP reinforcing bars is discussed in Bank et al. (2003).

Since FRP reinforcing bars are anisotropic and are manufactured from a variety of fibers and resin systems, they have different coefficients of thermal expansion (CTEs) in their longitudinal and transverse directions. The coefficients of thermal expansion of an FRP reinforcing bar can be obtained by use of conventional thermal mechanical analysis (TMA) according to CSA Annex M and JSCE E-536. It should be noted that due to the nature of the TMA test device used, small coupons are used and that this is not a full-section test method. In addition, only the longitudinal CTE is measured in the methods noted above. This method is equivalent to ASTM E 831.

No unique full-section test methods have been developed for testing FRP grids and gratings used as reinforcements for concrete. The test methods used for FRP bars are intended for use with FRP grids and gratings. According to the test methods, individual bars in the grid or grating should be cut from the grid and tested. This is not entirely appropriate, as grid intersections create regions of stress concentrations and have very different properties from those of nonintersection regions. In addition, grid bars are generally smooth and not specifically deformed for bond to the concrete. Grids and gratings acquire their bonding properties from the anchorage provided by the perpendicular grid members in the concrete. Although a standard test method has not been developed to determine the development length of a grid or grating bar, it is generally accepted that at least two perpendicular bars are required to fully develop a single grid bar (Bakht et al., 2000, p. 10). In addition, while splicing of individual FRP reinforcing bars by overlapping is reasonable, splicing of FRP grids and gratings by overlapping splices cannot generally be carried out (unless the grids are very thin and flexible) and specialized splicing methods are required for grids and gratings. In-house test methods developed for the NEFMAC FRP grid material are reviewed in Sugita (1993).

No test methods have been developed for FRP stay-in-place (SIP) formwork systems that are required to serve the dual purpose of forms for the uncured (i.e., wet) concrete and of reinforcements for the cured (i.e., hardened) concrete. Often, these FRP systems have quite complex cross-sectional shapes, and coupon-level testing is difficult. In addition, since they serve as tensile reinforcement for FRP reinforced concrete in much the same way as do FRP reinforcing bars, the extent of the bond and the development length of the FRP SIP form and the concrete needs to be determined. Since FRP SIP forms are bonded to the concrete on only one side, the determination of the bond and development length properties often require specialized in-house test methods to be developed that depend on the geometry of the FRP SIP form.

Full-Section Tests for FRP Precured Strips and Hand-Laid-Up Sheets and Fabrics A number of test methods have been developed to test FRP precured strips and FRP hand-laid-up sheets or fabrics (known as *wraps* when they partially or fully encircle a member) that are adhesively bonded to the surface of concrete or masonry structures. The full-section tests developed include those for testing the direct tensile pull-off bond strength of the cured FRP when bonded adhesively to the concrete, the longitudinal shear bond strength, and the development length of the cured FRP when bonded adhesively to the concrete, and tensile and tensile splice tests of the FRP itself (i.e., not bonded to the concrete).

The test methods developed for the tensile testing (ACI L.2, CSA Annex G, and JSCE E-541) are essentially modifications of the standard coupon tests used to test FRP unidirectional materials on the laminate level, such as ASTM D 3039. The key differences between the standard tensile test methods and those developed for FRP strips and fabrics lie in the methods of specimen preparation and in the techniques of data reduction.

Since FRP laid-up sheets and fabrics are fabricated in situ, a procedure (ACI L.2) is described to fabricate test panels (also known as *witness panels*) in the field at the time of actual fabrication of the FRP composite on the concrete surface. The procedure allows the fabricator to use a roller to "work out trapped air in the laminate" and the use of a flat cover plate when curing the test panel to create a smooth top surface (but not to apply significant pressure to the laminate). Coupons cut from the cured test panel are then tested to determine the in situ properties of the FRP hand laid-up materials. It is very important to note that this procedure requires the fabricator to cure the test panel at the same conditions as those of the FRP material. If postcuring is not used on the in-field applied FRP material, postcuring must not be used on the test panel. A procedure is also described for producing a laboratory test panel for a laid-up FRP material system. The laboratory procedure is similar to the field procedure, however, it allows the fabricator to forcibly push out excess resin using a flat-edged paddle, and furthermore, it allows the laminate to be cured under pressure and/or elevated temperature provided that "the engineer and material supplier . . . agree on a representative specimen fabrication process"[28] (ACI L.2). When precured FRP strips are tested, the strip must be tested at its full thickness. It is important to note that in all the methods described above, the laminate tested must be the full thickness of the actual laminate used in the field. Testing of individual unidirectional plies and calculation of effective properties are not permitted. This is in contradiction to the design procedures in ACI 440.2R-02. It is also important to note that the ACI L.2 test method is applicable only to unidirectional or bidirectional (cross-ply) FRP materials.

The second key difference between the full-section test method and the standard tensile test method lies in the methods allowed for data reduction to determine the effective tensile strength and stiffness of an FRP laminate. In

[28] When using test data from manufacturer specification sheets it is very important to know whether the test panel has been produced similar to the in situ material or has been in produced in laboratory-controlled conditions. Removing excess resin to increase fiber volume fraction and elevated temperature postcuring of an FRP laminate can have a significant influence on both the mechanical (e.g., strengths, stiffnesses) and physical (e.g., glass transition temperature, hardness) properties of the laminate.

addition to allowing the fabricator to obtain the properties based on the gross cross-sectional area of the coupon (which is standard for composite material test methods), an alternative data reduction method based on the net cross-sectional area of the fibers (only) in the coupon can be used.[29] The distinction between the way in which these data are reported and how they are used in design is of utmost importance, and a structural engineer must know if the tensile properties reported are for the FRP laminate or for the fibers only.

The test methods for overlap splices for FRP strips and hand-laid-up sheets and fabrics (ACI L.3, CSA Annex Q, and JSCE-E 542) are similar to those used to test laminates in single- or double-lap configurations. However, the key differences detailed above for specimen fabrication and data reduction are also applied to lap splice specimens.

Two types of test methods are provided to test the properties of the bond between the FRP laminate (precured or hand laid-up) and the concrete substrate. One is a direct tension test that can be used to measure tensile bond strength normal to the FRP laminate; the other is a shear and peeling test that can be used to measure shear bond strength and interfacial fracture energy parallel to the FRP laminate.

The direct tension pull-off test (ACI L.1, CSA Annex E, and JSCE-E 545) uses a standard pull-off test fixture to measure the bond strength between the FRP and the concrete. Modifications have been made to the standard pull-off test that define allowable types of failure modes when the tester is used to test laminates bonded to concrete. Allowable failure modes are cohesive failure with the laminate, cohesive failure in the concrete substrate, or interfacial failure at the interface of the adhesive and the concrete or the FRP. It is important to note that the ACI version of this method does not permit failure in the interfacial region. The results of this test are not actually used for design of FRP strengthening systems but rather, to perform tests on the concrete substrate to determine its suitability for strengthening. The design guides for FRP strengthening generally require that the concrete pull-off tensile strength in the surface layer to which the FRP is bonded exceeds 200 psi (1.4 MPa) (ACI 440.2R-02, 2002).

The other bond test is a single-lap shear test (CSA Annex P and JSCE E-543) in which the FRP is bonded to the concrete and then placed in longitudinal tension so as to shear it off the concrete to simulate the stress state in the FRP strengthening system when it is used as a tensile strengthening element. The test method depends on the determination of an effective bond length.[30] The results of the test can be used to determine bond stress, effective length, and interfacial fracture energy. However, they cannot generally be used

[29] In this method the area of the fibers is obtained from manufacturers' specification sheets.

[30] It has been shown that shear force transfer in a bonded laminate occurs only over a limited length, called the *effective length,* and that the bond stress can therefore not increase in an unlimited manner. The effective length is a function of properties of the concrete substrate, the FRP material, and the adhesive, and is usually determined by experiment.

at this time to predict either laminate-end debonding or midspan debonding of FRP laminates, due to the lengthwise-varying stress resultants, peeling stresses at the plate ends, and crack patterns that develop in beams when loaded in flexure.

Test methods have also been developed for evaluating the long-term fatigue and durability of FRP strengthening systems when used to strengthen concrete (JSCE-E 546, JSCE-E 547, JSCE-E 548, and JSCE-E 549). These test methods use essentially the same specimen types as the tensile tests (or bond tests in the case of JSCE-E 549) but expose the specimens to different accelerated conditioning environments to determine residual properties.

Full-Section Tests for FRP Profiles FRP pultruded profile sections commonly used in structural engineering include I-shaped beams, channels, angles, and square and rectangular tubes. Channels and angles are generally used in back-to-back configurations to create double-channel and double-angle sections that have coincident centroids and shear centers. The mechanical and physical properties of FRP materials used in these pultruded profiles are obtained from tests on coupons cut from the sections (i.e., on the laminate level) or from tests on test panels that are produced by pultrusion having nominally identical properties (Wang and Zurelck, 1994). Most pultrusion manufacturers report the properties of their pultruded materials for an entire class of structural shapes (e.g., I-beams, plates) and resin systems even though layup of the roving and mat layers may differ within a class. The properties of the thinner pultruded material shapes [$<\frac{1}{4}$ in. (6 mm)] are generally penalized at the expense of the thicker parts [$>\frac{1}{4}$ in. (6 mm)] in order to publish lower bound values for use in design. A designer may want to use test data to determine design values of the pultruded material for a specific part used in a design in lieu of the manufacturer-reported values. In general, pultruded profiles for braced frame and truss structures are designed using the properties obtained from the coupon test and appropriate theoretical methods as explained in the design chapters of this book. However, in certain situations it is often preferred to use full-section tests on individual profiles or on subassemblies of profiles to develop full-section properties or capacities for use in design. This is because there are situations in which a full-section property is easier to use in a design or when coupon property data cannot be used with confidence to predict the performance of complicated details.

Since designs of pultruded profiles are often controlled by serviceability and buckling criteria,[31] the bending, transverse shear, and torsional rigidities of the full section (EI, GA_s, and GJ) are needed in design (Roberts and Al-Ubaidi, 2002). Although theoretical methods are available to predict these properties from coupon data, this requires the designer to make a number of simplifying assumptions related to the profile's geometry, homogeneity, and

[31] This includes flexural (Euler) buckling, torsional buckling, lateral buckling, and local buckling.

anisotropy (Bank et al., 1995a). Full-section testing is a way to obtain an effective property for a profile that can be used in a stress-resultant theory.

The most widely used full-section test is the test for the *full-section modulus,* or *E-modulus* as it referred to by most manufacturers. In this test, the FRP profile is tested in three- or four-point bending, and an effective longitudinal modulus is determined from load and midspan deflection data (CEN Part 2 Annex D). However, due to the presence of shear deformation effects in pultruded profiles because of moderately high longitudinal-to-shear modulus ratios (4 to 6 for glass-reinforced profiles) and because pultruded profiles are generally used on shorter spans than steel beams, the results of the full-section bending test need to be interpreted with caution (Bank, 1989a,b). As noted in the CEN test method, the effective flexural modulus is a span-dependent property, due to the effect of shear deformations. To minimize the effects of the shear deformation, the CEN test method requires the profile to be tested over a span-to-depth ratio of at least 20. In addition, the measured value of the full-section modulus that is obtained from the test is multiplied by a factor of 1.05 to compensate approximately for the effects of shear deformation. However, this test does not yield the full-section transverse shear rigidity (GA_s) that is often needed in a design where a shear deformation beam theory is used to predict deflections.

CEN Part 2 Annex G provides a method to measure the bending rigidity (EI) and the shear rigidity (GA_s) of a profile section simultaneously by testing it on different span lengths and performing a data analysis. This method has been used by a number of researchers and manufacturers to obtain results that can be used in design, and had become a de facto standard prior to publication by CEN (Bank, 1989a,b; Roberts and Al-Ubaidi, 2002; Giroux and Shao, 2003). CEN Part 2 Annex G also provides a method to determine the full-section torsional stiffness of an FRP profile using a fixture that applies a uniform torsional moment to the profile and measures the rotation. It is important to note that the torsional stiffness thus obtained is the unrestrained or *nonwarping restrained torsional stiffness* (also known as the *Saint-Venant torsional stiffness*). In most structural engineering applications the ends of the structural members are restrained against warping and the warping torsional constant (EC_w) is needed for design. This can be measured in an indirect method using a nonuniform torsion test (Roberts and Al-Ubaidi, 2002). It should also be pointed out that singly symmetric open-cross-section FRP profiles have very low torsional rigidity and should not be used in situations where significant torsional moment carrying resistance is needed. When used as beams and columns, FRP profiles must be laterally braced to prevent lateral or torsional instability under bending or axial loads.

Full-section tests to determine the strengths and stiffness of connections in pultruded structures are not standardized at this time. Most manufacturers of pultruded profiles provide load tables for simple framing connections. These tables are based on full-section testing of subassemblies of profiles conducted using undisclosed in-house methods. Data are generally not provided on failure modes or deformations, nor are they correlated with coupon property data obtained from bearing tests. Neither is there a standard procedure for full-section testing of semirigid connections for pultruded frame structures that is needed to obtain strength and moment–rotation characteristics for these connections for use in the analysis of frames with semirigid connections. Single- and double-beam methods proposed by Bank et al. (1990) and Mottram and Zheng (1999a,b) have been used in a number of investigations.

3.4 RELEVANT STANDARD TEST METHODS FOR FRP COMPOSITES FOR STRUCTURAL ENGINEERS

3.4.1 American Society of Testing and Materials Test Methods

ASTM Standard Test Methods

C 1557 Standard Test Method for Tensile Strength and Young's Modulus of Fibers

D 149 Standard Test Method for Dielectric Breakdown Voltage and Dielectric Strength of Solid Electrical Insulating Materials at Commercial Power Frequencies

D 256 Standard Test Methods for Determining the Izod Pendulum Impact Resistance of Plastics

D 570 Standard Test Method for Water Absorption of Plastics

D 635 Standard Test Method for Rate of Burning and/or Extent and Time of Burning of Plastics in a Horizontal Position

D 638 Standard Test Method for Tensile Properties of Plastics

D 648 Standard Test Method for Deflection Temperature of Plastics Under Flexural Load in the Edgewise Position

D 695 Standard Test Method for Compressive Properties of Rigid Plastics

D 696 Standard Test Method for Coefficient of Linear Thermal Expansion of Plastics Between −30°C and 30°C with a Vitreous Silica Dilatometer

D 790 Standard Test Methods for Flexural Properties of Unreinforced and Reinforced Plastics and Electrical Insulating Materials

D 792 Standard Test Methods for Density and Specific Gravity (Relative Density) of Plastics by Displacement

D 953 Standard Test Methods for Bearing Strength of Plastics

D 1929 Standard Test Method for Determining Ignition Temperature of Plastics

D 2343 Standard Test Method for Tensile Properties for Glass Fiber Strands, Yarns, and Rovings Used in Reinforced Plastics

D 2344 Standard Test Method for Short-Beam Strength of Polymer Matrix Composite Materials and Their Laminates

D 2583 Standard Test Method for Indentation Hardness of Rigid Plastics by Means of a Barcol Impressor
D 2584 Standard Test Method for Ignition Loss of Cured Reinforced Resins
D 3039 Standard Test Method for Tensile Properties of Polymer Matrix Composite Materials
D 3171 Standard Test Method for Constituent Content of Composite Materials
D 3379 Standard Test Method for Tensile Strength and Young's Modulus of High-Modulus Single-Filament Materials
D 3410 Standard Test Method for Compressive Properties of Polymer Matrix Composite Materials with Unsupported Gage Section by Shear Loading
D 3846 Standard Test Method for In-Plane Shear Strength of Reinforced Plastics
D 3916 Standard Test Method for Tensile Properties of Pultruded Glass-Fiber-Reinforced Plastic Rod
D 4018 Standard Test Methods for Properties of Continuous Filament Carbon and Graphite Fiber Tows
D 4475 Standard Test Method for Apparent Horizontal Shear Strength of Pultruded Reinforced Plastic Rods by the Short-Beam Method
D 4476 Standard Test Method for Flexural Properties of Fiber-Reinforced Pultruded Plastic Rods
D 4762 Standard Guide for Testing Polymer Matrix Composite Materials
D 5083 Standard Test Method for Tensile Properties of Reinforced Thermosetting Plastics Using Straight-Sided Specimens
D 5379 Standard Test Method for Shear Properties of Composite Materials by the V-Notched Beam Method
D 5961 Standard Test Method for Bearing Response of Polymer Matrix Composite Laminates
D 6856 Standard Guide for Testing Fabric-Reinforced "Textile" Composite Materials
E 84 Standard Test Method for Surface Burning Characteristics of Building Materials
E 662 Standard Test Method for Specific Optical Density of Smoke Generated by Solid Materials
E 831 Standard Test Method for Linear Thermal Expansion of Solid Materials by Thermomechanical Analysis
E 1356 Standard Test Method for Assignment of the Glass Transition Temperatures by Differential Scanning Calorimetry or Differential Thermal Analysis
E 1640 Standard Test Method for Assignment of the Glass Transition Temperature by Dynamic Mechanical Analysis
E 2092 Standard Test Method for Distortion Temperature in Three-Point Bending by Thermal Mechanical Analysis

ASTM Standard Terminologies

C 162 Standard Terminology of Glass and Glass Products
C 904 Standard Terminology Relating to Chemical-Resistant Nonmetallic Materials
D 123 Standard Terminology Relating to Textiles
D 883 Standard Terminology Relating to Plastics
D 907 Standard Terminology of Adhesives
D 3878 Standard Terminology of Composite Materials
D 3918 Standard Terminology Relating to Reinforced Plastic Pultruded Products
E 6 Standard Terminology Relating to Methods of Mechanical Testing
E 631 Standard Terminology of Building Constructions

ASTM Standard Specifications

D 3917 Standard Specification for Dimensional Tolerance of Thermosetting Glass-Reinforced Plastic Pultruded Shapes

ASTM Standard Practices

C 581 Standard Practice for Determining Chemical Resistance of Thermosetting Resins Used in Glass-Fiber-Reinforced Structures Intended for Liquid Service
D 618 Standard Practice for Conditioning Plastics for Testing
E 122 Standard Practice for Calculating Sample Size to Estimate, with a Specified Tolerable Error, the Average for a Characteristic of a Lot or Process
E 632 Standard Practice for Developing Accelerated Tests to Aid Prediction of the Service Life of Building Components and Materials

3.4.2 Full-Section Test Methods for FRP Bars and Laminates

ACI Test Methods for FRP Bars for Concrete Structures

B.1 Test Method for Cross-Sectional Properties of FRP Bars
B.2 Test Method for Longitudinal Tensile Properties of FRP Bars
B.3 Test Method for Bond Strength of FRP Bars by Pullout Testing
B.4 Test Method for Transverse Shear Strength of FRP Bars
B.5 Test Method for Strength of FRP Bent Bars and Stirrups at Bend Locations
B.6 Accelerated Test Method for Alkali Resistance of FRP Bars
B.7 Test Method for Tensile Fatigue of FRP Bars
B.8 Test Method for Creep Rupture of FRP Bars

B.9 Test Method for Long-Term Relaxation of FRP Bars
B.10 Test Method for Performance of Anchorages of FRP Bars
B.11 Test Method for Tensile Properties of Deflected FRP Bars
B.12 Test Method for Determining the Effect of Corner Radius on Tensile Strength of FRP Bars
Appendix A: Anchor for Testing FRP Bars Under Monotonic, Sustained and Cyclic Tension

ACI Test Methods for FRP Laminates for Concrete and Masonry

L.1 Test Method for Direct Tension Pull-off Test
L.2 Test Method for Tension Test of Flat Specimen
L.3 Test Method for Overlap Splice Tension Test
Appendix B: Methods for Calculating Properties of Flat Specimen

CEN 13706 Test Methods for FRP Pultruded Profiles

CEN Normative Test Methods

Part 2 Annex A Visual Defects: Descriptions and Acceptance Levels
Part 2 Annex B Dimensional Tolerance for Pultruded Profiles
Part 2 Annex C Workmanship
Part 2 Annex D Determination of Effective Flexural Modulus
Part 2 Annex E Determination of the Pin Bearing Strength

CEN Informative Test Methods

Part 2 Annex F Recommended Test Methods for Particular Requirements
Part 2 Annex G Determination of Flexural Shear and Torsional Stiffness Properties

CSA S806 Standard Test Methods for FRP Bars and Laminates

CSA Normative Test Methods

Annex A Determination of Cross-Sectional Area of FRP Reinforcement
Annex B Anchor for Testing FRP Specimens Under Monotonic, Sustained and Cyclic Tension
Annex C Test Method for Tensile Properties of FRP Reinforcements
Annex D Test Method for Development Length of FRP Reinforcement
Annex E Test Method for FRP Bent Bars and Stirrups
Annex F Test Method for Direct Tension Pull-off Test
Annex G Test Method for Tension Test of Flat Specimens

CSA Informative Test Methods

Annex H Test Method for Bond Strength of FRP Rods by Pullout Testing
Annex J Test Method for Creep of FRP Rods
Annex K Test Method for Long-Term Relaxation of FRP Rods
Annex L Test Method for Tensile Fatigue of FRP Rods
Annex M Test Method for Coefficient of Thermal Expansion of FRP Rods
Annex N Test Method for Shear Properties of FRP Rods
Annex O Test Methods for Alkali Resistance of FRP Rods
Annex P Test Methods for Bond Strength of FRP Sheet Bonded to Concrete
Annex Q Test Method for Overlap Splice in Tension

Japan Society of Civil Engineers Standard Test Methods for Continuous Fiber Reinforcing Materials

JSCE-E-131 Quality Specification for Continuous Fiber Reinforcing Materials
JSCE-E 531 Test Method for Tensile Properties of Continuous Fiber Reinforcing Materials
JSCE-E 532 Test Method for Flexural Tensile Properties of Continuous Fiber Reinforcing Materials
JSCE-E 533 Test Method for Creep of Continuous Fiber Reinforcing Materials
JSCE-E 534 Test Method for Long-Term Relaxation of Continuous Fiber Reinforcing Materials
JSCE-E 535 Test Method for Tensile Fatigue of Continuous Fiber Reinforcing Materials
JSCE-E 536 Test Method for Coefficient of Thermal Expansion of Continuous Fiber Reinforcing Materials by Thermo Mechanical Analysis
JSCE-E 537 Test Method for Performance of Anchors and Couplers in Prestressed Concrete Using Continuous Fiber Reinforcing Materials
JSCE-E 538 Test Method for Alkali Resistance of Continuous Fiber Reinforcing Materials
JSCE-E 539 Test Method for Bond Strength of Continuous Fiber Reinforcing Materials by Pull-out Testing
JSCE-E 540 Test Method for Shear Properties of Continuous Fiber Reinforcing Materials by Double Plane Shear

Japan Society of Civil Engineers Standard Test Methods for Continuous Fiber Sheets

JSCE-E 541 Test Method for Tensile Properties of Continuous Fiber Sheets

JSCE-E 542 Test Method for Overlap Splice Strength of Continuous Fiber Sheets

JSCE-E 543 Test Method for Bond Properties of Continuous Fiber Sheets to Concrete

JSCE-E 544 Test Method for Bond Strength of Continuous Fiber Sheets to Steel Plate

JSCE-E 545 Test Method for Direct Pull-off Strength of Continuous Fiber Sheets with Concrete

JSCE-E 546 Test Method for Tensile Fatigue of Continuous Fiber Sheets

JSCE-E 547 Test Method for Accelerated Artificial Exposure of Continuous Fiber Sheets

JSCE-E 548 Test Method for Freeze-Thaw Resistance of Continuous Fiber Sheets

JSCE-E 549 Test Method for Water, Acid and Alkali Resistance of Continuous Fiber Sheets

PROBLEMS

3.1 Write a computer program to calculate the stiffness properties (E_x, E_y, $G_{xy} = E_s$, and ν_x) of a two-dimensional FRP composite given the properties of the fiber and matrix. Assume that both the matrix and the fiber are isotropic and use the material properties given below. Use the rule-of-mixtures micromechanics equations.

(a) Use your program to verify the results of Analysis Example 3.1.

(b) Plot the stiffness properties (E_x, E_y, $G_{xy} = E_s$, and ν_x) for an E-glass/vinylester composite and for a carbon–epoxy composite as a function of fiber volume fraction from 20 to 60%.

(c) Plot the longitudinal strength of the two FRP composites above as a function of fiber volume from 20 to 60%.

(d) Plot the fiber weight (or mass) fraction as a function of fiber volume fraction from 20 to 60%.

The following properties are given for the fibers and matrices. For the E-glass fiber:

E (psi)	10,500,000
ν	0.20
σ_f^{ult} (psi)	270,000
ρ (lb/in^3)	9.400×10^{-2}

For the carbon fiber:

E (psi)	33,000,000
ν	0.20
σ^{ult} (psi)	570,000
ρ (lb/in^3)	6.500×10^{-2}

For the vinylester matrix:

E (psi)	500,000
ν	0.38
σ^{ult} (psi)	11,000
ρ (lb/in^3)	4.570×10^{-2}

For the epoxy:

E (psi)	620,000
ν	0.34
σ^{ult} (psi)	11,300
ρ (lb/in^3)	4.570×10^{-2}

3.2 Write a computer program to calculate the off-axis stiffness coefficients [Q_{ij} (i,j = 1,2,6)] for a composite lamina with on-axis stiffness coefficients [Q_{ij} ($i,j = x,y,s$)] rotated at an angle θ from the structural axes (1,2,6) when given the engineering constants (E_x, E_y, E_s, ν_x) of the lamina. You can use any programming language. A high-level language/program such as MathCad, MatLab, or Excel is recommended.

(a) Use your program to verify the off-axis stiffness coefficients of the +10° carbon–epoxy lamina in Analysis Example 3.3.

(b) For the 51% volume fraction E-glass/polyester lamina in Example 3.2, provide a plot of the engineering constants (E_1, E_2, E_6, ν_{21}) as a function of the angle θ from 0 to 360°. Discuss how the properties vary relative to one another as the lamina (ply) is rotated through 360°.

(c) For the 23% continuous filament mat lamina in Analysis Example 3.2, provide a plot of the engineering constants (E_1, E_2, E_6, ν_{21}) as a function of the angle θ from 0 to 360°. Discuss how the properties vary relative to one another as the lamina (ply) is rotated through 360°.

3.3 Write a computer program to (1) calculate the [A–B–D] matrix for a three-ply laminate with arbitrary ply thicknesses and properties, and (2) calculate the effective in-plane and flexural engineering properties of the laminate. You can use any programming language. A high-level language/program such as MathCad, MatLab, or Excel is recommended.

(a) Use the program to verify the results of Analysis Example 3.2.

(b) Use the program to verify the results of Analysis Example 3.3.

3.4 Given a quasi-isotropic eight-layer $[0/90/+45/-45]_s$ laminate of E-glass/polyester with a 51% fiber volume fraction per layer. Use material properties from Analysis Example 3.2. Show that the laminate is isotropic for its in-plane stiffness but not isotropic for its flexural stiffness; that is, show that

$$E_1^0 = E_2^0 \qquad E_1^f \neq E_2^f$$

and

$$E_6^0 = \frac{E_1^0}{2(1 - \nu_1^0)} \qquad E_6^f \neq \frac{E_1^f}{2(1 + \nu_1^f)}$$

3.5 An FRP laminate is to be used to strengthen a concrete structure. The laminate is constructed of three plies of SikaWrap Hex 103C unidirectional carbon fiber fabric and SikaDur Hex 306 epoxy resin. The layup for the laminate is given as [0/90/0]. The longitudinal axis (the 1-direction) of the laminate is at 0°. Manufacturer specification sheets for the fabric–resin system are available at www.sikaconstruction.com. You are required to consult the specification sheets to obtain the data you need for the calculations to follow. Use the average values measured at room temperature for the lamina properties. In addition to the data provided in the specification sheets, you are given that the major Poisson ratio for a single composite ply of the SikaWrap Hex 103C unidirectional carbon fiber fabric with SikaDur Hex 306 epoxy resin is $\nu_x = 0.32$ (since this property is not reported by the manufacturer).

For the FRP laminate described, find the following:

(a) The on-axis stiffness matrix $[Q]$ for a single ply of the SikaWrap Hex 103C unidirectional carbon fiber fabric and SikaDur Hex 306 epoxy resin.

(b) The $[A]$ matrix for the three-ply laminate.

(c) The effective longitudinal in-plane (extensional) modulus of the laminate (in the 0° direction).

(d) The fiber volume fraction (assuming no voids in the composite).

3.6 Given a $\frac{1}{4}$-in.-thick pultruded material that will be used in a 6 × 6 in. wide flange pultruded profile (i.e., the flange width is 6 in. and the web depth is 6 in.). The FRP material consists of E-glass/polyester and consists of four 1.0-oz/ft^2 E-glass continuous filament mats and two layers of unidirectional 113 yield glass roving, as shown in Fig.

1-oz CFM, $t = 0.02$ in.
100-113 yield E-glass rovings, $t = 0.085$ in.
two 1-oz CFM, $t = 0.04$ in.
100-113 yield E-glass rovings, $t = 0.085$ in.
1-oz CFM, $t = 0.02$ in.

Figure P3.6 Pultruded material layup.

P3.6 (in a 6-in. width). The area of one 113 yield roving is given as 2.615×10^{-3} in^2. The isotropic properties of the E-glass/polyester resin matrix are as follows:

$$E_{\text{glass}} = 10.5 \times 10^6 \text{ psi} \qquad E_{\text{polyester}} = 4.7 \times 10^5 \text{ psi}$$

$$\nu_{\text{glass}} = 0.20 \qquad \nu_{\text{polyester}} = 0.38$$

$$\sigma^{\text{ult}} = 270 \text{ ksi} \qquad \sigma^{\text{ult}} = 10.4 \text{ ksi}$$

The in-plane engineering properties of the CFM layer are

$$E = 1.26 \times 10^6 \text{ psi} \qquad \nu = 0.35 \qquad X_t = 12.45 \text{ ksi}$$

Determine the following:

(a) The fiber volume fraction in the roving layer.

(b) The engineering stiffness properties (E_x, E_y, E_s, ν_x, ν_y) of the roving layer using the rule-of-mixtures equations.

(c) The engineering stiffness properties (E_x, E_y, E_s, ν_x, ν_y) of the CFM layer.

(d) The $[A]$ matrix of the layup.

(e) The $[D]$ matrix of the layup.

(f) The effective engineering in-plane extensional properties of the material: E_x^0, E_y^0, E_s^0, ν_x^0, ν_y^0. Compare these properties with those given by Strongwell and Creative Pultrusions for their E-glass/polyester pultruded profiles (see online specification sheets). Discuss the differences and similarities between the calculated and manufacturer-reported data.

(g) Provide an estimate of the longitudinal tensile strength of the material. Justify your answer with an explanation and show how you have calculated the strength.

3.7 Review ASTM test methods D 3039, D 638, and D 5083 for tensile testing of composites and plastics and list their similarities and differences in tabulated form.

3.8 Review ASTM test methods D 3171 and D 2548 for volume fraction testing and discuss the various methods. Explain why different methods are used for different types of composite materials.

3.9 Review ASTM test methods D 1356, D 1640, and D 648 for glass transition temperature and heat deflection temperature. Do these test methods measure the same physical or mechanical property for the fiber-reinforced polymer? Why are these properties important to know when using FRP composites in structural engineering?

3.10 To determine the four independent in-plane engineering stiffness constants (E_x, E_y, E_s, ν_x) of an orthotropic lamina of FRP composite material, longitudinal (for E_x, ν_x), transverse (for E_y), and in-plane shear (for E_s) tests are required. It is often difficult or impossible to perform transverse and in-plane shear testing due to the coupon size required for the testing relative to the size of the part from which it is cut or the complexity of the test itself. For example, a 4-in. wide by $\frac{1}{16}$-in. FRP strengthening strip is too narrow to comply with the requirements of ASTM D 3039 for tensile testing or too thin to be used in the special testing fixture for shear testing in accordance with ASTM D 5379. In addition, the special fixture needed for shear testing (called the Iosipescu fixture) requires time-consuming specimen preparation and strain gauging. It is, however, possible to obtain the four independent properties of an orthotropic lamina by using only the tensile test. To do this, one tests a coupon cut parallel to the major axis of orthotropy (the fiber direction) and another coupon cut at an angle (typically, 10 to 15°) from the major axis of orthotropy. In both tests the longitudinal and the transverse strains (parallel and perpendicular to the coupon edges) are measured as a function of applied load (or stress).

Show analytically how the experimental data obtained from these two tests can be used to calculate all four of the in-plane stiffness properties of the orthotropic lamina. (*Hint:* Use the expressions for engineering properties and the on- and off-axis compliance relationships.)

4 Design Basis for FRP Reinforcements

4.1 OVERVIEW

In this chapter we introduce the subject of FRP reinforcing bars (FRP *rebars*) for reinforced concrete structures such as beams and slabs. The properties of FRP reinforcing bars are reviewed and their behavior under different loads is discussed. The primary focus of this book is on glass fiber reinforced bars. Although the design procedures presented are applicable to FRP rebars made of other reinforcing fibers, such as carbon or aramid fibers, at present economical rebars of these types are not generally available for concrete reinforcing applications.

We then proceed to describe the basis for the design of concrete members reinforced with FRP bars. This basis is in accordance with American Concrete Institute publication ACI 440.1R-06, *Guide to the Design and Construction of Structural Concrete Reinforced with FRP Bars* (ACI, 2006). This document is compatible with ACI publication ACI 318-05, *Building Code Requirements for Structural Concrete and Commentary* (ACI, 2006). Throughout this chapter and subsequent chapters that describe in detail the procedures for the design of concrete structures with FRP rebars, it is assumed that the reader has reasonable familiarity with ACI 318-05 and that the reader is familiar with the design of concrete beams and slabs reinforced with conventional steel reinforcing bars. The equations and examples presented in what follows use U.S. standard units throughout since these are the familiar units for design in the United States, where the ACI codes are mainly used.

With regard to FRP reinforcing bars, one of the purposes of the book is to point out the key differences between design with steel bars and design with FRP bars. Extensive examples are not presented; rather, a limited number of detailed illustrative examples are presented. These examples generally follow those in ACI 440.1R-06 and are intended to allow the reader to analyze these examples critically and compare them to steel-reinforced design. After studying the book readers are challenged to open their favorite text on reinforced concrete design and attempt to compete the example problems in that text using FRP rebars rather than steel rebars. Only by understanding the differences between steel and FRP reinforced concrete design can readers truly get a feel for design with FRP bars.

4.2 INTRODUCTION

Fiber-reinforced plastic (FRP) reinforcing bars and grids have been produced for reinforcing concrete structures for over 30 years (Nanni, 1993b; ACI, 1996). FRP reinforcing bars have been developed for prestressed and non-prestressed (conventional) concrete reinforcement. In this section we consider only nonprestressed reinforcement for concrete structures. A review of design recommendations for FRP reinforcement for prestressed concrete structures can be found in Gilstrap et al. (1997). ACI guide 440.4R-04 (ACI, 2004b) provides design guidelines for prestressing concrete with FRP tendons.

Current FRP reinforcing bars (referred to as FRP rebars in what follows) and grids are produced commercially using thermosetting polymer resins (commonly vinylester and epoxy) and glass, carbon, or aramid reinforcing fibers. The bars are primarily longitudinally reinforced with volume fractions of fibers in the range 50 to 65%. FRP reinforcing bars are usually produced by a process similar to pultrusion and have a surface deformation or texture to develop bond to concrete. Photographs of typical FRP rebars are shown in Fig. 1.1.

A number of design guides and national standards are currently published to provide recommendations for the analysis, design, and construction of concrete structures reinforced with FRP rebars (JSCE, 1997; Sonobe et al., 1997; Bakht et al., 2000; CSA, 2002; ACI, 2006). This book follows the ACI 440.1R-06 guidelines. A number of industry groups, including the Market Development Alliance of the American Composite Manufacturers Association (ACMA), coordinate activities of FRP rebar producers in the United States. Activities of these organizations are closely coordinated with American Concrete Institute (ACI) Technical Committee 440-FRP Reinforcements. Research in the use of FRP reinforcements in concrete structures has been the focus of intense international research activity since the late 1980s. A biannual series of symposia entitled "Fiber Reinforced Plastics in Reinforced Concrete Structures" (FRPRCS) has been the leading venue for reporting and disseminating research results. Symposia in the series date back to 1993. FRPRCS-7 was held in Kansas City, Missouri, in November 2005.

4.3 PROPERTIES OF FRP REINFORCING BARS

Glass fiber reinforced vinylester bars, the most common commercially produced FRP rebars, are available from a number of manufacturers. Bars are typically produced in sizes ranging from $\frac{3}{8}$ in. (9 mm) in diameter to 1 in. (25 mm) in diameter (i.e., No. 3 to No. 8 bars). Bars have either a sand-coated external layer, a molded deformation layer, or a helically wound spiral fiber layer, to create a nonsmooth surface. The longitudinal strength of FRP rebars is bar size dependent, due to materials used in different-sized bars and due to shear lag effects. The strength of a glass FRP bar decreases as the diameter increases.

Figure 4.1 shows an FRP rebar cage produced for a concrete beam 8 × 12 in. × 9 ft long. The main bars are No. 7 glass FRP bars and the stirrups and top bars are No. 3 bars. FRP rebars cannot be bent and must be premanufactured with bends. Therefore, hooks at the ends of the beam for anchorage are produced as separate pieces and lap spliced to the main tension bars, as shown in Fig. 4.1.

FRP rebars are typically elastic and brittle such that the stress–strain relation in axial tension is linear elastic to failure. Figures 4.2 and 4.3 show the failure mode and the stress–strain curve for a coupon from a glass FRP rebar tested in tension. The failure mode for a full-section test of an FRP rebar potted in end anchors is shown in Fig. 4.4.

As noted, the ultimate tensile strength of FRP rebars decreases with bar diameter. Typical properties are given in Table 1.1 for glass fiber FRP rebars and carbon fiber FRP bars. It should be noted that the carbon fiber bars are typically used as prestressing tendons or *near-surface-mounted* (NSM) strengthening and not as conventional reinforcing bars due to cost considerations. In accordance with ACI 440.1R-06 recommendations for guaranteed tensile strength and longitudinal modulus,[1] guaranteed properties of glass–vinylester FRP reinforcing bars commercially manufactured in North America are listed in Table 4.1. As noted, the strength of the reinforcing bar decreases with the diameter of the bar, however, the longitudinal modulus does not change appreciably.

FRP rebars are considered to be transversely isotropic from a continuum mechanics perspective (Bank, 1993a). Theoretical equations used to predict the mechanical and physical properties of FRP rebars from the properties of fiber and resin constituents are provided in Chapter 3. Theoretical methods are not currently available to predict the bond properties and long-term du-

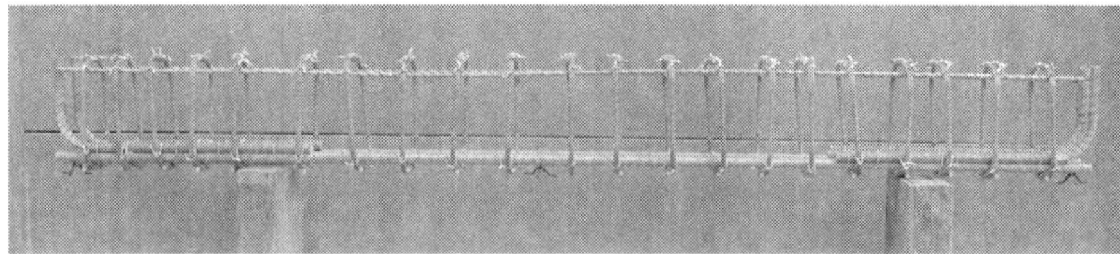

Figure 4.1 Reinforcing cage for a concrete beam made with FRP rebars. (Courtesy of Dushyant Arora.)

[1] To be discussed further in what follows.

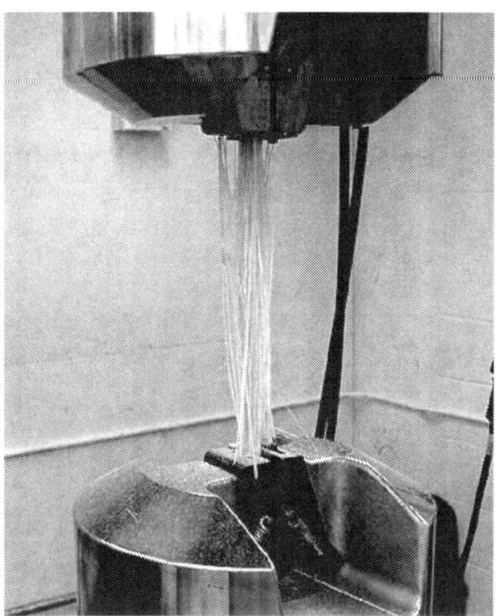

Figure 4.2 Failure of a tensile coupon cut from a No. 7 GRFP rebar. (Courtesy of Joshua Dietsche.)

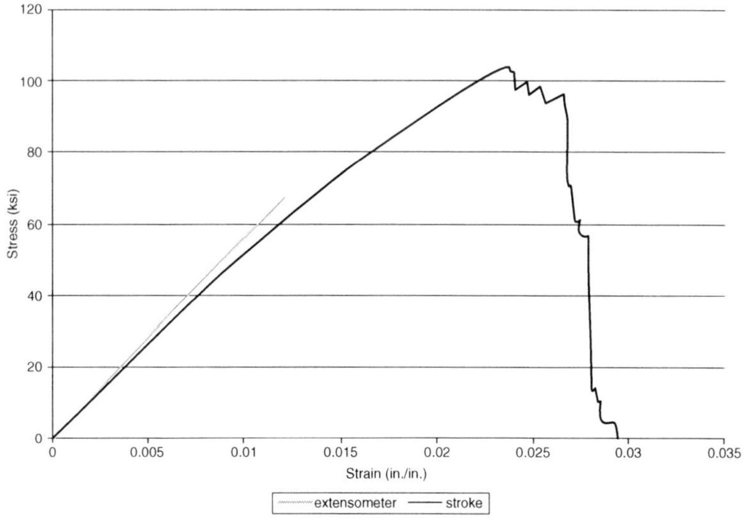

Figure 4.3 Tensile stress–strain curve for a glass FRP rebar coupon.

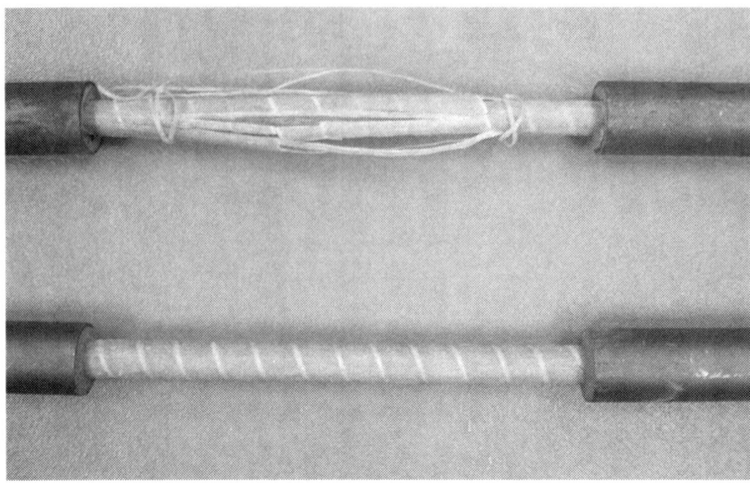

Figure 4.4 Pre- and posttested glass FRP rebars tested in end anchorages at full section.

rability characteristics of FRP rebars. Test methods for determining the properties of FRP rebars can be found in ACI 440.3R-04 (ACI, 2004a) and are described in Chapter 3.

FRP rebars should only be used at service temperatures below the glass transition temperature of the polymer resin system used in the bar. For typical vinylester polymers this is around 200°F (~95°C). The bond properties have been shown to be highly dependent on the glass transition temperature of the polymer (Katz et al., 1999).

TABLE 4.1 Guaranteed Properties of Glass FRP Rebars Produced in North America, 2005[a]

Nominal Bar Size No.	Nominal Diameter (in.)	Nominal Area (in.^{2})	Reported Measured Area (in.^{2})	Guaranteed Longitudinal Strength (ksi)	Guaranteed Longitudinal Modulus (Msi)
2	0.25	0.05	0.05	120–127	5.9–6.7
3	0.375	0.11	0.13	110–111	5.9–6.3
4	0.50	0.20	0.23	100–103	5.9–6.4
5	0.625	0.31	0.34	95–99	5.9–6.8
6	0.75	0.44	0.46	90–95	5.9–6.9
7	0.875	0.60	0.59	85	5.9
8	1.0	0.79	0.83	80–87	5.9–6.0
9	1.125	1.00	1.00	75	5.9
10	1.25	1.27	1.25	70	5.9

[a]Not all manufacturers supply all bar sizes. Manufacturers' current specifications must always be consulted for design calculations.

The coefficient of thermal expansion of an FRP rebar is not the same in the transverse (radial) direction as in the longitudinal direction (Gentry and Husain, 1999). The coefficient of thermal expansion may be close to an order of magnitude higher in the transverse direction of the bar due to its anisotropic properties (see typical properties in Table 1.1). This may cause longitudinal splitting in the concrete at elevated temperatures if insufficient cover is not provided. FRP rebars, especially those containing glass fibers, can fail catastrophically under sustained load at stresses significantly lower than their short-term static tensile strengths, a phenomenon known as *creep rupture* or *static fatigue.* The amount of sustained load on FRP rebars is therefore limited by design guides.

FRP reinforcing bars made of thermosetting polymers (e.g., vinylester, epoxy) cannot be bent in the field and must be produced by the FRP rebar manufacturer with "bends" for anchorages or for stirrups. The strength of the FRP rebar at the bend is substantially reduced and must be considered in the design. FRP rebars with thermoplastic polymer resins, which may allow them to be bent in the field, are in the developmental stages.

The compressive behavior of FRP rebars has not been studied adequately. It appears that the compressive properties of FRP bars may be lower than their tensile properties (Dietz et al., 2003). In addition, due to their low modulus, FRP rebars have a greater tendency to buckle than do steel rebars.

4.4 DESIGN BASIS FOR FRP-REINFORCED CONCRETE

In what follows the design basis of the American Concrete Institute presented in ACI 318-05 and ACI 440.1R-06 are followed with respect to design philosophy and load and resistance factors for design.[2] Similar design bases, conforming to the limit states design procedure, are recommended by many standards organizations and professional organizations for FRP reinforced concrete (JSCE, 1997; Sonobe et al., 1997; Bakht et al., 2000; CSA, 2002). ACI 440.1R-06 recommends the use of traditional methods of strain compatibility and equilibrium to determine internal forces in an FRP reinforced concrete section. This includes the assumption that there is no slip (i.e., local relative longitudinal displacement) between the rebars and the concrete in an FRP-reinforced section.

The resistance factors for flexural strength design of glass FRP reinforced members have been developed using the LRFD probability-based approach. A reliability factor, β, of at least 3.5 is obtained using the design methods presented in what follows. The calibration for the reliability factor was based on ASCE 7 load combination 2 ($1.2D + 1.6L$) and a ratio of dead load to

[2] The ultimate strength design procedure, a load and resistance factor design (LRFD) procedure followed by ACI 440.1R-06, is compatible with ACI 318-05, which uses the load factors of ASCE 7.

live load of 1:3. It is important to note that the flexural resistance factors presented in previous editions of ACI 440.1R (ACI, 2001, 2003) were not probabilistically calibrated, although they still provided safe designs. The resistance factor for shear strength design has not yet been probabilistically calibrated in ACI 440.1R-06.

4.4.1 Resistance Factors

The resistance factors for determination of the ultimate flexural capacity and ultimate shear capacity of transversely loaded members (i.e., beams, slabs, and beam-columns) reinforced with FRP rebars as listed in ACI 440.1R-06 are given below.[3]

Flexural capacity:

$$\phi = \begin{cases} 0.55 \text{ for an underreinforced beam section } (\rho_f < \rho_{fb}) \\ 0.65 \text{ for a substantially overreinforced beam section } (\rho_f > 1.4\rho_{fb}) \\ 0.3 + 0.25\rho_f/\rho_{fb} \text{ for a lightly overreinforced beam section} \\ \quad (\rho_{fb} < \rho < 1.4\rho_{fb}) \end{cases}$$

(ACI 440.1R-06:8-7)

Shear capacity:

$$\phi = 0.75 \qquad \text{(ACI 318-05 Sec. 9.3.2.3)}$$

In the equations above, ρ_f is the FRP reinforcement ratio and ρ_{fb} is the balanced FRP reinforcement ratio for a rectangular cross section:

$$\rho_f = \frac{A_f}{bd} \qquad \text{(ACI 440.1R-06: 8-2)}$$

The balanced FRP reinforcement ratio is given as

$$\rho_{fb} = 0.85\beta_1 \frac{f'_c}{f_{fu}} \frac{E_f \varepsilon_{cu}}{E_f \varepsilon_{cu} + f_{fu}} \qquad \text{(ACI 440.1R-06: 8-3)}$$

In these equations, A_f is the area of FRP reinforcement, b the width of the section, d the depth of the FRP reinforcement, β_1 a factor that depends on concrete strength and is given in ACI 318-05 (e.g., 0.85 for 4000-psi concrete), f'_c the cylinder compressive strength of the concrete, E_f the guaranteed

[3]Throughout this book, where equations are taken directly from ACI guides and codes, they are cited with the equation numbers or section numbers given in the source ACI documents.

longitudinal modulus of the FRP rebar, ε_{cu} the ultimate nominal compressive strain in the concrete (usually taken as 0.003), and f_{fu} the design longitudinal strength of the FRP rebar.

Only tensile FRP reinforcement is considered in the design of FRP-reinforced sections. If FRP reinforcement exists in the compression zone, it should not be considered in a determination of the strength of the section. Neither is it assumed to reduce the strength of the section. The resistance factors provided above for flexural design do not consider any beneficial or deleterious effect of FRP compression reinforcement on the capacity of the section.

Rationale for Flexural Resistance Factors The rationale for the use of the flexural resistance factors given above is based on the nature of the failure modes in over- and underreinforced RC sections with FRP rebars. When the section is underreinforced, the failure is due to rupture of the FRP bars in tension, which is sudden and catastrophic since FRP rebars currently available are brittle and linear elastic to failure. For this case, a resistance factor of $\phi = 0.55$ has been obtained. Note that this is less than the factor of $\phi = 0.65$ that is used for steel-reinforced concrete when brittle failure occurs per ACI 318-05.

When the section is overreinforced, the failure is due to concrete crushing. This failure mode, although brittle, is somewhat less brittle than the mode due to FRP rupture and is assigned a higher resistance factor of $\phi = 0.65$ when the overreinforced failure mode can be assured (i.e., the reinforcement ratio is significantly above the balanced ratio). This resistance factor is equal to the resistance factor for brittle failure due to concrete crushing for steel-reinforced concrete. When the section is slightly overreinforced, the resistance factor is assumed to vary from 0.65 down to 0.55 (equal to the underreinforced resistance factor), due to uncertainty in the actual failure mode that will occur. Note that the well-known flexural resistance factor of $\phi = 0.90$ that is used when ductile failure of a steel-reinforced concrete section is assured, when $\varepsilon_s > 0.005$, can never be used for FRP-reinforced concrete.

4.4.2 Minimum Reinforcement Requirements

A minimum amount of FRP flexural reinforcement is required to prevent sudden brittle failure in FRP underreinforced flexural members. A minimum amount of FRP shear reinforcement is required to prevent sudden brittle shear failure in FRP flexural members. In addition, to prevent cracking, a minimum amount of temperature and shrinkage reinforcement is required for FRP-reinforced slabs. Details are provided in later chapters.

4.4.3 Determination of Guaranteed Properties of FRP Rebars

The FRP bar strength (called the *guaranteed strength*) and strain to failure of FRP rebars (called the *guaranteed rupture strain*) are defined as the mean minus three standard deviations of a minimum of 25 test samples. The guaranteed design strength, f^*_{fu}, and design failure strain, ε^*_{fu}, are expected to be supplied by the manufacturer. FRP rebars must be tested according to the procedures detailed in ACI 440.3R-04 (see Chapter 3 for details).

The strength of an FRP rebar at a bend is highly dependent on the manufacturing processes. As noted previously, FRP rebars produced with thermosetting polymer resins such as polyester, vinylester, and epoxy must be bent during production prior to the resin curing. ACI 440.3R-04 provides two test methods for determining the effect of bends on the tensile strength of FRP bars. However, manufacturers at this time, are not required to report a guaranteed strength of FRP bars with bends, nor are there standard dimensions for bend radii at this time. An analytical equation for determining the strength of a bar with a bend is provided in ACI 440.1R-06.

$$f_{fb} = \left(0.05\,\frac{r_b}{d_b} + 0.3\right) f_{fu} \qquad \text{(ACI 440.1R-06:7-3)}$$

where f_{fb} is the tensile strength of the FRP rebar with a bend, r_b the inside radius of the bend, and d_b the diameter of the FRP rebar. A minimum inside radius-to-diameter ratio of 3 is required. Standard bend radii are reported by manufacturers and range from 2 to 3 in. for typical FRP rebars.

4.4.4 Design for Environmental Effects on FRP Rebars

The design strength, f_{fu} ,and design failure strain, ε_{fu} , are obtained from the guaranteed strength and guaranteed failure strain by multiplying them by an environmental reduction factor, C_E, which depends on the fiber type in the bar and the type of service intended, as shown in Table 4.2.

TABLE 4.2 Environmental Reduction Factors for FRP Rebars from ACI 440.1R-06

Exposure Condition	Fiber Type	Environmental Reduction Factor, C_E
Concrete not exposed to ground and weather	Carbon	1.0
	Glass	0.8
	Aramid	0.9
Concrete exposed to ground and weather	Carbon	0.9
	Glass	0.7
	Aramid	0.8

$$f_{fu} = C_E f_{fu}^* \qquad \text{(ACI 440.1R-06:7-2)}$$

The environmental reduction factor accounts for the fact that even though FRP rebars are not susceptible to conventional electropotential corrosion that affects metallic materials, they can nevertheless deteriorate in a variety of chemical environments, both alkaline and acidic. This deterioration is accelerated at elevated temperatures. Glass fibers are especially susceptible to corrosion in alkaline and neutral solutions, due to leaching of the fiber (Bank et al., 1995b). Aramid fibers do not leach like glass fibers but are known to absorb moisture and swell, leading to loss of mechanical and physical properties. Carbon fibers are the most chemically inert and are generally recommended when extreme exposure to chemical aqueous environments is anticipated. In addition, the polymer matrix is susceptible to deterioration in various chemical environments. Polymer resin producers and manufacturers of FRP products typically supply a corrosion-resistance guide that lists the various chemical exposures that are known to affect their materials.

The outer surface of the FRP rebars is the part of the bar that is typically most vulnerable to chemical attack. Once the outer layer has been breached, diffusion of chemical solutions to the interior of the bar can occur more readily. In some cases the exterior helical wrap or deformation layer on the bar can aid in protecting the outer surface, in other cases it has been shown that the outer layer can detach from the core and lead to separation of the exterior layer from the interior core of the bar (Bank et al., 1998). Scanning electron micrographs (SEMs) of deterioration at the surface of a smooth bar and deterioration at the exterior wrap and core interfaces is show in Figs. 4.5 and 4.6.

The FRP bar modulus (also called the *specified tensile modulus*), E_f, measured in the axial direction of the bar, is defined as the average modulus of bars tested in accordance with ACI 440.3R-04. No reduction of the measured modulus is required for either statistical or environmental reasons.

4.4.5 Special Considerations Regarding FRP Rebars

As with steel bars, the design of FRP-reinforced concrete is conducted using only the longitudinal properties (f_{fu}, ε_{fu}, E_f) of the bars. This assumes that the transverse properties of the FRP bars and the shear properties of the FRP bars, which are known to be significantly lower than the longitudinal properties of the bars due to the anisotropic nature of FRP materials, do not significantly influence the flexural behavior of an FRP-reinforced RC section. This has been verified in numerous research studies (Faza and Gangarao, 1993; Nanni, 1993b; Benmokrane et al., 1996a).

Since FRP reinforcing bars are linear elastic to failure, redistribution of moments and development of plastic hinges in statically indeterminate FRP reinforced concrete structures cannot be assumed to occur as in steel-

Figure 4.5 Deterioration at the bar surface following accelerated conditioning of a smooth FRP bar.

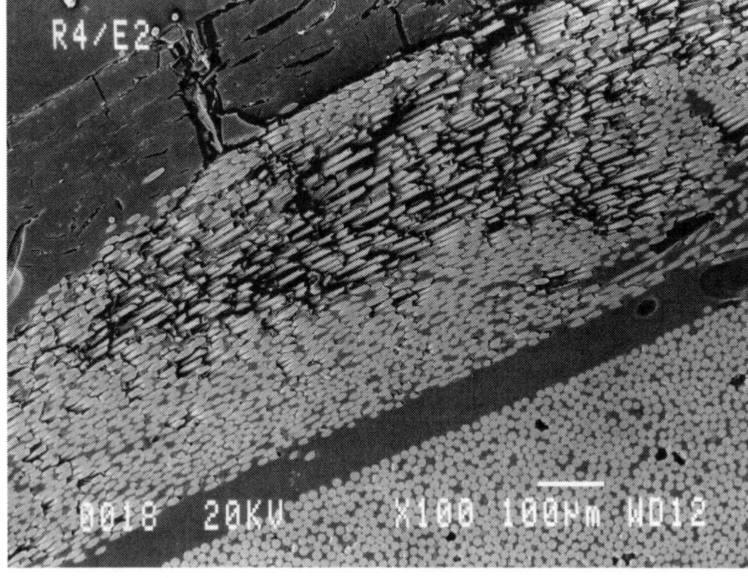

Figure 4.6 Deterioration at the exterior polymer surface following accelerated conditioning of an FRP bar.

reinforced concrete structures. Where FRP bars are used in layers the stress in each of the layers must be calculated separately to determine the moment capacity of the section. The outermost layer of bars must be used to determine the flexural capacity of the member when a beam fails due to FRP rupture. (In steel-reinforced concrete beams it is permissible to assume that the resultant tensile force in the bars acts through the centroid of the bar layers since all bars will be at their yield stress at ultimate failure.)

At this time, FRP reinforcing bars should not be use as compression reinforcement, as insufficient test data on the compression properties of FRP bars have been obtained. If FRP bars are placed in the compression zones of members, they should not be assumed to carry any compression load, and they should be confined adequately to prevent local bucking, which could lead to premature concrete compression failure. The use of shear-friction and strut-and-tie design methods (Nilsen et al., 2004) to determine the ultimate capacity of concrete structures reinforced with FRP bars has not been studied in sufficient detail at this time and is not discussed by ACI 440.1R-06.

4.4.6 Design for Serviceability

The design of FRP rebars for the serviceability limit state considers two primary serviceability conditions; deflection and cracking. Since FRP rebars typically have a lower modulus than that of steel rebars, the serviceability limit states (deflections and crack widths) can often control the design of FRP-reinforced concrete sections. Permissible service load deflections for FRP-reinforced concrete structures are generally taken as the same as those for all structures, and pertinent building code guidelines should be followed. Permissible crack widths for FRP-reinforced concrete are not the same as for conventional reinforced concrete structures, due to the superior corrosion resistance of FRP bars.

Limits on Crack Widths Limits on maximum crack width under service loads for FRP-reinforced concrete are recommended. The limits on crack widths are intended to limit the ingress of fluids that could degrade the FRP bars, and also to ensure an acceptable aesthetic appearance of an FRP-reinforced flexural member. Although perhaps not technically unsafe, large flexural cracks may cause public discomfort, due to a perceived risk of structural collapse. The maximum crack widths recommended by ACI 440.1R-06 are listed in Table 4.3 together with the historically recognized permissible crack widths for steel-reinforced concrete from ACI 318-95 (ACI, 1995). It is important to note that since ACI 318-99 (ACI, 1999) the explicit calculation of crack widths and limits on crack widths in reinforced concrete structures has not been required for design of steel-reinforced concrete. An indirect method based on the spacing of the reinforcing bars as a function of the bar stress at service loads and the concrete cover is now used for steel-reinforced concrete. However, due to the possibility of low flexural rigidity and the high

TABLE 4.3 Permissible Crack Widths for FRP and Steel RC Structures

Exposure	FRP-Reinforced Concrete [in. (mm)]	Steel-Reinforced Concrete [in. (mm)]
Exterior	0.020 (0.5)	0.013 (0.3)
Interior	0.028 (0.7)	0.016 (0.4)

strength of FRP-reinforced concrete members, ACI 440.1R-06 still recommends the explicit calculation of crack widths in concrete structures reinforced with FRP bars.

Limits on Service Load Stresses To prevent failure of FRP-reinforced beams at service loads due to creep rupture, fatigue, or other long-term environmental effects, the stress in FRP rebars under sustained service loads or fatigue loads is limited by ACI 440.1R-06 to a fraction of its design strength. These limits depend on the type of fiber in the FRP rebar. For glass FRP rebars, the limiting stress is 20% of the ultimate strength (i.e., $0.2f_{fu}$.) For aramid- and carbon-reinforced FRP bars, the limits are $0.3f_{fu}$ and $0.55f_{fu}$, respectively.

4.4.7 Temperature and Shrinkage Reinforcement in Slabs

In one-way slabs, a minimum amount of reinforcement is required perpendicular to the main bars to limit cracking parallel to the main bars that may develop due to thermal and shrinkage effects in the concrete.

A modified version of the ACI 318-05 equations for steel rebars is provided by ACI 440.1R-06. The minimum amount of temperature and shrinkage reinforcement required is

$$\rho_{f,ts} = 0.0018 \left(\frac{60{,}000}{f_{fu}}\right) \frac{E_s}{E_f} \qquad \text{(ACI 440.1R-06:10-1)}$$

but not less than 0.0014. The reinforcement ratio, $\rho_{f,ts}$, is computed over the gross concrete area, and E_s is the modulus of grade 60 steel rebar. The spacing of the FRP temperature and shrinkage reinforcement should not exceed three times the slab thickness, or 12 in.

PROBLEMS

4.1 Visit the following FRP rebar manufacturers'[4] Web sites and request or download copies of their FRP rebar property specification sheets. These will be needed for solving design problems in the chapters that follow.

[4]No endorsement of the manufacturers listed is implied. The reader is free to choose any manufacturer's products to use in the examples that follow, and is encouraged to collect similar data for locally available FRP rebars.

Hughes Brothers (Aslan 100, Aslan 200): www.hughesbros.com
Pultrall (V-Rod): www.pultrall.adsinc.ca *or* www.fiberglassrebar.com
Schöck (comBAR): www.schoeck.com
Pullwell (R-bar): www.pullwellpultrusions.com

4.2 For the following FRP rebars, determine[5] the ACI 440.1R-06 design strength, f_{fu}, the design failure strain, ε_{fu}, and the design longitudinal modulus, E_f, when the rebar is used in a concrete structure exposed to the ground. Also, indicate the weight per unit length and the current sales price of the rebar per foot of length.[6] Attach copies of the manufacturer's specification sheets to your homework.

(a) No. 3 Aslan 100 GFRP bar
(b) No. 3 V-Rod GFRP bar
(c) No. 8 Aslan 100 GFRP bar
(d) No. 8 V-Rod GFRP bar
(e) No. 3 Aslan 200 CFRP bar
(f) No. 3 V-rod CFRP bar
(g) $\phi 8$ comBAR GFRP bar
(h) $\phi 32$ comBAR GFRP bar
(i) $\phi 9$ R-bar GFRP bar
(j) $\phi 32$ R-bar GFRP bar

4.3 Derive the expression for the balanced reinforcement ratio for an FRP-reinforced beam given in ACI 440.1R-06 as

$$\rho_{fb} = 0.85\beta_1 \frac{f'_c}{f_{fu}} \frac{E_f \varepsilon_{cu}}{E_f \varepsilon_{cu} + f_{fu}}$$

4.4 Determine the balanced reinforcement ratio for the beams listed in Table P4.4 when they are used in interior construction. Provide a bar graph comparing the balanced reinforcement ratios for the beams and discuss why there are such significant differences in the balanced ratios for various types of reinforcement.

4.5 For the bars listed in the Problem 4.2, determine the strength of the bar at a bend. Assume a minimum bar radius at the bend equal to 3. Also, determine the percent decrease in bar strength at the bend for these bars.

4.6 Obtain strength, failure strain, stiffness, weight per foot, and cost per foot data for currently produced No. 3, No. 8, $\phi 8$, and $\phi 32$ steel and epoxy-coated steel reinforcing bars. Compare these data to those for the

[5] This requires the manufacturer guaranteed strength as defined by ACI 440.1R-06. If this is not provided, use the reported strength but be aware of the difference.
[6] You will need to contact a distributor or the manufacturer to obtain pricing.

TABLE P4.4 Properties of Beams

Beam	f'_c (psi)	Rebar Type
1	4000	No. 4 Aslan 100
2	4000	No. 6 Aslan 100
3	4000	No. 8 Aslan 100
4	4000	No. 3 Aslan 200
5	4000	No. 6 grade 60 steel
6	8000	No. 4 Aslan 100
7	8000	No. 6 Aslan 100
8	8000	No. 8 Aslan 100
9	8000	No. 3 Aslan 200
10	8000	No. 6 grade 60 steel

FRP bars obtained in Problem 4.2. Explain how the nominal strength, failure strain, and stiffness of a steel reinforcing bar are determined. List the numbers of the applicable ASTM test methods used to determine properties of steel and epoxy-coated steel reinforcing bars. How is the effect of the environment accounted for in the design of concrete structures reinforced with steel and epoxy-coated steel bars?

SUGGESTED ACTIVITIES

4.1 Obtain samples of 3- to 4-ft No. 3 or No. 4 GFRP rebars. Also, obtain equivalent-diameter steel rebars. Try to bend the bars to create a permanent 90° bend in the middle of the bar. Clamp one end in a vice and bend the other end by sliding a steel pipe over the free end to obtain some leverage. Push the FRP bar in only one direction and be careful, as the bar may break suddenly. Heat the FRP bar locally to see if this will help you to bend the FRP bar into the shape required. Now bend the FRP bar back and forth about 20 times, each time pushing a little farther. Record your observations with a digital camera. Write a 1000-word report on your qualitative study of the properties of FRP rebars and steel bars under flexural loads. (Make sure to take appropriate safety precautions when doing this activity. Coordinate this project with your laboratory manager and your safety engineer. Wear gloves, boots, long pants, and safety glasses. If heating, make sure that the room is ventilated and do not allow the bar to ignite.)

4.2 Obtain samples of small- and large-diameter (e.g., No. 3 and 7) glass FRP rebars. Cut short (1- to 2-in.) lengths of the bars using a circular saw with a masonry or diamond blade. Polish a cut end (to 200-grit paper) and observe it under an optical microscope. Observe fiber bundles, resin-rich areas, and surface deformations. Cut lengthwise sections of the bars and perform similar studies. Write a 1000-word report on your observations (including photomicrographs).

5 FRP Flexural Reinforcement

5.1 OVERVIEW

In this chapter the design of flexural members, such as beams and slabs, reinforced with FRP rebars is discussed. This includes procedures to determine the strength of either an over- or an underreinforced FRP concrete section according to ACI 440.1R-06 using the LRFD basis. We also provide procedures to design a flexural member for serviceability limit states, which include deflection, flexural crack width limitations, and maximum sustained stress on FRP bars under service loads. The design of flexural members to resist transverse shear forces is discussed in Chapter 6. The geometric details, bond, and development length of FRP reinforcing bars are discussed in Chapter 7.

The design procedures presented in this chapter follow ACI 440.1R-06, which is compatible with ACI 318-05. The reader is assumed to have familiarity with the design of flexural members with conventional steel reinforcing bars. This includes having an understanding of the design of the section for serviceability and includes an understanding of the issues of bond and development length. In this chapter the flexural design of a rectangular reinforced concrete section is discussed. The section may be a beam, which needs additional shear reinforcement to carry transverse loads, or it may be a slab, which may not need, and for constructability reasons should not have, shear reinforcement. In recent years, reinforcement of slabs has emerged as a large application area for FRP bars, particularly in highway bridge decks (Benmokrane et al., 2004). In the flexural design of a slab without additional shear reinforcement, special attention must be paid to the shear capacity discussed in Chapter 6. The amount of tensile reinforcement in a slab can depend on the shear capacity of the slab, something that does not occur in a steel-reinforced slab.

5.2 INTRODUCTION

Most structural engineers are relatively unfamiliar with the design of over-reinforced beam sections, since traditional steel-reinforced concrete design philosophy strongly encourages the use of underreinforced sections. This is to ensure a ductile failure mode in steel-reinforced concrete beams. The desired failure mode in a steel-reinforced beam is yielding of the tension steel, followed by eventual crushing of the concrete in the compression zone of the member. It is important for the reader to review the fundamentals of steel-reinforced concrete design and to understand that even though the Whitney stress block is used to determine the strength of an underreinforced section, the failure is nevertheless due to the steel yielding and depends upon the plastic strain capacity of steel reinforcing bars. To an experienced structural engineer this may be obvious, but engineering students need to come to appreciate this fundamental fact.

Since FRP reinforcing bars are linear elastic to failure when loaded in tension and fail in a brittle manner, a ductile steel-like failure does not occur in FRP-reinforced concrete. This is fundamentally different from steel-reinforced concrete and makes the choice between an over- and underreinforced section much less clear. In fact, as will be seen in what follows, failure by concrete crushing, which is generally avoided in steel-reinforced concrete design, is slightly preferred in FRP-reinforced concrete design. This is because confined concrete has some measure of postpeak large strain capacity, albeit at reduced stress levels. However, this large stain capacity is nothing like the large strain capacity of steel reinforcing bars. It will be seen in what follows that the commonly used Whitney rectangular compression stress block, which is so familiar to structural engineers who design underreinforced steel sections, will be used in the design of overreinforced FRP reinforced sections. This requires some new thinking and a different perspective on reinforced concrete design. Therefore, the reader is encouraged to review the fundamental derivation of the equations for steel-reinforced concrete design and to derive equations for the overreinforced concrete steel section using strain compatibility and equilibrium of the resultant forces in the section. This is discussed in most reinforced concrete textbooks, but usually not in great detail (Nilson et al., 2004).

Figure 5.1 shows the difference in the behavior of steel- and FRP-reinforced concrete beams based on the results of tests on six 98-in.-long (90-in. simple span) reinforced concrete beams, all with the same 8×12 in. cross section (Ozel et al., 2003). The FRP beams were reinforced with three No. 7 glass FRP bars (FRP 1 and 2), and the steel beams were reinforced with three No. 6 grade 60 bars (steel 1 and 2) and three No. 5 grade 60 bars (steel 3 and 4), respectively. The relative reinforcement ratios, ρ_f/ρ_{bf} and ρ/ρ_b, for the FRP- and steel-reinforced beams were 3.14, 0.50, and 0.35, respectively. The concrete strength was close to 5000 psi in all beams. The steel-reinforced beams failed due to yielding of the steel rebars followed by crushing of the compression zone in the concrete. The FRP-reinforced beams failed due to crushing of the concrete in the compression zone (FRP 2) or debonding of the stirrups in the tension zone (FRP 1) (Ozel et al., 2003). The FRP reinforcement cage details for the FRP-reinforced beams were shown in Fig. 4.1. The failure modes of the FRP-reinforced beams are shown in Figs. 5.4 and 6.1.

Notice the lower stiffness of the FRP-reinforced beams in the elastic range, even though their reinforcement ratios are much larger than those of the steel

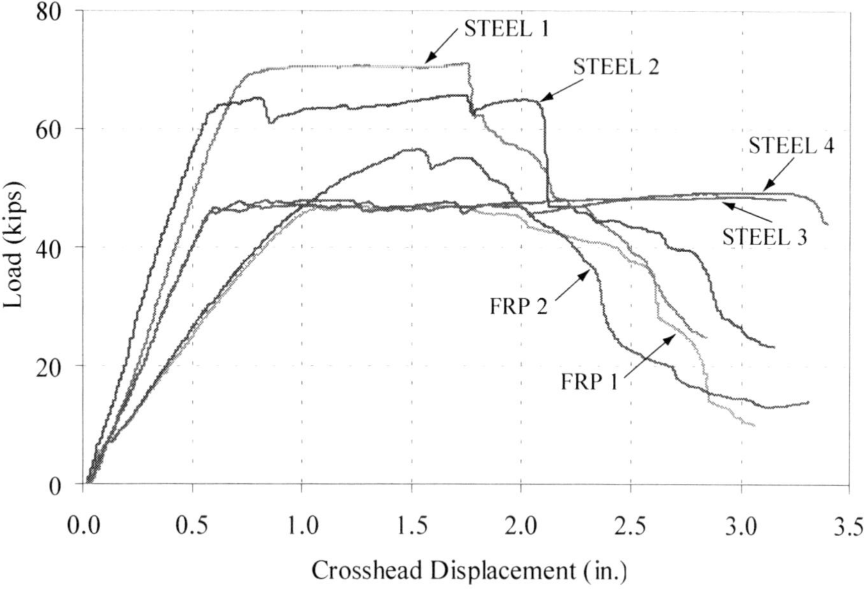

Figure 5.1 Comparison of behavior of FRP- and steel-reinforced beams.

beams. Also notice the elastic–plastic behavior of the steel-reinforced beams and their eventual failure at large deflections. Note also, however, that due to their lower flexural stiffness, the FRP-reinforced beams undergo large deflections that serve as a warning to the user. Recognize that the service loads that would typically be placed on this beam would be in the 30-kip range, and compare the deflection of the two beams in this range. Then compare the ultimate deflections of beams to the service load deflections of beams. One can see that the ratios of the ultimate deflections and service deflections are about two to three times greater in the steel-reinforced beams, even though the postpeak failure of the FRP-reinforced beam appears to have some measure of ductility. Note, however, that this data set shows the results of tests conducted in displacement control (as is usually the case in a laboratory), not in load control (as is usually the case in a real structure).

5.3 FLEXURAL STRENGTH OF AN FRP-REINFORCED SECTION

The proportioning of a member with FRP rebars[1] for flexural strength (or capacity) follows the ACI strength design method, a load and resistance factor

[1] Only singly reinforced rectangular sections are considered in this book. Since FRP bars are not used in compression, doubly reinforced sections are not designed. Procedures for T-beams can be developed following the procedures described in what follows.

design (LRFD) method. In this approach the factored nominal moment resistance, ϕM_n, of the member must be greater than the factored (ultimate) moment demand, M_u:

$$\phi M_n > M_u \tag{5.1}$$

According to ACI 440.1R-06, the nominal moment capacity of FRP-reinforced concrete members is determined in a fashion similar to that of a steel-reinforced section. However, since FRP rebars do not yield (i.e., are linear elastic to failure), the ultimate strength of the bar replaces the yield strength of the steel rebar, f_y, in the traditional reinforced concrete analysis procedure, which assumes equilibrium of forces and that plane sections remain plane. The design of either under- or overreinforced sections is permitted, but due to serviceability limits (deflections and crack widths), most glass FRP-reinforced flexural members will be overreinforced. The strains, stresses, and resulting section forces in the balanced condition are shown in Fig. 5.2.

The depth of the compression zone, which is equal to the distance from the extreme outer concrete surface to the neutral axis of the section, c_b, is given as

$$c_b = \frac{\varepsilon_{cu}}{\varepsilon_{cu} + \varepsilon_{fu}} d \qquad \text{(ACI 440.1R-06: 8-6}c\text{)}$$

and the balanced FRP reinforcement ratio (also given in Chapter 4) is

$$\rho_{fb} = 0.85\beta_1 \frac{f'_c}{f_{fu}} \frac{E_f \varepsilon_{cu}}{E_f \varepsilon_{cu} + f_{fu}} \qquad \text{(ACI 440.1R-06: 8-3)}$$

To determine the compressive force resultant in an FRP-reinforced section, it is assumed that the Whitney rectangular stress block (Whitney, 1937) can

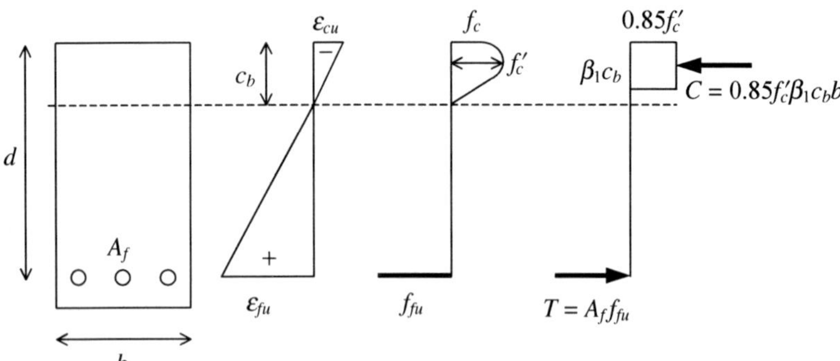

Figure 5.2 Strains, stresses, and section forces in the balanced condition of an FRP-reinforced beam.

be used to represent the nonlinear stress in the concrete at the balanced condition shown in Fig. 5.2. However, it is important to recognize that the stability of the section at the balanced state in the FRP-reinforced beam is not analogous to that of a steel-reinforced beam in the balanced state. In a steel-reinforced beam the balanced state is defined as when the reinforcing bar yields, not when it fails, which means that the section can continue to carry loads beyond this point since the steel will not rupture, there will be some stress redistribution in the concrete, and a gradual pseudoductile failure or an excessive deflection will occur. In the case of an FRP-reinforced beam, when the FRP bar fails there can be no redistribution of stresses in the section and a catastrophic collapse will occur.

In addition, in typical FRP-reinforced beams the neutral axis will be much higher up in the section than in a steel-reinforced beam at the balanced condition because of the large strain to failure in the FRP rebar relative to the yield strain in a steel bar. This means that the depth of the compression zone (or the distance from the extreme outer concrete surface to the neutral axis of the section), c_b, can be very small in an FRP-reinforced beam. Often, the compression zone will be contained entirely above the top bars of the section and will be unconfined by the shear stirrups, making it more susceptible to failure as soon as the concrete compressive strain is reached at the top of the section. Note that this also implies that the top bars in the section can provide some, albeit very limited, additional tensile reinforcement to the section.

5.3.1 Overreinforced Section

When $\rho_f > \rho_{fb}$ (and $c > c_b$), a section[2] will fail due to concrete crushing, and the nominal moment capacity is given in a fashion similar to that for a section reinforced with steel rebar (where the rebar has not yet reached its yield stress). The strains, stresses, and force resultants for the overreinforced condition are shown in Fig. 5.3.

The nominal moment capacity of the overreinforced section is given as

$$M_n = A_f f_f\left(d - \frac{a}{2}\right) \qquad \text{(ACI 440.1R-06: 8-4}a\text{)}$$

where

$$a = \frac{A_f f_f}{0.85 f'_c b} \qquad \text{(ACI 440.1R-06: 8-4}b\text{)}$$

and

[2] The term *section* is used, as the procedure can be used for any rectangular beam section and is applicable to a beam and a slab.

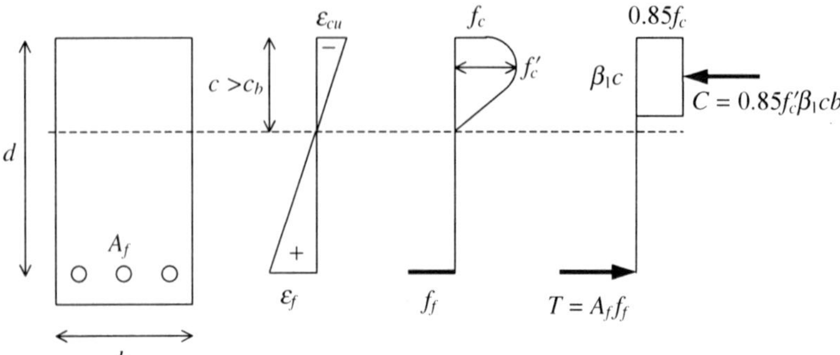

Figure 5.3 Strains, stresses, and force resultants in an overreinforced FRP beam section.

$$f_f = \sqrt{\frac{(E_f\varepsilon_{cu})^2}{4} + \frac{0.85\beta_1 f_c'}{\rho_f} E_f\varepsilon_{cu}} - 0.5E_f\varepsilon_{cu} \qquad \text{(ACI 440.1R-06: 8-4}d\text{)}$$

Here f_f is the stress in the FRP rebar at concrete compressive failure, and a is the depth of the Whitney stress block in the concrete. Figure 5.4 shows an overreinforced FRP reinforced beam that has failed due to concrete crushing.

5.3.2 Underreinforced Section

When $\rho_f < \rho_{fb}$ (and $c < c_b$), a section will fail due to rupture of the FRP rebars in tension before the concrete has reached its ultimate strain. Since the FRP reinforcement will not yield prior to rupturing, failure cannot be assumed to occur due to concrete compression failure following the rebar yielding (as in the case of a steel rebar). Consequently, the Whitney stress block cannot be assumed to exist in the concrete, and the compression force resultant and its location need to be calculated using the nonlinear stress distribution in the

Figure 5.4 FRP-reinforced beam failure due to concrete crushing (FRP 1). (Courtesy of Dushyant Arora.)

concrete. In addition, the depth of the neutral axis is not known in this case. The strains, stresses, and force resultants in an underreinforced section are shown in Fig. 5.5.

Calculation of the nominal moment capacity of an underreinforced section therefore requires use of the nonlinear stress–strain curve of the concrete, and this necessitates a numerical solution procedure, which is not suited to design calculations. To overcome this situation, the ACI 440.1R-06 guide recommends computing the approximate (and conservative) nominal flexural capacity in an underreinforced section using the approximate equation

$$M_n = A_f f_{fu}\left(d - \frac{\beta_1 c_b}{2}\right) \qquad \text{(ACI 440.1R-06: 8-6}b\text{)}$$

where c_b is the depth of the neutral axis at the balanced reinforcement ratio, given previously.

Since $c < c_b$, it follows that $\beta_1 c < \beta_1 c_b$ and that the moment arm calculated using $\beta_1 c_b$ will be less than the moment arm calculated using $\beta_1 c$, and therefore the term in parentheses will be greater than that for the actual condition.

Numerical solution using the nonlinear stress–strain characteristics of the concrete requires the use of a nonlinear stress–strain model for the concrete.[3] Two commonly used models are the Todeshini model (Todeshini et al., 1964) and the FIB model (FIB, 2001). To use these models, the nonlinear concrete stress–strain relation is converted to an equivalent rectangular stress block using two strain- and stress-dependent parameters, β_1 and γ (see ACI 440.2R-02). In the case of a balanced or overreinforced section, where the Whitney stress block applies, γ is 0.85 and β_1 depends on concrete strength according

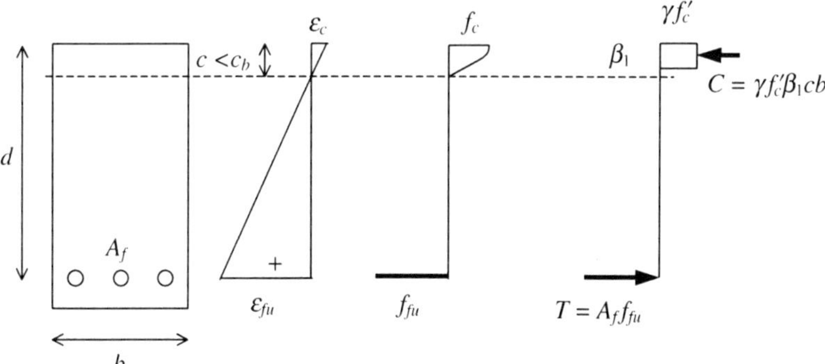

Figure 5.5 Strains, stresses, and force resultants in an underreinforced section.

[3] Note that the nonlinear calculation procedure is permitted by ACI 440.1R-06; however, the equations presented in this section are not provided in ACI 440 design guides.

to ACI 318-05 (e.g., 0.85 for 4000-psi concrete). The height and the depth of the equivalent rectangular stress block for the nonlinear models are, $\beta_1 c$ and $\gamma f'_c$, respectively.

The stress–strain curve of Todeshini et al. (1964) is given by a single function as

$$f_c = \frac{2f'_c(\varepsilon_c/\varepsilon'_c)}{1 + (\varepsilon_c/\varepsilon'_c)^2} \tag{5.2}$$

where ε'_c is the concrete strain at f'_c. The equivalent stress block parameters for the Todeshini model are given as (MBrace, 1998)

$$\beta_1 = 2 - \frac{4[(\varepsilon_c/\varepsilon'_c) - \tan^{-1}(\varepsilon_c/\varepsilon'_c)]}{(\varepsilon_c/\varepsilon'_c)\ln[1 + (\varepsilon_c/\varepsilon'_c)^2]} \tag{5.3}$$

$$\gamma = \frac{0.90\ \ln[1 + [\varepsilon_c/\varepsilon'_c)^2]}{\beta_1(\varepsilon_c/\varepsilon'_c)} \tag{5.4}$$

$$\varepsilon'_c = \frac{1.71 f'_c}{E_c} \tag{5.5}$$

The iterative solution consists of the following steps:

1. Assume a depth of the neutral axis of $c < c_b$ (i.e., $\rho_f < \rho_{fb}$).
2. Calculate the strain in the concrete, ε_c ($\varepsilon_f = \varepsilon_{fu}$ since failure is due to FRP rupture).
3. Calculate the parameters of the equivalent stress block, β_1 and γ.
4. Calculate the actual depth of the neutral axis, c, from the equilibrium equation

$$c = \frac{A_f f_{fu}}{\gamma f'_c \beta_1 b} \tag{5.6}$$

5. Compare the calculated c with the assumed c and iterate until they are equal.
6. Calculate the nominal moment with the final values of β_1 and c using

$$M_n = A_f f_{fu}\left(d - \frac{\beta_1 c}{2}\right) \qquad \text{(ACI 440.1R-06:8-6}a\text{)}$$

This numerical solution is accomplished readily using a simple spreadsheet computer program.

5.3.3 Minimum Flexural Reinforcement

A minimum amount of flexural reinforcement should be provided when the FRP-reinforced beam is designed to fail by FRP bar rupture to prevent failure at concrete cracking (i.e., $\phi M_n \geq M_{cr}$). The amount is given as

$$A_{f,min} = \frac{4.9\sqrt{f'_c}}{f_{fu}} b_w d \geq \frac{330}{f_{fu}} b_w d \qquad \text{(ACI 440.1R-06:8-8)}$$

5.4 DESIGN PROCEDURE FOR AN FRP-REINFORCED FLEXURAL MEMBER

The design procedure presented in what follows is applicable to FRP-reinforced beams having rectangular sections. It is applicable to any type of brittle linear-elastic FRP bar. For the sake of completeness, the procedure includes the steps for shear and serviceability design that are covered later in the book. However, the focus of this section and the design example that follows is on flexural strength design.

Step 1. Determine the design loads and the moment capacity required. Calculate the nominal and factored loads using ACI 318-05 load factors for the design. For a slab, calculate all loads per unit width (typically, 1 ft). Use 1.2 for dead loads and 1.6 for live loads (ASCE 7-02). For the self-weight, estimate the depth using the maximum deflection limits recommended by ACI 440.1R-06, shown in Table 5.1. For a beam, estimate the width as two-thirds of the beam height. Using the depths in Table 5.1 will lead to overreinforced section designs.

Step 2. Assume a bar size and obtain f^*_{fu} from the manufacturer's specifications. Assume a stirrup diameter (if stirrups are required) and a concrete cover depth ($2d_b$ is recommended) and calculate d. The bar size needs to be chosen a priori since the bar strength is a function of the bar size. The

TABLE 5.1 Minimum Thickness of FRP-Reinforced Flexural Members from ACI 440.1R-06[a]

	Support Conditions			
Element Type	Simply Supported	One End Continuous	Both Ends Continuous	Cantilever
Solid one-way slabs	$l/13$	$l/17$	$l/22$	$l/5.5$
Beams	$l/10$	$l/12$	$l/16$	$l/4$

[a] Based on a target maximum deflection of $l/240$.

procedure used to design steel-reinforced beams whereby the area of reinforcement is determined first based on the demand cannot be used easily.

Step 3. Calculate the balanced reinforcement ratio. Calculate the design strength of the FRP bar, f_{fu}, which depends on the service environment and is given by C_E. Then calculate the balanced reinforcement ratio, ρ_{fb}.

Step 4. Choose the numbers of bars. For an overreinforced design, select the number of bars such that $\rho_f > 1.0\rho_{fb}$. For an underreinforced design, select $\rho_f < 1.0\rho_{fb}$. For an overreinforced design, $\rho_f \sim 1.4\ \rho_{fb}$ is recommended, as these beams are more likely to be yield designs that also satisfy deflection and crack width serviceability criteria. Check the beam width to ensure that the number of bars selected can actually fit in the beam. Note that bend radii of FRP stirrups (see Chapter 7) are generally larger than those of steel stirrups. Calculate the actual reinforcement ratio $\rho_f = A_f/bd$.

Step 5. Calculate the nominal moment capacity of the section. For an overreinforced section, first calculate the stress in the FRP bars at concrete failure, f_f. Confirm that this stress is less than the design strength of the bar. Calculate the depth of the Whitney stress block, a, and then the nominal moment capacity, M_n. For an underreinforced beam, first calculate the depth of the neutral axis at the balanced condition, then calculate the depth of the Whitney stress block, a, and then the nominal moment capacity, M_n. For an underreinforced beam, check the minimum flexural reinforcement requirements.

Step 6. Calculate the flexural resistance factor. Determine the flexural resistance factor based on the actual FRP reinforcement ratio.

Step 7. Calculate the factored moment capacity. Calculate ϕM_n and compare with M_u. If $\phi M_n < M_u$, go to step 1 and repeat the design with different section dimensions and/or bar sizes and numbers (remember to use a new value of f_{fu} if you change the bar size). If $\phi M_n > M_u$, the beam design for strength capacity is completed. However, it may be unable to meet serviceability design criteria. If $\phi M_n >> M_u$, the beam is overdesigned for flexural capacity and may be able to be optimized. However, it is always advisable to check the serviceability design criteria before doing too much optimization. To meet the serviceability criteria, FRP beams are often well overdesigned (by as much as 100%) for flexural capacity.

Step 8. Design the beam for the serviceability limit state. Serviceability design is discussed in detail later in this chapter.[4]

Step 9. Check the section shear capacity and design the shear reinforcement. See Chapter 6 for shear design procedures.

Step 10. Detail the cross section. Draw a sketch of the cross section to scale, showing the bars and stirrups. Plot the strain and stress profiles at the

[4] Note that steps 8 and 9 are required in a general flexural design procedure; however, they are discussed separately later in the book.

ultimate condition to get a physical feel for the location of the neutral axis and the compression zone depth.

Design Example 5.1: Strength Design of an FRP-Reinforced Beam[5] Design a simply supported interior beam for a l = 11-ft span. Design the beam for the strength limit state.[6] Use f'_c = 4000-psi concrete and the following FRP manufacturer-guaranteed rebar properties: f^*_{fu} = 90 ksi, ε^*_{fu} = 0.014, E_f = 6500 ksi. (Assume to begin with that the bar strength is *not* a function of the bar diameter.) Design the beam for the following loads: live load: 400 lb/ft. Superimposed dead load: 208 lb/ft (does not include the beam self-weight, which must be added to this to obtain the total dead load on the beam).

SOLUTION

Step 1. Determine the design loads and the moment capacity required. The beam depth is estimated to be 12 in. (132/10 ≃ 13). Ideally, a beam depth larger than that suggested in Table 5.1 is actually recommended, since deflection limits may be greater than $l/240$ and the reinforcement ratio may be lower than assumed in developing the recommendations. The width is estimated to be 7 in., as this is the minimum width recommended for two No. 5 or two No. 6 main bars, assuming No. 3 stirrups and $1\frac{1}{2}$ in. of clear cover on all sides and that bend radii are used for steel bars (Wang and Salmon, 2002, p. 68). (See Chapter 7 for further discussion as to the applicability of this assumption.) This assumes that No. 3 FRP stirrups can be obtained with the same bend radii as steel bars (i.e., $2d_b$ in this case). This may not actually be the case in practice, and the designer should always check with FRP rebar suppliers as to their production capability.

The unfactored (service) loads on the beam are

$$w_{\text{LL}} = 400 \text{ lb/ft}$$

$$w_{\text{DL}} = 208 \text{ lb/ft} + 150 \text{ (lb/ft}^3\text{)}(7/12)(12/12) = 295.5 \text{ lb/ft}$$

The factored loads according to ACI 318-05 are

$$W_u = 1.2(87.5 + 208) + 1.6(400) = 994.6^7 \text{ lb/ft}$$

$$M_u = \frac{W_u l^2}{8} = \frac{994.6(11)^2}{8} = 15{,}043 \text{ lb-ft} = 15.04 \text{ kip-ft} = 180.5 \text{ kip-in.}$$

[5] Design of the serviceability limit states for this beam is discussed later in the chapter.

[6] This example follows the Beam Design Example given in Part 5 of ACI 440.1R-06.

[7] In the example problems in this book "extra" significant figures will often be used that are not typically "carried" in engineering calculations. This is to enable the reader to follow subsequent calculations. In actual engineering design, such numbers should be rounded.

Step 2. Assume a bar size and obtain f^*_{fu} from the manufacturer's specifications. Reinforcement bar sizes are also needed at this stage to estimate the effective depth, d. Assume reinforcement of two No. 6 main bars (since this is the largest bar permitted for a 7-in. beam width) and No. 3 stirrups: d_b = 0.75 in. and A_{bar} = 0.44 in^2. Note that the area of the bar given above is obtained from the nominal diameter of the bar and is a nominal area. The actual cross-sectional area of an FRP bar as determined by ACI 440.4R-03 test method B.1 by water displacement may be different (usually, larger) than that determined from the nominal diameter (see Table 4.1 for reported actual bar areas). Use of the nominal area is recommended for design calculations. Assume a clear cover of 1.5 in. The effective depth of the section is therefore

$$d = h - \text{cover} - \text{stirrup diameter} - 0.5(\text{bar diameter})$$
$$= 12 - 1.5 - 0.375 - 0.5(0.75) = 9.75 \text{ in.}$$

The guaranteed properties of the No. 6 glass FRP bar provided are given as f^*_{fu} = 90 ksi[8], ε^*_{fu} = 0.014, and E_f = 6500 ksi.

Step 3. Calculate the balanced reinforcement ratio. The design strength of the bar is determined using the environmental conditions specified and the type of fiber in the FRP bar. For a glass-reinforced FRP bar designed for interior exposure, C_E = 0.8 and

$$f_{fu} = C_E f^*_{fu} = 0.8(90 \text{ ksi}) = 72 \text{ ksi}$$

$$\varepsilon_{fu} = C_E \varepsilon^*_{fu} = 0.8(0.014) = 0.011$$

$$E_f = 6500 \text{ ksi (no reduction in the modulus is taken from the manufacturer-guaranteed value)}$$

The ultimate compressive strain in the concrete is taken as ε_{cu} = 0.003 per ACI 318-05.

$$\rho_{fb} = 0.85\beta_1 \frac{f'_c}{f_{fu}} \frac{E_f \varepsilon_{cu}}{E_f \varepsilon_{cu} + f_{fu}}$$
$$= 0.85(0.85)\left(\frac{4}{72}\right)\left[\frac{6500(0.003)}{6500(0.003) + (72)}\right] = 0.0086$$

[8] Note that this is a hypothetical example. In reality, the strength of the bar is a function of the bar diameter and cannot be assumed a priori. This tends to complicate the design of FRP-reinforced beams.

Step 4. Choose the numbers of bars. To achieve a compression-controlled failure that will give a beam that will probably satisfy the deflection limits given in Table 5.1 and used to size the beam in this example, choose a reinforcement ratio greater than $1.0\rho_{fb}$ for an overreinforced beam. If $1.4\rho_{fb}$ is chosen, the use of the higher resistance factor ($\phi = 0.65$) is allowed. Therefore, it is advisable to choose 1.4 ρ_{fb} as a starting point for the design.

$$\rho_f^{\text{reqd}} = 1.4\rho_{fb} = 1.4(0.0086) = 0.0120$$

$$A_f^{\text{reqd}} = \rho_f^{\text{reqd}} bd = 0.0120(7)(9.75) = 0.82 \text{ in}^2$$

$$A_{fb} = \rho_{fb} bd = 0.0086(7)(9.75) = 0.59 \text{ in}^2$$

Note that for two No. 5 bars, $A_f = 0.62$ in^2 would be sufficient for an overreinforced design; however, this is very close to the balanced condition and may lead to the beam not meeting serviceability criteria and perhaps not failing in the desired mode.[9] Therefore, two No. 6 bars are chosen as the main tensile reinforcement:

$$A_f = 0.88 \text{ in}^2$$

$$\rho_f = \frac{0.88}{7(9.75)} = 0.0129 = 1.5\rho_{fb}$$

Step 5. Calculate the nominal moment capacity of the section. Since the beam is overreinforced, the FRP bar will be at a stress less than its design strength, and the stress in the FRP bar at concrete compression failure is calculated as

$$f_f = \sqrt{\frac{(E_f\varepsilon_{cu})^2}{4} + \frac{0.85\beta_1 f'_c}{\rho_f} E_f\varepsilon_{cu}} - 0.5E_f\varepsilon_{cu}$$

$$= \sqrt{\frac{(6500)^2(0.003)^2}{4} + \frac{0.85(0.85)(4)(6500)(0.003)}{0.0129}}$$

$$- 0.5(6500)(0.003) = 57.06 \text{ ksi}$$

The depth of the Whitney rectangular compression block is

$$a = \frac{A_f f_f}{0.85 f'_c b} = \frac{0.88(57.06)}{0.85(4)(7)} = 2.110 \text{ in.}$$

The nominal moment capacity is

[9] The detailed calculation for this case is given in ACI 440.1R-06, Part 5.

$$M_n = A_f f_f\left(d - \frac{a}{2}\right) = 0.88(57.06)\left(9.75 - \frac{2.110}{2}\right)$$
$$= 436.6 \text{ kip-in.} = 36.4 \text{ kip-ft}$$

Step 6. Calculate the flexural resistance factor. Since $\rho_f > 1.4\ \rho_{fb}$, $\phi = 0.65$.

Step 7. Calculate the factored moment capacity.

$$\phi M_n = 0.65(436.6) = 283.8 \text{ kip-in.} > 180.5 \text{ kip-in. } (M_u)$$

This is a safe design, but the beam appears to be overdesigned by quite a large amount (i.e., 57%). However, serviceability criteria need to be checked to see if the beam is, in fact, overdesigned. This is done in the serviceability example that follows.

Discussion of Alternative Designs

Results for alternative designs for this beam are now presented to compare a number of different designs for the beam. In design there is no *right* answer. Many good designs are possible, and the choice will often be made on the basis of nonstructural issues such as geometrical constraints, costs, and availability of FRP bars and stirrups. All other items being equal, the choice will probably be made on a cost basis.

Option a: Alternative overreinforced beam design If two No. 5 bars had been chosen initially, the design procedure outlined above is used. It is assumed that the guaranteed strength does not increase for the smaller bar. The effective depth is calculated as $d = 9.81$ in. The nominal moment is calculated to be $M_n = 385.0$ kip-in. $= 32.08$ kip-ft. Since $\rho_f = 0.0090$ and $1.0\rho_{fb} < \rho_f < 1.4\rho_{fb}$, the resistance factor is calculated as

$$\phi = 0.3 + \frac{\rho_f}{4\rho_{fb}} = 0.3 + \frac{0.0090}{4(0.0086)} = 0.56$$

and the factored moment is

$$\phi M_n = 0.56(385.0) = 215.6 \text{ kip-in.} > 180.5 \text{ kip-in. } (M_u)$$

This is also a safe design and the beam appears to be well designed for strength (only 19% overdesigned for strength). However, this beam does not satisfy serviceability criteria (as noted in ACI440.1R-06, Part 5.)

Option b: Underreinforced beam design with the same beam height Consider an underreinforced design as an alternative. Assume that the beam is designed using five No. 3 main bars (A_f = 0.55 in^2) that are bundled in twos at the corners, stacked vertically. The two $\frac{3}{8}$-in. bundled bars are then equivalent in dimensions to one No. 6 bar, and the effective depth is d = 9.75 in., as before. The use of bundled FRP bars is not prohibited. (See provisions of ACI 318-05: Section 7.6.6.) Note that this assumes that the bar strength does not change and that the design strength for the No. 3 bars is f_{fu} = 72 ksi (f^*_{fu} = 90 ksi)). (What happens when the guaranteed bar strength is changed to f^*_{fu} = 110 ksi is discussed in what follows.) The actual reinforcement ratio for this configuration is

$$\rho_f = \frac{0.55}{7(9.75)} = 0.0081 < 0.0086 \quad (\rho_{fb} \text{ based on } f^*_{fu} = 90 \text{ ksi})$$

To design the underreinforced beam, the approximate method is used. The depth of the neutral axis at the balanced condition is calculated:

$$c_b = \frac{\varepsilon_{cu}}{\varepsilon_{cu} + \varepsilon_{fu}} d = \frac{0.003}{0.003 + 0.011}(9.75) = 2.089 \text{ in.}$$

and the nominal moment is

$$M_n = A_f f_{fu}\left(d - \frac{\beta_1 c_b}{2}\right) = 0.55(72)\left[9.75 - \frac{0.85(2.089)}{2}\right]$$
$$= 350.9 \text{ kip-in.} = 29.2 \text{ kip-ft}$$

Since the beam is underreinforced, the resistance factor, ϕ = 0.55, and the factored capacity is

$$\phi M_n = 0.55(350.9) = 190.0 \text{ kip-in.} > 180.5 \text{ kip-in.} \ (M_u)$$

The underreinforced beam is also safely designed from a strength design perspective, but may not meet serviceability criteria.

Option c: Underreinforced design using a deeper cross section. Consider an underreinforced beam using two No. 6 bars, but designed intentionally to be underreinforced. In this case, choose $\rho_{\text{reqd}} = \rho_{fb}$ = 0.0086. Assume that d = h − 2.25 in. and that b is 7 in., as before. $d_{\text{reqd}} = A_f/\rho_{\text{reqd}}\, b$ = 0.88/ 0.0086(7) = 14.6 in. → use 14.75 in. Therefore, h = 14.75 + 2.25 = 17.0 in.

Recalculate the design loads:

$$w_{\mathrm{LL}} = 400 \text{ lb/ft}$$

$$w_{\mathrm{DL}} = 208 \text{ lb/ft} + 150 \text{ (lb/ft}^3)(7/12)(17/12) = 332 \text{ lb/ft}$$

Factor the load according to ACI 318-05:

$$W_u = 1.2(332) + 1.6(400) = 1038.4 \text{ lb/ft}$$

$$M_u = \frac{W_u\, l^2}{8} = \frac{1038.4(11)^2}{8} = 15{,}706 \text{ lb-ft} = 15.71 \text{ kip-ft} = 188.5 \text{ kip-in.}$$

Determine the depth of the neutral axis in the balanced condition:

$$c_b = \frac{\varepsilon_{cu}}{\varepsilon_{cu} + \varepsilon_{fu}}\, d = \frac{0.003}{0.003 + 0.011}(14.75) = 3.161 \text{ in.}$$

The nominal moment capacity is

$$M_n = A_f f_{fu}\left(d - \frac{\beta_1 c_b}{2}\right) = 0.88(72)\left[14.75 - \frac{0.85(3.161)}{2}\right]$$

$$= 849.4 \text{ kip-in.} = 70.8 \text{ kip-ft}$$

The factored capacity is

$$\phi M_n = 0.55(849.4) = 467.2 \text{ kip-in.} > 188.5 \text{ kip-in. } (M_u)$$

Therefore, the design is safe, but the beam appears to be significantly overdesigned. As noted in the overreinforced case, the beam must still be checked for serviceability criteria. However, in this case the underreinforced beam will usually not have difficulty meeting the serviceability criteria, due to its large cross section. Therefore, the beam design may be optimized by choosing less reinforcement (e.g., two No. 5 bars) and decreasing the beam height.

For this underreinforced beam, check the minimum flexural reinforcement:

$$A_f = 0.88 \text{ in}^2$$

$$A_{f,\min} = \frac{4.9\sqrt{f'_c}}{f_{fu}}\, b_w d = \frac{4.9\sqrt{4000}}{72{,}000}(7)(14.75) = 0.44 \text{ in}^2$$

but not less than

$$\frac{330}{f_{fu}} b_w d = \frac{330}{72{,}000}(7)(14.75) = 0.47 \text{ in}^2$$

Therefore, the second condition controls, and A_f must be greater than 0.47 in^2. Since $A_f = 0.88$ in^2, the minimum flexural reinforcement has been provided for the underreinforced beam, and catastrophic, brittle failure at the cracking moment cannot occur.

Option d: Effect of higher bar strength for smaller bars Reconsider the beam designed in option b, with five No. 3 bars in a 7 × 12 in. cross section. Now assume that the guaranteed strength of the No. 3 bar is $f^*_{fu} = 110$ ksi, and determine the capacity of the beam.

The design strength of the bar is $f_{fu} = 0.8(110) = 88$ ksi, and the balanced reinforcement ratio must be recalculated:

$$\rho_{fb} = 0.85\beta_1 \frac{f'_c}{f_{fu}} \frac{E_f\varepsilon_{cu}}{E_f\varepsilon_{cu} + f_{fu}}$$

$$= 0.85(0.85)\left(\frac{4}{88}\right)\left[\frac{6500(0.003)}{6500(0.003) + (88)}\right] = 0.0060$$

Notice that this is significantly less than the ratio of 0.0086 obtained previously with $f_{fu} = 72$ ksi. Now the beam with five No. 3 bars ($\rho_f = 0.0081$) is actually overreinforced, and an overreinforced calculation needs to be made. Following the procedure above yields

$$f_f = \sqrt{\frac{(E_f\varepsilon_{cu})^2}{4} + \frac{0.85\beta_1 f'_c}{\rho_f} E_f\varepsilon_{cu}} - 0.5E_f\varepsilon_{cu}$$

$$= \sqrt{\frac{(6500)^2(0.003)^2}{4} + \frac{0.85(0.85)(4)(6500)(0.003)}{0.0081}}$$

$$- 0.5(6500)(0.003) = 74.23 \text{ ksi}$$

$$a = \frac{A_f f_f}{0.85 f'_c b} = \frac{0.55(74.23)}{0.85(4)(7)} = 1.715 \text{ in.}$$

$$M_n = A_f f_f\left(d - \frac{a}{2}\right) = 0.55(74.23)\left(9.75 - \frac{1.715}{2}\right) = 363.0 \text{ kip-in.}$$

$$= 30.25 \text{ kip-ft}$$

$$\phi = 0.3 + \frac{\rho_f}{4\rho_{fb}} = 0.3 + \frac{0.0081}{4(0.0086)} = 0.64$$

$$\phi M_n = 0.64(363.0) = 231.4 \text{ kip-in.} > 180.5 \text{ kip-in.} (M_u)$$

A summary of the design alternatives is presented in Table 5.2 for comparison purposes. It can be seen that all of the beams have sufficient strength to carry the design loads. The choice of which beam to use is made based on serviceability criteria described in what follows.

5.4.1 Design of FRP-Reinforced Bridge Deck Slabs

The use of FRP rebars in highway bridge decks is viewed as a promising method to increase the durability of highway bridges. A number of projects have been completed in North America in recent years in which FRP rebars have been used in place of epoxy-coated steel bars to reinforce bridge decks (Benmokrane et al., 2004). Other bridges have used combinations of bottom mat steel and top mat FRP bars (Bradberry and Wallace, 2003) or FRP grids (Steffen et al., 2003; Bank et al., 2006). Combinations of FRP bars, FRP grids, and FRP stay-in-place forms have also been used in full-scale highway bridge projects in recent years (Berg et al., 2006).

Bridge Deck Design Procedure In what follows a design procedure for FRP-reinforced bridge decks spanning transverse to the traffic direction that is compatible with ACI 440-1R.06 and the AASHTO *Standard Specification for Highway Bridges*[10] (AASHTO, 2002) is discussed. The loads, load factors, and other reinforcement requirements follow the AASHTO load factor design (LFD) method, and the resistance factors are taken from ACI 440.1R-06 in this design procedure.

In the AASHTO LFD procedure the bridge is designed for a dead load factor of DL = 1.3 and a live load factor of LL = 1.3(1.67) = 2.17. In addition, the nominal live load is subject to an impact factor that is given as

TABLE 5.2 Summary of FRP-Reinforced Beam Design Alternatives[a]

Beam	Bars	f^*_{fu} (ksi)	Dimensions, $b \times h$ (in.)	Over- or Underreinforced	M_n (kip-in.)	ϕM_n (kip-in.)
Original	Two No. 6	90	7×12	Over	436.6	283.8
Option a	Two No. 5	90	7×12	Over	385.0	215.6
Option b	Five No. 3	90	7×12	Under	350.9	193.0
Option c	Two No. 6	90	7×17	Under	834.5	458.9
Option d	Five No. 3	110	7×12	Over	363.0	231.4

[a]The moment demand for these designs: M_u = 180.5 kip-in (except Option C: M_u = 188.5 kip-in).

[10]Alternatively, the loads could be taken from the AASHTO *LRFD Bridge Design Specifications* (2005).

$$I = \frac{50}{L + 125} \tag{5.7}$$

where L is the span length in feet and I is the impact factor, which has a maximum of 30%. For bridge decks that span transversely between longitudinal girders, this implies that the nominal live load is multiplied by 1.3 prior to application of the load factors.

The live load moment in units of ft-lb per foot width of the slab (ft-lb/ft) is given by AASHTO for HS20 loading as

$$\frac{S + 2}{32} P_{20} \tag{5.8}$$

where S is the clear span length in feet perpendicular to the traffic (i.e., between the girders). If the slab is continuous over three or more supports (i.e., girders), the live load moment can be taken as 80% of the above (called a *continuity factor*). The concentrated wheel load for HS20 loading, P_{20} = 16,000 lb. The main slab top and bottom reinforcement transverse to the traffic direction is designed using the dead load of the slab, any additional dead load due to a possible future wearing surface (usually taken as 20 lb/ft^2), and the live load given above.

Reinforcement in the longitudinal traffic direction in the slab, called *distribution reinforcement,* is given as a percentage of the required main reinforcement:

$$\text{percentage} = \frac{220}{\sqrt{S}} \qquad \text{maximum } 67\% \tag{5.9}$$

This amount of distribution reinforcement must be placed in the middle half of the slab (i.e., in the positive moment region for the transverse reinforcement). Not less than 50% of this amount can be used in outer quarters of the slab (i.e., nearer the girders in the negative moment region for the main transverse reinforcement). The distribution reinforcement is usually distributed equally across the width of the slab and tied below the top main bars and above the bottom main bars.

When designing a steel rebar deck, the designer is not required to check for shear or bond, as these are assumed to be adequate for steel reinforcing bars designed for the dead and live loads stipulated. Neither is there a requirement to check for deflections or crack widths. In the case of FRP bars, these assumptions are not appropriate. Even if AASHTO loads and load factor procedures are used, the FRP deck must be checked for shear and deflections (and even bond). ACI 440.1R-06 provisions should be followed to determine shear strength (transverse shear and punching shear) and serviceability criteria (deflection, crack width, and creep rupture stress).

For the punching shear design, the tire footprint (AASHTO, 2002) is taken as a rectangle with an area $0.01P$ in^2. The ratio of the width to the length of the rectangle is taken as 2.5 (i.e., the footprint is wider in the direction perpendicular to the girders than in the direction parallel to the girders).

Bridge Parapet Design The bridge parapet, or more precisely the bridge guardrail, is an integral part of a reinforced concrete bridge deck slab. At this time a FRP-rebar-reinforced standard parapet design has not been crash tested and approved by the U.S. Federal Highway Administration according to NCHRP 350 crash test procedures. However, a reinforced concrete parapet with FRP reinforcing bars has been tested in recent years using pendulum impact and static test methods (El-Salakawy et al., 2003; Dietz et al., 2004). A FRP-reinforced guardrail has been used on an FRP-reinforced deck constructed in Canada. In many projects where FRP deck reinforcing has been used, conventional epoxy-coated steel reinforcing has been used for the parapet.

Design Example 5.2: FRP-Reinforced Bridge Deck Slab Design a bridge deck reinforced with GFRP rebars for a simply supported two-lane highway bridge 100 ft long by 42 ft wide. The bridge is to be designed for AASHTO HS20-44 loading according to the AASHTO *Standard Specification for Highway Bridges* (AASHTO, 2002). The bridge superstructure consists of prestressed concrete I-girders with rigid flanges that act compositely with the deck slab and are spaced at 7 ft on-center. The width of the top flange of the prestressed girder is 18 in. Assume that there is no overhang and that conventional steel-reinforced parapets will be used. The bridge deck is designed of concrete with $f'_c = 4000$ psi and has a minimum depth of 8 in. An instantaneous deflection limit of $L/1200 = 0.08$ inches is prescribed and maximum crack widths per ACI 440.1R-06 for exterior harsh exposure. A clear cover of 0.75 in. is used per ACI 318-05 recommendations for slabs.

SOLUTION

Step 1. Determine the design loads and the moment capacity required. For dead load calculations use 150 lb/ft^3 for concrete and a future wearing surface of 20 lb/ft^2. For live loads use the AASHTO formula plus 30% impact. The deck span is calculated from the face of the girder when rigid girders are used. The dead and live load moments are calculated per foot width of the deck.

$$S = 7(12) - 2(9) = 66 \text{ in.} = 5.5 \text{ ft}$$

$$w_{DL} = (150\ \text{lb/ft}^3)(8/12) + 20\ \text{lb/ft}^2 = 120\ \text{lb/ft}$$

$$m_{DL} = \frac{w_{DL}S^2}{8} = \frac{120(5.5)^2}{8} = 454\ \text{lb-ft/ft} = 0.45\ \text{kip-ft/ft}$$

$$m_{LL+I} = \frac{S+2}{32}P_{20}(1.3) = \frac{5.5+2}{32}16(1.3) = 4.88\ \text{kip-ft/ft}$$

Factored moments for simple spans:

$$M_u = 1.3(0.45) + 2.17(4.88) = 11.22\ \text{kip-ft/ft}$$

Multiplying by 0.8 for continuity gives the design demand:

$$M_u = 0.8(11.22) = 8.98\ \text{kip-ft/ft}$$

Step 2. Assume a bar size and obtain f^*_{fu} from the manufacturer's specifications. A glass FRP No. 5 bar with $f^*_{fu} = 95$ ksi and $E_f = 5920$ ksi is selected for a trial design. A bottom clear cover of 0.75 in. is used. For top bars (for a negative moment capacity over the girders) the top cover is usually taken as 2.5 in. for steel bars. For FRP bars this may be excessive. $2d_b$ is suggested for constructability reasons. Calculate the effective depth for the bottom bars for positive moment capacity.

$$\begin{aligned} d &= h - \text{cover} - 0.5(\text{bar diameter}) \\ &= 8 - 0.75 - 0.5(0.625) = 6.94\ \text{in.} \end{aligned}$$

Step 3. Calculate the balanced reinforcement ratio. For a glass FRP bar designed for exterior exposure, $C_E = 0.7$ and

$$f_{fu} = C_E f^*_{fu} = 0.7(95\ \text{ksi}) = 66.5\ \text{ksi}$$

$$\begin{aligned} \rho_{fb} &= 0.85\beta_1 \frac{f'_c}{f_{fu}} \frac{E_f\varepsilon_{cu}}{E_f\varepsilon_{cu} + f_{fu}} \\ &= 0.85(0.85)\left(\frac{4}{66.5}\right)\left[\frac{5920(0.003)}{5920(0.003) + (66.5)}\right] = 0.0092 \end{aligned}$$

Step 4. Choose the numbers of bars.

$$1.4\,\rho_{fb} = 1.4(0.0092) = 0.0129$$

Choose three No. 5 bars with $A_f = 0.93$ in^2 and $\rho_f = 0.0112$ (slightly less than 1.4 $\rho_{fb)}$.

Step 5. Calculate the nominal moment capacity of the section. The stress in the FRP bars at flexural failure is

$$f_f = \sqrt{\frac{(E_f\varepsilon_{cu})^2}{4} + \frac{0.85\beta_1 f'_c}{\rho_f} E_f\varepsilon_{cu}} - 0.5E_f\varepsilon_{cu}$$

$$= \sqrt{\frac{(5920)^2(0.003)^2}{4} + \frac{0.85(0.85)(4)(5920)(0.003)}{0.0122}}$$

$$- 0.5(5920)(0.003) = 59.4 \text{ ksi}$$

The depth of the Whitney rectangular compression block is

$$a = \frac{A_f f_f}{0.85 f'_c b} = \frac{0.93(59.4)}{0.85(4)(12)} = 1.35 \text{ in.}$$

The nominal moment capacity is

$$M_n = A_f f_f\left(d - \frac{a}{2}\right) = 0.93(59.4)\left(6.94 - \frac{1.35}{2}\right) = 346.09 \text{ kip-in./ft}$$

$$= 28.8 \text{ kip-ft/ft}$$

Step 6. Calculate the flexural resistance factor.

$$\phi = 0.3 + \frac{\rho_f}{4\rho_{fb}} = 0.3 + \frac{0.00112}{4(0.0092)} = 0.60$$

Step 7. Calculate the factored moment capacity.

$$\phi M_n = 0.60(28.2) = 17.3 \text{ kip-ft/ft} > 8.98 \text{ kip-ft/ft}$$

With the chosen reinforcement, the slab can easily meet the positive moment flexural demand. However, the slab needs to be checked for shear before finalizing the design. The required distribution reinforcement is 67% of the main reinforcement. Choose two No. 5 bars. Since the slab is overdesigned in flexure, it could be argued that too much distribution reinforcement is being provided in the FRP slab. However, it is common practice to have reasonably balanced bidirectional reinforcement in bridge deck slabs, where punching shear under the wheel loads is typically the failure mode observed.

The slab reinforcement proposed at this stage of the design for flexural capacity is three No. 5 (No. 5 at 4 in.) GFRP bars in the transverse direction at the top and bottom of the slab and a distribution reinforcement of two No. 5 (No. 5 at 6 in.) GFRP bars in the longitudinal direction. Note that the effective depth for the top bars may be less than the design depth assumed for the bottom bars, depending on the clear cover used for the top bars. A thicker clear cover is generally used for top bars in bridge decks reinforced with steel bars; however, it is not known if such a conservative approach is needed for FRP-reinforced decks. A clear top cover of 1.0 in. has been used on two FRP-reinforced decks built in Wisconsin (Berg et al., 2006; Bank et al., 2006).

This design can be compared with the published data for the FRP-reinforced bridge deck on the Morristown Bridge in Morristown, Vermont, constructed in 2002 (Benmokrane et al., 2004). A 9-in.-thick deck with top and bottom clear covers of 2.5 and 1.5 in., respectively (typical steel clear covers) were used. Girder spacing was 7 ft on center. The design was based on a maximum flexural crack width of 0.020 in. and called for No. 5 at 4 in. bottom main bars, No. 6 at 4 in. top main bars, and No. 5 at 6 in. distribution bars (top and bottom). For constructability reasons, the actual reinforcement used in the bridge was No. 6 at 4 in. main bars (top and bottom) and No. 6 at 6 in. distribution bars. A photograph of the FRP bars in place prior to concrete placement is shown in Fig. 5.6.

Figure 5.6 Glass FRP bars on the Morristown, Vermont, bridge in 2002. (Courtesy of Brahim Benmokrane.)

The punching shear design for the slab is considered in Chapter 6. The transverse shear capacity is often assumed to be satisfactory if the punching shear capacity is satisfactory. Deflection, crack width, and creep rupture checks must still be performed. However, since the slab is overreinforced and the AASHTO load factors are significantly larger than the ASCE 7 (i.e., ACI 318-05) load factors, the serviceability design should be sufficient. However, this must be verified with calculations in an engineering design, since the deflection limits imposed for bridges are more severe than in building-related applications.

5.5 SERVICEABILITY DESIGN OF FRP-REINFORCED BEAMS

For the serviceability limit state, both deflections and crack widths must be checked according to procedures detailed in ACI 440.1R-06. The designer is also required to check stresses under sustained service loads against creep rupture stress limits and fatigue stress limits. The strain and stress distributions and resultant forces in the cracked section at service loads are shown in Fig. 5.7.

5.5.1 Deflections Under Service Loads

Due to the lower modulus of glass FRP rebars relative to steel rebars, deflections in FRP-reinforced beams for equivalent reinforcement ratios to steel-reinforced beams will be much larger. Deflections should be determined under service loads for both immediate (i.e., short-term) and long-term sustained loads and compared with building code–permitted deflections. Similar to the case of steel-reinforced concrete members, an "effective" second moment of area based on the gross, I_g, and cracked, I_{cr}, cross section second moments is used to calculate deflections of an FRP-reinforced beam. A modified form of the Branson equation is used for FRP-reinforced beams to determine the ef-

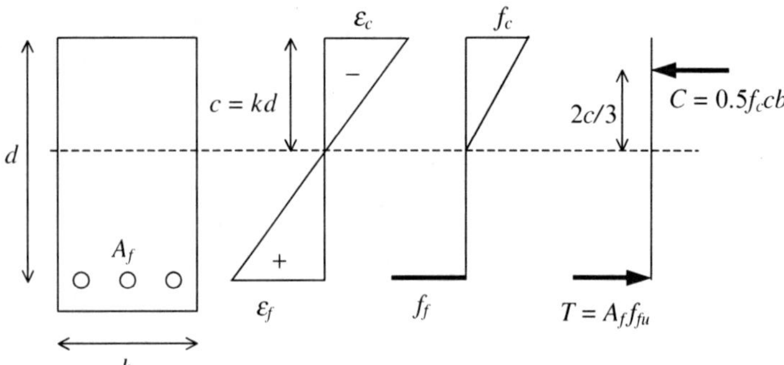

Figure 5.7 Strains, stress, and force resultants at service loads.

fective second moment of the section as a function of maximum applied service load moment, M_a, as follows:

$$I_e = \left(\frac{M_{cr}}{M_a}\right)^3 \beta_d I_g + \left[1 - \left(\frac{M_{cr}}{M_a}\right)^3\right] I_{cr} \leq I_g \qquad \text{(ACI 440.1R-06:8-13}a\text{)}$$

where M_{cr} is the moment at cracking and β_d is a reduction coefficient for FRP-reinforced beams and is given as

$$\beta_d = \frac{1}{5}\left(\frac{\rho_f}{\rho_{fb}}\right) \leq 1.0 \qquad \text{(ACI 440.1R-06:8-13}b\text{)}$$

As in conventional steel-reinforced concrete, the cracked (transformed) second moment of area is given as

$$I_{cr} = \frac{bd^3}{3}k^3 + \eta_f A_f d^2(1 - k)^2 \qquad \text{(ACI 440.1R-06:8-11)}$$

where $k = c/d$ is the ratio of the depth of the neutral axis to the effective depth of the section under service loads and η_f is the modular ratio for the FRP reinforcement,

$$k = \sqrt{(\rho_f \eta_f)^2 + 2\rho_f \eta_f} - \rho_f \eta_f \qquad \text{(ACI 440.1R-06:8-12)}$$

$$\eta_f = \frac{E_f}{E_c} \qquad (5.10)$$

The gross second moment of area is taken as

$$I_g = \frac{bh^3}{12} \qquad (5.11)$$

Alternatively, the properties of the cracked section under service loads can be calculated directly using the bimodular (composite beam) material formula (Gere and Timoshenko, 1997). The location of the neutral axis shown in Fig. 5.8, is given, by definition, by the equation

$$E_1 \int y_1 \, dA_1 + E_2 \int y_2 \, dA_2 = 0 \qquad (5.12)$$

Substituting $E_1 = E_c$, $E_2 = E_f$, and $y = c =$ the distance from the neutral axis to the top of the section, integrating, and rearranging gives the quadratic equation for c as

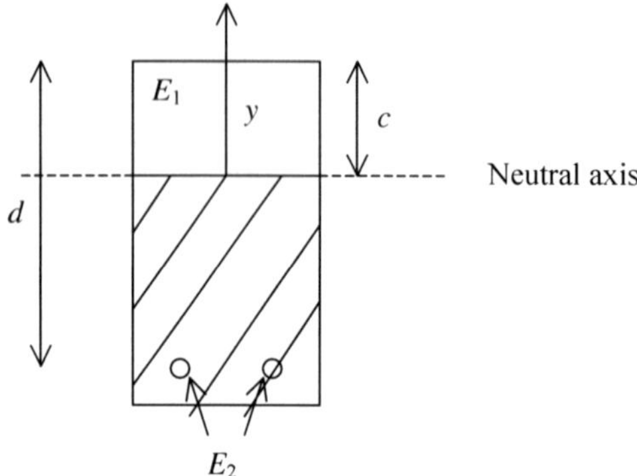

Figure 5.8 Cracked section dimensions and notation.

$$E_c b \frac{c^2}{2} + E_f A_f (c - d) = 0 \tag{5.13}$$

The flexural stiffness (EI) of the composite section is given as

$$\sum EI = E_1 I_1 + E_2 I_2 \tag{5.14}$$

and the cracked (transformed) second moment of area as

$$I_{cr} = \frac{\sum EI}{E_c} \tag{5.15}$$

The long-term deflection in an FRP-reinforced beam that includes the effects of creep and shrinkage of the concrete under sustained long-term service loads (i.e., the dead load and sustained live load) is calculated in a fashion similar to that of a steel rebar–reinforced beam and is given as

$$\Delta_{cp+sh} = 0.6\xi(\Delta_i)_{sus} \qquad \text{(ACI 440.1R-06:8-14)}$$

For FRP reinforcements where no compression reinforcement is used, $\xi = \lambda$. λ is the conventional ACI multiplier for additional deflection due to long-term effects in reinforced concrete members, and ξ is the time-dependent factor for sustained loads. $\xi = 2.0$ for sustained loads with a duration of five years as per ACI 318-05. The 0.6 coefficient is added for an FRP-reinforced beam to account for larger initial deflection in an FRP-reinforced beam and for the compressive creep of the concrete.

5.5.2 Flexural Cracking

The width of flexural cracks in beams reinforced with FRP rebars is limited by ACI 440.1R-06. The crack width for FRP-reinforced members may be calculated from

$$w = 2\frac{f_f}{E_f}\beta k_b\sqrt{d_c^2 + \left(\frac{s}{2}\right)^2} \qquad \text{(ACI 440.1R-06:8-9)}$$

where w is the maximum crack width in inches, f_f the service load stress in the FRP reinforcement in ksi, E_f the modulus of the FRP rebars in ksi, β the ratio of the distance between the neutral axis and the bottom of the section (i.e., the tension surface) and the distance between the neutral axis and the centroid of reinforcement (as shown in Fig. 5.9), d_c the thickness of the concrete cover from the tension face to the center of the closest bar, s the center-to-center bar spacing of the main FRP bars, and k_b a bond-related coefficient (Frosch, 1999). β is calculated from

$$\beta = \frac{h_2}{h_1} = \frac{h - kd}{d(1 - k)} \tag{5.16}$$

where the distances h_1 and h_2 are as shown in Fig. 5.9,

The stress in the FRP at service loads can be calculated from

$$f_{f,s} = \frac{m_{\mathrm{DL+LL}}\ \eta_f d(1 - k)}{I_{\mathrm{cr}}} \tag{5.17}$$

or

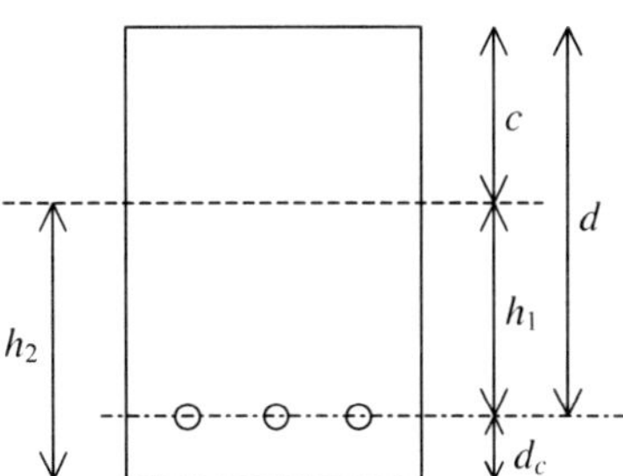

Figure 5.9 Cross-sectional dimensions used to calculate β.

$$f_{f,s} = \frac{m_{\mathrm{DL+LL}}\, E_f(d - c)}{\sum EI} \tag{5.18}$$

ACI 440.1R-06 recommends that the bond-related coefficient is taken as $k_b = 1.4$ for commercially produced FRP rebars. Lower values of k_b can be used if experimental data are available to support such a value. For steel bars, $k_b = 1.0$, which implies that FRP bars have a different bond behavior than that of steel bars.

5.5.3 Creep and Fatigue at Service Loads

FRP-reinforced concrete beams must be checked for possible failure due to creep rupture or fatigue under service loads. Creep rupture is checked with respect to all sustained service loads, and fatigue is checked with respect to all sustained loads plus the maximum loading induced in a fatigue loading cycle.

$$f_{f,\text{creep rupture}} = \frac{m_{\text{sus}}\eta_f d(1 - k)}{I_{\text{cr}}} \tag{5.19}$$

or

$$f_{f,\text{creep rupture}} = \frac{m_{\text{sus}}}{A_f d(1 - k/3)} \tag{5.20}$$

The stress limits for creep rupture and fatigue given by ACI 440.1R-06 are $0.2f_{fu}$, $0.3f_{fu}$, and $0.55f_{fu}$ for glass, aramid, and carbon FRP bars, respectively.

5.6 DESIGN PROCEDURE FOR SERVICEABILITY

The design for serviceability is usually performed after the strength design for reinforced concrete members. The design typically involves checking to see that the section and reinforcement selected can meet the code-mandated serviceability requirements. Serviceability design in the LRFD approach is performed by setting all the load factors to 1.0 and all the resistance factors to 1.0. Therefore, the loads are the nominal service loads, and the calculated effects (deflections and stresses) are the actual effects (i.e., they are not modified by any factors). The approach is therefore the same approach as that used to calculate deflections and service-level stresses in the old working stress design (WSD) method (the equivalent of the allowable stress design method). A similar approach is used for FRP-reinforced sections. The procedure for the serviceability check is as follows:

Step 1. Determine the service loads and moments. The service moments due to dead loads, live load, and the sustained dead and live loads are calculated.

Step 2. Calculate the properties of the uncracked and cracked transformed section. The cracked section properties can be calculated by the transformed section method or the composite mechanics method. Note, however, that the properties of the cracked section with the reinforcement transformed to the equivalent properties of the concrete are used in calculating the FRP-reinforced beam deflection using the modified Branson equation.

Step 3. Calculate the immediate (short-term) and long-term deflection of the section. Using the modified Branson equation, calculate the immediate deflection due to the full dead and live loads. Then calculate the long-term deflection due to creep and shrinkage under the sustained loads only. Compare the calculated deflections with the code-permitted short- and long-term deflections.

Step 4. Calculate the width of the cracks due to the flexural load. The flexural crack width is determined and the value is compared with the ACI 440.1R-06 code-permitted maximum crack widths for either interior or exterior exposures given in Table 4.3.

Step 5. Calculate the axial stress in the FRP bars under long-term sustained service and fatigue loads. The stresses in the FRP bars under sustained loads are compared with the ACI 440.1R-06-permitted maxima to prevent failure due to creep rupture or fatigue.

Step 6. Modify the design if the serviceability limits are exceeded. If any of the three serviceability limits are exceeded, the section must be redesigned. Since the redesign may require changing the size of the FRP rebars, the strength design needs to be rechecked with the new section and bar properties.

Design Example 5.3: Serviceability Design for an FRP-Reinforced Beam Check the serviceability of the 12×7 in. overreinforced beam in Design Example 5.2. The geometric and material properties of the section are: simply supported beam, 11-ft span; 4000-psi concrete; FRP bar design strength, $f_{fu} = 72$ ksi; two No. 6 main bars; $A_f = 0.88$ in^2; $d = 9.75$ in.; $w_{LL} = 400$ lb/ft; $w_{\mathrm{DL}} = 295.5$ lb/ft (includes beam self-weight). Additionally, it is given that 20% of the live load is sustained and that the maximum permitted deflection under long-term sustained loads is $L/240$. The maximum crack width permitted for interior exposure is 0.028 in. (often referred to as 28 mils[11]). The maximum allowable stress in the glass FRP bars under sustained load is $0.2f_{fu} = 0.2(72) = 14.4$ ksi.

[11] 1 mil = 1/1000 in.

SOLUTION

Step 1. Determine the service loads and moments.

$$m_{\text{DL}} = \frac{w_{\text{DL}} l^2}{8} = \frac{(87.5 + 208)\,(11)^2}{8} = 4469.4 \text{ lb-ft} = 53.6 \text{ kip-in.}$$

$$m_{\text{LL}} = \frac{w_{\text{LL}} l^2}{8} = \frac{400(11)^2}{8} = 6050 \text{ lb-ft} = 72.6 \text{ kip-in.}$$

$$m_{\text{DL+LL}} = m_S = 126.2 \text{ kip-in.}$$

Step 2. Determine the modular ratio and section (cracked and gross) properties.

Modular ratio:

$$\eta_f = \frac{E_f}{E_c} = \frac{6{,}500{,}000}{57{,}000\sqrt{4000}} = 1.8$$

Neutral-axis depth ratio, $k = c/d$:

$$\begin{aligned} k &= \sqrt{(\rho_f \eta_f)^2 + 2\rho_f \eta_f} - \rho_f \eta_f \\ &= \sqrt{[0.0129(1.8)]^2 + 2(0.0129)(1.8)} - 0.0129(1.8) = 0.194 \end{aligned}$$

Cracked transformed second moment of area:

$$\begin{aligned} I_{\text{cr}} &= \frac{bd^3}{3} k^3 + \eta_f A_f d^2 (1 - k)^2 \\ &= \frac{7(9.75)^3}{3} (0.194)^3 + 1.8(0.88)(9.75)^2(1 - 0.194)^2 = 114 \text{ in}^4. \end{aligned}$$

Gross second moment of area:

$$I_g = \frac{bh^3}{12} = \frac{7(12)^3}{12} = 1008 \text{ in}^4$$

Alternatively, calculate directly using the bimodular (composite beam) material formula (see Fig. 5.8). Use the equation

$$E_1 \int y_1 \, dA_1 + E_2 \int y_2 \, dA_2 = 0$$

with

$$E_1 = E_c = 57{,}000\sqrt{4000} = 3605 \text{ ksi}$$

$$E_2 = E_f = 6500 \text{ ksi}$$

$y = c$ (distance from the neutral axis to the top of the section)

$$3605(7)c\,\frac{c}{2} + 6500[-(9.75 - c)](0.88) = 0$$

$$12{,}617.5c^2 + 5720c - 55{,}770 = 0$$

Solving the quadratic equation and taking the positive root yields

$$c = 1.888 \text{ in.} \quad \text{or} \quad k = \frac{c}{d} = \frac{1.888}{9.75} = 0.194 \text{ in.}$$

The flexural stiffness (EI) of the composite section is given as

$$\sum EI = E_1 I_1 + E_2 I_2$$

$$3{,}605{,}000\left[\frac{7(1.888)^3}{3}\right] + 6{,}500{,}000(0.88)(9.75 - 1.888)^2$$
$$= 4.102 \times 10^8 \text{ lb-in}^2$$

and

$$I_{cr} = \frac{\sum EI}{E_c} = \frac{4.102 \times 10^8}{3{,}605{,}000} = 113.8 \text{ in}^4 \approx 114 \text{ in}^4$$

Note that this is the same answer as obtained before using the transformed section method and ACI equations.

The tensile strength of the concrete

$$f_r = 7.5\sqrt{f'_c} = 7.5\sqrt{4000} = 474.3 \text{ psi}$$

Calculate the cracking moment:

$$M_{cr} = \frac{2f_r I_g}{h} = \frac{2(474.3)(1008)}{12} = 79{,}689 \text{ lb-in.}$$

$$= 79.7 \text{ kip-in.} < 126.2 \text{ kip-in. } (m_S)$$

The check that the cracking moment is smaller than the total service load moment is an important check for FRP-reinforced beams. This means that the beam is cracked under service loads. If this is not the case, the beam is uncracked under service loads, and the full gross second moment, I_g, is used in all the serviceability calculations (i.e., deflection, crack widths, and sustained stresses).

Step 3. Calculate the immediate (short-term) and long-term deflection of the section. Calculate the effective second moment of area (modified Branson formula). First, calculate the FRP reduction coefficient.

$$\beta_d = \frac{1}{5}\left(\frac{\rho_f}{\rho_{fb}}\right) = \frac{1}{5}\left(\frac{0.0129}{0.0086}\right) = 0.30 \leq 1.0$$

$$I_e = \left(\frac{M_{cr}}{M_{DL+LL}}\right)^3 \beta_d I_g + \left[1 - \left(\frac{M_{cr}}{M_{DL+LL}}\right)^3\right] I_{cr} \leq I_g$$

$$I_e = \left(\frac{79.7}{126.2}\right)^3 (0.30)(1008) + \left[1 - \left(\frac{79.7}{126.2}\right)^3\right](114) = 160.8 \text{ in}^4$$

The immediate deflection under the total (uniformly distributed) service load is

$$\Delta_i = \frac{5(w_{DL} + w_{LL})l^4}{384 E_c I_e} = \frac{5(295.5 + 400)(11)^4}{384(3605)(160.8)}\left(\frac{(12)^3}{(1000)}\right) = 0.395 \text{ in.}$$

The long-term sustained load (i.e., the cause of the creep deflection) is

$$w_{sus} = w_{LL(sus)} + w_{DL} = 0.2(400) + (208 + 87.5) = 375.5 \text{ lb/ft}$$

$$(\Delta_i)_{sus} = \Delta_i \frac{w_{sus}}{w_{DL} + w_{LL}} = 0.395\left(\frac{375.5}{695.5}\right) = 0.213 \text{ in.}$$

The deflection due to the long-term sustained loads is

$$\Delta_{(cp+sh)} = 0.6\xi(\Delta_i)_{sus} = 0.6(2)(0.213) = 0.256 \text{ in.}$$

The total deflection at a time in the future (greater than five years) is the sum of the instantaneous deflection due to the live load only plus the long-term deflection due to the sustained loads:

$$\Delta_{i(\text{LL})} = \Delta_i \frac{w_{\text{LL}}}{w_{\text{DL}} + w_{\text{LL}}} = 0.395 \left(\frac{400}{695.5}\right) = 0.227 \text{ in.}$$

$$\begin{aligned}\Delta_{\text{long term}} &= \Delta_{i(\text{LL})} + \Delta_{\text{cp+sh}} = 0.227 + 0.256 \\ &= 0.483 \text{ in.} < 0.55 \text{ in.} \ (\Delta_{\text{allowable}} = L/240)\end{aligned}$$

The beam therefore meets the long-term deflection limit. However, it is somewhat overdesigned for deflection limits (as it was for strength).

Step 4. Calculate the width of the cracks due to the flexural load. To calculate the width of the flexural cracks, the center-to-center spacing of the main bars is needed. For two No. 6 bars, this is $s = 2.5$ in. (see Chapter 7 for details). The bond-related coefficient, $k_b = 1.4$, as recommended by ACI 440.1R-06 since no additional data are provided to reduce this number. The stress in the bars at service and the β ratio are, as follows: Using the transformed crack second moment yields

$$f_f = \frac{m_{\text{DL+LL}}\, \eta_f d(1 - k)}{I_{\text{cr}}} = \frac{126.1(1.8)(9.75)(1 - 0.194)}{114} = 15.7 \text{ ksi}$$

or using the composite flexural stiffness gives us

$$f_f = \frac{m_{\text{DL+LL}} E_f\,(d - c)}{\sum EI} = \frac{126.1(6500)(9.75 - 1.89)}{4.102 \times 10^5} = 15.7 \text{ ksi}$$

$$\beta = \frac{h_2}{h_1} = \frac{h - kd}{d(1 - k)} = \frac{12 - 0.194(9.75)}{9.75(1 - 0.194)} = 1.286$$

Substituting in the expression for the flexural crack width gives

$$\begin{aligned}w &= 2\frac{f_f}{E_f}\beta k_b \sqrt{d_c^2 + \left(\frac{s}{2}\right)^2} \\ &= 2\left(\frac{15.7}{6500}\right)(1.286)(1.4)\sqrt{(2.25)^2 + \left(\frac{2.5}{2}\right)^2} \\ &= 0.024 \text{ in.} < 0.028 \text{ in.} \ (w_{\text{allowable}})\end{aligned}$$

Therefore, the beam meets crack width requirements.

TABLE 5.3 Summary of Flexural Design for a 7 × 12 Section with Two No. 6 Bars

Criterion	Designed	Required (or Permitted)	Overdesign Factor
Flexural strength	283.8 kip-in.	180.5 kip-in.	1.57
Maximum long-term deflection	0.483 in.	0.550 in.	1.14
Crack width	0.024 in.	0.028 in.	1.17
Creep rupture stress	8.46 ksi	14.4 ksi	1.70

Step 5. Calculate the axial stress in the FRP bars under long-term sustained service and fatigue loads.

$$m_{\text{sus}} = \frac{375.5(11)^2}{8} = 5679 \text{ lb-ft} = 5.68 \text{ kip-ft} = 68.16 \text{ kip-in.}$$

$$f_f = \frac{m_{\text{sus}}\, \eta_f d(1 - k)}{I_{cr}} = \frac{68.16(1.8)(9.75)(1 - 0.194)}{114}$$

$$= 8.46 \text{ ksi} < 14.4 \text{ ksi } (0.2\, f_{fu})$$

Therefore, long-term creep rupture of the FRP bars is not a concern.

The flexural design of a 7 × 12 in. beam with two No. 6 main bars is summarized in Table 5.3. It can be seen that the beam is optimized from a serviceability perspective but is quite overdesigned with regard to flexural strength and creep rupture stress. This will generally be the case with glass FRP bars that have a relatively low longitudinal modulus but high strength. It should be also recalled that this beam is overreinforced. It is important to note that the deflection or the crack width limits often control design. Therefore, it is advisable to check the serviceability limits before optimizing the design for strength or starting the design from the serviceability calculation.

PROBLEMS

Note to the reader: The analysis and design homework problems provided here address the capacity and serviceability of reinforced concrete beams and slabs subjected to flexural loads. Additional design problems for beams and slabs that include flexure (capacity and serviceability), shear, and detailing in an integrated fashion are given in Chapter 7.

5.1 The expression for the nominal moment capacity for an overreinforced FRP-reinforced section with two layers of tension reinforcement is

$$M_n = A_{f1}f_{f1}\left(d_1 - \frac{\beta_1 c}{2}\right) + A_{f2}f_{f2}\left(d_2 - \frac{\beta_1 c}{2}\right)$$

where A_{f1} and A_{f2} are the areas of the FRP reinforcements in the first (lowest) and second rows of the reinforcement. f_{f1} and f_{f2} the stresses at concrete compression failure in the reinforcements, and d_1 and d_2 the effective depths for the two layers. The layers have moduli E_{f1} and E_{f2}, respectively. Derive the expression for the depth of the neutral axis, c, for use in the expression above and provide expressions for calculating f_{f1} and f_{f2}.

5.2 An approximate expression for the nominal moment capacity for an underreinforced FRP-reinforced section with two layers of tension reinforcement (assuming that the lowest layer ruptures first) is

$$M_n = A_{f1}f_{fu1}\left(d_1 - \frac{\beta_1 c_b}{2}\right) + A_{f2}f_{f2}\left(d_2 - \frac{\beta_1 c_b}{2}\right)$$

where A_{f1} and A_{f2} are the areas of the FRP reinforcements in the first (lowest) and second rows of the reinforcement, f_{fu1} the design strength of the FRP reinforcement in the lowest layer, f_{f2} is the stress in the second layer at failure of the section, and d_l and d_2 are the effective depths for the two layers. The layers have moduli E_{f1} and E_{f2}, respectively. Provide an equation for the neutral axis at the balanced condition, c_b, and an equation for f_{f2} for use in the expression above.

5.3 Prove that the following two equations, which are used to calculate the stress in the FRP reinforcement under sustained service loads, are equivalent:

$$f_{f,\text{sus}} = \frac{m_{\text{sus}}}{A_f d(1 - k/3)} \qquad f_{f,\text{sus}} = \frac{m_{\text{sus}} n_f d(1 - k)}{I_{\text{cr}}}$$

5.4 Consider the FRP-reinforced concrete beams listed in Table P5.4.[12] The beams are to be used in interior construction, and a clear cover of 1.5 in. is to be used for all bars. Assume that No. 3 GFRP stirrups are provided and that shear is not critical. Use one layer of main bars. Determine the nominal moment capacity, M_n, and the factored moment capacity, ϕM_n, for the beams. Comment on the effect of high-strength concrete on the beam capacities for both over- and underreinforced beams. Provide a bar graph comparing the moment capacities (nominal and factored) of the eight beams.

[12] Instructors may assign only selected beams. A choice of both under- and overreinforced beams is recommended.

TABLE P5.4 Properties of FRP-Reinforced Beams for Analysis

Beam	f'_c (psi)	Bars	b (in.)	h (in.)
1	4000	4 No. 4 Aslan 100	12	18
2	4000	4 No. 6 Aslan 100	12	18
3	4000	4 No. 8 Aslan 100	12	18
4	4000	4 No. 4 Aslan 200	12	18
5	8000	4 No. 4 Aslan 100	12	18
6	8000	4 No. 6 Aslan 100	12	18
7	8000	4 No. 8 Aslan 100	12	18
8	8000	4 No. 4 Aslan 200	12	18

5.5 For the beams listed in Table P5.4, determine the load-carrying capacity (w_{DL}, w_{LL}, and $w_{DL} + w_{LL}$) per unit length (lb/ft) based on the factored moment capacity of the beam (determined in Problem 5.4). Assume that the beams are simply supported, uniformly loaded beams with an 18-ft span and that the ratio of dead load to live load is 1:3. Compare the load-carrying capacities of the eight beams in a bar graph. Also, compare the load-carrying capacity to the beam self-weight (per linear foot) and to the load that causes cracking in the beam. Discuss your observations.

5.6 For the load-carrying capacities determined for the FRP beams in Table P5.4, design grade 60 epoxy-coated steel-reinforced beams at $0.4\rho_b$ (and width-to-height ratios of approximately 1 to 1.5) according to ACI 318-05. Consider only moment capacity in your design of the steel beams. Compare the costs of the FRP beams to the steel-reinforced beams. Consider main bars only (assume straight bars with no bends) in your cost calculations. Assume epoxy-coated steel bars at \$0.45 per pound. For concrete, use \$75 per cubic yard for 4000 psi and \$100 per cubic yard for 8000 psi. Assume the following costs for straight FRP bars: GRPF No. 4, \$0.65 per foot; GRFP No. 6, \$1.25 per foot; GRFP No. 8, \$1.80 per foot, CFRP No. 4, \$5.50 per foot. Provide a comparison table showing FRP versus steel beam size, main bars, and costs.

5.7 Consider the beam in Design Example 5.1. Design the beam using carbon FRP bars instead of glass FRP bars. Use the same 7 × 12 in. cross section as in the design example. Consider an overreinforced design and an underreinforced design. Bundle bars if larger CFRP bar diameters are not available. The guaranteed properties for Nos. 3 and 4 CFRP rebars are $f^*_{fu} = 300$ ksi and $E^*_f = 18{,}000$ ksi. Consider only the flexural strength in this problem. Compare your results to those of the design example.

5.8 Consider the beam in Design Example 5.1. Design the beam using 8000-psi concrete instead of 4000-psi concrete. Consider an overreinforced design and an underreinforced design. Compare your results to those of the design example.

5.9 Consider the underreinforced beams in Design Example 5.1 (options b and c). Redesign the beams using the nonlinear concrete stress–strain relations instead of the approximate method given in ACI 440.1R-06 (and in the design example). Use equations (5.3) to (5.5) to determine the rectangular stress block parameters and use the iterative procedure described in Section 5.3.2. Compare your results to those of the design example. Comment on the applicability of the ACI 440.1R-06 approximate procedure. Is it conservative?

5.10 Determine the nominal and the factored moment capacities (per foot of width) of the one-way FRP-reinforced concrete slabs listed in Table P5.10. Use a 0.75-in. clear cover for slabs and do not use shear reinforcement.

5.11 Consider the beams in Table P5.4 (except for beams 1 and 5[13]). The beams are used in interior construction and loaded with a uniformly distributed superimposed dead load of 100 lb/ft and a uniformly distributed live load of 900 lb/ft over a simply supported span of 18 ft. Assume that 20% of the live load is sustained on the beam. Determine **(a)** the maximum midspan instantaneous deflection under short-term loading, **(b)** the maximum midspan deflection under long-term loading, **(c)** the maximum crack width, and **(d)** the sustained stress in the FRP bars. To calculate the bar spacing, s, in the crack width equation, assume a clear cover of 1.5 in. to the side of the beam and that the strirrup has a bend radius of $3d_b$. Compare the maximum deflections to $L/360$, and the crack widths and maximum sustained stresses in the bars to maximums permitted by ACI 440.1R-06. Indicate which of the beams satisfy the serviceability requirements.

5.12 Consider Design Example 5.1. Perform the serviceability checks (instantaneous and long-term deflection, crack width, and FRP stress un-

TABLE P5.10 Properties of FRP-Reinforced Slabs for Analysis

Slab	f_c (psi)	Bars	h (in.)
1	4000	No. 5 Aslan 100 at 4 in. on center	7
2	4000	No. 6 Pultrall GFRP at 6 in. on center	7
3	4000	No. 4 Aslan 200 CFRP at 4 in. on center	7
4	5000	No. 5 Aslan 100 at 4 in. on center	9
5	5000	No. 6 Pultrall GFRP at 6 in. on center	9

[13] Beams 1 and 5 do not have the moment capacity to carry the loads stipulated. Verify as an exercise.

der sustained loads) for design options a, b, c, and d. (Be aware that the section may not necessarily be cracked under the service loads.) Provide a summary of your results by expanding Table 5.2 and adding to the table the deflection, crack width, and FRP stresses under sustained loads. Indicate which beam designs satisfy both strength and serviceability requirements and which do not.

5.13 Consider the CFRP-reinforced beams designed in Problem 5.7. Perform the serviceability checks (instantaneous and long-term deflection, crack width, and FRP stress under sustained loads) for these beams. Compare the behavior of the CFRP over- and underreinforced beams with the GFRP over- and underreinforced beams (Design Example 5.1 and option a).

5.14 Consider the FRP-reinforced one-way slabs listed in Table P5.10. The slabs are to be used as simply supported one-way floor slabs spanning 12 ft between beams. They are loaded with a superimposed dead load of 30 lb/ft^2 (which acts in addition to the self-weight of the slab) and a live load of 100 lb/ft^2. Determine the maximum midspan deflection after 20 years of service, the maximum crack width, and the stress in the FRP due to sustained loads on the slabs.

5.15 Design a GFRP-reinforced beam to carry a dead load of 500 lb/ft (not including the beam self-weight) and a live load of 1500 lb/ft for a built-in (fixed ends) beam in interior construction on a 20-ft span. Use 5000-psi concrete. Design the beam for positive and negative moment capacity, a long-term deflection of less than $L/360$, a maximum crack width of less than 28 mils, and a sustained stress of less than $0.2f_{fu}$. Assume that the top bars can be sufficiently developed and anchored at the fixed supports.

5.16 Design a GFRP-reinforced normal-weight concrete slab for an interior floor system in a commercial building. Design for a superimposed dead load of 20 lb/ft^2 and a live load of 125 lb/ft^2 (10% sustained). Use 5000-psi concrete. The slab spans continuously over beams (nonintegral supports) spaced at 10 ft on center. Assume one-way slab action and use ACI coefficients for slab negative and positive moments (ACI 318-05 8.3.3). Recall that FRP bars are not effective as flexural reinforcements in compression. Design the slab for both flexural strength and serviceability. Limit long-term deflections to $L/360$. Provide temperature and shrinkage FRP reinforcement in the transverse direction. Calculate the FRP reinforcement cost per square foot. Consider only flexural design.

5.17 Design a bridge deck slab using GFRP rebars for a steel girder highway bridge with a girder spacing of 8 ft on center. The girder top flange is 8 in. wide. Use 4000-psi concrete and use a 1-in. clear cover for the bottom bars and a 1.5-in. cover for the top bars. Use the AASHTO standard specifications for the slab moments. Perform serviceability checks of the slab assuming an effective width strip $E = 8S/(S + 2)$ for a single concentrated wheel load (S is the effective span perpendicular to the girders, and one-way action of the slab is assumed). Design the distribution reinforcement. Provide a reinforcement diagram for the slab. Determine the cost of the FRP reinforcement per square foot. Compare this to the cost of a traditional 8-in.-deep epoxy-coated steel bridge deck with main (transverse to traffic) bars of No. 5 at 7.5 in. and distribution (parallel to traffic) bars of No. 4 at 7.5 in.

5.18 Design a simply supported GFRP-reinforced beam to carry a dead load of 800 lb/ft (not including the beam self-weight) and a live load of 2500 lb/ft in interior construction on a 20-ft span. Use 5000-psi concrete. For architectural reasons, the beam height is limited to 24 in. Use a double layer of FRP tension reinforcement and use maximum No. 10 bars (see Problem 5.1). Design the beam for moment capacity, a long-term deflection of less than $L/300$, a maximum crack width of less than 28 mils, and a sustained stress of less than $0.2f_{fu}$.

6 FRP Shear Reinforcement

6.1 OVERVIEW

In this chapter we discuss the design of concrete flexural members subjected to transverse shear forces and that have FRP main tension reinforcing bars. This includes the determination of the shear capacity of the concrete itself when the section is reinforced with FRP main bars, and the design of additional shear reinforcement in the form of stirrups if this additional reinforcement is needed. The flexural members that fall into the scope of this chapter are slabs, which will have no additional shear reinforcement, and beams, which will usually have additional shear reinforcement. As discussed in detail, the shear capacity of the concrete itself when FRP main bars are used is significantly less than when conventional steel bars are used. This is a significant fact that has a major impact on the design of both beams and slabs reinforced with FRP bars (even though additional shear reinforcement is not used in slabs.) We do not address the design of torsion of reinforced concrete members.

The design procedures presented in this chapter follow ACI 440.1R-06, which is compatible with ACI 318-05. As noted in Chapter 4, the design follows the LRFD basis for strength with a resistance factor of $\phi = 0.75$. This is the same factor that is used for steel-reinforced concrete according to ACI 318-05. The reader is assumed to have familiarity with the procedures used for shear design of concrete flexural member using steel reinforcing bars.

6.2 INTRODUCTION

The shear strength of a reinforced concrete section is traditionally assumed to come from a number of different material and geometric properties of the section. It is also assumed that a critical diagonal shear crack develops in the section. In a simply-supported beam the critical shear crack typically initiates from a flexural crack at the bottom of the beam and propagates toward the top surface, but does not reach the top surface except at the ultimate state. The shear resistance at this time is provided by (1) the shear strength of the portion of the concrete section at the top of the beam that is not cracked by the diagonal crack; (2) the resistance provided by the friction between the two sides of the beam on either side of the critical crack, known as *aggregate interlock;* (3) the vertical component of shear force that is carried in the main tensile reinforcement, which tends to prevent the shear crack from displacing vertically at the bottom of the beam, known as *dowel action;* (4) the vertical component of the force provided by any additional reinforcing bars, if they exist, that are placed along the length of the beam in the cross-sectional plane, known as *shear reinforcement;* and (5) in deep beams, the additional thrust provided by the compression arch that develops in the section roughly parallel to the shear cracks, known as *arching action.* No analytical methods are currently available, even for steel-reinforced concrete, to quantitatively determine the magnitude of any of these individual contributions, and the shear design of reinforced concrete members is still largely an empirical endeavor.

Current reinforced concrete design practice is to lump all the contributions to shear capacity, except for the contribution from the added shear reinforcement, in what is known as the *concrete contribution to shear capacity* and given the symbol V_c. The contribution due to the additional shear reinforcement is given the symbol V_s. The value of V_c has traditionally been determined based on empirical data obtained from tests of beams without shear reinforcement but with steel main reinforcement. The value of V_s has been obtained by assuming that the steel stirrups yield locally where they cross the critical diagonal shear crack. The force V_s is then obtained simply as the vertical component of this tensile force in the bar, which is given by the bar area multiplied by the yield strength of the bar.[1]

The model for shear capacity of a reinforced concrete member described above is problematic if it is applied to the shear capacity of FRP-reinforced sections and the empirical approach above is not necessarily applicable to FRP-reinforced sections. Considering the five contributors to the shear capacity of a steel-reinforced section described above, the following can be noted (in order): (1) The depth of the neutral axis in the FRP-reinforced section is typically much less than in a steel-reinforced beam, so the contribution from the concrete itself to shear resistance is lower in the FRP-reinforced beam; (2) the width of the critical diagonal crack in an FRP-reinforced beam is likely to be wider than in a steel-reinforced beam since the flexural stiffness of the FRP-reinforced beam will be lower than its steel counterpart (for the same load-carrying capacity); (3) the transverse shear force resisted by the FRP main bars will be less than that resisted by steel bars since FRP bars are anisotropic and even for the same area will have lower shear rigidity than that of steel bars; (4) FRP stirrups do not yield, so the stress in the FRP stirrup cannot be calculated in the simple manner in which the force in the steel stirrup is calculated (the force that will develop in the FRP bar will depend on the displacement of the bar, which in turn will depend on how much the shear crack opens at the location where the FRP bar crosses it; this is not a trivial calculation); and (5) arching action should

[1] This is known as the *truss model for shear.* A 45° inclination of the critical shear crack is traditionally assumed.

not be affected by the FRP bar provided that the compression zone in the top of the beam stays intact (larger shear cracks and vertical displacements should not affect arching action but may well affect user perception, but arching action is applicable only to a section with special aspect ratios and cannot be relied on in long beams ($a/d > 6$).

In addition to the above, the development of flexural cracking in a beam will depend on the bond coefficient of the main bars to the concrete, which as noted in Chapter 5, is larger than in steel-reinforced concrete (at least at service loads). Few flexural cracks usually lead to fewer shear cracks. Fewer cracks mean that each crack is wider and that the critical shear crack in an FRP-reinforced beam will be wider than in a steel-reinforced beam, leading to less shear resistance in the section (for many of the reasons above).

The comparisons above serve to support the opinion that the shear resistance of an FRP-reinforced section designed for the same nominal moment-carrying capacity as a steel-reinforced section will have a much reduced concrete shear capacity than that of an equivalent steel-reinforced beam. These observations have been supported by tests on FRP-reinforced beams and slabs that have failed in shear (Michaluk et al., 1998; Yost et al., 2001; Tureyen and Frosch, 2002; El-Sayed et al., 2005). The research identifies the lower longitudinal modulus of glass FRP bars as being the primary factor leading to the lower shear capacity of concrete reinforced with FRP bars.

Figure 6.1 shows an FRP-reinforced beam that failed initially in shear due to debonding of the FRP shear stirrups. The two U-shaped FRP stirrups used in the FRP-reinforced beams are shown in Fig. 6.2 alongside a steel stirrup used for the steel beams. The FRP stirrups were anchored in the tension zone, which is not permitted. However, these were the only size of stirrups that the manufacturer could provide at the time. Notice the large diagonal shear crack on the left side of the beam where failure occurred. Compare this with Fig. 5.4, where the beam failed due to compression failure at a higher load than the beam shown in Fig. 6.1.

Figure 6.1 Shear failure of an FRP-reinforced beam. (Courtesy of Dushyant Arora.)

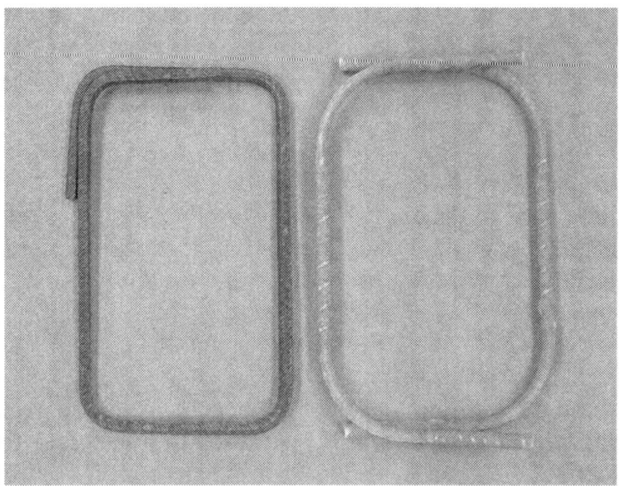

Figure 6.2 Steel and FRP stirrups in test beams.

6.3 SHEAR DESIGN OF AN FRP-REINFORCED CONCRETE SECTION

The LRFD design for shear strength for a reinforced concrete member subjected to flexural loading requires that the factored nominal shear resistance, ϕV_n, of the member be greater than the factored (ultimate) shear demand, V_u:

$$\phi V_n > V_u \tag{6.1}$$

The nominal shear capacity, V_n, of an FRP-reinforced beam according to ACI 440.1R-06 is given as

$$V_n = V_c + V_f \tag{6.2}$$

where V_c is the concrete contribution to the shear capacity as defined in ACI 440.1R-06 (to be defined in what follows), and V_f the contribution to the shear capacity from the FRP stirrups (hence the subscript f). It is important to note that the definition of V_c in ACI 440.1R-06 is different from the traditional definition of V_c in ACI 318-05.

6.3.1 Concrete Contribution to Shear Capacity

The equation to determine the concrete contribution to the shear capacity when FRP bars are used as main bars has evolved with the three editions of

ACI440.1R. In the first two editions, the concrete contribution to the shear capacity of the FRP-reinforced section was given as

$$V_{c,f} = \frac{A_f E_f}{A_s E_s} V_c \tag{6.3}$$

where $V_{c,f}$ was used to denote the shear capacity in the concrete beam reinforced with FRP bars and V_c was defined in the traditional manner as the "conventional" shear capacity of the concrete, usually taken as

$$V_c = 2\sqrt{f'_c}\, b_w d \tag{6.4}$$

The ratio of the axial stiffness of the FRP bars, $A_f E_f$, to the axial stiffness of an equivalent area of steel bars, $A_s E_s$, was used to "decrease" the traditional concrete contribution.

In ACI 440.1R-06, the symbol $V_{c,f}$ is no longer used as an equation that accounts directly for the axial stiffness of the main reinforcing bars is used to find the shear contribution of the concrete. This new equation can be used for both FRP bars and steel bars and hence the subscript f is dropped. The equation given in ACI 440.1R-06 is

$$V_c = 5\sqrt{f'_c}\, b_w c \qquad \text{(ACI 440.1R-06:9-1)}$$

or

$$V_c = \frac{5k}{2}\left(2\sqrt{f'_c}\, b_w d\right) \qquad \text{(ACI 440.1R-06:9-1a)}$$

where b_w is the width of the beam web and c is the depth of the neutral axis in the cracked elastic section as defined for the serviceability limit state flexural behavior in Chapter 5 and given there as

$$c = kd$$

$$k = \sqrt{(\rho_f \eta_f)^2 + 2\rho_f \eta_f} - \rho_f \eta_f$$

$$\eta_f = \frac{E_f}{E_c}$$

As can be seen in the second form of the equation for V_c, it is the traditional equation multiplied by 2.5k. Since the FRP bars have less axial stiffness than that of steel bars, the value of k will be less than that of steel bars and the contribution to the shear capacity will be less than in the case of steel bars.

The equation is empirical and was developed by fitting to test data (Tureyen and Frosch, 2003).

6.3.2 Shear Capacity of FRP Stirrups

The shear capacity of the FRP stirrups, V_f, is given by equations similar to those given for steel stirrups. The equation, which assumes a 45° inclination of the critical shear crack to the longitudinal axis of the beam, is for vertical stirrups,

$$V_f = \frac{A_{fv} f_{fv} d}{s} \qquad \text{(ACI 440.1R-06:9-2)}$$

for inclined stirrups,

$$V_f = \frac{A_{fv} f_{fv} d}{s} (\sin \alpha + \cos \alpha) \qquad \text{(ACI 440.1R-06:9-5)}$$

and for continuous spirals,

$$V_f = \frac{A_{fv} f_{fv} d}{s} \sin \alpha \qquad \text{(ACI 440.1R-06:9-6)}$$

where A_{fv} is the total area of the stirrups across the shear crack, f_{fv} the stress in the FRP stirrups, s the stirrup spacing measured along the beam axis (also called the *pitch*), and α the angle between the beam axis and the inclined stirrup. The use of bent-up main bars as shear reinforcement is not recommended, due to the strength reduction at the bend in the bar. In addition, bent-up bars are not generally available, and their use in reinforced concrete has not been tested. Inclined stirrups, if used, should be closed or U-shaped stirrups, similar to vertical stirrups.

As indicated previously, FRP stirrups will not yield like steel stirrups. Neither do they rupture when the beam fails in shear, and therefore the design strength of an FRP stirrup cannot be used in place of the yield stress in the case of the steel stirrup. The stress in the FRP stirrup is limited by its tensile strain. The tensile strain in the stirrup is limited to prevent the diagonal shear crack from opening too much, thereby reducing the effectiveness of the concrete contribution to the shear strength. In addition, a large shear crack can lead to overall instability of the section, due to rigid-body rotations (as can be seen in Fig. 6.1). The strength of the FRP stirrup, f_{fv}, is limited by the smaller of

$$f_{fv} = 0.004E_f \leq f_{fb} \qquad \text{(ACI 440.1R-06:9-3)}$$

where f_{fb} is the tensile strength of the FRP rebar with a bend (given in Chapter 4),

$$f_{fb} = \left(0.05\,\frac{r_b}{d_b} + 0.3\right) f_{fu} \qquad \text{(ACI 440.1R-06:7-3)}$$

where r_b is the inside radius of the bend and d_b is the diameter of the FRP rebar. A minimum inside radius-to-diameter ratio of 3 is required for FRP bars according to ACI 440.1R-06.

Figure 6.3 is a close-up of bends in Nos. 7 and 3 FRP bars from different manufacturers. Due to the manufacturing method used to produce FRP bars with bends, the corner of the bar at the bend is "flattened" and the thickness of the bar is smaller in the bend region. Internal fibers on the exterior of the bend are longer than those on the interior of the bend. As the bar is stretched, the corner of the bar experiences through-the-thickness radial stresses, which cause premature failure of the linear–elastic material in the FRP bar. This is quite unlike a steel stirrup with a bend, where the permanent set due to plastic deformation in the corner region causes the bend. For elastic–plastic materials such as steel, the yield stress in the corner region is not reduced and no reduction is taken for bends in steel bars.

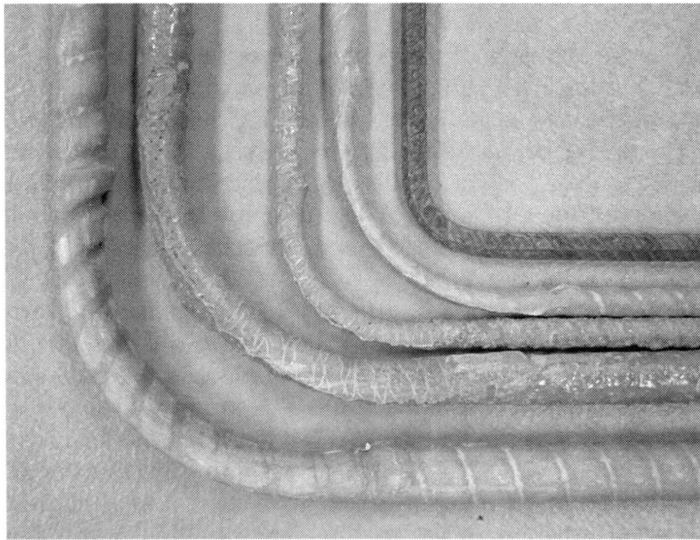

Figure 6.3 Close-up of bends in Nos. 7 and 3 FRP bars and No. 3 steel bar shown for comparison.

6.3.3 Punching Shear Capacity in Slabs

The punching shear capacity of concrete slabs reinforced with FRP bars in two orthogonal directions can be calculated from

$$V_c = 10\sqrt{f'_c}\, b_0 c \qquad \text{(ACI 440.1R-06:9-8)}$$

where b_0 is the perimeter of critical section evaluated at $d/2$ away from the point load (or support). To calculate the reinforcement ratio, the average area and average depth of the reinforcement in both directions in the tensile reinforcement layer only (if there are two layers) are used (Ospina, 2005).

6.4 LIMITS ON SHEAR REINFORCEMENT AND SHEAR STRENGTHS FOR SHEAR DESIGN

Similar to steel-reinforced beams, the maximum spacing permitted for FRP stirrups is given as

$$s_{\max} = \frac{d}{2} \quad \text{or} \quad 24 \text{ in.} \qquad (6.5)$$

This condition often determines the minimum amount of shear reinforcement that needs to be provided in FRP-reinforced beams that are appropriately designed for flexural capacity. This condition ensures that the assumed shear crack at 45° will be intercepted by at least one stirrup. The tail length of 90° hooks in the stirrups should be at least $12d_b$. Hooks must be anchored in the compression zone of the member. This is particularly important to recall for continuous members.

In addition, a minimum amount of shear reinforcement is required to prevent brittle shear failure of the beam when $V_u > \phi V_c/2$. The amount (which is given by the same formula as that for steel-reinforced beams) is given as

$$A_{fv,\min} = \frac{50 b_w s}{f_{fv}} \qquad \text{(ACI 440.1R-06:9-7)}$$

To prevent shear crushing failure, the shear capacity of the concrete is limited to

$$V_c \leq 8\sqrt{f'_c}\, b_w d \qquad (6.6)$$

6.5 DESIGN PROCEDURE FOR FRP SHEAR REINFORCEMENT

Step 1. Determine the shear demand. The shear demand is calculated in the same way as in ACI 318-05 and is calculated using the same loading as that used for flexural design of the member.

Step 2. Determine the concrete shear capacity with FRP rebars. The concrete shear capacity is calculated using the ACI 440.1R-06 formula. This calculation has the most significant effect on the design of FRP members for transverse loads (both shear and flexure) and is the most significant difference between the steel and FRP design procedures for concrete members for shear.

Step 3. Check if additional shear reinforcement is required. According to ACI 318-05, the following are checked: If $V_u \geq 0.5\phi V_c$, minimum shear reinforcement is required to be provided; if $V_u \geq \phi V_c$, additional shear reinforcement is required to be designed.

Step 4. Design the shear reinforcement. FRP shear reinforcement is designed using strain limits to determine the maximum permissible stress in the FRP stirrups. The designer should confirm that FRP stirrups of the sizes anticipated in the design can actually be obtained from a manufacturer. Consider varying the stirrup spacing along the beam length to economize the stirrup costs, if possible.

Step 5. Determine the factored shear strength of the beam and compare it with the demand. A resistance factor of $\phi = 0.75$ is used per ACI 318-05 to determine the factored shear strength of the member. If the demand is not met, redesign the beam or slab for shear capacity. For slabs in which no additional shear reinforcement is provided, this may require a complete flexural redesign of the slab, which could lead to a thicker slab or more flexural reinforcement and even a change in the failure mode designed. This is significantly different from the design of slabs with steel bars, where shear does not usually control the flexural design (or thickness) of commonly designed slabs.

Step 6. Check the limits on reinforcement and concrete strength. Check the minimum amount of (or maximum spacing for) the stirrups. If maximum spacing controls, the stirrups can be redesigned using a smaller-diameter bar for the stirrups if available.

Step 7. Detail the stirrups. Use closed loops with 90° hooks with a minimum tail length of $12d_b$. Place the hooks in the compression zone of the beam (i.e., the top for a simply supported beam). (For further discussion on detailing of FRP rebars, see Chapter 7.) Draw a sketch of the final design to scale (including the bend radii) to make sure that the beam size is sufficient to fit the FRP stirrups and that the main bar spacing is as needed for bond development length calculations (see Chapter 7).

Design Example 6.1: Beam Design for Shear Strength Design shear reinforcement for the overreinforced beam discussed in Design Example 5.1. The geometric and material properties are: simply supported beam, 11-ft span; 4000-psi concrete; FRP bar design strength, f_{fu} = 72 ksi; two No. 6 main bars; A_f = 0.88 in^2; d = 9.75 in. (No. 3 FRP strirrups were assumed in the flexural design); w_{LL} = 400 lb/ft (20% sustained); w_{SDL} = 208 lb/ft (superimposed dead load); w_{DL} = (150 lb/ft^3)(7/12)(12/12) = 87.5 lb/ft (self-weight); W_u = 1.2(87.5 + 208) + 1.6(400) = 1094 lb/ft.

SOLUTION

Step 1. Determine the shear demand. For a simply supported beam, calculate V_u at a distance d from the support:

$$V_u = W_u\left(\frac{l}{2} - d\right) = 994.6\left(\frac{11}{2} - \frac{9.75}{12}\right) = 4662 \text{ lb} = 4.66 \text{ kips}$$

Step 2. Determine the concrete shear capacity with FRP rebars. The properties of the elastic section were determined previously as follows:

$$\begin{aligned} k &= \sqrt{(\rho_f\eta_f)^2 + 2\rho_f\eta_f} - \rho_f\eta_f \\ &= \sqrt{[0.0129(1.8)]^2 + 2(0.0129)(1.8)} - 0.0129(1.8) = 0.194 \end{aligned}$$

Therefore,

$$c = kd = 0.194(9.75) = 1.89 \text{ in.}$$

and

$$V_c = 5\sqrt{f'_c}b_w c = 5\sqrt{4000}(7)(1.89) = 4184 \text{ lb} = 4.18 \text{ kips}$$

Notice that the concrete contribution to shear resistance of concrete beam with FRP bars is significantly less than for conventional steel bars. For steel bars the contribution would have been calculated using the ACI 318-05 equation as

$$V_c = 2\sqrt{f'_c}\, b_w d = 2\sqrt{4000}(7)(9.75) = 8633 \text{ lb} = 8.6 \text{ kips}$$

Step 3. Check if additional shear reinforcement is required. If $V_u \geq 0.5\phi V_c$,

$$4.66 \text{ kips} \geq 0.5(0.75)(4.18) = 1.57 \text{ kips}$$

$\therefore$minimum shear reinforcement required

If $V_u \geq \phi V_c$,

4.66 kips $\geq$ 0.75(4.18) = 3.140 kips

$\therefore$additional shear reinforcement required

Step 4. Design the shear reinforcement. Assume No. 3 FRP stirrups. A_{fv} = 2(0.11) = 0.22 in^2. The design strength of the FRP stirrups is the lesser of

$$0.004E_f = 0.004(6500) = 26 \text{ ksi}$$

or

$$f_{fb} = \left(0.05\,\frac{r_b}{d_b} + 0.3\right)f_{fu} = [0.05(3.0) + 0.3](72) = 32.4 \text{ ksi}$$

Therefore, f_{fv} = 26.0 ksi. Note that in the calculation of the strength of a bar with the bend, r_b/d_b = 3 has been assumed and the strength of the No. 3 bar is not increased (i.e., taken as 72 ksi). If the bar strength were increased to f_{fu} = 88 ksi (for f^*_{fu} = 110 ksi), the strength of the bar with the bend would increase. It most cases this will not control since the bar strength as required for the strain limit on the FRP bar will control. Similarly, if a larger r_b/d_b were used, as is most likely to be the actual case for FRP bars with bends, the strength of the bar with the bend will increase.

The stirrup spacing required is given by the lesser of

$$s = \frac{\phi A_{fv} f_{fv} d}{V_u - \phi V_c} = \frac{0.75(0.22)(26)(9.75)}{4.66 - 0.75(4.18)} = 27.4 \text{ in.}$$

or

$$s = \frac{d}{2} = \frac{9.75}{2} = 4.9 \text{ in.}$$

Therefore, use s = 4.5 in.

Step 5. Determine the factored shear strength of the beam and compare it with the demand. The shear capacity provided by the FRP stirrups is

$$V_f = \frac{A_{fv} f_{fv} d}{s} = \frac{0.22(26)(9.75)}{4.5} = 12.39 \text{ kips}$$

The shear capacity of the concrete is

$$V_c = 4.18 \text{ kips}$$

and

$$V_n = V_c + V_f = 4.18 + 12.39 = 16.57 \text{ kips}$$

$$\phi V_n = 0.75(16.57) = 12.43 \text{ kips} > 4.66 \text{ kips } (V_u)$$

The beam is well overdesigned for shear since the maximum spacing controls.

Step 6. Check the minimum limits on reinforcement and concrete strength.

$$A_{fv}^{\min} = \frac{50bs}{f_{fv}} = \frac{50(7)(4.5)}{26{,}000} = 0.06 \text{ in}^2 < 0.22 \text{ in}^2 \rightarrow \text{OK}$$

$$V_c \leq 8\sqrt{f'_c}\, bd = 8\sqrt{4000}(7)(9.75) = 34{,}532 \text{ lb}$$

$$= 34.5 \text{ kips } (>V_c = 4.18) \rightarrow \text{OK}$$

Step 7. Detail the stirrups. Use closed loops with 90° hooks with a minimum tail length of $12d_b = 12(0.375) = 4.5$ in. Place the hooks in the compression zone of the beam (i.e., the top for the simply supported beam given).

Design Example 6.2: Punching Shear Design for a Bridge Deck Slab The bridge deck in Design Example 5.2 is revisited to determine if the punching shear capacity of the slab with three No. 5 GFRP main bars and two No. 5 GFRP distribution bars is sufficient.

SOLUTION

Step 1. Determine the punching shear demand. The punching shear load is the factored P_{20} wheel load plus the impact multiplied by the appropriate AASHTO live load factors and calculated as

$$V_u = 2.17(16.0)(1.3) = 45.14 \text{ kips}$$

Step 2. Determine the concrete shear capacity with FRP rebars. For punching shear under the wheel load, only the bottom reinforcement mat is regarded as active. The average depth and the average reinforcement ratios of the transverse and longitudinal bars need to be calculated. For the main reinforcement (No. 5 at 4 in.) in the transverse direction (superscript T in what follows), we have previously calculated, $d^T = 6.94$ in. and $\rho_f^T = 0.0112$. For the distribution reinforcement (No. 5 at 6 in.) in the longitudinal direction (superscript L in what follows), we obtain the following: $d^L = 6.31$ in. and $\rho_f^L = 0.0082$. The average values are $d = 6.63$ in. and $\rho_f = 0.0097$.

The distance to the neutral axis of the cracked section of the slab is found using the average effective depth and the average reinforcement ratio as follows: The modular ratio is

$$\eta_f = \frac{E_f}{E_c} = \frac{5{,}920{,}000}{57{,}000\sqrt{4000}} = 1.64$$

The neutral-axis depth ratio, $k = c/d$,

$$k = \sqrt{(\rho_f\eta_f)^2 + 2\rho_f\eta_f} - \rho_f\eta_f$$
$$= \sqrt{[0.0097(1.64)]^2 + 2(0.0097)(1.64)} - 0.0097(1.64) = 0.163$$

$$c = kd = 0.163(6.63) = 1.08 \text{ in.}$$

The punching shear perimeter is measured at a distance of $d/2$ from the tire footprint. The tire area is

$$0.01P_{20} = 0.01(16{,}000) = 160 \text{ in}^2$$

The width of the load patch is therefore 20 in. and the length is 8 in. The punching shear perimeter is found as

$$b_0 = 2[(20 + 6.63) + (8 + 6.63)] = 82.5 \text{ in.}$$

and the nominal punching shear capacity of the slab is

$$V_n = V_c = 10\sqrt{f_c'}b_0c = 10\sqrt{4000}(82.5)(1.08) = 56{,}370 \text{ lb} = 56.4 \text{ kips}$$

The factored punching shear capacity is

$$\phi V_n = 0.75(56.4) = 42.3 \text{ kips} < 45.1 \text{ kips } (V_u)$$

The factored nominal capacity of the slabs is *less* than the punching shear demand and the slab does not satisfy the loading requirements. The slab bottom reinforcement should be increased slightly for punching shear capacity. The top reinforcement does not need to be redesigned. Increasing the bottom bars to three No. 5 per foot of width (No. 5 at 4 in.), bars in the longitudinal direction will meet the requirements. Note, however, that the AASHTO LFD factors were originally developed to be compatible with the "old" ACI 318-99 resistance factor for shear of $\phi = 0.85$. In that case, the design above is satisfactory.

The slab should also be checked for transverse shear capacity at the edge of the girders. To do this calculation, the shear force per unit width of the slab under the HS 20 wheel load is required. The AASHTO approximate method for determining the moment per foot width of the slab does not provide a value for the shear force per foot width (or does not provide the effective width for the wheel load). The AASHTO LRFD bridge design specification provides an equation to calculate the distribution width for deck slabs which can be used to determine the shear force per unit width for this calculation.

PROBLEMS

Note to the reader: The analysis and design homework problems provided here address the capacity of reinforced concrete beams and slabs subjected to transverse and punching shear loads. Additional design problems for beams and slabs that include flexure (capacity and serviceability), shear, and detailing in an integrated fashion are given in Chapter 7.

6.1 For the beams listed in Table P5.4,[2] determine, V_c, the concrete contribution to the shear capacity, and V_f, the FRP stirrup contribution to the shear capacity, if No. 3 Aslan 100 GFRP vertical stirrups are provided at the maximum spacing of $d/2$. Determine the nominal shear capacity, V_n, and the factored shear capacity, ϕV_n, of the beams. Determine the shear capacities of the beams if they are reinforced with grade 60 No. 6 steel main bars (determine an appropriate d for interior construction with steel bars) and grade 60 No. 3 steel stirrups at $d/2$. Compare the shear capacities of the FRP- and steel-reinforced beams.

6.2 For the values of the factored shear capacities calculated in Problem 6.1, determine the load-carrying capacity (w_{DL}, w_{LL}, and $w_{DL} + w_{LL}$) per unit length (lb/ft) for the FRP-reinforced beams. Assume that the beams are simply supported uniformly loaded beams with an 18-ft span and that the ratio of dead load to live load is 1:3. Calculate the shear demand at a distance d from the face of the support (assumed to be at the nominal span length). Compare the load-carrying capacities of the eight beams in a bar graph. Also, compare the load-carrying capacities to the load-carrying capacities for the beams based on their flexural capacities (Problem 5.5). Discuss your observations.

6.3 Consider the FRP-reinforced one-way slabs listed in Table P5.6. Determine the nominal shear capacity, V_n, and the factored shear capacity, ϕV_n, of each slab. Check if this shear capacity is sufficient if the slabs

[2] Instructors may assign only selected beams from the table.

are used as simply supported one-way floor slabs spanning 12 ft between beams and loaded with a superimposed dead load of 30 lb/ft^2 (which acts in addition to the self-weight of the slab) and a live load of 100 lb/ft^2.

6.4 Consider the beams in Design Example 5.1, options a to d. Design the shear reinforcement for the beams assuming No. 3 GRFP stirrups with the nominal properties used in Design Example 6.1. Determine the actual shear capacity of the beams if minimum shear reinforcement requirements control. Calculate the cost of the FRP shear reinforcement for each beam. Use \$0.40 per foot for No. 3 GFRP bar in addition to \$1.00 per bend (i.e., five bends per closed stirrup with 90° hooks).

6.5 Consider the beams in Table P5.4. The beams are used for interior construction and are loaded with a uniformly distributed superimposed dead load of 100 lb/ft and a uniformly distributed live load of 900 lb/ft over a simply supported span of 18 ft. Design the shear reinforcement to carry the loads indicated. Assume that the flexural capacity of these beams is sufficient for both strength and serviceability. Calculate the cost of the FRP shear reinforcement for the beams. Use \$0.40 per foot for No. 3 GFRP bar and add the cost of bends at \$1.00 per bend.

6.6 Consider the beams in Table P5.4. The beams are used for exterior construction and are loaded with a uniformly distributed superimposed dead load of 400 lb/ft and a uniformly distributed live load of 1500 lb/ft over a simply supported span of 18 ft. Design the shear reinforcement to carry the loads indicated. Assume that the flexural capacity of these beams is sufficient for both strength and serviceability. Consider No. 4 GFRP bars for the vertical stirrups. Use minimum shear reinforcement where appropriate in the interior of the span to minimize the cost of the stirrups. Sketch the stirrup layout along the beam length.

6.7 Consider beams 3 and 4 in Table P5.4. The beams are used for exterior construction and are loaded with a uniformly distributed superimposed dead load of 400 lb/ft and a uniformly distributed live load of 1200 lb/ft over a simply supported span of 18 ft. Design the shear reinforcement to carry the loads indicated. Assume that the flexural capacity of these beams is sufficient for both strength and serviceability. Use No. 2 GFRP rectangular spiral reinforcement for the beams. Determine the required pitch of the spiral reinforcement.

6.8 Consider the slabs in Table P5.10 with reinforcement as indicated used in both directions (i.e., two-way slab action). Determine the punching shear capacity of the slab when used in a reinforced concrete building frame with typical square 20 $\times$ 20 in. reinforced concrete columns.

6.9 Consider the FRP-reinforced highway bridge deck discussed in Design Examples 5.2 and 6.2. Determine the transverse shear capacity (as opposed to the punching shear capacity) of the deck parallel to the edges of the longitudinal prestressed girders. Assume an effective width strip $E = 8S/(S + 2)$ for a single concentrated wheel load (S is the effective span perpendicular to the girders, and one-way action of the slab is assumed).

6.10 Consider a beam with a double layer of main FRP tensile reinforcement such as that discussed in Problems 5.1, 5.2, and 5.17. Discuss how the use of a double layer of main bars will affect the shear capacity of the FRP-reinforced beam.

7 FRP Reinforcement Detailing

7.1 OVERVIEW

In this chapter we discuss the subject of the detailing of FRP rebars for use in concrete structures. This includes a discussion of the geometric issues that need to be considered when using FRP bars and the quantitative issues that need to be addressed to ensure that FRP bars are sufficiently anchored in the concrete to transfer the required forces from the bars into the concrete. The chapter follows the procedures of ACI 440.1R-06 and the code equations for calculating the development length for FRP bars at different locations along the length of the reinforced concrete member. Since FRP bars are only used in tension, no information is provided on development and anchorage for compression reinforcement.

The design procedures presented in this chapter follow ACI 440.1R-06, which is compatible with ACI 318-05. It is assumed that readers are familiar with the subject of bond and development in steel reinforcing bars and with the procedures provided in ACI 318-05 for the calculation of development length for steel reinforcing bars.

7.2 INTRODUCTION

Detailing of reinforcing bars refers to the determination of the dimensions (but not sizes), shapes, and positions of the bars in cross section. The strength and serviceability calculations described previously are used to determine the cross-sectional areas (and bar sizes) of the tension (main) and shear reinforcing bars in a beam or slab; however, these calculations do not give the designer the lengths of the bars, the spacing of the bars, or the location of the bars in the section. (However, recommended geometric spacing limits are used to decide on the number of bars possible in a section.) Neither do these equations check whether the bars are sufficiently anchored or embedded in the concrete at all sections along the length to ensure that the strengths calculated can be achieved. This falls to *detailing,* which should more rightly be called *rebar length, shape, and position design.* It is important to note that this is a very important part of the design of a reinforced concrete structure and is not simply a construction-related item as the term *detailing* could imply.

In addition to the geometric calculations performed to detail the locations of the FRP bars in the section and along the length of the member, the primary calculations that are performed are those related to what is traditionally known as *development and splices of reinforcement,* covered in Section 12 of ACI 318-05. From the early days of research into the use of FRP bars as reinforcements for concrete, it was recognized that anchorage of the reinforcement in the concrete and the transfer of stresses between the FRP bar and the concrete were likely to be defining characteristics of FRP bars and their design. Since FRP bars require a high percentage of longitudinal fiber to develop their tensile strength and tensile modulus, the key question was how to create a suitable surface in the resin layer on the outside of the bar that would be able to transfer the forces from the concrete and mobilize the longitudinal fibers in the interior of the bar. First-generation FRP bars used a surface layer that contained relatively coarse sand particles and helically wrapped fibers. These were followed by bars that had more controlled helical wraps and a much finer silica sand coating or with bars having a sand coating only. Some manufacturers produced bars with helical wraps and no coatings. Bars were also produced that had a molded surface with protruding deformations that were reinforced with fine silica particles. Many of these products are no longer available on the market. Figure 7.1 shows images of FRP rebars tested by Katz (2000) that are representative of the bars produced in the 1990s.

Research was conducted on the durability of the surface deformation layer to determine if the failure of the bond layer due to environmental degradation, high temperature, or cyclic loading would lead to failure of the bar itself or lead to the bar not being able to mobilize its core of unidirectional fiber reinforcements (e.g., Nanni et al., 1995; Bank et al., 1998; Katz et al., 1999; Katz, 2000). The other major topic of research related to the bonding of FRP bars focused on characterizing the effect of bar mechanical properties on the

Figure 7.1 Surface textures of various FRP bars of the 1990s. (Courtesy of Amnon Katz.)

local constitutive behavior of FRP bars when pulled from concrete (Chaallal and Benmokrane, 1993; Cosenza et al., 1997; Focacci et al., 2000). The objective of this work was to try to determine analytically the embedment length that was required to develop the bar tensile stress.

Since the 1990s, the research and development for new types of glass FRP bars has slowed, and two types of bars are available in North America: bars with helically wound surface fibers that create a small undulating bar surface together with a light silica sand coating (Fig. 7.2) and bars with a sand-coated layer only (Fig. 7.3). Both of these bars have been shown to develop a satisfactory bond to concrete.

7.3 GEOMETRIC DETAILS

Although nominally the same as steel bars in terms of bar number sizes, FRP rebars have different geometric features that should be accounted for by the designer. This applies particularly to the available bend radii for bent bars used as strirrups and as hooks. Bent bars are produced in a separate manufacturing process and are not made from straight bars as are steel bent bars. They are therefore really "bars with bends" rather than "bent bars." Samples of FRP bars with bends are shown in Fig. 7.4.

As noted previously, FRP bars are made from thermosetting resins that cannot be bent after they have cured, nor can they be heated and bent. A thermosetting resin will not soften when heated and then re-formed into a different shape upon cooldown [as can be done with a thermoplastic resin (see Chapter 2 for more on materials)]. Therefore, all bent FRP bars must be specially manufactured for the job at the manufacturer's plant. A selection of bent bars ready for shipment from a manufacturer's plant is shown in Fig. 7.5.

Close coordination with the bar manufacturer is required during the design phase of the job. This is unlike the design of a steel-reinforced member, where it can be assumed that any size bar of bent bar can be obtained. In addition, at this time, due to the way in which bent bars are manufactured, it is not possible to obtain bars with bends of other than 90°. Therefore, if FRP bars

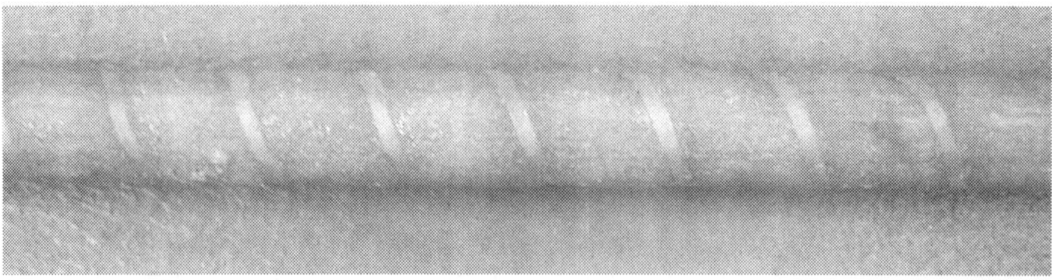

Figure 7.2 Close-up of glass FRP bar with helical wrap and fine sand coating.

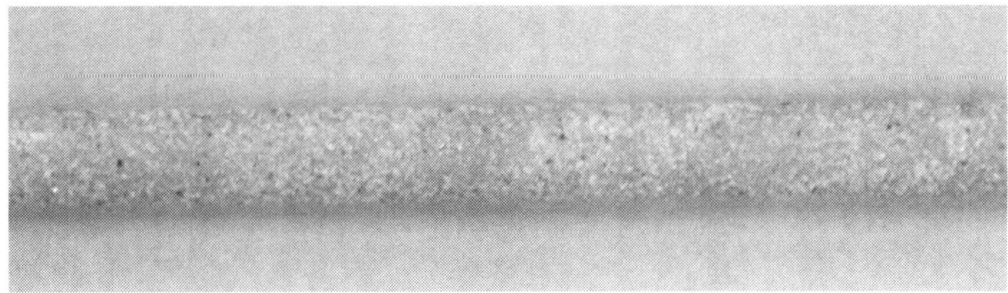

Figure 7.3 Close-up of glass FRP bar with sand coating only.

with other bends are required, the FRP bar manufacturer should be consulted well in advance of the design of the concrete member. It is also not possible to obtain closed-loop stirrups of all sizes. All stirrups must be preordered and manufactured to specific dimensions.

Typical inside bend radii reported by an FRP bar manufacturer and the ratio r_b/d_b needed to calculate the strength of the bar with the bend are given in Table 7.1. The smaller the ratio, the larger the reduction in bar strength; the larger the ratio, the more difficult it is to position bars at the horizontal tangential corner of the stirrup as required for a single layer of bottom bars in a beam, for example. Since Nos. 3 and 4 bars are typically used as stirrups in FRP-reinforced beams, the large inside radii of these bars mean that the first "corner" bar is positioned farther from the edge of the beam than in a steel bar-reinforced beam, and standard tables for the number of bars permitted in a given width should be used with care.

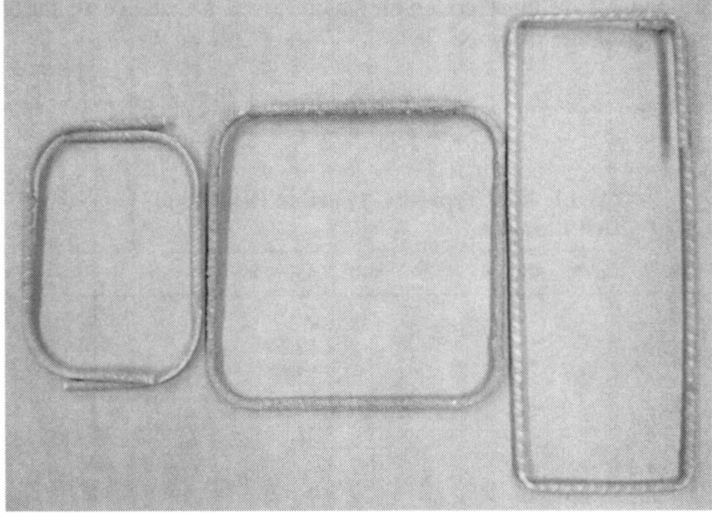

Figure 7.4 Samples of FRP bars with bends.

Figure 7.5 FRP rebars with bends at manufacturer's plant ready for shipping. (Courtesy of Hughes Brothers.)

7.3.1 Calculation of Bar Spacing

The side distance for an FRP corner bar is calculated in the same way as for a steel bar. However, the condition that the inside radius be $3d_b$ or greater needs to be enforced. Figure 7.6 shows the geometry of the corner bar and the parameters A, B, and C (clear cover, stirrup diameter, and inside stirrup radius) that are used to calculate this distance.

For example, a beam with a No. 3 stirrup with an inside bend radius of 2.125 in. and a side clear cover of 1.5 in. gives a distance to the center of the corner bar (edge distance) of

$$A + B + C = 1.5 + 0.375 + 2.125 = 4.0 \text{ in.}$$

TABLE 7.1 Typically Available Bend Radii for FRP Bars

Bar Number	Inside Radius (in.)	r_b/d_b
2	1.5	6.0
3	2.125	5.67
4	2.125	4.25
5	2.25	3.6
6	2.25	3.0
7	3.0	3.43
8	3.0	3.0

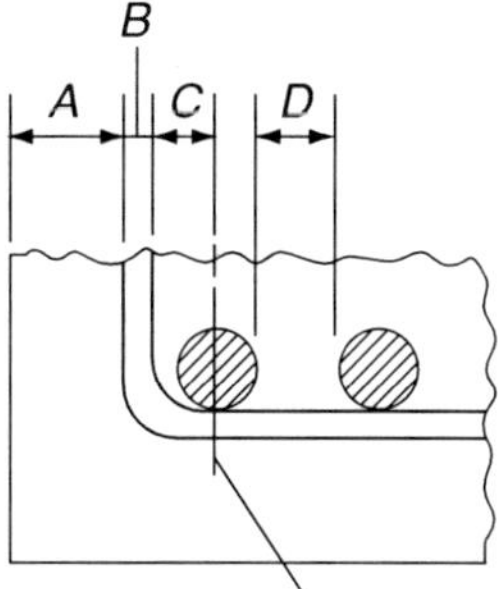

Figure 7.6 Corner bar side distance calculation parameters. (From Wang and Salmon, 2002, with permission from John Wiley & Sons.)

For a minimum clear (*D* in Fig. 7.6) spacing of d_b or 1.0 in. for the main bars, the minimum width of a beam with two No. 6 bars and No. 3 stirrups is therefore

$$\begin{aligned} \text{minimum width} &= \text{edge distance} + \tfrac{1}{2}d_b + d_b \text{ or 1.0-in. space} + \tfrac{1}{2}d_b \\ &\quad + \text{edge distance} = 4 + 0.375 + 1.0 + 0.375 + 4 \\ &= 9.75 \text{ in.} \end{aligned}$$

For a steel bar this minimum is 7.0 in., since for a No. 3 steel stirrup the inside radius is only twice the stirrup diameter (i.e., 0.75 in.) and $A + B + C = 2.625$ in.

Notice that the 7-in. beam width used in the illustrative examples presented in Chapters 5 and 6 is possible only if it is assumed that a No. 3 stirrup with an inside radius of $3d_b$ is available and that the outside clear cover is reduced to 1.125 in., and a minimum clear spacing between the main bars is used. In this case,

$$A + B + C = 1.125 + 0.375 + 3(0.375) = 2.625$$

and the beam width is

$$\begin{aligned} &\text{edge distance} + \tfrac{1}{2}d_b + d_b \text{ or 1.0-in. space} + \tfrac{1}{2}d_b + \text{edge distance} \\ &\quad = 2.625 + 0.375 + 1.0 + 0.375 + 2.625 = 7.0 \text{ in.} \end{aligned}$$

7.4 BOND STRENGTH OF FRP BARS

The transfer of forces between the FRP reinforcing bar and the concrete depends on the development of sufficient bond strength between the FRP bar and the surrounding concrete. Fundamentally, the bond stress can be determined using the equilibrium equation

$$l_e \pi d_b u = A_{f,\mathrm{bar}} f_f \qquad \text{(ACI 440.1R-06:11-1)}$$

which states that the average bond stress (or longitudinal shear stress), u, on the outer surface of the bar over an embedment length, l_e, must be equal to the tensile stress in the bar, f_f, that acts on the cross-sectional area of that bar, $A_{f,\mathrm{bar}}$. However, the equilibrium equation above is not used in the simple form given since the bond stress over the length l_e varies, and more important, it may not be able to be mobilized over the full length l_e due to premature failure of the concrete surrounding the bar or due to the bar pulling out of the concrete. Therefore, an alternative form is provided in ACI 318-05 for steel bars and in ACI 440.1R-06 for FRP bars.

One of the fundamental assumptions of reinforced concrete behavior is that strain compatibility exists between the reinforcing bar and the concrete. This means that it is assumed that "perfect" bond exists and that no slip between the reinforcing bar and the concrete occurs when the tensile forces are transferred from the bar to the concrete by shear stresses (or bond stresses) on the surface of the bar. For this to occur, there must be a sufficient amount of the bar embedded in the concrete, l_e, so that all the tensile stress in the bar can be transferred into the concrete. If insufficient length is available, the amount of tensile force required is not transferred into the concrete and the element can fail, since the internal equilibrium of tensile and compressive forces required for flexure cannot be maintained. However, it is not always possible to transfer all of the tensile force in the bar into the concrete for a given embedment length, since it may not be possible to develop sufficient bond strength, due either to splitting of the concrete surrounding the bar or because the bar pulls out of the concrete (as noted previously). Based on an analysis of beam test data by Wambeke and Shield (2006) in which both splitting and pull-out failures occurred, ACI 440.1R-06 recommends the maximum (effective) stress achievable in an FRP bar based on bond failure be taken as

$$f_{fe} = \frac{\sqrt{f'_c}}{\alpha}\left(13.6\frac{l_e}{d_b} + \frac{C}{d_b}\frac{l_e}{d_b} + 340\right) \leq f_{fu} \qquad \text{(ACI 440.1R-06:11-3)}$$

where C is the lesser of (1) the distance from the center of the bar to the nearest outer concrete surface in the tension zone or (2) half the on-center spacing of the bars (side by side), and α is the bar location factor. It is taken as 1.0 for bars that are in the bottom 12 in. of the formwork when the concrete is cast and as 1.5 when the bars are more than 12 in. above the bottom of the formwork when the beam is cast (known as *top bars*). ACI 440.1R-06 further recommends that the term C/d_b not be taken larger than 3.5 and that the minimum embedment length, l_e, be at least 20 bar diameters, or $20d_b$.

It is seen from the equation above that the effective design strength of the FRP rebar can be less than the maximum design strength of the bar based on tensile failure. This is unique to FRP bars and can complicate the design of flexural members since the bar strength is needed in the design process, and usually the details of the reinforcement are not precisely known at the time of the initial flexural design. This does not happen in the design of steel-reinforced concrete, in which the bar will always be able to achieve its yield stress and will not fail prematurely due to bond failure provided that it is appropriately detailed (i.e., it has sufficient embedment length or anchorage). Therefore, in steel design, detailing is in fact "only" detailing, and it occurs after the flexural design is completed. This is due primarily to the fact that FRP bars have a very high strength relative to their modulus compared with steel bars.

In the flexural design, sections that are controlled by the bar not being able to achieve its design strength due to bond failure are referred to in 440.1R-06 as *bond-critical sections*. In this case the nominal moment capacity should be reevaluated using the lower bar-effective design strength. If the bond strength controls, the design of the beam and its capacity are obtained using the equations for an underreinforced beam, and a resistance factor of ϕ = 0.55 is recommended (even if the section was originally designed as an overreinforced section) since bond failure by splitting is regarded as a brittle failure mode. Alternatively, the beam can be redesigned as an overreinforced beam using the lower bond-critical bar strength, and the resistance factor for an overreinforced beam can be used, $0.55 \leq \phi \leq 0.65$.

7.5 DEVELOPMENT OF STRAIGHT FRP BARS

The development length of a reinforcing bar is the length that is required to anchor the bar in the concrete in order to develop the tensile stress in the bar that is required for internal moment equilibrium at any section. In steel-reinforced concrete, in which underreinforced beams are most common, the bar stress that needs to be developed at the maximum moment location is the yield stress of the bar.

In FRP design, however, where overreinforced beams will most commonly be designed (primarily for serviceability, as discussed in Chapter 5), the bar is not stressed to its maximum design strength at the maximum moment. Therefore, in FRP-reinforced beams the development length of the bar is determined from the actual stress, f_{fr}, that the bar will experience at the failure of the beam at the location of the maximum moment. This stress is either (1) the design strength of the bar for underreinforced beams, f_{fu}, or (2) the actual stress in the bar for overreinforced sections, f_f, or (3) the effective bond critical design stress in the bar for both over- and underreinforced sections, f_{fe}.

The development length, l_d, required is then given as

$$l_d = \frac{\alpha(f_{fr}/\sqrt{f'_c}) - 340}{13.6 + C/d_b} d_b \qquad \text{(ACI 440.1R-06:11-6)}$$

All other ACI 318-05 provisions related to the development of bars at inflection points or at bar cutoffs, for steel bars are all also required for FRP bars. FRP bars should extend a distance d or $12d_b$, whichever is greater, beyond points of inflection or where terminated when not required for moment capacity.

The provision for the development of FRP bars at simple supports is

$$l_d \leq \frac{\phi M_n}{V_u} + l_a \qquad \text{(ACI 440.1R-06:11-7)}$$

Note that the resistance factor is added to the nominal moment in the numerator of the first term. When tension bars are confined by the compressive reaction at the simple support, they have less tendency to cause splitting and the development length can be increased to

$$l_d \leq 1.3\frac{\phi M_n}{V_u} + l_a \qquad (7.1)$$

For simply supported beams, at least one-third of the main tension bars must extend over the support.

7.6 DEVELOPMENT OF HOOKED FRP BARS

For hooked bars the basic development length is given as a function of the FRP rebar ultimate strength and the concrete strength. For FRP rebars with ultimate strengths in the range 75 to 150 ksi (the typical range for glass FRP rebars), the development length of a hooked bar, l_{bhf}, is given as

$$l_{bhf} = \frac{f_{fu}}{37.5}\frac{d_b}{\sqrt{f'_c}} \qquad \text{(ACI 440.1R-06:11-5)}$$

and not less than $12d_b$ or 9 in. In addition, the tail length of the hooked bar, l_{thf}, should be greater than $12d_b$ or 9 in., and the radius of the bend should not be less than $3d_b$. Values of l_{bhf} for FRP bars with strengths less than 75 ksi or greater than 150,000 ksi are given in ACI 440.1R-06. A photograph of hooks on the ends of No. 7 FRP bars is shown in Fig. 7.7.

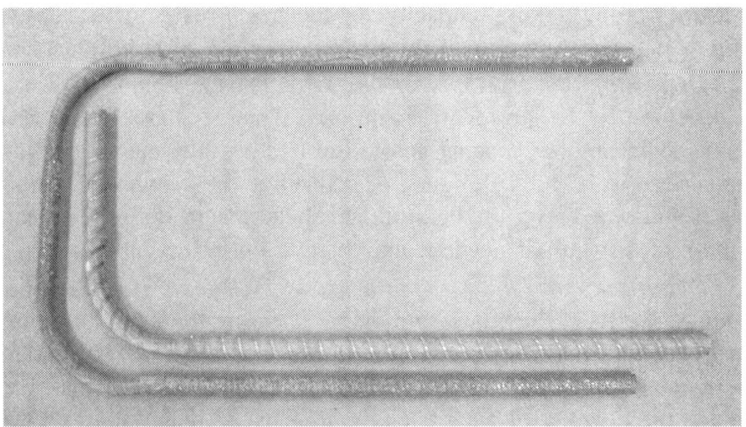

Figure 7.7 FRP short bars (J-hooks) with hooked ends.

As can be seen, the "long" leg of the bar is quite short. This is not because it has been cut to length. At the present time it is not actually possible to obtain long glass FRP bars with end hooks; they cannot be produced by the existing manufacturing technology. Only long straight bars or short pieces of bars with bends can be produced. These are commonly referred to as *J-hooks.* They are spliced to longitudinal bars at their ends, where a hooked anchorage is required, as shown in Fig. 4.1. This can significantly crowd the end of the beam region and may lead to splitting. It is therefore recommended that the structural geometry be designed to enable straight FRP bars to be used wherever possible.

7.7 LAP SPLICES FOR FRP BARS

Tension lap splices for FRP rebars are based on recommendations for steel rebars and limited test data. ACI 440.1R-06 recommends that all FRP bar splices be designed as class B lap splices and have a length of $1.3l_d$.

7.8 DESIGN PROCEDURE TO DETAIL FRP BARS IN A BEAM

Step 1. Determine the bond-critical effective strength of the FRP main bar. The bond-critical effective strength of the FRP bar, f_{fe}, is a function of the embedment length available. This will depend on where along the length of the beam the nominal design moment is calculated. For a simply supported beam with the maximum moment at midspan (such as a uniformly loaded beam), the available embedment length will be half the beam span. Check to see that the bond-critical effective strength is more than that used to determine the flexural capacity for the beam (over- or underreinforced design). If it is not, the main reinforcement must be redesigned.

Step 2. Determine the development length of the FRP main bars. Calculate the development length for the main bars. This depends on the bar bottom or side cover and bar spacing across the width of the member. The section details need to be fully designed to determine these numbers. This should have been done during the flexural and shear design stages. The bend radii of the FRP stirrups (if needed) may have a big effect on this step.

Step 3. Detail the bars and provide a list of FRP materials. Provide a longitudinal sketch of the concrete member showing the main bar lengths and stirrup details (if stirrups are required). Develop a list of FRP materials for the member designed. Identify each bar by length, size, and geometry. Provide the quantity of each bar type in a list of materials.

Design Example 7.1: Detailing of FRP Bars in a Beam Assume that the two No. 6 main bars in the 7 × 12 in. beam in Design Example 5.1 are extended to the end of the span over the supports, which are 8-in.-wide CMU walls. Assume that a cover of $2d_b$ is required at the end of the beam. Detail the main bars.

SOLUTION

Step 1. Determine the bond-critical effective strength of the FRP main bar. To determine the bond-critical design strength of the FRP bar, first find the distance, C. The spacing calculation is based on a No. 3 stirrup with an inside radius of $3d_b$, an outside clear cover of 1.125 in., and a bar clear spacing of 1.0 in. (discussed previously) In this case the distance to the closest bar is either (1) the distance to the side of the beam, $A + B + C = 1.125 + 0.375 + 3(0.375) = 2.625$, or (2) the distance to the bottom on the beam, $1.5 + 0.375 + 0.75/2 = 2.2$, or (3) half the center-to-center spacing of the bars, $(1.0 + 0.75)/2 = 0.875$ (controls).

The available length for embedment from the point of the maximum moment is half the beam length,

$$l_e = \frac{l}{2} = \frac{11(12)}{2} = 66 \text{ in.}$$

The bar location factor, $\alpha = 1.0$, since the bars are bottom bars.

$$\frac{C}{d_b} = \frac{0.875}{0.75} = 1.167 < 3.5 \quad \rightarrow \text{OK}$$

$$f_{fe} = \frac{\sqrt{f'_c}}{\alpha}\left(13.6\frac{l_e}{d_b} + \frac{C}{d_b}\frac{l_e}{d_b} + 340\right)$$

$$= \frac{\sqrt{4000}}{1.0}\left(13.6\frac{66}{0.75} + \frac{0.875}{0.75}\frac{66}{0.75} + 340\right) = 103{,}689 \text{ psi}$$

$$= 103.7 \text{ ksi} > 72 \text{ ksi } (f_{fu})$$

Therefore, the bar is not bond critical, and the flexural capacity determined previously for the beam can be achieved. If f_{fe} had been less than f_{fu}, the beam would have needed to be redesigned for flexural capacity using the lower bar design strength.

Step 2. Determine the development length of the FRP main bars. The beam is overreinforced and the stress in the bar at failure is $f_f = 57.06$ ksi. The bar stress in the development length equation is the lesser of (1) the bond-critical strength, 103.6 ksi, (2) the bar design strength, 72 ksi, and (3) the bar stress at failure, 57.06 ksi. Therefore, use $f_{fr} = 57.06$ ksi.

$$l_d = \frac{\alpha(f_{fr}/\sqrt{f'_c}) - 340}{13.6 + C/d_b}d_b = \frac{(1.0)[57{,}060/\sqrt{4000}] - 340}{13.6 + 0.875/0.75}(0.75) = 28.6 \text{ in.}$$

Check that the development length is less than that allowed for a simple support. Assume that $l_a = 0$. Since the bars extend over the support and they are in the bottom of the section, they are confined and the length available for development is

$$1.3\frac{\phi M_n}{V_u} + l_a = (1.3)\frac{0.65(436.6)}{4.66} + 0 = 79.2 \text{ in.} > 28.6 \text{ in.} \rightarrow \text{OK}$$

PROBLEMS

7.1 Determine the bond-critical effective FRP bar strength, f_{fe}, for the FRP bars listed in Table P7.1 for the embedment lengths specified. Assume that the bars are spaced at the minimum on-center bar spacing permitted (i.e., minimum of 1-in. clear spacing, or d_b) and that this controls the bar-edge spacing parameter, C.

TABLE P7.1 Bar Bond Stress and Development Length Analysis

No.	Bar Type	f'_c (psi)	Location	l_e
1	No. 4 GFRP	4000	Bottom	$20d_b$
2	No. 6 GFRP	4000	Bottom	$20d_b$
3	No. 8 GFRP	4000	Bottom	$20d_b$
4	No. 10 GFRP	4000	Bottom	$20d_b$
5	No. 4 GFRP	4000	Bottom	$40d_b$
6	No. 6 GFRP	4000	Bottom	$40d_b$
7	No. 8 GFRP	4000	Bottom	$40d_b$
8	No. 10 GFRP	4000	Bottom	$40d_b$
9	No. 4 GFRP	8000	Bottom	$20d_b$
10	No. 6 GFRP	8000	Bottom	$20d_b$
11	No. 8 GFRP	8000	Bottom	$20d_b$
12	No. 10 GFRP	8000	Bottom	$20d_b$
13	No. 4 GFRP	4000	Top	$20d_b$
14	No. 6 GFRP	4000	Top	$20d_b$
15	No. 8 GFRP	4000	Top	$20d_b$
16	No. 10 GFRP	4000	Top	$20d_b$
17	No. 4 GFRP	4000	Bottom	$20d_b$
18	No. 4 GFRP	4000	Bottom	$40d_b$

7.2 Determine the required development length, l_d, for the bars listed in Table P7.1 assuming they are used in underreinforced FRP reinforced beams and that they have sufficient embedment length so as not to be bond critical.

7.3 Consider the GFRP-reinforced test beams (FRP1 and FRP2) shown in Fig. 4.1, whose load deflection results are shown in Fig. 5.1. The beams were reinforced with three No. 7 GFRP Aslan 100 main bars and No. 3 Aslan GFRP stirrups at 5 in. on center. The stirrups were two-part horizontal U shapes with inside corner radii of 3.25 in. (shown on the left side of Fig. 7.4). The beam had an 8 × 12 in. cross section and was fabricated with 5000-psi design strength concrete (5880-psi actual 28-day strength). The bottom and side clear covers were 1 in. The beam was 98 in. long and was loaded in four-point bending over a 90-in. span at its third points (i.e., the moment and shear spans all equal 30 in.). The failure mode of beam FRP 1 is shown in Fig. 6.1, and the failure mode of FRP 2 is shown in Fig. 5.4. As noted in the text, FRP 1 failed in shear, whereas FRP 2 failed in flexure, due to concrete crushing. You are required to use the theory presented in the text to critically evaluate the results of the experiments on FRP 1 and FRP 2.

(a) Determine the nominal moment and nominal shear capacities of the beams predicted and compare with the experimental results (note that Fig. 5.1 shows the total load on beam and that the deflection due to the beam self-weight is not included).

(b) Construct the load–deflection curves for the beams from zero to the maximum predicted load at 5-kip intervals. Plot the prediced load–deflection curves for the beams and compare these to the experimentally measured curves (plot the experimental curves approximately by taking measurements from Fig. 5.1).

(c) Critically evaluate the results and comment on the ability of the theory to predict the beam behavior in the service load range and in the ultimate state (including the failure modes). Were the J-hooks required for the main bars?

(d) Perform the same exercise for the steel-reinforced beams and comment on the accuracy of the predictions for the steel-reinforced beams. (The steel beams had closed loop No. 3 steel stirrups at 5 in. on center.)

7.4 Calculate the costs of the GFRP- and steel-reinforced beams in Problem 7.3. Assume uncoated steel (black steel) bars at a cost of $0.35 per pound. For concrete use $75 per cubic yard for 5000-psi design strength. For FRP bar costs, use data provided in the problem sections of Chapters 5 and 6 and interpolate if necessary (or obtain pricing from the bar manufacturer). Determine the strength/cost, stiffness/cost, and energy absorbed[1]/cost ratios for the GFRP and steel beams. Compare the results.

7.5 Design a 28-ft-long two-span continuous FRP-reinforced beam having a rectangular cross section. Use 5000-psi concrete and Aslan 100 GFRP rebars (main bars, stirrups, and hooks if necessary). The beam depth is limited to 20 in. You are given the following design loads and serviceability limits: dead load 400 lb/ft, live load 900 lb/ft (20% sustained), maximum long-term deflection $L/360$.

(a) Design the beam for ultimate (flexure, shear) and service limit states (deflection control, crack width control, creep rupture control). Identify the failure modes of the beam. Explain your design choices where necessary.

(b) Provide a bar layout diagram showing bar cutoffs and stirrup layout. Show cross sections in the positive moment spans and over the middle support. Provide a list of FRP rebar materials, giving the number, size, and length of each bar type and a small sketch of each bar.

7.6 Design a 75-ft-long continuous GFRP-reinforced beam having a rectangular cross section to support an overhead 10-ton gantry crane in a highly corrosive chemical plant. The continuous beam is supported on 10-in.-wide column brackets at 15 ft on center. The overhead crane

[1] Assume that the energy absorbed by the beams is equal to the area under the load–deflection curves.

gantry weights 1000 lb and is 30 ft wide. The crane has a lift capacity of 10 tons (20 kips). The gantry travels along the FRP-reinforced concrete beams (typically called *crane rails*). The beams have an additional superimposed dead load of 50 lb/ft of mechanical equipment. Due to space limitations, the width of the beam is limited to 12 in. maximum. A durable 6000-psi concrete is specified. The maximum long-term deflection of the beam is limited to 0.20 in. according to the crane mechanical specifications. Design the crane rail. Determine the section size, the main and shear reinforcement, and detail the FRP reinforcing bars for the negative and positive moments. Check the FRP bars for long-term sustained stress and for fatigue stresses. Show a diagram of the beam over the six supports (brackets). Indicate where bars are terminated along the length, if required. Provide a bill of materials for the beam, listing the quantity, size, and type of all FRP rebars to be used to reinforce the beam. Explain all your design assumptions and justify your design decisions as necessary.

SUGGESTED FRP-REINFORCED CONCRETE STRUCTURAL DESIGN PROJECTS

The following design projects are suggested for students in a composites for construction design class. The types of projects selected are those in which the noncorrosive properties of FRP reinforcements are known to be beneficial and could result in service life savings over traditional steel-reinforced concrete. The design project should preferably be done in groups of two or three students. Students should be given 4 to 6 weeks to compete the design project (which therefore needs to be assigned early in the semester). The final deliverable should be a design proposal[2] that includes the problem statement, scope, codes and specifications, loads, materials, design calculations, design drawings, and a cost analysis. A presentation of the design proposal should be given in class. It is recommended that the current conditions of the existing structure and any corrosion-related problems be discussed and that owners and designers be invited to form part of the project jury to obtain feedback from industry on the designs presented.

7.1 *FRP-reinforced concrete pedestrian bridge.* Design an FRP-reinforced concrete bridge to replace an existing pedestrian bridge/elevated walkway (originally concrete, steel, or timber). Students should scout the local area (campus, downtown, etc.) to find a pedestrian bridge with 20- to 40-ft spans and use this bridge as the basis for the design. If possible, the original plans of the structure should be obtained from the university

[2] Items listed are at the discretion of the instructor.

physical facilities department (or equivalent). Design the concrete slab and support beams (girders, stringers) as required, depending on the type of structural system used. Use the AASHTO *Guide Specification for Design of Pedestrian Bridges* (1997) for loads and dynamic requirements or other applicable codes.

7.2 *FRP-reinforced concrete retaining wall.* Design an FRP-reinforced concrete retaining wall (or foundation wall) to replace an existing retaining wall. Students should scout the local area (campus, downtown, etc.) to find a small retaining wall structure 8 to 12 ft high and use this wall as the basis for the design. If possible, obtain the original plans of the structure from the university physical facilities department (or equivalent). Design the wall and the foundation as required using FRP reinforcements.

7.3 *FRP-reinforced concrete beams in a wastewater treatment plant.* Visit your local wastewater treatment plant[3] and identify any cast-in-place reinforced concrete structures (preferably beam and slab structures, which are usually near the settling tanks). Consult with the operator of the facility and discuss any corrosion-related problems at the facility. Redesign a concrete beam or flat slab member using FRP reinforcements.

7.4 *FRP-reinforced precast double T-beams.* Prestressed and precast double-T floor beams are frequently used in the construction of precast parking garages. Scout your local area and identify a parking structure of this type. Obtain typical plans and specifications for the double-T beams (available in textbooks and other resources). Redesign the nonprestressed reinforcement in the flanges and the shear reinforcement in the webs of such a member using FRP-reinforced concrete. Be sure to address the issue of the steel flange inserts (shear connectors) that are used to attach the flanges together to create the floor diaphragm (typically, by on-site welding). Consider how to use FRP materials for this purpose.

[3] The environmental engineering faculty may be able to be of assistance in recommending a local facility and contacts.

8 Design Basis for FRP Strengthening

8.1 OVERVIEW

This chapter provides an introduction to the subject of FRP strengthening materials, systems, and methods for strengthening and retrofitting reinforced concrete structures such as beams, slabs, and columns. The properties of FRP strengthening systems are reviewed and their behavior under various loads is discussed. Since FRP strengthening systems are used both to increase the strength of reinforced concrete members and to increase their ductility, the subject of strengthening is more complex than that of reinforcing with FRP rebars.

To apply a strengthening system to a concrete member, the surface must be suitably prepared. This important aspect of strengthening of concrete structures is not covered in this book, as it is not directly related to the design of the strengthening system but rather, to the constructability. It is assumed that the reader is familiar with these procedures.[1] Part of a strengthening design often includes evaluation of the current strength of an existing structure. Often, the original plans are no longer available, and the structure may have deteriorated since its original construction. This can present one of the most challenging engineering aspects of designing a strengthening system. Determination of the existing capacity of a deteriorated concrete structure is not covered in this book.

In this chapter we describe the basis for the design of concrete members strengthened with FRP materials. This basis is in accordance with the American Concrete Institute publication ACI 440.2R-02, *Guide for the Design and Construction of Externally Bonded FRP Strengthening Systems for Strengthening Concrete Structures* (ACI, 2002). This document is compatible with ACI publication ACI 318-99, *Building Code Requirements for Structural Concrete and Commentary* (ACI, 1999). It is very important to note that the current ACI FRP strengthening design guide is not compatible with ACI 318-05 (or ACI 318-02) provisions and is not compatible with ASCE 7-02 (ASCE, 2002) load factors. In particular, the load factors used with ACI 440.2R-02 are the "old" ACI load factors, the most common of these being 1.4 for the

[1] See Chapter 2 for discussion and guidance on this topic.

dead load and 1.7 for the live load. The resistance factors presented in what follows for the design of FRP strengthening systems must therefore be used only with the 1.4 and 1.7 load factors.[2]

Throughout this chapter and the chapters that follow that describe in detail the procedures for the design of concrete structures with FRP strengthening systems, it is assumed that the reader has reasonable familiarity with ACI 318-99 and is familiar with the design of concrete beams and slabs reinforced with conventional steel reinforcing bars. The equations and examples that follow use U.S. standard units since these are the familiar units for design in the United States, where the ACI codes are mainly used.

8.2 INTRODUCTION

Fiber-reinforced-plastic (FRP) reinforcing systems for strengthening structurally deficient concrete structural members and for repairing damaged or deteriorated concrete structures have been used since the mid-1980s. The first applications involved beams strengthened to increase their flexural capacity using high-strength lightweight fiber-reinforced epoxy laminates bonded to the soffits of the beams (e.g., Meier and Kaiser, 1991; Saadatmanesh and Ehsani, 1991; An et al., 1999; Triantafillou et al., 1992; Meier, 1995). The method is a modification of the method using epoxy-bonded steel plates to strengthen concrete beams which has been in use since the mid-1960s. The FRP systems were shown to provide significant benefits over the steel plates in constructibility and durability. Work was conducted on strengthening of concrete columns to enhance their axial capacity, shear capacity, and ductility, primarily for seismic loadings (e.g., Fardis and Khalili, 1981; Katsumata et al., 1988; Priestley et al., 1992; Seible et al., 1997). This method is a modification of one using steel jackets to strengthen concrete columns. This was followed closely by work on shear strengthening of beams (e.g., Chajes et al., 1995; Khalifa et al., 1998; Triantafillou, 1998). A review of the state of the art on the subject can be found in Bakis et al. (2002) and Teng et al. (2001). The method has also been used to strengthen masonry and timber structures; however, applications of this type are not covered in this book. In recent years a variant of the FRP strengthening method called the *near-surface-mounted* (NSM) *method* has been developed. In this method the FRP strip is adhesively bonded into a slot (or groove) that is saw-cut to a shallow depth [0.5 to 1.0 in. (13 to 25 mm)] in the concrete.

Current FRP strengthening systems for concrete fall into two popular types. One type consists of factory-manufactured (typically, unidirectional pul-

[2] ACI Committee 440 is currently in the process of revising ACI 440.2R to be compatible with ACI 318-05. The reader should keep this in mind when using future editions of the ACI guide in conjunction with this book.

truded) laminates (also known as *strips* or *plates*) of carbon- or glass-reinforced thermosetting polymers (epoxy or vinylester) bonded to the surface of the concrete using an epoxy adhesive. The manufactured laminates typically have a volume fraction of fibers in the range 55 to 65% and are cured at high temperatures (typically, >300°F) but bonded in the field at ambient temperatures. The other type consists of layers (or plies) of unidirectional sheets or woven or stitched fabrics of dry fibers (glass, carbon, or aramid) that are saturated in the field with a thermosetting polymer (epoxy or vinylester) which simultaneously bonds the FRP laminate (thus formed) to the concrete. These *formed-in-place* or *laid-up* FRP systems typically have a fiber volume fraction between 20 and 50% and are cured at ambient temperatures in the field. Figure 8.1 shows a concrete beam with a bonded precured strip, and Fig. 8.2 shows a concrete beam with a laid-up sheet.

As reviewed in Chapter 1, a number of design guides and national standards are currently published that provide recommendations for the analysis, design, and construction of concrete structures strengthened with FRP materials (AC 125, 1997; FIB, 2001; JSCE, 2001; ACI, 2002; TR 55, 2004). Manufacturers of FRP strengthening systems for concrete typically provide their own design and installation guides for their proprietary systems. Since the performance of the FRP strengthening system is highly dependent on the adhesive or saturating polymer used, the preparation of the concrete surface prior to application of the FRP strengthening system, and the field installation and construction procedures, manufacturers typically certify "approved contractors" to ensure that their systems are designed and installed correctly. In addition, code guidance is provided to ensure that FRP strengthening systems are installed appropriately (AC 187, 2001; ACI, 2002; TR 57, 2003; NCHRP, 2004).

Research into the use of FRP strengthening systems for concrete structures has been the focus of intense international research activity since the early

Figure 8.1 Beam with a bonded precured FRP strip.

Figure 8.2 Beam with a laid-up FRP sheet.

1990s. A biannual series of symposia entitled "Fiber Reinforced Plastics in Reinforced Concrete Structures" (FRPRCS) has been the leading venue for reporting and disseminating research results. Recent symposia in the series, dating back to 1993, were held in Singapore in July 2003 (Tan, 2003) and in Kansas City in November 2005 (Shield et al., 2005).

8.3 PROPERTIES OF FRP STRENGTHENING SYSTEMS

Carbon fiber–reinforced epoxy laminates (or strips) are most commonly used in the adhesively bonded products. Depending on the type of carbon fiber used in the strip, different longitudinal strengths and stiffness are produced. Strips are typically thin [less than 0.100 in. (2.5 mm)] and are available in a variety of widths [typically, 2 to 4 in. (50 to 100 mm)]. Since the strips are reinforced with unidirectional fibers, they are highly orthotropic and have very low properties in the transverse and through-the-thickness directions. Narrow strips and small-diameter FRP rebars are also used in the NSM strengthening method.

Manufacturers typically only report properties in longitudinal directions and report very few data on physical properties. The strips are bonded to the concrete with a compatible adhesive that is supplied by the strip manufacturer. Typical properties of strips are given in Table 1.2. It is important to note that the properties shown for the strips are the properties of the FRP composite, not the properties of the fibers alone.

In the types of products for FRP strengthening that consist of dry fibers in sheet or fabric form with compatible polymer-saturating resins, there is a greater array of products available that depend on fiber type and sheet or fabric architecture. In this group of products, a unidirectional highly orthotropic carbon fiber tow sheet is produced by a number of manufacturers and is often used in strengthening applications. The individual carbon tows in the sheet are held together by a very light polymeric binder (or a light stitching), are supplied on a wax paper backing, and have a typical weight of $\frac{1}{2}$ oz/ft^2 (~150 g/m^2). Sheets are typically 10 to 40 in. (250 to 1000 mm) wide and may be applied in multiple layers with different orientations. Common fabric materials in this group are woven or stitched glass or carbon fiber materials, typically consisting of 12 to 24 oz/yd^2 (360 to 720 g/m^2) materials with a variety of weaves, which can give the fabric properties from highly orthotropic to square symmetric. Carbon fiber tow fabrics and hybrid fabrics (with more than one fiber type) are also available. Fabrics are typically much thicker than tow sheets and can also be used in multiple layers. Because of the wide variety of products available and their different thicknesses, it is not easy to compare their properties directly. In addition, the fibers must be used with a compatible resin system applied with a controlled volume fraction in order to achieve an FRP composite with measurable properties. In the case of sheet and fabric materials, manufacturers typically report the mechanical properties of the dry fibers and the thickness (or area) of the fibers. It is important to note that when reported in this fashion, the properties are not the properties of the FRP composite. Also, if the fabric has bidirectional or multidirectional fibers, only those fibers in the direction that is to be strengthened must be used to determine the fiber area for strengthening. Properties of some commonly available unidirectional fiber sheet and fiber fabric materials are listed in Table 1.3.

The performance of the FRP strengthening system is highly influenced by the properties of the adhesive layer in the case of the bonded strip and by the properties of the saturating polymeric resin in the case of the sheets and fabrics. The interface between the FRP composite and the concrete substrate transfers the loads from the concrete to the FRP composite. In the case of flexural, shear, or axial tensile strengthening, this load transfer is primarily in shear, and the properties and quality of the interface bond between the FRP composite and the concrete effect load transfer into the FRP strengthening system. Such applications are termed *bond-critical applications.* In the case of axial compressive strengthening, where the role of the strengthening system is to confine the lateral expansion of the cracked concrete, the interface bond is not as critical as long as the FRP strengthening system is in intimate contact with the concrete and is wrapped around the concrete continuously so as to provide a required confining pressure. Such applications are termed *contact critical.* It is important to note, however, that in either type of system the polymer resin still plays a critical role in the FRP composite in transferring load to the fibers and protecting the fibers. A poorly applied sheet or fabric having a low fiber volume fraction (<20%) and a relatively thick uneven thickness polymer layer with a large void content (>3%) will not provide a durable FRP strengthening system, even though the amount of fiber placed in the wrap may be according to design requirements.

Since most FRP strengthening systems depend on curing of polymer adhesives or resins at ambient temperature, the cured glass transition temperatures of these components may be quite low [120 to 180°F (~50 to 80°C)]. Since the effectiveness of the strengthening system depends to a great degree on the stiffness of the adhesive or resin used to bond the FRP composite to the concrete, an FRP strengthening system can become unserviceable at even moderately high temperatures (~100°F). In certain cases where the fibers are sufficiently anchored away from the region subjected to the high temperature, the FRP strengthening system may still be effective. Designers should always be aware of the glass transition temperature of the FRP composite or adhesive they are using in a design. In the event of a fire, the integrity of an FRP strengthening system may be severely compromised in a very short time unless it is protected by insulation. This is currently the subject of ongoing research studies (Bisby et al., 2005).

8.4 DESIGN BASIS FOR FRP STRENGTHENING SYSTEMS

In what follows the design basis of the American Concrete Institute presented in ACI 318-99 and ACI 440.2R-02 are followed with respect to design philosophy and load and resistance factors for design.[3] Similar design bases, conforming to the limit states design procedure, are recommended by many standards organizations and professional organizations for FRP-reinforced concrete (JSCE, 2001; FIB, 2001; CSA, 2002; TR 55, 2004).

8.4.1 Resistance Factors

For the design of concrete members with FRP strengthening systems, the ACI 440.2R.-02 recommends use of the traditional resistance factor, ϕ, which is a function of the intended use (or the dominant stress) of the structural member. The resistance factor is applied to the nominal capacity of the member to obtain the factored capacity for use in the strength design method. For FRP strengthening systems, ACI 440.2R-02 also recommends the use of a strength reduction factor (similar to a material partial safety factor), ψ_f, which is applied to the FRP system only. The strength reduction factor is applied to the stress in the FRP material to obtain the nominal capacity of the structural member. Consequently, the factored (ultimate) capacity of a concrete member depends on the member resistance factor and on the FRP strength reduction factor. The following factors are provided by ACI 440.2R-02 for design of FRP-strengthened concrete structures.

[3]The ultimate strength design procedure, a load and resistance factor design (LRFD) procedure, is followed by ACI 440.2R-02; however, at this time, resistance factors are not probabilistically based. Load factors are those recommended by ACI 318-99 for all concrete structures.

Flexural capacity (for tensile strengthening) (ACI 440.2R-02:9-5):

$$\phi = \begin{cases} 0.9 & \text{for a ductile failure when the member fails after steel yielding and } \varepsilon_s > 0.005 \\ 0.7 & \text{for a brittle failure when the member fails prior to steel yielding and } \varepsilon_s < \varepsilon_{sy} \\ 0.7 + \dfrac{0.2(\varepsilon_s - \varepsilon_{sy})}{0.005 - \varepsilon_{sy}} & \text{for a semiductile failure when the member fails after steel yielding and } \varepsilon_s > \varepsilon_{sy} \text{ but } \varepsilon_s < 0.005 \end{cases}$$

$$\psi_f = 0.85 \text{ for FRP bond-critical designs}$$

Shear capacity (for shear strengthening) (ACI 318-99):

$$\phi = 0.85$$

$$\psi_f = \begin{cases} 0.85 & \text{for FRP bond-critical designs} \\ 0.95 & \text{for FRP contact-critical designs} \end{cases}$$

Axial compressive capacity (for confinement) (ACI 318-99):

$$\phi = \begin{cases} 0.75 & \text{for steel spiral reinforcement} \\ 0.70 & \text{for steel tied reinforcement} \end{cases}$$

$$\psi_f = 0.95 \text{ for FRP contact-critical designs}$$

8.4.2 Guaranteed Properties

The tensile strength of an FRP strengthening system (called the *ultimate tensile strength*) and tensile strain to failure of FRP strengthening systems (called the *ultimate rupture strain*) are defined by ACI 440.2R-02 as the mean minus three standard deviations of a minimum of 20 test samples. The ultimate strength, f^*_{fu}, and ultimate rupture strain, ε^*_{fu}, are expected to be supplied by the manufacturer. FRP strengthening systems must be tested according to the procedures detailed in in ACI 440.3R-04 (see Chapter 3 for details).

8.4.3 Environmental Effects

The design strength, f_{fu}, and design failure strain, ε_{fu}, are obtained from the ultimate strength and ultimate rupture failure strain by multiplying them by an *environmental reduction factor,* C_E, which depends on the type of FRP system used and the type of intended service of the structure, as shown Table 8.1. The FRP system tensile modulus (also called the *specified tensile modulus*), E_f, measured in the axial direction of the fiber is defined as the average

TABLE 8.1 Environmental Reduction Factors for FRP Strengthening Systems per ACI 440.2R-02

Exposure Condition	FRP System Type	Environmental Reduction Factor, C_E
Interior (FRP system in an indoor environment)	Carbon–epoxy	0.95
	Glass–epoxy	0.75
	Aramid–epoxy	0.85
Exterior (FRP system exposed to the elements)	Carbon–epoxy	0.85
	Glass–epoxy	0.65
	Aramid–epoxy	0.75
Especially aggressive environments (chemical exposure, wastewater)	Carbon–epoxy	0.85
	Glass–epoxy	0.50
	Aramid–epoxy	0.70

modulus of the test data in accordance with ACI 440.3R-04. No reduction to the measured modulus is taken, for either statistical or environmental reasons.

The environmental reduction factor accounts for the fact that even though FRP systems are not susceptible to conventional electropotential corrosion that affects metallic materials, they can nevertheless deteriorate in various chemical environments, both alkaline and acidic. This deterioration is accelerated at elevated temperatures. Glass fibers are especially susceptible to corrosion in alkaline and neutral solutions, due to leaching of the fiber (Bank et al., 1995b). Aramid fibers do not leach like glass fibers but are known to absorb moisture and swell, leading to loss of mechanical and physical properties. Carbon fibers are the most chemically inert and are generally recommended when extreme exposure to chemical aqueous environments is anticipated. In addition, the polymer matrix is also susceptible to deterioration in different chemical environments. However, carbon fibers are not electrochemically inert and can cause a phenomenon known as *galvanic corrosion* to occur when they are in contact with metals, as discussed in Chapter 1. This is not likely to be a problem in FRP-strengthened concrete since the FRP does not come into contact with the steel reinforcing bars. However, if metallic anchor bolts or plates are used to anchor a carbon FRP system, this can be a consideration.

8.4.4 Limits of Strengthening

Although flexural strengthening over 100% of original design strength is possible to achieve, ACI 440.2R-02 limits the amount of strengthening to prevent catastrophic failure of the concrete member in the event of loss of, or damage to, the strengthening system (due to vandalism or environmental degradation). The ACI recommends that the strengthened member still have sufficient original factored capacity (i.e., discounting the strengthening system) to resist a substantial portion of the future load on the strengthened member, given as

$$1.2D + 0.85L \tag{8.1}$$

where D is the dead load effect[4] and L is the live load effect on the strengthened structure. To account for fire, additional restrictions are placed on the factored capacity of the FRP-strengthened structured. In this case the portion of the new load on the strengthened member is given as

$$1.0D + 1.0L \tag{8.2}$$

The maximum permissible strengthening of a section is a function of the ratio of future live load to future dead load on the strengthened beam. If the original factored moment capacity of a beam, $(\phi M_n)_{\text{orig}}$, is known the maximum service moment due to the future dead load, $m_{\text{DL(max)}}$, and the maximum service load moment due to the future live load, $m_{\text{LL(max)}}$, can be obtained from

$$1.2m_{\text{DL(max)}} + 0.85m_{\text{LL(max)}} = (\phi M_n)_{\text{orig}} \tag{8.3}$$

Given the load ratio on the strengthened beam,

$$R = \frac{m_{\text{LL(max)}}}{m_{\text{DL(max)}}} \tag{8.4}$$

gives[5]

$$(\phi M_n)_{\max} = (\phi M_n)_{\text{orig}} \frac{1.4 + 1.7R}{1.2 + 0.85R} \tag{8.5}$$

and the maximum permissible percentage increase in factored moment capacity is

$$\%_{\max} = \frac{0.2 + 0.85R}{1.2 + 0.85R} \times 100 \tag{8.6}$$

When a concrete member is strengthened to increase its capacity in a selected mode (e.g., flexure) the member must be checked to ensure that the capacities in other failure modes (e.g., shear) are not exceeded. If this is the case, the

[4] Such as bending moment, shear force, and axial force.

[5] This does not necessarily mean that the resistance factor, ϕ, will be the same for the original and the strengthened sections.

strengthening should be decreased or the secondary capacity needs to be enhanced with its own strengthening system.

8.4.5 Limits on Stresses in FRP Strengthening Systems at Service Loads

The stress level in the FRP at service loads is limited by the creep rupture properties of different FRP materials and by their fatigue resistance. ACI 440.2R-02 limits the service-level stresses in an FRP strengthening system in strengthened reinforced concrete members according to the fiber type used in the strengthening system (Table 8.2).

8.4.6 Compression Strengthening in Flexural Members

At this time, FRP systems should not be used as compression strengthening of flexural members. If FRP strengthening systems are placed in the compression zones of flexural members, they should not be assumed to carry any compression load, and they should be adequately attached, to prevent local bucking, which could lead to premature failure.

8.5 DEFLECTIONS IN FRP-STRENGTHENED STRUCTURES

For deflections in flexural members where stresses are in the service range, the contribution of the FRP strengthening system is typically small. Flexural deflections in the service range can be estimated by use of an effective composite-section second moment-of-area (I_e) analysis where the tensile contribution of the FRP is added to the contribution of the steel reinforcing. In the inelastic range after the primary reinforcing steel has yielded, the contribution of the FRP strengthening to the postyield stiffness can be considerable and should be accounted for in inelastic analysis.

8.6 FRP STRENGTHENING SYSTEM AREA CALCULATIONS

It is extremely important to note that the method of determining the tensile force resultant in an FRP strengthening system depends on the type of system

TABLE 8.2 Maximum Permissible Service Load Stress in an FRP Strengthening System

Stress Type	Glass FRP	Aramid FRP	Carbon FRP
Sustained stress plus maximum stress under cyclic load	$0.20f_{fu}$	$0.30f_{fu}$	$0.55f_{fu}$

used. In bonded strip systems, the ultimate force is obtained from the strength of the FRP composite and the gross cross-sectional area of the strip. In dry fiber systems, the ultimate force is obtained from the strength of the fibers and the thickness of the net area of the fibers. The designer must know if the strength (and stiffness) reported for an FRP strengthening system is for the FRP composite (gross cross section) or for the fibers alone (net fiber cross section). Both methods of calculation are permitted at this time by the ACI 440.2R-02 guide.

PROBLEMS

8.1 Visit the following FRP strengthening system manufacturers'[1] Web sites and request or download copies of their FRP precured strip, sheet, or fabric property specification sheets. These will be needed for solving the problems that follow.

Sika (SikaWrap Hex fabrics and Carbordur strips): www.sikaconstruction.com

Fyfe Company (Tyfo fabrics and strips): www.fyfeco.com

Degussa (WaboMBrace sheets and strips): www.wbacorp.com

VSL (V-Wrap): www.vsl.net

Hughes Brothers (Aslan 400 strips): www.hughesbros.com

S&P Clever Reinforcement Company (S&P sheets and laminates): www.sp-reinforcement.ch

8.2 For the following FRP fabric strengthening systems, determine the ACI 440.2R-2 nominal design strength, f_{fu}, the nominal design failure strain, ε_{fu}, and the longitudinal modulus, E_f, when the strengthening system is used in an exterior nonaggressive environment. Also indicate the thickness of a single ply of this system, the weight of the fabric in oz/yd^2, and whether the design is based on the laminate properties or on the fiber properties. Attach copies of the specification sheets to your homework.

(a) SikaWrap Hex 100G with Sikadur Hex 100 epoxy.

(b) Tyfo SEH 51 with Tyfo S epoxy.

(c) WaboMbrace CF530 with WaboMbrace epoxy encapsulation resin.

(d) VSL V-Wrap C200 with epoxy material type 1.

(e) S&P C-sheet 640 with S&P resin epoxy 55/50.

[1] No endorsement of the manufacturers listed is implied. The reader is free to choose any manufacturer's products to use in the examples that follow, and is encouraged to collect similar data for locally available strengthening products.

8.3 For the following FRP precured strip strengthening systems, determine the ACI 440.2R-02 nominal design strength, f_{fu}, the nominal design failure strain, ε_{fu}, and the longitudinal modulus, E_f, when the strengthening strip is used in an interior environment. Also, indicate the available thicknesses and widths of a single strip of this system and the adhesives that are specified by the manufacturer for use with this strip. Attach copies of the specification sheets to your homework.

(a) Sika Carbodur laminates

(b) Tyfo UG composite laminate strip system

(c) Aslan 400 laminates

(d) S&P laminates CFK 150/2000

8.4 Consider equation (8.6), which gives the maximum permissible flexural strengthening of a member based on its original capacity and the future live load/dead load ratio. Use this equation and show graphically the maximum permissible strengthening as a function of the future live load/dead load ratio. Plot a graph of the maximum percentage increase in strengthening permitted in an FRP-strengthened section as a function of the live load/dead load ratio from 0.5 to 5. Discuss what can be learned from this graph about the types of structures that are most beneficial to strengthen with FRP materials.

8.5 Develop an equation in a form similar to equation (8.6) that can be used to determine the maximum strengthening permissible when an FRP-strengthened structure needs to be designed for possible fire loads [per equation (8.2)]. Plot a graph of the maximum percentage increase of strengthening permitted in this type of FRP-strengthened section as a function of the live load to dead load ratio from 0.5 to 5, and comment on the results.

SUGGESTED ACTIVITIES

8.1 Obtain small (4 × 4 in. to 6 × 6 in.) swatches of FRP strengthening sheets and fabrics. Request samples from the vendors of the FRP strengthening systems in Problem 8.1 or obtain samples from regular suppliers of fiber fabrics (glass and carbon, primarily) in other parts of the composites industry (see Problem 2.1). If necessary, small sample swatches can be purchased. Ideally, obtain tow sheets or fabrics ranging from about 5 to 24 oz/yd^2 (150 to 720 g/m^2). Try to obtain thin tow sheets, woven fabrics, and stitched fabrics.

(a) Determine the areal weight of the fabrics received and examine the structure of the weave or stitching.

(b) Practice handling the fabric materials: folding, cutting, and rolling. Write a two-page report describing the properties and the fabrics you have investigated and describing the difference between tow sheets, woven fabrics, and stitched fabircs.

8.2 Obtain small (4- to 6-in.) lengths of precured unidirectional carbon fiber–reinforced FRP strengthening strips (or laminates). Request samples from the vendors of the FRP strengthening systems listed in Problem 8.1.

(a) Examine the strip for flatness and note any lengthwise or widthwise curvatures.

(b) Experiment with various hand and power saws to cut the strip in the longitudinal and transverse directions. Consult manufacturers' recommendations on how to cut strips in the field.

(c) Manually bend (in an attempt to break) a short longitudinal strip (about 6 in. long) and a short transverse strip (about 2 in. long). Notice the difference in flexural behavior in the two directions. Write a brief report on your oversvations.

8.3 Download a copy of the 2004 National Cooperative Highway Research Program (NCHRP) report (number 514), entitled *Bonded Repair and Retrofit of Concrete Structures Using FRP Composites* from http://trb.org/. Write a six-page double-spaced essay that describes what you learned from the report about repair and strengthening of concrete with FRP composites.

9 FRP Flexural Strengthening

9.1 OVERVIEW

This chapter deals with the design and strengthening of reinforced concrete flexural members, such as beams, slabs, walls, and columns with precured FRP strips and plates, and laid-up FRP sheet and fabric systems. Only non-prestressed concrete flexural members that are reinforced with conventional steel reinforcing bars are considered. The design procedures presented in this chapter follow ACI 440.2R-02, which is compatible with ACI 318-99. For the purposes of this chapter, it is assumed that the reader is familiar with both the fundamentals and details of the design of concrete flexural members reinforced with conventional steel reinforcing bars.

The examples presented in this chapter and those to follow are based on the examples in ACI 440.2R-02 and are intended to allow the reader to analyze these examples critically and to consider design alternatives not presented in the ACI guide. The material in the chapter covers strength design and serviceability design. For pedagogical and practical reasons, these are separated into two parts. In reality, strength and serviceability design are not separate and must be considered as equally important limit states. When dealing with FRP materials, this is always important to keep in mind.

9.2 INTRODUCTION

Flexural strengthening of reinforced concrete beams, slabs, walls, and columns is achieved by attaching an FRP strengthening system (bonded strip or saturated dry fabric) to the tension face (or a portion thereof) of a flexural member to increase the effective tensile force resultant in the member and thereby increase the moment capacity of the member.[1] Typical applications are shown in Figs. 9.1 and 9.2.

Flexural strengthening with FRPs is analogous to adding steel strengthening strips (or plates) to the tension face member. However, two fundamental differences exist. First, the FRP strengthening system behaves in a linear elastic fashion and does not yield, and second, the FRP strengthening system

[1] In what follows, only singly reinforced rectangular sections are discussed. For doubly reinforced and nonrectangular sections, similar derivations are possible.

Figure 9.1 Flexural (and shear) strengthening of a reinforced concrete beam. (Courtesy of Fyfe Company.)

Figure 9.2 Flexural strengthening of a reinforced concrete slab. (Courtesy of Sika Corporation.)

is more susceptible than steel plate systems to detachment (debonding or delamination) failures. Since the steel plates themselves will yield at a similar strain to the internal steel reinforcing, the stress level in the steel strengthening system is limited and epoxy-bonded steel plates are typically anchored with steel bolts at their ends (and also often along their length).

In FRP strengthening with FRP systems having ultimate tensile strengths exceeding 300 ksi and stiffness exceeding 40 Msi (see Tables 1.2 and 1.3), the stress level in the FRP can be significantly higher than that in steel strengthening systems. Consequently, the strength enhancement provided by the FRP can be much greater than that of steel systems. If the internal steel reinforcing in the strengthened RC beam yields before the FRP strengthening system fails, which is desirable, the FRP system will "pick up" the load beyond the yield load of the internal steel. The large tensile forces in the FRP strengthening systems and the large post-steel-yielding deflections in the member can lead to the FRP strengthening system delaminating (or debonding) from the concrete. Therefore, the maximum stress and strain in the FRP usually cannot be attained, and they are limited in the design procedure to prevent this type of failure from occurring.

A typical load–midspan deflection plot of an FRP strengthened concrete beam is shown in Fig. 9.3 (Alkhrdaji et al., 2000). Initially, the beam behaves like a conventional reinforced concrete beam with internal steel reinforcing bars carrying the majority of tensile force in the section. At some point (generally seen as a change in slope of the load–deflection curve) the internal steel yields and cannot carry additional tensile force. This additional tensile force is carried by the FRP system and an increase in the load capacity of

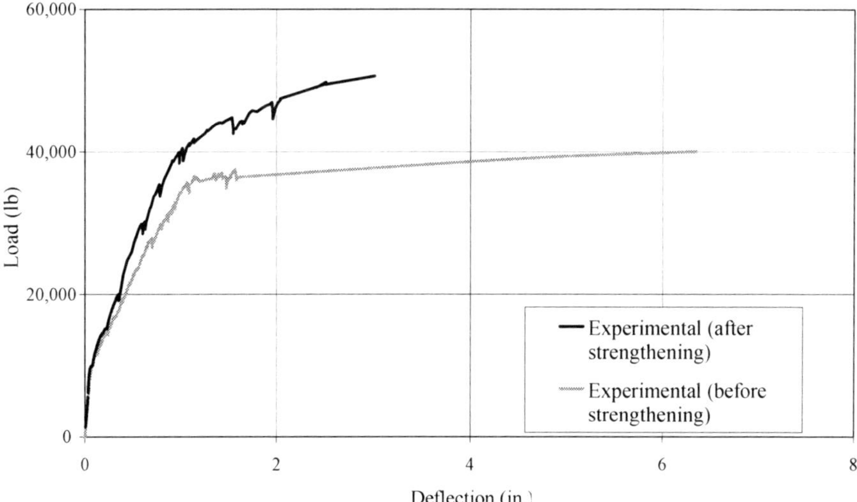

Figure 9.3 Load–deflection response of an FRP-strengthened concrete beam. (Courtesy of A. Nanni and R. Mayo.)

the member is obtained. Eventually, the FRP-strengthened beam fails. The various failure modes are discussed below.

Notwithstanding the problem of debonding noted previously, the analysis and design of FRP flexural strengthening systems according to the ACI 440.2R-02 guide is based on the traditional assumptions of reinforced concrete design. These assumptions are compatibility of strain through the depth of the section, and equilibrium of the tension and compression force resultants at any point along the length of the member. This implies that there is no relative axial displacement (or slip) between the FRP system and the concrete (i.e., axial strain compatibility exists at the interface) and that the adhesive layer does not deform in shear when it transfers the forces from the concrete to the FRP strengthening system. These assumptions are assumed to hold even after the internal steel has yielded and the concrete cover layer between the tensile steel and the FRP system is significantly cracked.

Strengthening of members in flexure is achieved only if there is sufficient additional compressive capacity in the concrete to allow for the increase in internal moment in the section. Therefore, flexural strengthening is most suitable for concrete members that are lightly to moderately reinforced, having steel reinforcement in the range 20 to 40% of the balanced steel (unstrengthened) reinforcement ratio. This reinforcement ratio is not uncommon in reinforced concrete members, especially in older structures. In addition, strengthening can be achieved with bonded FRP strengthening systems only if the concrete in the bottom cover layer to which the FRP system is bonded is in suitable condition to permit transfer of the forces into the section above the internal reinforcing. In many older reinforced concrete members in which the internal steel reinforcing has corroded and the cover concrete has spalled, the cover concrete is generally not in good enough condition to perform this function and the concrete needs to be repaired before the FRP strengthening system can be applied.

9.3 FLEXURAL CAPACITY OF AN FRP-STRENGTHENED MEMBER

To begin with, it is important to understand that the key to determining the moment capacity of an FRP flexural strengthened member is to understand the flexural failure modes of the system and to account for them in the design process.[2] These include rupture of the FRP strengthening system, detachment of the FRP strengthening system (due to a variety of delamination or de-

[2] As noted previously, flexural strengthening can cause the strengthened member to be vulnerable to shear failure. Shear strengthening and shear failure modes of transversely loaded flexural members are discussed in Chapter 10.

bonding modes), or compressive failure of the concrete (Arduini and Nanni, 1997; Spadea et al., 1998). All of these modes can occur either before or after the internal steel has yielded. The most desirable mode of failure is concrete compressive failure after the internal steel has yielded with the FRP strengthening system still attached. This is often difficult to achieve, and the mode of FRP detachment (or less frequently, rupture) at large deflections after the internal steel has yielded but prior to the concrete crushing often occurs. The strengthening system can detach in a number of ways. The system can delaminate from the concrete substrate (due to failure in the concrete, the adhesive layer, or in the FRP laminate itself) either at the ends (due to high peeling and shear stresses) or in the interior of the beam due to flexural and shear cracks in the beam at large deflections (Buyukozturk et al., 2004). The various failure modes are shown schematically in Fig. 9.4.

Analytical methods to predict the various detachment (also known as *delamination* or *debonding*) failure modes are still not fully developed,[3] and the ACI guide does not provide an explicit procedure to determine whether or not a detachment failure will occur. Instead, the ACI guide limits the strain permitted in the FRP strengthening system to ensure that none of the detachment failure modes [modes (d) to (g) in Fig. 9.4] is likely to occur. The maximum effective tensile strain in the FRP strengthening system, ε_{fe}, is obtained by multiplying the design ultimate rupture strain, ε_{fu}, by an empirically obtained bond-dependent coefficient, κ_m, which is a function of the stiffness and thickness of the FRP system and defined as

$$\kappa_m = \begin{cases} \dfrac{1}{60\varepsilon_{fu}}\left(1 - \dfrac{nE_ft_f}{2{,}000{,}000}\right) \leq 0.90 & \text{for} \quad nE_ft_f \leq 1{,}000{,}0000 \text{ lb/in.} \\ \dfrac{1}{60\varepsilon_{fu}}\left(\dfrac{500{,}000}{nE_ft_f}\right) \leq 0.90 & \text{for} \quad nE_ft_f > 1{,}000{,}000 \text{ lb/in.} \end{cases}$$

(ACI 440.2R-02:9-2)

where n is the number of layers (or plies) of FRP strips or sheets or fabrics, E_f the longitudinal tensile modulus of the FRP composite in the case of strips or the longitudinal modulus of the fibers in the strengthening direction in the case of sheets or fabrics, and t_f the thickness of an individual strip in the case of FRP strips or the net thickness of the fibers in a single sheet or fabric in the case of sheets or fabrics.

Since κ_m must be less than 0.9, the strain in the FRP is never allowed to reach the ultimate rupture strain in the ACI design procedure. Therefore, theoretically, the failure mode of FRP rupture [mode (a) in Fig. 9.4] can never

[3] See Teng at al. (2001) for a detailed discussion of existing analytical models to predict the loads at which detachment will occur.

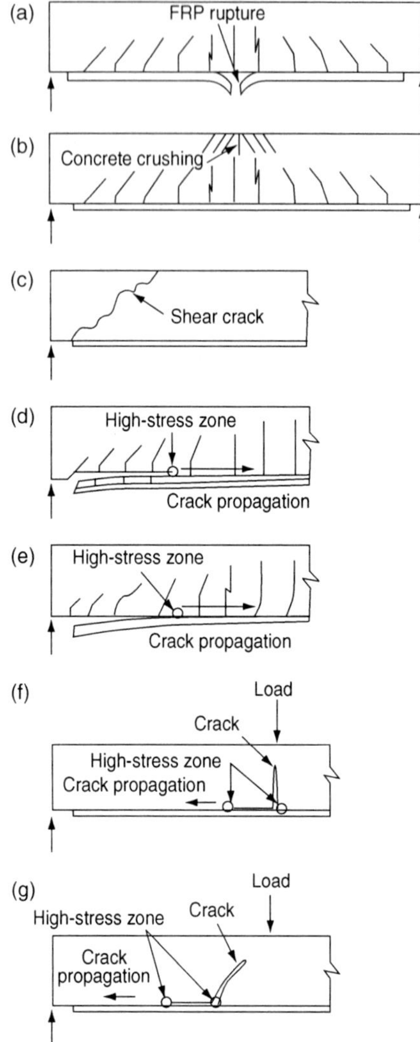

Figure 9.4 Failure modes of flexurally strengthened beams. (From Teng et al., 2001, with permission from John Wiley & Sons.)

occur, according to the ACI. According to the ACI 440.2R-02 guide, only two failure modes are assumed to occur for the purposes of design calculations: compressive failure of the concrete (called *mode 1* in what follows) and failure of the FRP strengthening system (called *mode 2* in what follows). For each of these design failure modes, the stress in the FRP, the internal steel, and the concrete are required to determine the ultimate bending capacity of the section.

9.3.1 Stress in the FRP Strengthening System

Concrete compression failure can occur either after (called *mode 1a*) or before (called *mode 1b*) the internal steel has yielded with the FRP system intact and still attached. This is the failure mode of an overreinforced strengthened concrete beam. In this case, the effective strain in the FRP at failure is obtained from the assumed linear variation of the strain through the depth of the section shown in Fig. 9.5. The effective strain in the FRP is given as

$$\varepsilon_{fe} = \varepsilon_{cu}\frac{h - c}{c} - \varepsilon_{bi} \qquad \text{(ACI 440.2R-02:9-3)}$$

where ε_{fe} is the effective strain in the FRP at ultimate failure of the member, ε_{cu} the ultimate compressive strain in the concrete, c the depth of the neutral axis, h the depth of the section (or more accurately, the distance to the centroid of the FRP material), and ε_{bi} the existing tensile strain in the concrete substrate at the location of the FRP strengthening system.

FRP failure can occur either after (called *mode 2a*) or before (called *mode 2b*) the internal steel has yielded, with the concrete in the compression zone below its compression strength and ultimate failure strain ($\varepsilon_{cu} = 0.003$). Here *failure* means any of the possible detachment modes or even FRP rupture if supplemental anchorages are used to prevent detachment failures. This is the failure mode of an underreinforced strengthened concrete beam. In this case, the effective strain in the FRP at failure is obtained from

$$\varepsilon_{fe} = \kappa_m \varepsilon_{fu} \qquad \text{(ACI 440.2R-02:9-3)}$$

where ε_{fu} is the design strength of the FRP strengthening system. The effective stress, f_{fe}, in the FRP is the stress in the FRP at failure (either in mode 1 or in mode 2) and is linearly related to the effective strain as

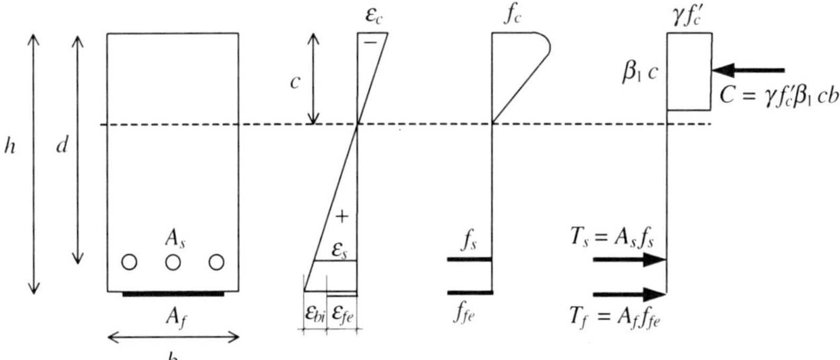

Figure 9.5 Strain, stresses, and force resultants in strengthened section at the ultimate state.

$$f_{fe} = E_f \varepsilon_{fe} \qquad \text{(ACI 440.2R-02:9-4)}$$

9.3.2 Strain in the Internal Reinforcing Steel

For both modes, the strain in the internal reinforcing steel is related linearly to the strain in the FRP at failure as

$$\varepsilon_s = (\varepsilon_{fe} + \varepsilon_{bi}) \frac{d - c}{h - c} \qquad \text{(ACI 440.2R-02:9-8)}$$

and the stress in the internal steel, which must be less than or equal to the yield stress of the steel, is given as

$$f_s = E_s \varepsilon_s \leq f_y \qquad \text{(ACI 440.2R-02:9-9)}$$

9.3.3 Neutral-Axis Depth

The depth of the neutral axis for a singly reinforced section,[4] c, is obtained from force equilibrium in the x-direction as shown in Fig. 9.5.

$$\sum F_x = 0 \rightarrow T_s + T_f - C = 0 \qquad (9.1)$$

where

$$T_s = A_s f_s \qquad (9.2)$$

$$T_f = A_f f_{fe} \qquad (9.3)$$

$$C = \gamma f'_c \beta_1 c b \qquad (9.4)$$

Substituting in the equilibrium equation yields

$$c = \frac{A_s f_s + A_f f_{fe}}{\gamma f'_c \beta_1 b} \qquad \text{(ACI 440.2R-02:9-10)}$$

where A_s is the area of the tensile steel, f_s the stress in the steel at failure, d the depth of the steel reinforcing, β_1 the depth ratio of an equivalent rectangular stress block, A_f the area of the FRP strip or the fibers in a dry fiber system, γ the intensity of an equivalent rectangular stress block (also know as the concrete stress resultant factor), and b the width of the section. The nominal moment capacity of the section is then found from

[4] For doubly reinforced sections the compression steel can be added following the same procedure. If the compression steel yields, additional failure modes are possible.

$$M_n = A_s f_s\left(d - \frac{\beta_1 c}{2}\right) + \psi_f A_f f_{fe}\left(h - \frac{\beta_1 c}{2}\right) \qquad \text{(ACI 440.2R-02:9-11)}$$

9.3.4 Existing Substrate Strain

Since the FRP strengthening system is attached to the existing deflected structure, strains will already exist in the steel and concrete prior to the FRP being attached. These existing strains must be accounted for in the design process if the member is not shored up during the strengthening operation. Neglecting the existing strain in the concrete will lead to unconservative designs. To illustrate the strain conditions in the strengthened beam, the strains in the section can be depicted as a superposition of strains from before and after the FRP is attached, as shown in Fig. 9.6. The three strain states depicted are (1) the initial state: before the FRP is attached, considering *only* the exiting load at the time the FRP is attached; (2) the strengthened state: after the FRP has been attached, considering *only* the supplemental loads that are applied after the FRP has been attached (a hypothetical situation); and (3) the final state: after the FRP has been attached with the existing and supplemental loads. Notice how the strain in the initial state is added to the strain that develops in the FRP strengthening system to give the final strain state at the underside of the beam. This final strain at the underside of the beam must be compatible with the strain in the concrete in the final state. Notice, too, how the neutral axis shifts position in the three states.

In ACI 440.2R-02 the existing strain at the substrate, $\varepsilon_{b(1)}$, is identified as ε_{bi} and can be found from the properties of the transformed cracked section[5] as

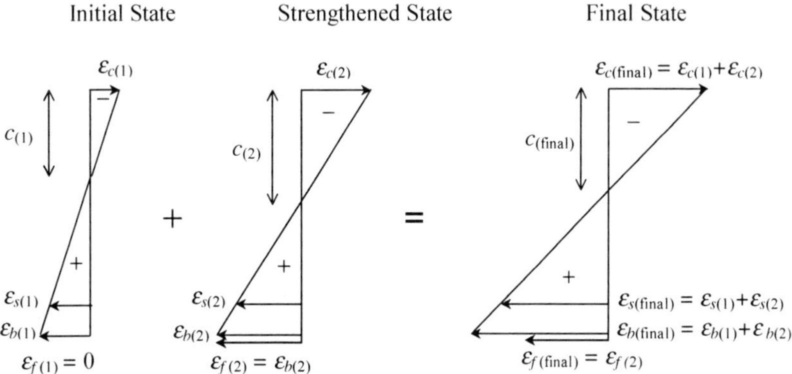

Figure 9.6 Superposition of strains.

[5] Throughout the chapters that discuss FRP strengthening the elastic (cracked) properties of the original (i.e., unstrengthened) section are identified with a subscript 1 and the properties of the strengthened section are identified with a subscript 2.

$$\varepsilon_{bi} = \frac{m_1(h - k_1 d)}{(I_{cr})_1 E_c} \tag{9.5}$$

where m_1 is the service load moment in the beam at the time the FRP system is attached (typically, only the dead load on the beam). The remaining terms are the usual properties of the transformed cracked section of the unstrengthened beam.

9.4 DETERMINATION OF FAILURE MODES AND FLEXURAL CAPACITY

The solution to the preceding equations yields the depth of the neutral axis, the stresses and strains in all the materials (concrete, steel, and FRP), and the nominal moment capacity of the strengthened section. The correct solution to the equations depends on correctly identifying the failure mode of the section. As indicated previously, there are two primary failure modes, each with two variations, as follows:

Mode 1a: Concrete crushing after steel yields
Mode 1b: Concrete crushing before steel yields
Mode 2a: FRP failure after steel yields
Mode 2b: FRP failure before steel yields

To arrive at a solution, one of the four failure modes above must first be assumed. Thereafter, the location of the neutral axis, c, for this assumed failure mode is determined either by a trial-and-error method or by solving a quadratic equation. The correct value of c is the unique solution that gives the stresses and strains in the materials that are compatible with the assumed failure mode. Each of the four modes is discussed in detail below.

It is important to point out that in most practical strengthening designs it is generally accepted that the failures should be in mode 1a or mode 2a; that is, failure occurs in the concrete or the FRP after the internal tension steel has yielded. This is because most FRP strengthening systems have lower moduli and higher strains to failure than the internal reinforcing steel bars, and they are located a similar distance from the neutral axis of the section (i.e., d or h). To utilize the large strain capacity of FRP strengthening systems effectively, the strain in the steel at failure of the strengthened system needs to be much larger than the yield strain of steel, which is 0.00207 for grade 60 rebar. However, for very high modulus strengthening systems, that have lower strains to failure, such as ultrahigh-modulus carbon systems, substantial strengthening can be achieved without the steel yielding before failure (modes 1b and 2b). In addition, for specific geometries and FRP strengthening configurations it may not be possible to achieve steel yielding before failure (becausc of FRP detachment, or an existing high reinforcement ratio, or a very low strength concrete). Also, in some designs the driving design requirement, generally when excessive live loads are a concern, is to reduce the strain in the internal steel at service loads. In these cases, failure after steel yielding is not a desirable failure mode. For all the reasons mentioned above, failure modes 1b and 2b are also covered in full detail in what follows, even though most FRP strengthened members will tend not to fail in these modes.

9.4.1 Mode 1a: Concrete Crushing After Steel Yields

In this mode the steel yields, and this is followed by concrete crushing with the strain in the FRP below its effective rupture strain. Therefore, the steel stress is set to its yield stress and the Whitney stress block parameters are applicable for the concrete. The depth of the neutral axis is found by substituting the following terms in the equilibrium equation:

$$T_s = A_s f_y \tag{9.6}$$

$$T_f = A_f f_{fe} = A_f E_f \left(\varepsilon_{cu} \frac{h - c}{c} - \varepsilon_{bi} \right) \tag{9.7}$$

$$C = 0.85\, f'_c \beta_1 cb \tag{9.8}$$

$T_s + T_f - C = 0$ and rearranging gives the quadratic equation for c as

$$(0.85 f'_c \beta_1 b)c^2 - A_s f_y c - A_f E_f [\varepsilon_{cu}(h - c) - \varepsilon_{bi} c] = 0 \tag{9.9}$$

which can be written in the form

$$Ac^2 + Bc + C = 0 \tag{9.10}$$

with

$$A = 0.85 f'_c \beta_1 b \tag{9.11}$$

$$B = -A_s f_y + A_f E_f (\varepsilon_{cu} + \varepsilon_{bi}) \tag{9.12}$$

$$C = -(A_f E_f \varepsilon_{cu} h) \tag{9.13}$$

With the value of c obtained, the strains in the steel and FRP are calculated and checked to ensure that the steel strain exceeds the steel yield strain, ε_{sy}, and that the FRP strain is less than $\kappa_m \varepsilon_{fu}$.

$$\varepsilon_s = \varepsilon_{cu}\frac{d-c}{c} \geq \varepsilon_{sy} \tag{9.14}$$

$$\varepsilon_{fe} = \varepsilon_{cu}\frac{h-c}{c} - \varepsilon_{bi} \leq \kappa_m\varepsilon_{fu} \qquad \text{(ACI 440.2R-02:9-3)}$$

If the two conditions above are satisfied, the steel stress is taken as f_y and the FRP stress is taken as

$$f_{fe} = E_f\varepsilon_{fe} \qquad \text{(ACI 440.2R-02:9-4)}$$

The nominal moment capacity is obtained from

$$M_n = A_s f_y\left(d - \frac{\beta_1 c}{2}\right) + \psi_f A_f f_{fe}\left(h - \frac{\beta_1 c}{2}\right) \qquad \text{(ACI 440.2R-02:9-11)}$$

and the factored moment capacity is given as

$$\phi M_n \tag{9.15}$$

where $\phi = 0.7 + 0.2(\varepsilon_s - \varepsilon_{sy})/(0.005 - \varepsilon_{sy})$ and depends on the value of the strain in the steel at the time of concrete crushing failure.

9.4.2 Mode 1b: Concrete Crushing Before Steel Yields

In this mode, the concrete is crushed before the steel yields with the strain in the FRP below its effective rupture strain. Therefore, the Whitney stress block parameters are applicable for the concrete. The depth of the neutral axis is found by substituting the following terms in the equilibrium equation:

$$T_s = A_s f_s = A_s E_s \varepsilon_s = A_s E_s \varepsilon_{cu}\frac{d-c}{c} \tag{9.16}$$

$$T_f = A_f f_{fe} = A_f E_f\left(\varepsilon_{cu}\frac{h-c}{c} - \varepsilon_{bi}\right) \tag{9.17}$$

$$C = 0.85 f'_c \beta_1 cb \tag{9.18}$$

$T_s + T_f - C = 0$ and rearranging gives the quadratic equation for c as

$$(0.85 f'_c \beta_1 b)c^2 - A_s E_s \varepsilon_{cu}(d-c) - A_f E_f[\varepsilon_{cu}(h-c) - \varepsilon_{bi}c] = 0 \tag{9.19}$$

which can be written in the form

$$Ac^2 + Bc + C = 0 \tag{9.20}$$

with

$$A = 0.85 f'_c \beta_1 b \tag{9.21}$$

$$B = A_s E_s \varepsilon_{cu} + A_f E_f (\varepsilon_{cu} + \varepsilon_{bi}) \tag{9.22}$$

$$C = -(A_s E_s \varepsilon_{cu} d + A_f E_f \varepsilon_{cu} h) \tag{9.23}$$

With the value of c obtained, the strains in the steel and FRP are calculated and checked to ensure that the steel strain is less than the steel yield strain, ε_{sy}, and that the FRP strain is less than $\kappa_m \varepsilon_{fu}$.

$$\varepsilon_s = \varepsilon_{cu} \frac{d - c}{c} \le \varepsilon_{sy} \tag{9.24}$$

$$\varepsilon_{fe} = \varepsilon_{cu} \frac{h - c}{c} - \varepsilon_{bi} \le \kappa_m \varepsilon_{fu} \qquad \text{(ACI 440.2R-02:9-3)}$$

If the two conditions above are satisfied, the steel stress is taken as

$$f_s = E_s \varepsilon_s \qquad \text{(ACI 440.2R-02:9-9)}$$

the FRP stress is taken as

$$f_{fe} = E_f \varepsilon_{fe} \qquad \text{(ACI 440.2R-02:9-4)}$$

the nominal moment capacity is obtained from

$$M_n = A_s f_s \left(d - \frac{\beta_1 c}{2} \right) + \psi_f A_f f_{fe} \left(h - \frac{\beta_1 c}{2} \right) \qquad \text{(ACI 440.2R-02:9-11)}$$

and the factored moment capacity is given as

$$\phi M_n \tag{9.25}$$

where $\phi = 0.7$ since the steel has not yielded at the time of concrete crushing failure.

9.4.3 Mode 2a: FRP Failure After Steel Yields

In this mode the steel yields, and following this the FRP fails with the strain in the concrete below its ultimate strain. Therefore, the steel stress is set to its yield stress and the FRP stress is set to its effective FRP failure strength, $f_{fe} = \kappa_m f_{fu}$. Since the concrete has not reached its ultimate compressive strain, the Whitney stress block parameters should technically not be used and the stress block parameters, β_1 and γ, should be obtained using a nonlinear analysis. This is the same as the case of the underreinforced section with FRP bars described in Chapter 5. The nonlinear stress–strain law described in Chapter 5 can be used.

However, ACI 440.2R-02 permits the use of the Whitney stress block parameters in lieu of conducting a nonlinear iterative analysis since experimental results indicate that the strain in the concrete is typically close to the ultimate strain, when the FRP fails in a well-designed strengthening system. (In a well-designed strengthening system the section is close to the inelastic balanced strengthening ratio, described in detail below). Nevertheless, the strain in the concrete at failure should be checked and compared with the concrete ultimate strain, ε_{cu} (i.e., 0.003), to verify that the concrete is close to compression failure. If this is not the case, the nonlinear analysis should be considered. The depth of the neutral axis is found by substituting the following terms in the equilibrium equation:

$$T_s = A_s f_y \tag{9.26}$$

$$T_f = A_f \kappa_m f_{fu} \tag{9.27}$$

$$C = 0.85 f'_c \beta_1 cb \tag{9.28}$$

$T_s + T_f - C = 0$ and rearranging gives the linear equation for c as

$$(0.85 f'_c \beta_1 b)c - A_s f_y - A_f \kappa_m f_{fu} = 0 \tag{9.29}$$

which gives

$$c = \frac{A_s f_y + A_f \kappa_m f_{fu}}{0.85 f'_c \beta_1 b} \tag{9.30}$$

With the value of c obtained, the strains in the concrete must be calculated and checked to ensure that the concrete strain is less than the concrete ultimate strain, ε_{cu}, and the strain in the steel must be calculated to ensure that it has indeed yielded.

$$\varepsilon_c = (\varepsilon_{fe} + \varepsilon_{bi}) \frac{c}{h - c} \leq \varepsilon_{cu} \tag{9.31}$$

$$\varepsilon_s = (\varepsilon_{fe} + \varepsilon_{bi}) \frac{d - c}{h - c} \geq \varepsilon_{sy} \tag{9.32}$$

If the conditions above are satisfied, the steel stress is taken as f_y and the FRP stress is taken as

$$f_{fe} = E_f \varepsilon_{fe} = \kappa_m f_{fu} \tag{9.33}$$

The nominal moment capacity is obtained as

$$M_n = A_s f_y \left(d - \frac{\beta_1 c}{2} \right) + \psi_f A_f f_{fe} \left(h - \frac{\beta_1 c}{2} \right) \qquad \text{(ACI 440.2R-02:9-11)}$$

and the factored moment capacity is given as

$$\phi M_n \tag{9.34}$$

where $\phi = 0.7 + 0.2(\varepsilon_s - \varepsilon_{sy})/(0.005 - \varepsilon_{sy})$ and depends on the value of the strain in the steel at the time of FRP failure.

9.4.4 Mode 2b: FRP Failure Before Steel Yields

In this mode, the FRP fails before the steel yields with the strain in the concrete below its ultimate strain. Therefore, the FRP stress is set to the effective FRP failure strength, $f_{fe} = \kappa_m f_{fu}$. Since the concrete has not reached its ultimate compressive strain, the Whitney stress block parameters should technically not be used and the stress block parameters, β_1 and γ, should be obtained using a nonlinear analysis. As discussed for mode 2a, ACI 440.2R-02 permits the use of the Whitney stress block parameters in lieu of conducting a nonlinear iterative analysis since experimental results indicate that the strain in the concrete is typically close to the ultimate strain when the FRP fails in a well-designed strengthening system. Nevertheless, the strain in the concrete at failure should be checked and compared with the concrete ultimate strain, ε_{cu} (i.e., 0.003), to verify that the concrete is close to compression failure.[6] If this is not the case, the nonlinear analysis should be considered. The depth of the neutral axis is found by substituting the following terms in the equilibrium equation.

$$T_s = A_s f_s = A_s E_s \varepsilon_s = A_s E_s (\varepsilon_{fe} + \varepsilon_{bi}) \frac{d - c}{h - c} \tag{9.35}$$

$$T_f = A_f f_{fe} = A_f \kappa_m f_{fu} = A_f E_f \varepsilon_{fe} \tag{9.36}$$

$$C = 0.85 f'_c \beta_1 c b \tag{9.37}$$

$T_s + T_f - C = 0$ and rearranging gives the quadratic equation for c as

[6] This can be assumed to be 0.002 or greater if the ultimate strain is taken to be 0.003 according to ACI 318.

$$(0.85f'_c\beta_1 b)c(h - c) - A_sE_s(\varepsilon_{fe} + \varepsilon_{bi})(d - c) - A_fE_f\varepsilon_{fe}(h - c) = 0 \quad (9.38)$$

which can be written in the form

$$Ac^2 + Bc + C = 0 \quad (9.39)$$

with,

$$A = 0.85f'_c\beta_1 b \quad (9.40)$$

$$B = -[0.85\beta_1 f'_c bh + A_sE_s(\varepsilon_{fe} + \varepsilon_{bi}) + A_fE_f\varepsilon_{fe}] \quad (9.41)$$

$$C = A_sE_s(\varepsilon_{fe} + \varepsilon_{bi})d + A_fE_f\varepsilon_{fe}h \quad (9.42)$$

With the value of c obtained, the strains in the concrete must be calculated and checked to ensure that the concrete strain is less than the concrete ultimate strain, ε_{cu}, and the strain in the steel must be calculated to ensure that it is less than the steel yield strain, ε_{sy}.

$$\varepsilon_c = (\varepsilon_{fe} + \varepsilon_{bl})\frac{c}{h - c} \leq \varepsilon_{cu} \quad (9.43)$$

$$\varepsilon_s = (\varepsilon_{fe} + \varepsilon_{bi})\frac{d - c}{h - c} \leq \varepsilon_{sy} \quad (9.44)$$

If the two conditions above are satisfied, the steel stress is taken as

$$f_s = E_s\varepsilon_s \quad \text{(ACI 440.2R-02:9-9)}$$

the FRP stress is taken as

$$f_{fe} = E_f\varepsilon_{fe} = \kappa_m f_{fu} \quad (9.45)$$

the nominal moment capacity is obtained from

$$M_n = A_sf_s\left(d - \frac{\beta_1 c}{2}\right) + \psi_f A_f f_{fe}\left(h - \frac{\beta_1 c}{2}\right) \quad \text{(ACI 440.2R-02:9-11)}$$

and the factored moment capacity is given as

$$\phi M_n \quad (9.46)$$

where $\phi = 0.7$ since the steel has not yielded at the time of FRP failure.

9.5 BALANCED CONDITION

A balanced condition can be defined for the FRP strengthened section in a fashion similar to the balanced condition for a typical reinforced concrete section. The balanced condition for the strengthened section is defined as the condition at which concrete compression failure and FRP failure occur simultaneously. Depending on the strain in the steel at which this simultaneous failure occurs, two balanced conditions are possible: defined herein as the *elastic balanced condition,* in which the tension steel has not yielded when the balanced strain condition is reached, and the *inelastic balanced condition,* in which the steel has yield prior to the balanced strain condition being reached. As discussed previously, for practical strengthening designs, the inelastic balanced condition is the balanced condition of principal interest. It is the only one that is presented in what follows; however, the elastic balanced condition can be derived following a similar procedure.

The inelastic balanced strain condition is obtained by setting the concrete failure strain to $\varepsilon_{cu} = 0.003$ and the FRP failure strain to $\varepsilon_{fe} = \kappa_m \varepsilon_{fu}$ and checking that the steel strain $\varepsilon_s \geq \varepsilon_{sy}$. Using this condition the *inelastic balanced reinforcement ratio* can be defined as

$$\rho_{fb} = \frac{A_{fb}}{bd} \tag{9.47}$$

Since the FRP failure strain is a function of the bond coefficient, κ_m, a specific number of layers, n, of the strengthening system must be assumed to determine the balanced reinforcement ratio. That is, different balanced ratios may exist for the same beam, depending on the number of layers of FRP strengthening system used. This is not a significant obstacle in design given that most practical flexural strengthening systems typically use between one and three layers.

The depth of the neutral axis at the balanced condition is found from strain compatibility as

$$c_b = \frac{\varepsilon_{cu} h}{\varepsilon_{cu} + \kappa_m \varepsilon_{fu} + \varepsilon_{bi}} \tag{9.48}$$

The resulting section forces (see Fig. 9.5) at the inelastic balanced condition are given as

$$C = 0.85 f'_c \beta_1 c_b b \tag{9.49}$$

$$T_s = A_s f_y \tag{9.50}$$

$$T_f = A_{fb} \kappa_m f_{fu} \tag{9.51}$$

Substitution into the equilibrium equation yields

$$\sum F_x = 0 \rightarrow T_s + T_f - C = 0 \tag{9.52}$$

and solving for A_{fb} gives

$$A_{fb} = \frac{0.85 f'_c \beta_1 bh \varepsilon_{cu}}{\kappa_m f_{fu}(\varepsilon_{cu} + \kappa_m \varepsilon_{fu} + \varepsilon_{bi})} - \frac{A_s f_y}{\kappa_m f_{fu}} \tag{9.53}$$

and

$$\rho_{fb} = 0.85 \frac{f'_c \beta_1}{\kappa_m f_{fu}} \frac{\varepsilon_{cu}}{\varepsilon_{cu} + \kappa_m \varepsilon_{fu} + \varepsilon_{bi}} \frac{h}{d} - \frac{f_y \rho_s}{\kappa_m f_{fu}} \tag{9.54}$$

Provided that the steel has yielded,

$$\varepsilon_s = (\varepsilon_{fe} + \varepsilon_{bi}) \frac{d - c_b}{h - c_b} \geq \varepsilon_{sy} \tag{9.55}$$

It is very important to note that unlike in steel reinforced sections, it is not always possible to achieve a balanced condition, due to the wide range of properties of different FRP strengthening systems and the existing dimensions of the concrete section. In particular, it is important to reemphasize that to obtain strengthening the compression force resultant in the section must increase from the unstrengthened (existing) case. For this to happen, the neutral axis of the strengthened section must move downward from its original position (i.e, $c_{b(\text{new})} > c_{\text{unstrengthened}}$). If the strain to failure in the FRP is high and the modulus is relatively low (such as for a glass FRP system), the opposite may occur (i.e., the neutral axis may move upward), which is not possible and means that the balanced condition cannot be achieved for the number of plies of the FRP system chosen. If the calculation for the balanced reinforcement ratio gives a rational result (i.e., $c_{b(\text{new})} > c_{\text{unstrengthened}}$), the balanced reinforcement ratio can be useful in determining the optimal area of FRP reinforcement required for a given strengthening demand.

In addition, if the balanced ratio can be found for a particular FRP system, it is a useful to tool for designing the FRP system to achieve a desired failure mode (i.e., concrete crushing or FRP failure) since:

If $A_f < A_{fb}$, the FRP failure mode (mode 2a) will control and $c < c_b$.
If $A_f > A_{fb}$, the concrete failure mode (mode 1a) will control and $c > c_b$.

9.6 DETAILING FOR FLEXURAL STRENGTHENING

At the strip ends two empirical checks are required by ACI 440.2R-02 to detail the FRP strip to ensure that premature failure does not occur in the beam prior to it achieving its design flexural strength. These are:

1. A strength check at the strip end or inflection point to prevent delamination due to anchorage failure. According to ACI 440.2R-02, the FRP flexural strengthening system must extend a distance d beyond the point corresponding to the cracking moment under factored loads for a simply supported beam.
2. A strength check at the strip end to prevent shear failure of the beam at the point of FRP strip termination. ACI 440.2R-02 requires that the factored shear force at the FRP strip termination point be less than two-thirds of the concrete shear capacity at this point.

9.7 DESIGN PROCEDURE FOR A FLEXURALLY STRENGTHENED CONCRETE MEMBER

To design an FRP flexural strengthening system for a reinforced concrete member to achieve a required moment capacity (demand), M_u, for future loads, the following steps should be followed:

Step 1. Determine the flexural and shear capacity of the original beam. The original capacity must be known prior to performing a strengthening design. The geometric and material properties of the existing beam are needed for this step. If existing plans and specifications are available, they can be consulted, but the designer needs to check that these accurately reflect the properties of the as-built member that will be strengthened. In many strengthening cases original plans or specifications no longer exist or can no longer be found, and the designer needs to obtain the data from the existing structure using field measurements, material tests, and engineering judgment.

Step 2. Determine the section properties of the original beam and the existing strain on the soffit. These properties are needed in subsequent calculations. It is usual to assume that the beam or slab soffit has been cracked by prior loads and that only the dead load will be acting when the strengthening system is applied. If other loads besides dead loads are on the member when it is to be strengthened or a shoring system is used, the designer needs to determine the strain on the soffit for these conditions.

Step 3. Determine current and future design loads and capacities and check the ultimate strengthening limits. Based on the given strengthening requirement the future required flexural and shear capacities (demands) need to be determined to determine if in fact strengthening is needed. In some cases strengthening may not actually be needed. This is because the member may have been overdesigned for the original loads due either to minimum reinforcement requirements (especially for shear reinforcement) or due to sizes of available bars and detailing requirements (especially for flexural reinforcement). In this step the maximum code-permitted strengthening limits for flexural strengthening should be checked to give the designer a feel for how much strengthening can be obtained and how this compares with the demand strengthening.

Step 4. Select an FRP strengthening system and determine its design properties. Numerous different FRP strengthening systems are available. The designer should be aware of which systems can readily be obtained and if a certified installer is available at the project location. Designs with different FRP systems should be considered. It is advisable to do a design with precured strips and laid-up sheets and to consider both carbon and glass systems. The unit costs of the various systems (both material and installation costs) should be obtained in this step of the design process.

Step 5. Estimate the amount of strengthening required and configure the FRP system. To begin the design, an estimate of the amount of strengthening can be obtained. An estimate of the required FRP area can be found as follows:

(i) Determine the additional factored moment that needs to be carried by the FRP, $\Delta\phi M_n$. Assume $\phi = 0.9$ and determine $\Delta M_n = (\Delta\phi M_n)/\phi$.

(ii) Approximate the stress in the FRP at failure as 80% of the effective strength (i.e., $\kappa_m f_{fu}$) with $\kappa_m = 0.8$. Since the maximum allowable stress in the FRP is 90% of the design strength, this assumes that failure due to concrete crushing may occur. Also, it assumes that a designer would like to try to stress the FRP system as much as possible to obtain an economically efficient design.

(iii) Approximate the depth of the compression block in the concrete as equal to c in the unstrengthened section, recognizing that it must be greater than a since the neutral axis will move downward after the strengthening has been applied.

(iv) The additional nominal moment carried by the FRP is given as

$$\Delta M_n = \psi_f A_f f_{fe}\left(h - \frac{c}{2}\right) \tag{9.56}$$

which gives the *estimated* required area of FRP strengthening as

$$A_f^{\text{reqd}} = \frac{\Delta M_n}{\psi_f f_{fe}}(h - c/2) \tag{9.57}$$

Or, using $A_f = nt_f w_f$, the approximate number of layers of the chosen FRP system is found as

$$n = \frac{\Delta M_n}{\psi_f t_f w_f f_{fe}(h - c/2)} \tag{9.58}$$

Step 6. Determine the inelastic balanced reinforcement ratio. Determine the inelastic balanced ratio for various numbers of layers of the system chosen, always ensuring that steel yielding has occurred. If a specific failure mode is desired, choose the number of layers and area of FRP to achieve this failure. To ensure the desired failure mode, it is recommended either 10% more (concrete failure) or less (FRP failure) FRP than the balanced FRP ratio be used.

Step 7. Select a trail design. Based on the data collected in steps 5 and 6, select a trail design.

Step 8. Determine the capacity and suitability of the section with the trial FRP design. At a very minimum, the flexural capacity for the strengthened member must exceed the flexural demand. However, the suitability of the design should be evaluated. If the FRP system is overdesigned, another iteration should be performed to reduce the amount of FRP used (provided that this does not have a negative impact on constructability). A design whereby failure is by the opposite failure mode from the one chosen in the trail design should be attempted and a cost comparison made between the two designs. A different FRP system may need to be considered in this case since different FRP systems may be not be susceptible to the same failure modes.

Step 9. Detail the FRP flexural strengthening system. (a) Check the strength at the strip end or inflection point to prevent delamination due to anchorage failure. (b) Check the strength at the strip end to prevent shear failure of the beam at the point of FRP strip termination.

Design Example 9.1: FRP Flexural Strengthening of a Beam An existing simply supported reinforced concrete beam in the interior of a building spans 24 ft and is constructed of 5000-psi concrete with three No. 9 grade 60 main bars.[7] Shear reinforcement is provided by No. 3 grade 60 stirrups at 10.5 in. on center. The beam was originally designed for the following loads:

$$w_{DL} = 1.0 \text{ kip/ft}$$

$$w_{LL} = 1.2 \text{ kip/ft}$$

$$W_u = 1.4(1.0) + 1.7(1.2) = 3.44 \text{ kip/ft}$$

[7] This example has the same dimensions and properties as the design example given in Section 14.3 of ACI 440.2R-02 but uses a different solution procedure and discusses design-related issues in more detail.

Due to a change in the use of the structure, the live load carrying–capacity of the beam needs to be increased by 50%. You are required to design an FRP strengthening system for the beam. The dimensions of the existing beam section are given in Fig. 9.7.

SOLUTION

Step 1. Determine the flexural and shear capacity of the original beam. The flexural capacity of the existing beam:

$$\rho_b = 0.85 \frac{f'_c}{f_y} \beta_1 \left(\frac{87{,}000}{87{,}000 + f_y} \right)$$

$$= 0.85 \left(\frac{5000}{60{,}000} \right) (0.80) \left(\frac{87{,}000}{87{,}000 + 60{,}000} \right) = 0.0335$$

$$\rho_s = \frac{A_s}{bd} = \frac{3(1.0)}{12(21.5)} = 0.0116$$

$$\frac{\rho_s}{\rho_b} = \frac{0.0116}{0.0335} = 0.34 \rightarrow \text{i.e., } 34\% \text{ of balanced}$$

Note that the existing beam is underreinforced and is a good candidate for strengthening (less than 40% of the balanced steel).

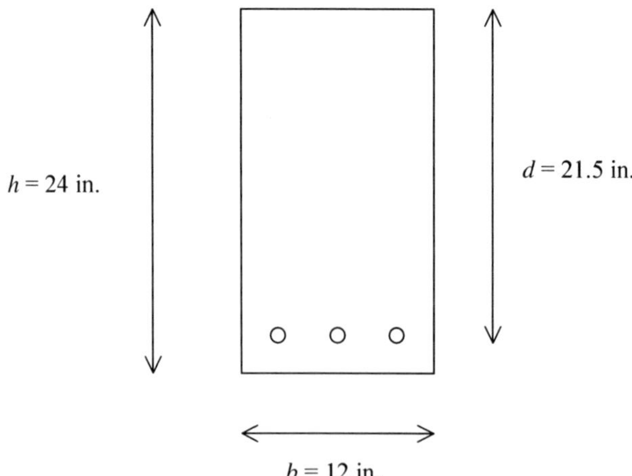

Figure 9.7 Beam cross-sectional dimensions for Example 9.1.

$$a = \frac{A_s f_y}{0.85 f'_c b} = \frac{3.0(60{,}000)}{0.85(5000)(12)} = 3.53 \text{ in.}$$

$$c = \frac{a}{\beta_1} = \frac{3.53}{0.80} = 4.41 \text{ in.}$$

$$M_n = A_s f_y\left(d - \frac{a}{2}\right) = 3.0(60{,}000)\left(21.5 - \frac{3.53}{2}\right)$$

$$= 3552 \text{ lb-in.} = 296 \text{ kip-ft}$$

and since the beam is underreinforced, $\phi = 0.90$[8] and $\phi M_n = 0.9(296) =$ 266 kip-ft.

Based on the existing loads, the original flexural design demand was

$$M_u = \frac{3.44(24)^2}{8} = 247.7 \text{ kip-ft} < 266 \text{ kip-ft}$$

and the original beam design was satisfactory for the existing loads. Note that this is only a check.

The shear capacity of the existing beam,

$$V_n = V_c + V_s = 2\sqrt{f'_c}bd + \frac{A_{sv} f_y d}{s}$$

$$= 2\sqrt{5000}(12)(21.5) + \frac{2(0.11)(60{,}000)(21.5)}{10.5}$$

$$= 36{,}487 + 27{,}029 = 63{,}516 \text{ lb} = 63.5 \text{ kips}$$

and the factored shear capacity is

$$\phi V_n = 0.85(63.5) = 54.0 \text{ kips}$$

Based on the existing loads, the original shear design demand was

$$V_u = W_u\left(\frac{\ell}{2} - d\right) = 3.44\left(\frac{24}{2} - \frac{21.5}{12}\right) = 35.12 \text{ kips} < 54.0 \text{ kips}$$

and the beam design is satisfactory for the existing loads. Note that this is only a check.

[8] Recall that ACI 440.2R-02 is based on ACI 318-99 load factors and resistance factors.

Step 2. Determine the section properties of the original beam and the existing strain on the soffit. Determine the existing strain on the underside (soffit) of the concrete beam prior to the attachment of the FRP strengthening. Calculate the strain based on the cracked section with only the dead load acting on the beam. To do this, the section properties of the beam are needed. It can generally be assumed that the existing beam has cracked under its prior service loading; however, this should be verified for the sake of completeness. However, even if it is found by calculation that the beam was not cracked under the service loads, it is advisable and conservative to assume that the beam was previously cracked in service in order to determine the existing strain on the substrate. Calculate the cracking moment:

$$f_r = 7.5\sqrt{f'_c} = 7.5\sqrt{5000} = 530.3 \text{ psi}$$

$$I_g = \frac{b\,h^3}{12} = \frac{12(24)^3}{12} = 13{,}824 \text{ in}^4$$

$$M_{cr} = \frac{2f_r I_g}{h} = \frac{2(530.3)(13{,}824)}{24} = 610{,}906 \text{ lb-in.} = 50.9 \text{ kip-ft}$$

$$m_{\mathrm{DL+LL}} = \frac{w_{\mathrm{DL+LL}} L^2}{8} = \frac{(1.0 + 1.2)(24)^2}{8} = 158.4 \text{ kip-ft} > 50.9 \text{ kip-ft}$$

Therefore, the section is cracked. Calculate the cracked section properties.

Method 1: Using the transformed section[9]

$$E_c = 57{,}000\sqrt{f'_c} = 57{,}000\sqrt{5000} = 4{,}030{,}508 \text{ psi} = 4030 \text{ ksi}$$

$$\eta_s = \frac{E_s}{E_c} = \frac{29{,}000{,}000}{4{,}030{,}000} = 7.2$$

$$k_1 = \sqrt{(\rho_s\eta_s)^2 + 2\rho_s\eta_s} - \rho_s\eta_s$$

$$= \sqrt{[0.0116(7.2)]^2 + 2(0.0116)(7.2)} - 0.0116(7.2) = 0.334$$

$$(I_{\mathrm{cr}})_1 = \frac{bd^3}{3} k_1^3 + \eta_s A_s d^2(1 - k_1)^2$$

$$= \frac{12(21.5)^3}{3}(0.334)^3 + 7.2(3.0)(21.5)^2(1 - 0.334)^2 = 5909 \text{ in}^4$$

[9] Throughout the chapters that discuss FRP strengthening the elastic (cracked) properties of the original (i.e., unstrengthened) section are identified with a subscript 1 and the properties of the strengthened section are identified with a subscript 2.

$$m_{\text{DL}} = \frac{w_{\text{DL}}L^2}{8} = \frac{1.0(24)^2}{8} = 72 \text{ kip-ft} = 864 \text{ kip-in.}$$

$$\varepsilon_{bi} = \frac{m_{\text{DL}}(h - k_1 d)}{(I_{\text{cr}})_1 E_c} = \frac{864[24 - 0.334(21.5)]}{5909(4030)} = 0.000611$$

Method 2: Using composite mechanics Following the procedure described in Chapter 5, the distance from the top of the section to the neutral axis is found from

$$E_c b \frac{c_1^2}{2} + E_s A_s [c_1 - d] = 0$$

$$4.030(12)\frac{(c_1)^2}{2} + 29(c_1 - 21.5)(3.0) = 0$$

$$24.18(c_1)^2 + 87c_1 - 1870.5 = 0$$

Solving the quadratic equation and taking the positive root gives

$$c_1 = 7.18 \text{ in.}$$

and

$$k_1 = \frac{c_1}{d} = \frac{7.18}{21.5} = 0.334 \text{ in.}$$

The flexural stiffness $(EI)_1$ of the original section is given as

$$(EI)_1 = \sum EI = E_1 I_1 + E_2 I_2 = E_c \frac{b(c_1)^3}{3} + E_s A_s (d - c_1)^2$$

$$= + 4{,}030{,}000\left[\frac{12(7.18)^3}{3}\right] + 29{,}000{,}000(3.0)(21.5 - 7.18)^2$$

$$= 2.38 \times 10^{10} \text{ lb-in}^2$$

and

$$\varepsilon_{bi} = \frac{m_{\text{DL}}(h - k_1 d)}{(EI)_1} = \frac{864[24 - 0.334(21.5)]}{2.38 \times 10^{10}} = 0.000611$$

which is the same result as obtained above using the transformed section method. In what follows it will be seen that the composite section method is easily adapted to the strengthened section and is recommended for use with the strengthened section.

Note that the transformed cracked second moment can be found from the above from

$$I_{cr} = \frac{(EI)_1}{E_c} = \frac{2.38 \times 10^{10}}{4{,}030{,}000} = 5906 \text{ in}^4 \approx 5909 \text{ in}^4$$

Step 3. Determine the current and future design loads and capacities and check the ultimate strengthening limit. The beam needs to be strengthened to carry the following future loads:

$$w_{DL} = 1.0 \text{ kip/ft}$$

$$w_{LL} = 1.2\ (1.5) = 1.8 \text{ kip/ft}$$

$$W_u = 1.4(1.0) + 1.7(1.8) = 4.46 \text{ kip/ft}$$

The future design demand $M_u = 4.46(24)^2/8 = 321.1$ kip-ft > 266 kip-ft, and therefore flexural strengthening is required for the future loads. This is equal to a percentage increase of 20.7% in the moment-carrying capacity. The future shear demand $V_u = 4.46(24/2 - 21.5/12) = 45.5$ kips < 54.0 kips, and therefore the existing shear capacity of the beam is sufficient to carry the future loads.

Check the strengthening limit for future limit-factored loads and compare with the existing unstrengthened capacity:

$$W_{\text{limit}} = 1.2(1.0) + 0.85(1.8) = 2.73 \text{ kip/ft}$$

$$M_{\text{limit}} = \frac{2.73(24)^2}{8} = 196.6 \text{ kip-ft} < 266 \text{ kip-ft}$$

Therefore, the beam can be strengthened to carry the future loads without exceeding the ACI mandated limit on the maximum strengthening enhancement allowed.[10]

The maximum permissible strengthened moment capacity allowed for this beam can also be calculated if desired as follows:

$$(\phi M_n)_{\max} = (\phi M_n)_{\text{org}} \frac{1.4 + 1.7R}{1.2 + 0.85R}$$

Assuming that the live load-to-dead load ratio will remain the same for the maximum strengthening (i.e., $R = 1.8/1.0 = 1.8$), we obtain

[10] In the general case, the designer must ensure that the limiting condition is met for the shear capacity as well as the flexural capacity.

$$(\phi M_n)_{\max} = 266\left[\frac{1.4 + 1.7(1.8)}{1.2 + 0.85(1.8)}\right] = 434.6 \text{ kip-ft} > 321.1 \text{ kip-ft}$$

and the maximum permissible % increase over the original capacity (which is not necessarily the original loads) is

$$\%_{\max} = \frac{0.2 + 0.85R}{1.2 + 0.85R} \times 100$$

$$\%_{\max} = \frac{0.2 + 0.85(1.8)}{1.2 + 0.85(1.8)} \times 100 = 63.4\% > 20.7\%$$

Step 4. Select an FRP strengthening system and determine its design properties. Consider a strengthening system with the following ply properties:

$$t_f = 0.04 \text{ in.} \qquad f_{fu}^* = 90 \text{ ksi} \qquad E_f = 5360 \text{ ksi}$$

$$w_f = 12 \text{ in.} \qquad \varepsilon_{fu}^* = 0.017 = 1.7\%$$

Note that this is a hypothetical FRP system and does not represent properties of any known actual FRP strengthening system. The properties provided are similar to those of a glass FRP strengthening system, although the ACI 440.2R-02 identifies this as a carbon FRP system.

Typically, only the longitudinal properties of an FRP strengthening system are given. However, when using a multiple fiber orientation strengthening layup, these properties cannot be used because a multidirectional laminate will have significantly different longitudinal properties than the unidirectional properties, as explained in Chapter 3. In this case, the effective longitudinal properties of the multidirectional layup in the direction of the required strengthening must first be determined using the methods for multidirectional laminates described in Chapter 3. Recall that all four in-plane stiffnesses and all five in-plane strengths of the individual orthotropic plies are needed to obtain the in-plane properties of a multidirectional laminate.

Determine the design strength and design rupture strain of the FRP system selected:

$$C_E = 0.95 \qquad \text{for carbon–epoxy in interior use}$$

$$f_{fu} = C_E f_{fu}^* = 0.95(90) = 85.5 \text{ ksi}$$

$$\varepsilon_{fu} = C_E \varepsilon_{fu}^* = 0.95(0.017) = 0.0162$$

Step 5. Estimate the amount of strengthening required and configure the FRP system. The additional moment that needs to be carried by the FRP is

$$\Delta\phi M_n = 321 - 266 \text{ kip-ft} = 55 \text{ kip-ft}$$

$$\Delta M_n = \frac{55}{0.9} = 61.1 \text{ kip-ft} = 733.20 \text{ kip-in.}$$

Approximate the stress in the FRP at failure as 80% of the effective strength (i.e., $\kappa_m f_{fu}$) with $\kappa_m = 0.8$. Since the maximum allowable stress in the FRP is 90% of the design strength, this assumes that failure due to concrete crushing may occur. Also, it assumes that a design would like to try to stress the FRP system as much as possible to obtain an economically efficient design,

$$f_{fe} = 0.8\kappa_m f_{fu} = 0.8(0.8)(85.5) = 54.72 \text{ ksi}$$

Approximate the depth of the compression block in the concrete as equal to c in the unstrengthened section because it will be greater than a with the neutral axis moving downward after the strengthening has been applied,

$$c = 4.41 \text{ in.}$$

The additional nominal moment carried by the FRP is given as

$$\Delta M_n = \psi_f A_f f_{fe}\left(h - \frac{c}{2}\right)$$

which gives the *estimated* required area of FRP strengthening as

$$A_f^{\text{reqd}} = \frac{\Delta M_n}{\psi_f f_{fe}(h - c/2)} = \frac{773.2}{0.85(54.72)(24 - 4.41/2)} = 0.763 \text{ in}^2$$

Or, using $A_f = nt_f w_f$, we find the approximate number of layers of the chosen FRP system with a 12-in. width[11] (i.e., the maximum width possible since the beam is 12 in. wide) as

[11] The designer should also consider the case of the FRP strengthening system not being the full width of the beam. A fully optimized solution will consider both different widths and different number of layers.

$$n = \frac{\Delta M_n}{\psi_f t_f w_f f_{fe}(h - c/2)} = \frac{773.20}{0.85(0.04)(12)(54.72)(24 - 4.41/2)}$$
$$= 1.6 \quad \therefore \text{ try 2 plies}$$

Step 6. Determine the inelastic balanced reinforcement ratio. First, determine the unit stiffness of the FRP strengthening system and the bond-dependent coefficient, κ_m. The procedure is demonstrated for two layers of the FRP system chosen,

$$nE_f t_f = 2(5{,}360{,}000)(0.04) = 428{,}800 \leq 1{,}000{,}000 \text{ lb/in.}$$

and therefore,

$$\kappa_m = \frac{1}{60\varepsilon_{fu}}\left(1 - \frac{nE_f t_f}{2{,}000{,}000}\right) = \frac{1}{60(0.0162)}\left(1 - \frac{428{,}800}{2{,}000{,}000}\right)$$
$$= 0.818 \leq 0.90$$

and

$$c_b = \frac{\varepsilon_{cu} h}{\varepsilon_{cu} + \kappa_m \varepsilon_{fu} + \varepsilon_{bi}} = \frac{0.003(24)}{0.003 + 0.818(0.0162) + 0.00061} = 4.27 \text{ in.}$$

Observe that 4.27 in. < 4.41 in., which is the depth of the neutral axis in the original unstrengthened state. Therefore, an area of FRP using two layers of the system chosen will not produce the balanced condition. In addition, it will not be possible to achieve FRP failure for two layers of this system, regardless of the area chosen. The section will fail due to concrete compression. To demonstrate this point, the balanced reinforcement area is calculated for this case.

First, confirm that the steel has yielded for this situation:

$$\varepsilon_{sy} = \frac{f_y}{E_s} = \frac{60{,}000}{29 \times 10^6} = 0.00207$$

$$\varepsilon_s = (\varepsilon_{fe} + \varepsilon_{bi})\frac{d - c}{h - c}$$
$$= [0.818(0.0162) + 0.00061]\frac{21.5 - 4.27}{24 - 4.27} = 0.0121 > 0.00207$$

Therefore, the steel has yielded and the inelastic balanced area can be found from

$$A_{fb} = \frac{0.85 f'_c \beta_1 b h \varepsilon_{cu}}{\kappa_m f_{fu}(\varepsilon_{cu} + \kappa_m \varepsilon_{fu} + \varepsilon_{bi})} - \frac{A_s f_y}{\kappa_m f_{fu}}$$

$$= \frac{0.85(5000)(0.80)(12)(24)(0.0003)}{0.818(85{,}500)[0.003 + 0.818(0.0162) + 0.00061]}$$

$$- \frac{3.0(60{,}000)}{0.818(85{,}500)}$$

$$= -0.083 \text{ in}^2 < 0 \rightarrow \text{not possible!}$$

Clearly, a negative area of FRP is not possible and the solution is invalid. The balanced condition cannot be achieved with two layers of the system chosen. Stated another way, if the neutral axis must move upward (i.e., the compression force resultant in the concrete is reduced from its original value), an additional compression force resultant is required in the section, which can only be supplied by the FRP—which is not possible!

At this stage in the design process, it is useful to determine the balanced FRP area for likely numbers of layers of the chosen reinforcement system. As discussed previously, this is useful in understanding how the failure modes relate to the FRP system area and the number of layers. As a result, specific failure modes can be selected based on the foregoing study. Table 9.1 shows the results of the calculations for one to five layers of the system chosen.

Step 7. Select a trail design. Based on the approximate calculation of $A_f^{\text{reqd}} = 0.76$ in^2 of step 5 and the balanced calculations of step 6, two possible trail designs *could* be considered: Option a: two layers of the FRP system with an estimated width of 9.5 in. (concrete failure mode), and Option b: four layers of the FRP system with an estimated width of 4.75 in. (FRP failure mode). Option a is considered first.

TABLE 9.1 Balanced FRP Areas and Width for Various Numbers of Layers

Layers, n	Unit Stiffness, $nE_f t_f$ (lb/in.)	Bond Coefficient, κ_m	Depth of NA, c_b (in.)	Steel Strain, ε_s	Balanced FRP Area, A_{fb} (in^2)[a]	Balanced FRP Width, w_{fb} (in.)
1	214,400	0.900	3.95	0.0133	—[b]	—
2	428,800	0.818	4.27	0.0121	—[b]	—
3	643,200	0.698	4.83	0.0104	0.28	2.36
4	857,600	0.588	5.48	0.0087	0.87	5.44
5	1,072,000	0.480	6.32	0.0072	1.90	9.50

[a] If $A_f < A_{fb}$, the failure mode is FRP failure; if $A_f > A_{fb}$, the failure mode is concrete failure.
[b] Since $c_b < c_{\text{unstrengthened}}$, a balanced condition is not possible.

Option a: Two layers of the FRP system chosen with an estimated width of 9.5 in. (concrete failure mode). Since it is known that for any amount of FRP area used, the failure mode will be concrete failure for two layers, a trial design with the maximum possible FRP for two layers (i.e., two layers each 12 in. wide[12]) is selected. This gives an area of FRP reinforcement $A_f = nt_f w_f = 2(0.04)(12) = 0.96$ in^2, which is greater than the estimate of 0.76 in^2 and should therefore be a feasible design.

Step 8. Determine the capacity and suitability of the section with the trial FRP design. The failure mode (mode 1a or 1b) must be determined to determine the moment capacity of the strengthened section. This requires determining the depth of the neutral axis and checking that the strains in the concrete, steel, and the FRP are compatible with the assumed failure mode. This can be determined directly from the equilibrium equations of the section as described in Section 9.4.1.

Start by assuming failure mode 1b,[13] concrete compression failure before the steel has yielded. The characteristic equation to be solved for this assumed failure mode is

$$(0.85f'_c\beta_1 b)c^2 - A_s E_s \varepsilon_{cu}(d - c) - A_f E_f[\varepsilon_{cu}(h - c) - \varepsilon_{bi} c] = 0$$

which can be written in the form

$$Ac^2 + Bc + C = 0$$

with,

$$A = 0.85f'_c\beta_1 b = 0.85(5)(0.80)(12) = 40.80 \text{ kip/in.}$$

$$B = A_s E_s \varepsilon_{cu} + A_f E_f(\varepsilon_{cu} + \varepsilon_{bi}) = 3.0(29{,}000)(0.003)$$
$$+ 0.96(5360)(0.003 + 0.00061) = 279.58 \text{ kips}$$

$$C = -(A_s E_s \varepsilon_{cu} d + A_f E_f \varepsilon_{cu} h) = -[3.0(29{,}000)(0.003)(21.5)$$
$$+ 0.96(5360)(0.003)(24)] = -5981.98 \text{ kip-in.}$$

Solving the quadratic equation and taking the positive root gives

[12] Note that this decision was made in hindsight after determining that two layers 9.5 in. wide would not provide the required moment capacity.

[13] Mode 1b is checked first in this example to demonstrate the solution procedure. Typically, mode 1a would be checked before mode 1b in an actual design and analysis procedure.

$$c = 9.15 \text{ in.}$$

Now, check if, in fact, the steel has not yielded (i.e., $\varepsilon_s < \varepsilon_{sy} = 0.00207$ for $f_y = 60$ ksi). If it has yielded, the solution above is not valid:

$$\varepsilon_s = \varepsilon_{cu}\frac{d-c}{c} = 0.003\left(\frac{21.5 - 9.15}{9.15}\right) = 0.0405 > 0.00207$$

Therefore, the steel has in fact yielded and the solution for c is not valid.

Now assume failure mode 1a, concrete crushing after the steel has yielded. The characteristic equation for this case is

$$(0.85f'_c\beta_1 b)c^2 - A_s f_y c - A_f E_f[\varepsilon_{cu}(h - c) - \varepsilon_{bi}c] = 0$$

which can be written in the form

$$Ac^2 + Bc + C = 0$$

with

$$A = 0.85f'_c\beta_1 b = 0.85(5)(0.80)(12) = 40.80 \text{ kip/in.}$$

$$B = A_s f_y + A_f E_f(\varepsilon_{cu} + \varepsilon_{bi}) = -3.0(60)$$
$$+ 0.96(5360)(0.003 + 0.00061) = -161.42 \text{ kips}$$

$$C = -(A_f E_f \varepsilon_{cu} h) = -0.96(5360)(0.003)(24) = -370.48 \text{ kip-in.}$$

Solving the quadratic equation and taking the positive root gives

$$c = 5.58 \text{ in.}$$

Note that this value is greater than 4.27 in, which is the depth of the neutral axis that would be required to develop a balanced section if it were possible. This confirms, once again, that the failure mode will be concrete compression since $c > c_b$.

Now check the strains in the steel and the FRP,

$$\varepsilon_s = \varepsilon_{cu}\frac{d-c}{c} = 0.003\left(\frac{21.5 - 5.58}{5.58}\right) = 0.008 > 0.00207 \rightarrow \text{OK}$$

Steel has yielded as assumed and

$$\varepsilon_{fe} = \varepsilon_{cu}\frac{h - c}{c} - \varepsilon_{bi} = 0.003\left(\frac{24 - 5.58}{5.58}\right) - 0.00061$$

$$= 0.00929 < \kappa_m\varepsilon_{fu} = 0.818(0.0162) = 0.0133$$

Since $\varepsilon_{fe} < \kappa_m\varepsilon_{fu}$, the concrete failure mode is once again confirmed; the FRP will not reach its effective design strain at failure. The stress in the FRP is found as

$$f_{fe} = E_f\varepsilon_{fe} = 5360(0.00929) = 49.79 \text{ ksi} < f_{fu} = 85.50 \text{ ksi}$$

Now determine the nominal moment capacity,

$$a = \beta_1 c = 0.80(5.58) = 4.46 \text{ in.}$$

$$M_n = A_s f_y\left(d - \frac{\beta_1 c}{2}\right) + \psi_f A_f f_{fe}\left(h - \frac{\beta_1 c}{2}\right)$$

$$= 3.0(60)\left(21.5 - \frac{4.46}{2}\right) + 0.85(0.96)(49.79)\left(24 - \frac{4.46}{2}\right)$$

$$= 3468.6 + 884.7$$

$$= 4352 \text{ kip-in.} = 362 \text{ kip-ft}$$

Note that the fraction of the nominal moment carried by the FRP is 884.7/3453 = 0.256, or 25.6%. The moment carried by the steel is 3468.6 kip-in. = 289.1 kip-ft. This is less than the moment of 296 kip-ft carried by the steel in the unstrengthened case since the neutral axis has been moved downward in the strengthened beam (i.e., $c = 5.58$ in. versus $c = 4.41$ in.).

Determine the factored moment capacity. Since the strain in the steel at failure is greater than 0.005, $\phi = 0.90$ and

$$\phi M_n = 0.90(362.0) = 326.5 \text{ kip-ft} > 321 \text{ kip-ft}$$

Therefore, the assumed FRP strengthening system meets the desired future moment demand. The width of the FRP could be reduced slightly since the strengthening system is slightly overdesigned in this case (by 5.5 kip-ft or 1.7%). It can be shown that two layers of FRP 10.6 in. wide ($A_f = 0.85$ in^2) will exactly meet the demand on 321 kip-ft. However, a 12-in.-wide FRP system, which is the same width as the beam itself, is much easier to apply in the field than a system that is slightly narrower than the beam. Therefore, a 12-in.-wide FRP system should be chosen for constructability reasons.

It is also important to note that the approximate design of step 5 (with a 9.5-in. width) would not have produced adequate moment capacity. The approximate calculation should be used as a guide, and detailed calculations must always be performed. A comparison of the approximate design of step

5 with the final design of step 7 is shown in Table 9.2. In addition, both serviceability and detailing checks are required for the system before it can be recommended for use.

Note on the effect of existing strain on the substrate: Neglecting the effect of the existing strain, ε_{bi}, in the concrete is not conservative. In the example given, if the existing strain were neglected in the design, the following results would be obtained:

$$A = 40{,}800 \text{ lb/in.}$$

$$B = -166{,}170 \text{ lb} \qquad (\textit{Note:} \text{ Only the } B \text{ term changes})$$

$$C = -370{,}480 \text{ lb-in.}$$

which gives $c = 5.67$ in. > 5.58 in., $\varepsilon_{fe} = 0.00970 > 0.00929$, $f_{fe} = 52.0$ ksi > 49.8 ksi, and $M_n = 4383$ kip-in. = 365 kip-ft > 362 kip-ft, in comparison to the values obtained when the existing strain was considered. That is, the calculation predicts a larger nominal moment capacity without the existing strain. Therefore, it is not conservative to neglect the existing strain on the substrate. However, it is also seen that the difference is very small in this particular example.

Step 9. Detail the FRP flexural strengthening system. Details of the FRP strengthening system must shown on the plans and in the specifications for an FRP strengthening systems. At the strip ends two empirical checks are required by ACI 440.2R-02 to assist with this detailing:

TABLE 9.2 Comparison of Estimated and Detailed Design Calculations

Design Method	FRP Area, A_f (in^2)	Depth of NA, c (in.)	Effective Design Stress, f_{fe} (ksi)	Factored Moment Capacity, ϕM_n (kip-ft)
Approximate (A_f obtained from estimated c and f_{fu})	0.76	4.41	54.72	≈321
Detailed (based on A_f estimated from approximate method)	0.76	5.37	52.30	316.5 < 321 NG
Detailed (based on final area selected)	0.96	5.58	49.79	326.5 > 321 OK
Detailed (based on area to exactly achieve the moment demand)	0.85	5.48	51.07	321

1. According to ACI 440.2R-02, the FRP flexural strengthening system must extend a distance d beyond the point corresponding to the cracking moment under factored loads for a simply supported beam.

$$M_{cr} = \frac{2f_r I_g}{h} = \frac{2(530.3)(13{,}824)}{24} = 610{,}906 \text{ lb-in.} = 50.9 \text{ kip-ft}$$

$$w_{\mathrm{DL}} = 1.0 \text{ kip/ft}$$

$$w_{\mathrm{LL}} = 1.2(1.5) = 1.8 \text{ kip/ft}$$

$$W_u = 1.4(1.0) + 1.7(1.8) = 4.46 \text{ kip/ft}$$

$$M_u = \frac{4.46(24)^2}{8} = 321.1 \text{ kip-ft} \quad \text{(for comparison)}$$

Determine the distance, x, along the beam from the support at which the moment is equal to the cracking moment, as shown in the free-body diagram in Fig. 9.8.

$$M_x = M_{\mathrm{cr}} = Rx - W \rightarrow \frac{4.46x^2}{2} - \frac{4.46(24)x}{2} + 50.9 = 0$$

$$x = 0.992 \text{ ft} = 11.9. \text{ in} < d \text{ (21.5 in.)}$$

Since insufficient length is available to anchor the FRP strip, a supplementary anchorage system is required.

2. A capacity check is required at the strip end to prevent shear failure of the beam at the point of FRP strip termination. Check that the factored shear force at the FRP strip termination point is less than two-thirds of the concrete shear capacity at this point.

Assuming, conservatively, that the FRP strip terminates at the support,

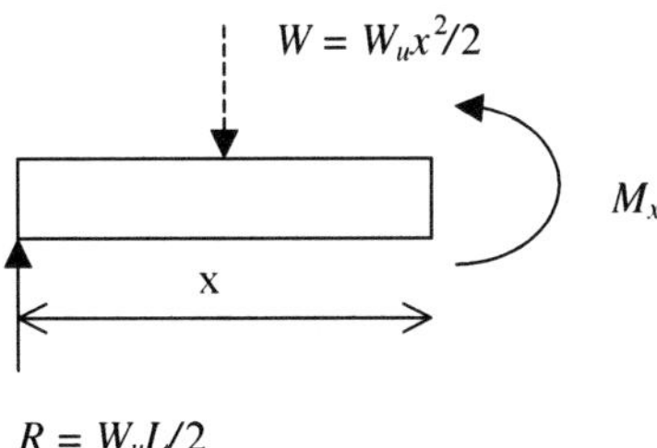

Figure 9.8 Free-body diagram of beam at distance x from support.

$$V_u = R = 4.46(12) = 53.5 \text{ kips}$$

Determine the concrete shear capacity:

$$V_c = 2\sqrt{f'_c}\, bd = 2\sqrt{5000}\,(12)(21.5) = 36{,}486 \text{ lb} = 36.5 \text{ kips}$$

$$\tfrac{2}{3}V_c = 24.3 \text{ kips} < 53.5 \text{ kips}$$

This check also requires that the FRP be anchored at the support.

A FRP U-wrap at the beam ends is generally used for this purpose (Spadea et al. 1998). Alternatively, proprietary fiber-anchors can be used to anchor the FRP strip Teng et al. (2001). In both cases, the strip should be extended as close as possible to the support. The length of the U-wrap should be at least equal to the depth of the beam.

The U-wrap is detailed as follows: Provide a U-wrap of width equal to the depth of the beam (24 in. in this case) and use a minimum of two layers. If possible, use the same FRP material system as used for the FRP flexural reinforcement. Apply the U-wrap FRP reinforcement at the same time as the FRP flexural reinforcement (i.e., before the adhesive for the flexural reinforcement has cured). Extend the U-wrap into the compression zone of the concrete. The corners of the beam should be rounded at the edges where the U-wrap is used ($\frac{1}{2}$ in. minimum radius). Figure 9.9 shows the detailing design.

If precured FRP laminates are used for the flexural strengthening, a preformed FRP L shape (see Chapter 10) or a wet-layup system can be used to anchor the strip and prevent shear failure of the beam. Steel anchors, steel cover plates, or steel angles can also be used (Hollaway and Garden, 1998). It is advisable to attach the steel plates or angles with drilled expansion bolts

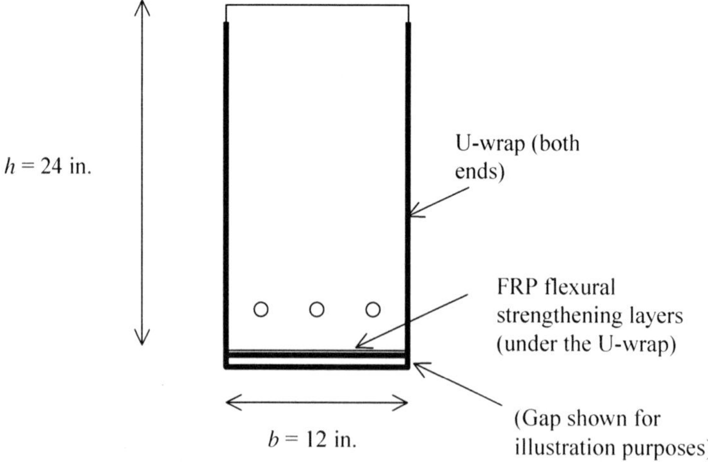

Figure 9.9 Details of FRP U-wrap at the end of the beam to anchor the FRP flexural strengthening system.

and to use a filler material (cementitious or resin based) to ensure good contact between the FRP and the cover plate. Holes for the expansion blots should be drilled through the FRP after it has cured.

Option b: Four layers of the FRP system chosen with an estimated width of 4.75 in. (FRP failure mode). It has been determined (see Table 9.1) that the balanced condition will occur with four layers of the FRP system selected for a width of the FRP system equal to $w_f = 5.44$ in. and a total FRP area of $A_f = 0.87$ in^2. A trial design is performed with this amount of FRP reinforcement, which is greater than the original estimate of 0.76 in^2 (4.75 in. width) based on the approximate calculation. However, recall that this estimate was shown in Table 9.2 to be rather inaccurate. For this design the theory predicts that the FRP will fail at exactly the same instant as the concrete will fail. If the FRP failure mode is to be guaranteed, an area of FRP smaller than 0.87 in^2, consisting of four layers, should be selected.

The unit stiffness of the FRP strengthening system and the bond-dependent coefficient, κ_m, for the four layers are determined as follows (data were shown in Table 9.1):

$$nE_f t_f = 4(5{,}360{,}000)(0.04) = 857{,}600 \leq 1{,}000{,}000 \text{ lb/in.}$$

and therefore

$$\kappa_m = \frac{1}{60\varepsilon_{fu}}\left(1 - \frac{nE_f t_f}{2{,}000{,}000}\right) = \frac{1}{60(0.0162)}\left(1 - \frac{857{,}600}{2{,}000{,}000}\right)$$
$$= 0.588 \leq 0.90$$

and

$$\varepsilon_{fe} = \kappa_m \varepsilon_{fu} = 0.588(0.0162) = 0.0095$$
$$f_{fe} = \kappa_m f_{fu} = 0.588(85.5) = 50.27 \text{ ksi}$$

The failure mode (mode 2a or 2b) must be determined to determine the moment capacity of the strengthened section. This requires finding the depth of the neutral axis and checking that the strains in the concrete, steel, and the FRP are compatible with the assumed failure mode. This can be determined directly from the equilibrium equations of the section.

Start by assuming failure mode 2b,[14] FRP failure before the steel has yielded. The characteristic equation to be solved for this assumed failure mode is

[14] Mode 2b is checked first in this example to demonstrate the solution procedure. Typically, mode 2a would be checked before mode 2b in an actual design and analysis procedure.

$$(0.85f'_c\beta_1 b)c(h - c) - A_s E_s(\varepsilon_{fe} + \varepsilon_{bi})(d - c) - A_f E_f \varepsilon_{fe}(h - c) = 0$$

which can be written in the form

$$Ac^2 + Bc + C = 0$$

with

$$A = 0.85f'_c\beta_1 b = 0.85(5)(0.80)(12) = 40.80 \text{ kips/in.}$$

$$B = -[0.85f'_c\beta_1 bh + A_s E_s(\varepsilon_{fe} + \varepsilon_{bi}) + A_f E_f \varepsilon_{fe}]$$

$$= -[0.85(5)(0.80)(12)(24) + 3.0(29{,}000)(0.0095 + 0.00061)$$

$$+ 0.87(5{,}360)(0.0095)]$$

$$= -1930.10 \text{ kips}$$

$$C = A_s E_s(\varepsilon_{fe} + \varepsilon_{bi})d + A_f E_f \varepsilon_{fe} h$$

$$= 3.0(29{,}000)(0.0095 + 0.00061)(21.5) + 0.87(5{,}360)(0.0095)(24)$$

$$= 18{,}901.76 \text{ kip-in.}$$

solving the quadratic equation and taking the positive root gives,

$$c = 13.86 \text{ in.}$$

Now check if, in fact, the steel has not yielded (i.e., $\varepsilon_s < \varepsilon_{sy} = 0.00207$ for $f_y = 60$ ksi). If it has yielded, the solution above is not valid!

$$\varepsilon_s = (\varepsilon_{fe} + \varepsilon_{bi})\frac{d - c}{h - c} = (0.0095 + 0.00061)\left(\frac{21.5 - 13.86}{24 - 13.86}\right)$$

$$= 0.00762 > 0.00207$$

Therefore, the steel has in fact yielded and the solution for c is not valid!

Now assume failure mode 2a, FRP failure after the steel has yielded. The characteristic equation for this case is linear, not quadratic, and is

$$(0.85f'_c\beta_1 b)c - A_s f_y - A_f \kappa_m f_{fu} = 0$$

which gives

$$c = \frac{A_s f_y + A_f \kappa_m f_{fu}}{0.85f'_c\beta_1 b} = \frac{3.0(60) + 0.87(0.588)(85.5)}{0.85(5)(0.80)(12)} = 5.484 \text{ in.}$$

which is equal to the value of c_b found previously since the area of the FRP chosen is exactly equal to that which gives the balanced condition.

Now check that the steel has yielded:

$$\varepsilon_s = (\varepsilon_{fe} + \varepsilon_{bi})\frac{d - c}{h - c} = (0.0095 + 0.00061)\left(\frac{21.5 - 5.484}{24 - 5.484}\right)$$

$$= 0.0087 \geq \varepsilon_{sy} = 0.00207$$

Therefore, the steel has yielded and the assumed failure mode is applicable. The strain in the concrete is

$$\varepsilon_c = (\varepsilon_{fe} + \varepsilon_{bi})\frac{c}{h - c} = (0.0095 + 0.00061)\left(\frac{5.484}{21.5 - 5.484}\right)$$

$$= 0.003 = \varepsilon_{cu}$$

This confirms that a balanced failure occurs for the chosen value of A_f.

The nominal moment capacity is obtained as

$$\begin{aligned} M_n &= A_s f_y\left(d - \frac{\beta_1 c}{2}\right) + \psi_f A_f f_{fe}\left(h - \frac{\beta_1 c}{2}\right) \\ &= 3.0(60)\left(21.5 - \frac{0.8(5.484)}{2}\right) \\ &\quad + 0.85(0.87)(50.27)\left[24 - \frac{0.8(5.484)}{2}\right] \\ &= 4285 \text{ kip-in.} = 357 \text{ kip-ft} \end{aligned}$$

and the factored moment capacity is given as

$$\phi M_n = (0.9)(357) = 321.3 \text{ kip-ft} > 321 \text{ kip-ft}$$

which meets the strengthening demand.

Discussion

A comparison of the results for options a and b is presented in Table 9.3. However, if option a with a width of 10.6 In is chosen (see pg. 259) the identical capacity can be achieved with the same area of FRP for either two or four layers of the material (small differences are due to numerical round-off). However, the failure is controlled by different mechanisms. If slightly less than the balanced area was chosen for the four-layer system, FRP failure would control. However, the balanced condition was chosen to demonstrate this special case. This is not surprising if one observes that the overreinforced

TABLE 9.3 Comparison of the results of Example 9.1 Options *a* and *b*

Option	Layers	Width, w_f (in.)	FRP Area, A_f (in^2)	FRP Stress at Failure, f_{fe} (ksi)	Depth of NA, c (in.)	Factored Moment Capacity, ϕM_n (kip-ft)	Failure Mode
a	2	12.0	0.96	49.79	5.58	326.5	Concrete
9.1	2	10.6	0.85	51.07	5.477	320.9 (~ 321)	Concrete (and FRP)
9.2	4	5.4	0.87	50.27	5.484	321.3 (~ 321)	FRP (and concrete)

case with concrete failure or the underreinforced case with FRP failure use the identical equation for determining the nominal moment shown previously:

$$M_n = A_s f_y \left(d - \frac{\beta_1 c}{2}\right) + \psi_f A_f f_{fe} \left(h - \frac{\beta_1 c}{2}\right) \qquad \text{(ACI 440.2R-02:9-11)}$$

9.8 SERVICEABILITY OF FRP-STRENGTHENED FLEXURAL MEMBERS

Under service loads the ACI 440.2R-02 guide requires the designer to check that the stresses in the existing steel and the FRP do not exceed specific limits to prevent failure of the strengthened beam due to creep and fatigue effects. In the service load range, a cracked elastic section analysis is used to determine the service load stresses. This is similar to conventional service load analysis of reinforced concrete members with an additional requirement. In the strengthened state the contribution of the FRP strengthening system is included in the cracked elastic section properties.

No specific requirements are given for deflections or crack control under service loads. The appropriate cracked section properties are used to determine service load deflections in FRP-strengthened beams. In general, the contribution of the FRP strengthening to the stiffness of the section in the service load range is rather small.

9.8.1 Cracked FRP Strengthened Section

To analyze the stresses and strains in the section that has been strengthened, the properties of the cracked section with the FRP strengthening system attached are required. The strain distribution, stress distribution, and force resultants in the service load range are shown in Fig. 9.10.

To calculate the properties of the cracked FRP-strengthened section, the transformed section method may be used or the composite section method may be used.

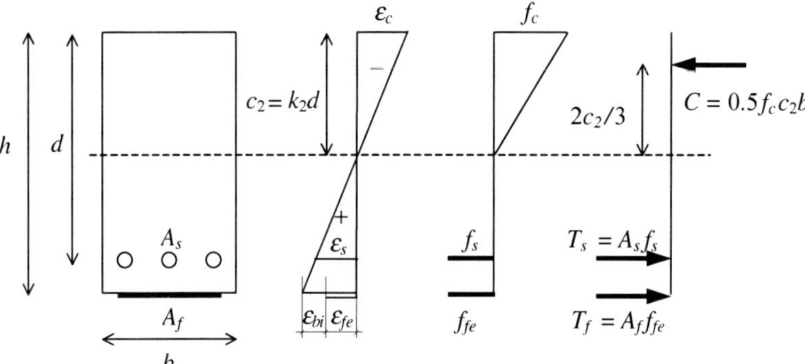

Figure 9.10 Strains, stresses, and force resultants in the service load range.

Method 1: Using the Transformed Section Since three different materials are present in the section, both the steel and the FRP must be transformed to the modulus of the concrete and two modular ratios are defined, $\eta_s = E_s/E_c$ and $\eta_f = E_f/E_c$, the steel-to-concrete modular ratio and the FRP-to-concrete modular ratio, respectively. The neutral-axis depth ratio, $k_2 = c_2/d$, is then given as

$$k_2 = \sqrt{(\rho_s\eta_s + \rho_f\eta_f)^2 + 2\left(\rho_s\eta_s + \rho_f\eta_f\frac{h}{d}\right)} - (\rho_s\eta_s + \rho_f\eta_f) \quad (9.59)$$

and

$$(I_{\text{cr}})_2 = \frac{bd^3}{3}k_2^3 + \eta_s A_s d^2(1 - k_2)^2 + \eta_f A_f d^2\left(\frac{h}{d} - k_2\right)^2 \quad (9.60)$$

Method 2: Using Composite Section Mechanics Following the procedure described in Chapter 5, the distance from the top of the section to the neutral axis, shown in Fig. 9.11, is found from

$$E_c b\frac{(c_2)^2}{2} + E_s A_s(c_2 - d) + E_f A_f(c_2 - h) = 0 \quad (9.61)$$

Solving the quadratic equation and taking the positive root gives the value of c_2. The flexural stiffness (EI) of the strengthened section is given as

$$(EI)_2 = \sum EI = E_c I_c + E_s I_s + E_f I_f = E_c\frac{b(c_2)^3}{3} + E_s A_s(d - c_2)^2 + E_f A_f(h - c_2)^2 \quad (9.62)$$

and

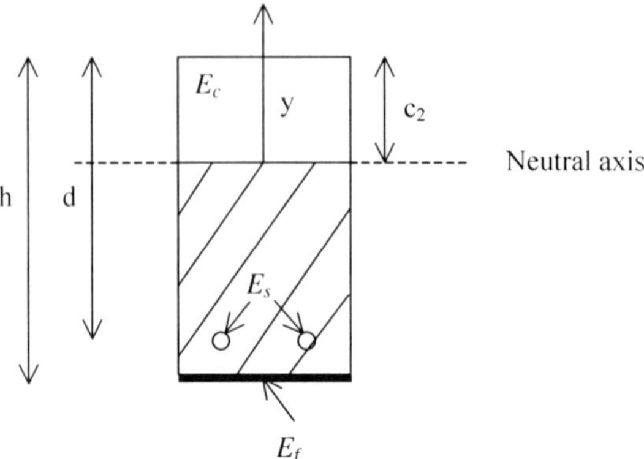

Figure 9.11 Properties of a strengthened section for determination of the neutral axis.

$$(I_{cr})_2 = \frac{(EI)_2}{E_c} \tag{9.63}$$

9.8.2 Service-Level Stress in the Internal Steel Reinforcing Bars

ACI 440.2R-02 limits the service-level stresses in the internal steel reinforcing bars in the strengthened reinforced concrete member as follows:

$$f_{s,s} \leq 0.80 f_y \tag{9.64}$$

This is to prevent yielding of the internal steel and to decrease the likelihood of fatigue failure of the internal steel rebars under service loads in the strengthened section. When the internal steel yields, the flexural stiffness of the beam decreases significantly and large, undesirable deflections will occur in the structure at service loads.

Since the steel internal reinforcing bars are subjected to stress and strain both before and after the FRP strengthening system is applied, the stress and strain is composed of two distinct stages, as shown in Fig. 9.6. The strains in these two stages of the loading are denoted as follows:

ε_{s1}: the strain in the steel due to the existing loads (usually, the dead loads only) at the time that the FRP is applied

ε_{s2}: the strain in the steel due to the supplemental applied loads after the FRP has been attached

In the *unstrengthened beam* (stage 1 of the loading),

$$\varepsilon_{s1} = \varepsilon_{b1}\frac{d - c_1}{h - c_1} \tag{9.65}$$

$$\varepsilon_{b1} = \frac{m_1(h - c_1)}{(I_{cr})_1 E_c} = \frac{m_1(h - c_1)}{(EI)_1} \tag{9.66}$$

and the corresponding stress is

$$f_{s1} = E_s \varepsilon_{s1} \tag{9.67}$$

or given directly as

$$f_{s1} = \frac{E_s m_1(d - c_1)}{(EI)_1} \tag{9.68}$$

where m_1 is the moment due to the existing loads at the time the FRP is attached and c_1 is the depth of the neutral axis in the unstrengthened beam. Also, recall that ε_{b1} is equal to ε_{bi}, which has previously been described as the strain in the substrate when the FRP is attached.

In the *strengthened beam* (stage 2 of the loading),

$$\varepsilon_{s2} = \varepsilon_{b2}\frac{d - c_2}{h - c_2} \tag{9.69}$$

$$\varepsilon_{b2} = \frac{m_2(h - c_2)}{(I_{cr})_2 E_c} = \frac{m_2(h - c_2)}{(EI)_2} \tag{9.70}$$

and the corresponding stress is

$$f_{s2} = E_s \varepsilon_{s2} \tag{9.71}$$

or given directly as

$$f_{s2} = \frac{E_s m_2(d - c_2)}{(EI)_2} \tag{9.72}$$

where m_2 is the moment due to the supplemental loads after the FRP has been attached and c_2 is the depth of the neutral axis in the strengthened beam. Therefore, the total stress in the internal steel at the service loads (i.e., the loads that existed on the structure at the time that the FRP was attached plus the loads that are applied to the structure after the FRP was attached[15]) is given by

$$f_{s,s} = f_{s1} + f_{s2} \tag{9.73}$$

or it can be determined directly from

$$f_{s,s} = \frac{[(m_1 - m_2) + \varepsilon_{bi}A_fE_f(h - k_2d/3)](d - k_2d)E_s}{A_sE_s(d - k_2d/3](d - k_2d) + A_fE_f[h - k_2d/3)(h - k_2d)}$$

(ACI 440.2R-02:9-12)

where $k_2 = c_2/d$ is the ratio of the depth of the neutral axis to the effective depth of the strengthened section (denoted as c_2 above).

9.8.3 Service-Level Stresses in the FRP Strengthening System

The loading combination for this FRP stress check is that due to all sustained loads on the structure and the maximum load effect of the cyclic loads on the structure that could lead to fatigue failure. The stress limits for the FRP strengthening system are given in Table 8.2. Since the FRP system is subjected to loading only in stage 2, after the FRP has been applied, the moment used in computing the strains and stresses is equal to the sustained load on the beam after the load has been applied, in addition to the load causing the fatigue moment on the beam, designated previously as m_2.[16] The strains in the FRP are therefore given as

$$\varepsilon_{f2} = \varepsilon_{b2} \tag{9.74}$$

$$\varepsilon_{b2} = \frac{m_2(h - c_2)}{(I_{cr})_2\, E_c} = \frac{m_2(h - c_2)}{(EI)_2} \tag{9.75}$$

and the corresponding stress is

$$f_{f2} = E_f\varepsilon_{f2} \tag{9.76}$$

or given directly as

[15]Note that this assumes that the load that existed on the beam when the FRP was attached remains on the beam after the FRP has been attached.

[16]It is possible that the moment m_2 that is used to determine the stress in the steel in stage 2 will not be the same as the sustained load plus maximum moment due to the cyclic load, but usually these are assumed to be the same.

$$f_{f2} = \frac{E_f m_2(d - c_2)}{(EI)_2} \tag{9.77}$$

Therefore, the stress, $f_{f,s}$, in the FRP at service is given as

$$f_{f,s} = f_{f2} \tag{9.78}$$

This stress can be calculated directly as

$$f_{f,s} = \frac{E_f m_2(h - c_2)}{(EI)_2} \tag{9.79}$$

Alternatively, the stress in the FRP can be determined directly from the stress in the steel,

$$f_{f,s} = f_{s,s}\frac{E_f}{E_s}\frac{h - k_2 d}{d - k_2 d} - \varepsilon_{bi}E_f \qquad \text{(ACI 440.2R-02:9-13)}$$

where $k_2 = c_2/d$ is the ratio of the depth of the neutral axis to the effective depth of the strengthened section (denoted as c_2 above), and the steel stress, $f_{s,s}$, is computed for the full service load moment $m_1 + m_2$.

9.9 LOAD–DEFLECTION RESPONSE OF FRP-STRENGTHENED FLEXURAL MEMBERS

The load–deflection response of an FRP-strengthened flexural member needs to be determined to ensure that serviceability deflection criteria are not violated by the strengthened beam under the increased loads. The load–deflection response of an FRP-strengthened member can be divided into a number of stages (regimes). Only the increasing (positive stiffness) load–deflection response to failure of the strengthened beam is considered in this section. The post-peak load–deflection response is not discussed and is beyond the scope of this book. At a first approximation, the increasing load–deflection response can be divided into three significant stages: (1) the linear elastic deflection at the time the FRP is applied, (2) the linear elastic deflection after the FRP has been applied up to the time the internal steel yields (assuming that the internal steel yields before the member fails in flexure), and (3) the nonlinear inelastic deflection from the time the steel yields until the FRP-strengthened beam fails due to concrete crushing or FRP failure. The flexural stiffness, $(EI)_i$, in each of these stages is used to determine the load–deflection response in an incremental fashion.

The flexural stiffnesses, $(EI)_1$ and $(EI)_2$ and the cracked transformed second moments $(I_{cr})_1$ and $(I_{cr})_2$ for stages (1) and (2) were derived previously. It is assumed that the beam is cracked prior to loading in the first stage and therefore that there is no precracking stage in the load–deflection response of an FRP-strengthened beam (Arduini and Nanni, 1997; Buyukotzurk and Hearing, 1998). The effective second moment of a concrete beam for deflection is generally taken to be that given by the classical Branson equation,

$$I_e = \left(\frac{M_{cr}}{M_a}\right)^3 I_g + \left[1 - \left(\frac{M_{cr}}{M_a}\right)\right] I_{cr} \leq I_g \tag{9.80}$$

The Branson equation can be assumed to be a reasonable approximation for the first and second stages of the load–deflection response since the beam stiffness in these stages is not affected significantly by the FRP (El-Mihilmy and Tedesco, 2000). It is also assumed that the response of the cracked section either before or after the FRP has been applied is linear elastic. This implies that the concrete modulus remains constant and equal to its initial value up to yielding of the internal steel. For low steel reinforcement ratios for which flexural strengthening is usually most effective, this assumption is reasonable.

The flexural stiffness in the third nonlinear inelastic stage after the internal steel has yielded can be approximated as a linear elastic stage for the purposes of obtaining an estimate of the deflection at ultimate load. Since in this stage the concrete stress–strain response is nonlinear and the steel is possibly hardening, it is most appropriate to approach this with a full incremental nonlinear analysis. However, some simplifying assumptions can be made: (1) After the steel has yielded, the steel modulus is set to zero (i.e., perfectly elastoplastic response); (2) after the steel has yielded, the concrete modulus is taken as 50% of its initial modulus (this is reasonable for the range of concrete strain from 0.001 to 0.002, which is the concrete strain for most of this stage),[17] and (3) after the steel yields strain, the compatibility between the FRP and the concrete substrate remains in effect even though the cover concrete may be cracked significantly.

The flexural stiffness $(EI)_3$ and the cracked transformed second moment $(I_{cr})_3$ for the third stage can be found using the composite mechanics method following the procedure described for the second stage. The distance from the top of the section to the neutral axis, c_3, is found from

$$\frac{E_c}{2} b \frac{(c_3)^2}{2} + E_f A_f (c_3 - h) = 0 \tag{9.81}$$

Solving the quadratic equation and taking the positive root gives the value of c_3. The flexural stiffness $(EI)_3$ of the strengthened section is given as

[17] In reality, this will not make much of a difference in the nonlinear range since the concrete contribution to the stiffness is quite small.

$$(EI)_3 = \sum EI = E_c I_c + E_f I_f = \frac{E_c}{2}\frac{b(c_3)^3}{3} + E_f A_f (h - c_3)^2 \quad (9.82)$$

and

$$(I_{cr})_3 = \frac{(EI)_3}{(E_c/2)} \quad (9.83)$$

Design Example 9.2: Serviceability of an FRP-Strengthened Beam Consider the same geometry and loading as the beam discussed earlier in this chapter in option a of the strength design. Determine (a) The stresses in the steel and the FRP under service loads, and (b) the midspan deflection of the strengthened beam under full dead and live loads. The pertinent properties of the beam and FRP system are:

$$f_{fu} = C_E f_{fu}^* = 0.95(90) = 85.5 \text{ ksi} \qquad w_f = 12 \text{ in.}$$

$$\varepsilon_{fu} = C_E \varepsilon_{fu}^* = 0.95(0.017) = 0.0162 \qquad \varepsilon_{bi} = 0.00061$$

$$E_f = 5360 \text{ ksi} \qquad E_c = 4030 \text{ ksi}$$

$$t_f = 0.04 \text{ in.} \qquad E_s = 29{,}000 \text{ ksi}$$

SOLUTION

(a) *Stresses in the steel and the FRP under service loads* Determine the unfactored service loads. The dead load that acts on the beam at the time the FRP is attached,

$$w_{DL} = 1.0 \text{ kip/ft}$$

The dead load plus the new live load that act on the FRP strengthened beam,

$$w_{DL} = 1.0 \text{ kip/ft} \qquad w_{LL} = 1.8 \text{ kip/ft}$$

Determine the service loads and moments for the original and strengthened sections.

Original section:

$$m_1 = m_{DL} = \frac{w_{DL}L^2}{8} = \frac{1.0(24)^2}{8} = 72 \text{ kip-ft} = 864{,}000 \text{ lb-in.}$$

Strengthened section:

$$m_2 = m_{LL} = \frac{w_{LL}L^2}{8} = \frac{1.8(24)^2}{8} = 129.6 \text{ kip-ft} = 1{,}555{,}200 \text{ lb-in.}$$

$$m_1 + m_2 = 2{,}419{,}200 \text{ lb-in.}$$

Determine the properties of the unstrengthened and strengthened sections. The properties of the original unstrengthened section were determined as follows:

$$c_1 = 7.18 \text{ in.} \qquad E_c = 4030 \text{ ksi}$$

$$k_1 = 0.334 \qquad (EI)_1 = 2.38 \times 10^{10} \text{ lb-in}^2$$

$$(I_{cr})_1 = 5909 \text{ in}^4$$

The properties of the strengthened section are calculated using method 2, composite mechanics. The distance from the top of the section to the neutral axis is

$$E_c b \frac{(c_2)^2}{2} + E_s A_s(c_2 - d) + E_f A_f(c_2 - h) = 0$$

$$4.030(12)\frac{(c_2)^2}{2} + 29(3.0)(c_2 - 21.5) + 5.36(0.96)(c_2 - 24) = 0$$

$$24.18(c_2)^2 + 92.15c_2 - 1993.49 = 0$$

Solving the quadratic equation and taking the positive root gives

$$c_2 = 7.37 \text{ in.} \quad \text{and} \quad k_2 = \frac{c_2}{d} = 0.343 \text{ in.}$$

(Compare with original section of values of $c_1 = 7.18$ in. and $k_1 = 0.334$.)

The flexural stiffness $(EI)_2$ of the section is obtained from

$$(EI)_2 = \sum EI = E_c I_c + E_s I_s + E_f I_f = E_c \frac{b(c_2)^3}{3}$$

$$+ E_s A_s(d - c_2)^2 + E_f A_f(h - c_2)^2$$

$$= (4.030 \times 10^6)\left(\frac{12(7.37)^3}{3}\right) + (29 \times 10^6)(3.0)(21.5 - 7.37)^2$$

$$+ (5.360 \times 10^6)(24 - 7.37)^2$$

$$= 2.52 \times 10^{10} \text{ lb-in}^2 > 2.38 \times 10^{10} \text{ lb-in}^2$$

which gives

$$(I_{cr})_2 = \frac{(EI)_2}{E_c} = \frac{2.52 \times 10^{10}}{4{,}030{,}000} = 6253 \text{ in}^4 > 5909 \text{ in}^4$$

It is interesting to note that the flexural stiffness has only increased by 5.9% in the elastic range due to the addition of the FRP. This implies that the FRP has a small effect on the deflections of the beam under the service load in this example.

Determine the stresses in the steel under service loads. Determine the strains and stress in the steel when the FRP is attached (stage 1) and after the FRP has been attached (stage 2).

Stage 1:

$$\varepsilon_{s1} = \varepsilon_{bi}\frac{d - c_1}{h - c_1} = 0.00061\left(\frac{21.5 - 7.18}{24 - 7.18}\right) = 0.00052$$

and the corresponding stress is

$$f_{s1} = E_s\varepsilon_{s1} = (29{,}000{,}000)(0.00052) = 15{,}080 \text{ psi}$$

This can also be calculated directly from

$$f_{s1} = \frac{E_s m_1(d - c_1)}{(EI)_1}$$

$$= \frac{29{,}000{,}000(864{,}000)(21.5 - 7.18)}{2.38 \times 10^{10}} = 15{,}076 \approx 15{,}080 \text{ psi}$$

Stage 2:

$$f_{s2} = \frac{E_s m_2(d - c_2)}{(EI)_2}$$

$$= \frac{29{,}000{,}000(1{,}555{,}200)(21.5 - 7.37)}{2.52 \times 10^{10}} = 25{,}367 \text{ psi}$$

Therefore, the stress in the steel at service loads is

$$f_{s,s} = f_{s1} + f_{s2} = 15{,}076 + 25{,}367 = 40{,}443 \text{ psi} = 40.4 \text{ ksi}$$

or calculate directly from

$$f_{s,s} = \frac{[(m_1 + m_2) + \varepsilon_{bi}A_fE_f(h - k_2d/3)](d - k_2d)E_s}{A_sE_s(d - k_2d/3)(d - k_2d) + A_fE_f(h - k_2d/3)(h - k_2d)}$$

$$k_2d = c_2 = 7.37 \text{ in.}$$

$$m_{\text{DL}} + m_{\text{LL}} = 864 + 1555.2^{18} = 2419.2 \text{ kip-in.}$$

$$f_{s,s} = \frac{[2419.2 + 0.00061(0.96)(5360)(24 - 7.37/3)](21.5 - 7.37)29{,}000}{3(29{,}000)(21.5 - 7.37/3)(21.5 - 7.37) + 0.96(5360)(24 - 7.37/3)(24 - 7.37/3)}$$

$$= 40.4 \text{ ksi}$$

Check the stress limits for the steel.

$$f_{s,s} \leq 0.80f_y$$

Substituting values gives

$$f_{s,s} = 40.4 \text{ ksi} < 0.8f_y = 0.8(60) = 48 \text{ ksi}$$

Therefore, the stress in the steel as service loads is OK.

Determine the stresses in the FRP under service loads.

$$f_{f,s} = \frac{E_fm_2(h - c_2)}{(EI)_2} = \frac{5360(1{,}555{,}200)(24 - 7.37)}{2.52 \times 10^{10}} = 5.50 \text{ ksi}$$

or calculate directly from

$$f_{f,s} = f_{s,s}\frac{E_f}{E_s}\frac{h - k_2d}{d - k_2d} - \varepsilon_{bi}E_f$$

$$= 40.4\left(\frac{5360}{29{,}000}\right)\left(\frac{24 - 7.37}{21.5 - 7.37}\right) - 0.00061(5360)$$

$$= 5.51 \text{ ksi}$$

Check the stress limits for the FRP. For the carbon–epoxy FRP strengthening system chosen, the limiting stress is

[18] In the example in ACI440.2R-02, this has been rounded to 130 kip-ft = 1560 kip-in., and hence the total moment is given as 2424 kip-in. This explains the slight difference in the final values.

$$f_{f,s} < 0.55 f_{fu}$$

Substituting gives

$$5.51 \text{ ksi} < 50 \text{ ksi} = 0.55(85.0)$$

Therefore, the stress in the FRP at service loads is OK.

(b) *Midspan deflection of the strengthened beam under full dead and live loads* Determine the flexural stiffnesses and cracked transformed second moments for the three stages of load–deflection response. For stages 1 and 2 these have been determined previously as

$$(EI)_1 = 2.38 \times 10^{10} \text{ lb-in}^2 \qquad (I_{cr})_1 = 5909 \text{ in}^4 \qquad E_c = 4030 \text{ ksi}$$

$$(EI)_2 = 2.52 \times 10^{10} \text{ lb-in}^2 \qquad (I_{cr})_2 = 6253 \text{ in}^4 \qquad M_{cr} = 610{,}906 \text{ lb-in.}$$

The flexural stiffness $(EI)_3$ of the section is obtained from

$$\begin{aligned} &\frac{E_c}{2} b \frac{(c_3)^2}{2} + E_f A_f (c_3 - h) \\ &= \frac{4.030}{2}(12)\frac{(c_3)^2}{2} + 5.360(0.96)(c_3 - 24) \\ &= 12.09(c_3)^2 + 5.15c_3 - 123.49 = 0 \end{aligned}$$

Solving the quadratic equation and taking the positive root gives

$$c_3 = 2.990 \text{ in.}$$

Note how shallow the compression block is in the third stage!

The flexural stiffness $(EI)_3$ of the strengthened section is given as

$$\begin{aligned} (EI)_3 = \Sigma\, EI &= \frac{E_c}{2} I_c + E_f I_f \\ &= \frac{E_c}{2}\frac{b(c_3)^3}{3} + E_f A_f (h - c_3)^2 \\ &= \frac{(4.030 \times 10^6)(12)(2.990)^3}{2(3)} \\ &\quad + (5.36 \times 10^6)(0.96)(24 - 2.990)^2 \\ &= 2.49 \times 10^9 \text{ lb-in}^2 \end{aligned}$$

and

$$(I_{cr})_3 = \frac{(EI)_3}{E_c/2} = \frac{(2.49 \times 10^9)}{4.030 \times 10^6/2} = 1236 \text{ in}^4$$

The load–deflection curve can now be constructed using either the values of the flexural stiffness, $(EI)_i$, calculated above or using the Branson formula with the values of $(I_{cr})_3$. Both methods are used in the example.

Method 1: Using the theoretical flexural stiffness To construct the piecewise linear load–deflection curve, the applied moments and the maximum deflection are required at the beginning and end of each stage. For stage 1, prior to the FRP being applied, the applied load is 1 kip/ft = 83.3 lb/in., and the maximum moment is

$$m_1 = \frac{w_1 \ell^2}{8} = \frac{83.3(24 \times 12)^2}{8} = 864{,}000 \text{ lb-in.}$$

The deflection corresponding to this moment is

$$\delta_1 = \frac{5}{384} \frac{w_1 \ell^4}{(EI)_1} = \frac{5}{384} \left(\frac{83.3(228)^4}{2.38 \times 10^{10}} \right) = 0.314 \text{ in.}$$

For stage 2 the moment (and distributed load) in the strengthened beam when the steel yields is first calculated using

$$f_{s2} = \frac{E_s m_y (d - c_2)}{(EI)_2}$$

rearranged as

$$m_y = \frac{(f_y - f_{s1})(EI)_2}{E_s(d - c_2)}$$

$$= \frac{(60{,}000 - 15{,}076)(2.52 \times 10^{10})}{(29{,}000{,}000)(21.5 - 7.37)} = 2{,}762{,}732 \text{ lb-in.}$$

and

$$w_y = \frac{8m_y}{\ell^2} = \frac{8(2{,}762{,}732)}{(288)^2} = 266.5 \text{ lb/in.}$$

$$\delta_y = \frac{5}{384} \frac{w_y \ell^4}{(EI)_2} = \frac{5}{384} \frac{266.5(228)^4}{2.52 \times 10^{10}} = 0.9473 \text{ in.}$$

At yielding of the internal steel, the total moment applied, total load applied, and total midspan deflection of the beam are 3,626,732 lb-in., 349.8 lb/in. (4.12 kip/ft), and 1.261 in., respectively.

The deflection at the design service loads occurs in stage 2 and is computed using the equations above but substituting the service load, which is 1.8 kips/ft (150 lb/in.), and corresponds to a moment of 1,555,200 lb-in. This deflection is equal to

$$\delta_2 = \frac{5}{384}\frac{w_2\ell^4}{(EI)_2} = \frac{5}{384}\left(\frac{150(228)^4}{2.52 \times 10^{10}}\right) = 0.533 \text{ in.}$$

and the total service load deflection is equal to 0.847 in. This corresponds to a deflection ratio of

$$\frac{L}{\delta} = \frac{288}{0.847} = 340 < 360$$

which implies that the deflection of the strengthened beam would exceed a typical allowable deflection ratio of 360 (using the constant cracked section properties).

For stage 3, the nominal moment at ultimate failure (without the FRP reduction factor, ψ_f) and the ultimate deflection are calculated using the approximated properties of the third stage as follows:

$$\begin{aligned} m_{\max} &= 3.0(60{,}000)\left(21.5 - \frac{4.46}{2}\right) + 0.96(49{,}790)\left(24 - \frac{4.46}{2}\right) \\ &= 3468.6 + 1040.6 = 4{,}509{,}200 \text{ lb-in.} \end{aligned}$$

This can be compared with $M_n = 4{,}352{,}000$ lb-in determined previously in the strength analysis of this beam. The moment increment in the third stage is therefore equal to

$$\Delta m_3 = 4{,}509{,}200 - 3{,}626{,}731 = 883{,}171 \text{ lb-in.}$$

which corresponds to a load increment of

$$\Delta w_3 = \frac{8\Delta m_3}{\ell^2} = \frac{8(883{,}171)}{(288)^2} = 85.2 \text{ lb/in.}$$

and a deflection increment of

$$\Delta\delta_3 = \frac{5}{384}\frac{\Delta w_3\ell^4}{(EI)_3} = \frac{5}{384}\left(\frac{85.2(228)^4}{2.49 \times 10^9}\right) = 3.065 \text{ in.}$$

At ultimate load the total moment applied, total load applied, and total midspan deflection of the beam are 4,509,200 lb-in., 435 lb/in. (5.22 kips/ft), and 4.326 in., respectively. The cumulative predicted load–deflection behavior of the beam, using the assumption of a constant cracked section stiffness in each stage, is shown in Fig. 9.12.

Method 2: Using the Branson-modified flexural stiffness When the Branson formula approach is used, the steps in the procedure are very similar except that the flexural stiffness $(EI)_i$ in each stage is replaced by $(E_cI_e)_i$. Since I_e is a function of the moment applied at any given instant, the lines in the load–deflection plots will no longer be straight lines in any of the stages but rather, will be curves. These curves can be constructed using a speadsheet program and incrementing the moment in each stage. In the interests of brevity, only the key points in the load–deflection curve (i.e., FRP attachment, service load, steel yielding, and ultimate load) will be computed in this example.

Stage (1):

$$(I_{cr})_1 = 5909 \text{ in}^4$$

At FRP attachment:

$$(I_e)_1 = \left(\frac{M_{cr}}{M_a}\right)^3 I_g + \left[1 - \left(\frac{M_{cr}}{M_a}\right)^3\right](I_{cr})_1$$

$$= \left(\frac{610{,}906}{864{,}000}\right)^3 (13{,}824) + \left[1 - \left(\frac{610{,}906}{864{,}000}\right)^3\right](5909) = 8717 \text{ in}^4$$

$$\delta_1 = \frac{5}{384}\frac{w_1\ell^4}{E_c(I_e)_1} = \frac{5}{384}\left[\frac{83.3(288)^4}{(4.03 \times 10^6)(8717)}\right] = 0.212 \text{ in.}$$

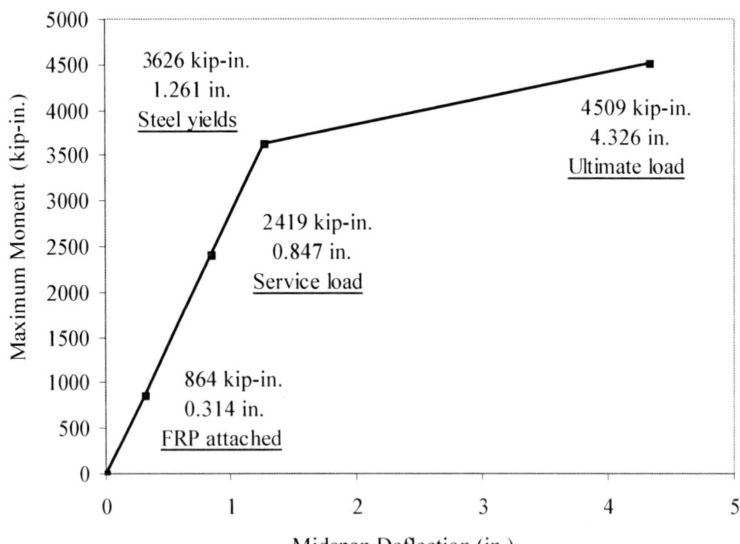

Figure 9.12 Load–deflection curve using the theoretical flexural stiffness.

Stage (2)

$$(I_{cr})_2 = 6253 \text{ in}^4$$

At service load:

$$(I_e)_{2(\text{service})} = \left(\frac{M_{cr}}{M_a}\right)^3 I_g + \left[1 - \left(\frac{M_{cr}}{M_a}\right)^3\right](I_{cr})_2$$

$$= \left(\frac{610{,}906}{2{,}419{,}200}\right)^3 (13{,}824) + \left[1 - \left(\frac{610{,}906}{2{,}419{,}200}\right)^3\right](6253) = 6375 \text{ in}^4$$

$$\delta_2 = \frac{5}{384}\frac{w_2\ell^4}{E_c(I_e)_2} = \frac{5}{384}\left[\frac{150(288)^4}{(4.03 \times 10^6)(6375)}\right] = 0.5230 \text{ in.}$$

Therefore, $\delta_{\text{service}} = 0.212 + 0.5230 = 0.7350$ in., and

$$\frac{L}{\delta} = \frac{288}{0.7350} = 392 > 360 \qquad \text{OK!}$$

At steel yielding:

$$(I_e)_{2(\text{final})} = \left(\frac{M_{cr}}{M_a}\right)^3 I_g + \left[1 - \left(\frac{M_{cr}}{M_a}\right)^3\right] I_{cr}$$

$$= \left(\frac{610{,}906}{3{,}626{,}732}\right)^3 (13{,}824) + \left[1 - \left(\frac{610{,}906}{3{,}626{,}732}\right)^3\right](6253) = 6319 \text{ in}^4$$

$$\delta_{2(\text{final})} = \frac{5}{384}\frac{w_2\ell^4}{E_c(I_e)_2} = \frac{5}{384}\left[\frac{116.5(288)^4}{(4.03 \times 10^6)(6{,}319)}\right] = 0.410 \text{ in.}$$

Therefore, $\delta_{\text{yielding}} = 0.212 + 0.5230 + 0.410 = 1.145$ in.

Stage (3)

$$(I_{cr})_3 = 1236 \text{ in}^4$$

$$I_{e3(\text{final})} = \left(\frac{M_{cr}}{M_a}\right)^3 I_g + \left[1 - \left(\frac{M_{cr}}{M_a}\right)^3\right](I_{cr})_3$$

$$= \left(\frac{610{,}906}{4{,}509{,}200}\right)^3 (13{,}824) + \left[1 - \left(\frac{610{,}906}{4{,}509{,}200}\right)^3\right](1236) = 1267 \text{ in}^4$$

$$\delta_{3(\text{final})} = \frac{5}{384}\frac{w_3\ell^4}{E_c(I_e)_3} = \frac{5}{384}\left[\frac{85.2(288)^4}{(2.02 \times 10^6)(1267)}\right] = 2.990 \text{ in.}$$

Therefore, $\delta_{\text{ultimate}} = 0.212 + 0.5230 + 0.410 + 2.990 = 4.135$ in.

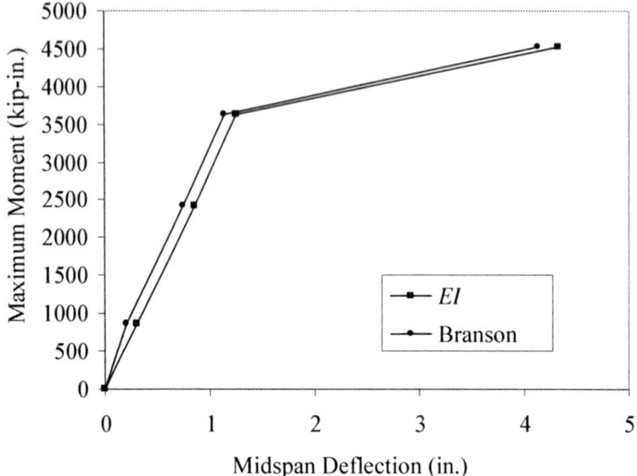

Figure 9.13 Comparison between theoretical and modified Branson load–deflection curves.

The load–deflection response using the Branson *EI* method is shown together with the theoretical *EI* method in Fig. 9.13. It can be seen that there is a small difference between the predictions using the Branson equations and the cracked section *EI*. However, this small difference in the service load can be significant in meeting serviceability design objectives (e.g., a $L/360$ deflection criterion), as shown in this example.

PROBLEMS

9.1 For the FRP strengthening systems[19] in Table P9.1,[20] calculate the bond-dependent coefficient, κ_m, for one, two, three, and four layers (plies) of the system.

9.2 The FRP strengthening systems in Table P9.1 are considered for strengthening an interior reinforced concrete beam with the following properties and loads: L = 16 ft, b = 10 in., h = 16 in., d = 13.5 in., f'_c = 3500 psi, f_y = 40,000 psi, two No. 8 bars, dead load= 400 lb/ft, and live load = 800 lb/ft. Consider one, two, three, and four layers

[19] Instructors may assign selected systems for homework problems.

[20] Properties reported in this table are design properties provided by manufacturers in their current (2006) online specification sheets and do not necessarily conform to the definition of the guaranteed property values in ACI 440.2R-02. They are assumed to be guaranteed properties for the purposes of these problems. For actual design, the user should obtain current guaranteed properties from manufacturers. These data can be compared with those obtained by the reader for Problems 8.2 and 8.3.

TABLE P9.1 FRP Strengthening Systems

No.	Fiber	FRP Strengthening System	f^*_{fu} (ksi)	E^*_f (Msi)	t_f (in.)
1	Glass	SikaWrap Hex 100G (laminate property)	81.0	3.55	0.040
2	Glass	Tyfo SEH 51 (laminate property)	66.7	3.03	0.050
3	Carbon	WaboMbrace CF130 (fiber property)	550.0	33.0	0.0065
4	Carbon	Sika Carbodur laminate (laminate property)	406.0	23.9	0.047
5	Glass	Tyfo UG composite laminate (laminate property)	130.0	6.0	0.055

of the given system. Assume that only the dead load (and self-weight) act on the beam when the FRP strengthening system is applied. Determine if the inelastic balanced reinforcement ratio can be achieved, and if so, its value. Why is the situation of not being able to achieve the balanced ratio not an issue with regard to traditional steel reinforcing?

9.3 Derive an expression [analogous to equation (9.54)] for the "elastic" balanced reinforcement ratio for an FRP-strengthened beam when the tension steel has not yielded in the section when a balanced condition is achieved.

9.4 Given an interior simply supported reinforced concrete beam with the following properties and loads: $L = 16$ ft, $b = 10$ in., $h = 16$ in., $d = 13$ in., $f'_c = 3500$ psi, $f_y = 40{,}000$ psi, two No. 8 bars, dead load = 300 lb/ft, and live load = 500 lb/ft. The beam is to be strengthened to increase its live-load carrying capacity. Consider the following strengthening alternatives: two layers each 10 in. wide, of the FRP sheet or fabric systems listed in Table P9.1 (systems 1 to 3) or two strips each 4 in. wide placed side by side of the FRP laminate systems in Table P9.1 (systems 4 and 5). Assume that only the dead load (and self-weight) act on the beam when the FRP strengthening system is applied. Determine **(a)** the nominal and factored moment capacity of the strengthened beam, and **(b)** the live-load capacity of the beam when it is strengthened. Compare the strengthening capabilities of the various systems.

9.5 Determine the factored moment capacity of the beam in Design Example 9.1 if five layers each 9.5 in. wide, of the FRP system are used. Find the required area and width of FRP needed to exactly achieve the moment demand with five layers of the FRP system. Compare this with the data presented in Table 9.3, which gives the results for two or four layers of the FRP system chosen. Can the moment demand be met with three layers of FRP system? What should the width of the

FRP be in this case, and what will the failure mode be? Compare the results for two, three, four, and five layers, and discuss the implications for the designer using FRP strengthening systems. What system would you ultimately choose as the best system to meet the strength demand? Do not consider serviceability and detailing or the influence of the effect of layers on possible shear failure of the strengthened beam. Justify your choice.

9.6 A simply supported reinforced concrete beam in the interior of a residential building spans 20 ft. It has a 12 × 18 in. cross section and is reinforced with three No. 7 grade 60 steel bars and No. 3 steel stirrups at 7.5 in. on center. It was designed with a clear cover of 1.5 in. The beam is supported on 12-in.-wide supports. $f'_c = 3500$ psi. The beam was originally designed to carry a live load of 400 lb/ft and a dead load of 600 lb/ft (not including its self-weight). The building has been sold and the new owner wants to convert the residential space to office space, which requires that the live load on the beam be increased. You are required to investigate a flexural strengthening system using Sika CarboDur Type S 1012 pultruded strips bonded to the beam with Sikadur 30 epoxy paste. For the five strengthening designs shown in Table P9.6, determine **(a)** the strengthened ultimate moment capacity (and % increase), **(b)** the failure mode, and **(c)** the new allowable live load that can be applied to the beam (and % increase). Follow ACI 440.2R-02 guidelines.

What can you conclude about the various strengthening designs? Are more strips necessarily better? Which design would you recommend to the owner to achieve the maximum live load increase and, why? (You do not need to consider serviceability limits for this exercise.)

9.7 Reconsider the beam described in Problem 9.6. Consider the same strengthening problem, but investigate the five cases using a Tyfo UG (unidirectional glass) laminate strip and Tyfo TC epoxy. Compare the

TABLE P9.6 Strengthening System Layouts

Design Case	Number of Strips	Strip Layout
1	1	Along centerline
2	2	Side by side parallel to centerline
3	2	Double layer along centerline[a]
4	3	Three strips alongside each other (in one layer)
5	4	Two double-layer strips side by side

[a]Precured strips are not usually used in multiple layers, due to their thickness. This example is for illustration purposes.

results using the glass laminate to those using carbon laminate. Discuss the differences and your observations.

9.8 A 5-in.-thick reinforced concrete slab in a building frame spans continuously over reinforced concrete beams 10 in. wide and 12 ft apart. $f'_c = 4000$ psi. The slab is reinforced with No. 4 grade 60 bars at 12 in. on center for positive and negative moments. It was designed for residential purposes to carry a superimposed dead load of 20 lb/ft^2 and a live load of 60 lb/ft^2. The building owner would like to increase the live-load capacity to 150 lb/ft^2 to use the space as office space. Design an FRP strengthening system using one of the FRP systems given in Table P9.1. (Note that the FRP does not have to cover the entire surface and can be provided in strips.) Perform only the design for strength. (Assume that serviceability is not a problem and that the beams have adequate capacity to support the new loads.) Provide strengthening for both the positive and negative moment regions. Detail the strengthening system and show termination points for a typical interior two-bay width (i.e., 24 ft). Discuss how you would detail the FRP strengthening system on the top surface of the floor.

9.9 A simply supported reinforced concrete beam in the interior of a residential building spans 18 ft. It has a 16 in. wide × 24 in. high cross section and is reinforced with three No. 9 grade 60 steel bars and No. 3 steel stirrups at 8 in. on center. Assume that the beam was designed with an effective depth of 21.5 in. $f'_c = 4000$ psi. The original dead load/live load ratio (including the beam self-weight) was 1:2, and this load ratio is maintained for the future loads. A consultant has designed a strengthening system for this beam to increase its load-carrying capacity. The consultant has recommended strengthening the beam by bonding two layers of 10-in.-wide sheets of WaboMBrace CF130 carbon fiber sheets and WaboMBrace epoxy saturant. The sheets are bonded to the underside of the beam and extended to the edges of the supports. You have been called in as an expert in FRP strengthening of concrete structures to peer-review the design. You are not told what the desired strengthening objective was; this is something the owner wants to verify with your check. The owner also wants to verify that the design is in compliance with the ACI 440.2R-02 design guide with regard to the ultimate capacity calculations, failure modes, and strengthening limits (including stresses in the steel and FRP and maximum deflection of the strengthened beam under future service loads). You are required to:

(a) Determine the failure mode of the strengthened beam.

(b) Determine the ultimate moment capacity of the strengthened beam.

(c) Determine the percent increase in the ultimate capacity.

(d) Determine the future (strengthened) dead- and live-load carrying capacity of the beam.

(e) Determine whether the strengthening is within ACI capacity (strength enhancement) strengthening limits.

(f) Determine whether or not shear strengthening is required for future loads.

(g) Determine the value of the maximum possible strengthened ultimate moment capacity the owner could obtain according to the ACI ultimate strengthening limits.

(h) Determine whether the strengthening is within ACI serviceability strengthening limits and explain why these limits are placed on strengthening.

(i) Provide a cost estimate to the owner for the strengthening system.[21]

9.10 Derive the folllowing equation (given in ACI 440.2R-02 without the subscript 2 on k) for the stress in the steel in the FRP-strengthened member at service loads:

$$f_{s,s} = \frac{[(m_1 + m_2) + \varepsilon_{bi} A_f E_f (h - k_2 d/3)](d - k_2 d) E_s}{A_s E_s (d - k_2 d/3)(d - k_2 d) + A_f E_f (h - k_2 d/3)(h - k_2 d)}$$

9.11 For the five strengthening systems evaluated in Problem 9.4, perform the required serviceability checks and detail the FRP systems.

(a) Check the service-level stresses in the steel and compare with the ACI 440.2R-02 permitted value,

(b) Check the service-level stresses in the FRP strengthening system and compare with the ACI 440.2R-02 permitted value,

(c) Check the maximum midspan deflection under service loads using the Branson formula and compare with $L/360$.

(d) Detail the strengthening systems.

9.12 For the five strengthening system configurations in Problems 9.6 and 9.7, perform the required serviceability checks.

(a) Check the service-level stresses in the steel and compare with the ACI 440 2R-02 permitted value.

(b) Check the service-level stresses in the FRP strengthening system and compare with the ACI 440 2R-02 permitted value.

(c) Check the maximum midspan deflection under service loads using the Branson formula and compare with $L/360$.

(d) Detail the strengthening systems.

9.13 A simply supported flat-slab bridge built in 1930 needs to be strengthened to increase its live-load capacity for current highway loads. The

[21] Contact the FRP system supplier for current pricing. Consider both material and installation costs.

bridge has a span of S = 23 ft, a width of 26 ft, a 20-in.-thick concrete slab, and a 6-in. asphalt overlay (assume that the asphalt overlay is 130 lb/ft^3). The bridge is reinforced with 1920s-era 1.0-in. square ribbed reinforcing bars at 6 in. on center with a yield strength of 33 ksi. The bottom clear cover is 1 in. and the 1920s concrete strength is f'_c = 2500 psi. The bridge has been rated by the local department of transportation and given an HS17 inventory load rating. Since it is rated below HS20, it is regarded as structurally deficient. The local authority would like to increase the load rating to HS25 (i.e., a 45-ton tractor-trailer, single wheel load, P_{25} = 20 kips, not including the impact factor of 30%) for future highway loads. To accomplish this, the nominal moment capacity of the bridge must be increased by 30%. You are required to design a strengthening system for the bridge using WaboMBrace CF130 carbon fiber sheets and WaboMBrace epoxy saturant according to ACI 440.2R-02 and the AASHTO Standard Specification (AASHTO, 2002)[22]. You need to perform the following as part of your design:

(a) Determine the current nominal moment capacity and current ultimate capacity (kip-ft/ft) and also the future (after the strengthening) nominal and ultimate capacities (kip-ft/ft) of the bridge.

(b) Determine the current and future allowable wheel load (live load) for the bridge. Assume that the wheel load is a single concentrated load applied at midspan over an effective width $E = 4 + 0.06S$ (where S is the span in feet) of the slab. (Do not forget to include the dead-load moment in this calculation.)

(c) Compare this to the wheel load for an HS25 truck (i.e., 20 kips). Note that an additional impact factor of 30% is applied to the live load to account for dynamic effects [i.e., the HS25 required wheel load capacity is 20(1.3) = 26 kips]. Does the increase in moment capacity you obtained meet the strengthening objectives (i.e., can the bridge carry a HS25 truck load)?

(d) Show a sketch of the strengthening system configuration. *Note:* You do not have to cover the entire surface; you can apply the sheet material in strips of width, w_f. Clear spacing of the strips should be less than two times the strip width and less than h.

[22] See Chapter 5 for AASHTO dead and live load factors, and consult the specifications.

10 FRP Shear Strengthening

10.1 OVERVIEW

This chapter deals with the design of shear strengthening systems for reinforced concrete flexural members loaded by transverse loads, such as beams, slabs, walls, and columns with precured FRP strips and plates and laid-up FRP sheet and fabric systems. The strengthening of reinforced concrete members subjected to torsion is not discussed. Only nonprestressed concrete members that are reinforced with conventional steel reinforcing bars are considered. For the purposes of this chapter, it is assumed that the reader is familiar with both the fundamentals and details of the shear design of concrete members reinforced with conventional steel reinforcing bars.

Although shear strengthening has been carried out in practice for about as long as flexural strengthening, the understanding of the behavior of members strengthened in shear is less developed than that for flexural strengthening. This parallels our understanding of the subject of shear behavior of reinforced concrete members in general, as noted in the discussion related to the shear design of concrete members with FRP rebars. The subject of shear strengthening is additionally complicated by the fact that design of the shear strengthening depends on how the FRP strengthening system encompasses, or wraps around, the concrete member. Therefore, when performing a shear strengthening design it is imperative that a structural engineer has a good knowledge of the particular FRP strengthening system that will be used by the contractor (or that will be specified in the specifications). Not all systems can be wrapped around a concrete member, and not all concrete members have their four sides accessible for the FRP to be applied.

The design procedures presented in this chapter follow ACI 440.2R-02, which is compatible with ACI 318-99. The examples presented in this chapter are based on the examples in ACI 440.2R-02 and are intended to allow the reader to analyze these examples critically and consider design alternatives not presented in the ACI guide. In recent years there have been significant applications of FRP shear strengthening systems to shear-strengthen deficient highway bridge columns in seismically active zones. ACI 440.2R-02 does not specifically address the issues of strengthening of columns. It has been shown that the axial force on a column can influence the shear capacity and the lateral displacement capacity. A method due to Priestley and colleagues, for the design of shear strengthening systems for columns that accounts for the effect of axial loads is presented at the end of this chapter.

10.2 INTRODUCTION

FRP strengthening systems can be used to increase the shear capacity of concrete beams, columns, and walls. FRP strengthening systems are applied to the webs of beams (or vertical sides of columns and walls) and function in a fashion similar to that of internal steel shear reinforcements such as stirrups, hoops, or ties. The FRP strengthening system is attached to the exterior surface of the concrete member in the transverse plane of the member and resists the transverse shear force resultant (or lateral force in a column or wall), as shown in Figs. 10.1 and 10.2. The FRP strengthening system adds to the shear resistance of the member, which traditionally is derived from the concrete itself and from the internal shear reinforcement.

In a general sense, an FRP shear strengthening system is similar in appearance on the exterior of the concrete member to an FRP flexural strengthening system (except for the orientation of the fibers, as explained in what follows). However, from a structural mechanics and constructability point of view, FRP shear strengthening is significantly more complex than FRP flexural strengthening. The reasons for this are based on the mechanics of shear strengthening itself and on the shapes and geometries of concrete members relative to the geometries and properties of FRP strengthening materials.

In the case of FRP flexural strengthening of a beam, column, or even a wall (typically, for out-of-plane strengthening) the FRP strengthening system is geometrically compatible with the member since the primary orientation of

Figure 10.1 FRP Shear strengthening of a beam. (Courtesy of Structural Group.)

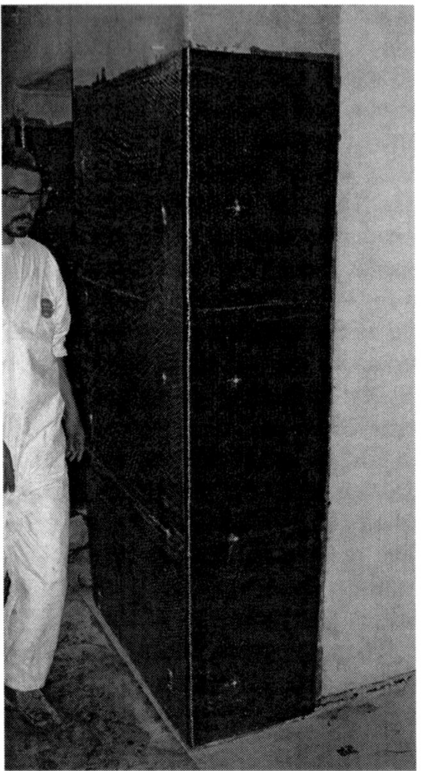

Figure 10.2 FRP shear strengthening of a wall with openings. (Courtesy of Racquel Hagan.)

the FRP strengthening system is in the lengthwise direction of the member. Therefore, a long precured strip or a unidirectional fabric tow sheet can readily be extended along the length of the member to provide flexural reinforcement. In addition, the bending moment along a flexural member varies along the length of the member, having points of maximum moment, minimum moment, and zero moment (at inflection points) along the length. Therefore, the FRP strengthening system is exposed to a varying axial force along its length, and any unit length portion of the adhesive system is only responsible for carrying the moment increment over that unit length. Wherever possible, the FRP strengthening is terminated and anchored at or near these regions of zero moment in the member, which reduces the potential for peeling anchorage failures at the ends of the strip. Added to this, a flexural strengthening system is typically attached at the most advantageous position on the cross section of the member, that is, as far from the neutral axis as possible. Finally, the structural behavior of reinforced concrete flexural members is well understood, and analytical models for flexural behavior have been verified over many years. It has been shown (as detailed in earlier chapters) that these models can be extended to FRP strengthening in a relatively straightforward manner provided that the force can be transferred between the FRP strengthening system and the concrete in a controlled fashion.[1]

Consider now the case of shear strengthening of a reinforced concrete member. In the case of a beam or a column, the structural depth of the member is much smaller than the length. Since the transverse shear force acts parallel to the structural depth of the member, the primary direction of the FRP reinforcement needs to be placed in this direction (the precise angle with respect to the beam axis is explained in what follows). Therefore, the FRP reinforcement can act only over relatively short lengths (and finite widths) on the sides of the member, which, as noted, is not very deep[2] and does not provide much room for anchorage of the FRP system. In addition, unlike the case of FRP flexural strengthening, where the force varies along its length, the force (or stress) in the FRP shear strengthening is constant through the depth of the section at any location along the length.[3]

To prevent debonding of the FRP shear strengthening system at the top and bottom of the sides of the member, it is possible to wrap the FRP strengthening system around the section, either partially (often called a *three-sided U-wrap*) or completely around the beam (often called a *four-sided wrap*), as shown schematically in Fig. 10.3. When the FRP system cannot be wrapped around either the top or the bottom of the section, the system is called a *two-*

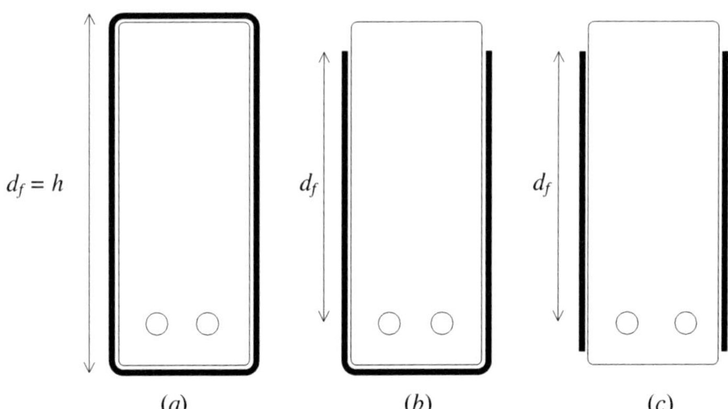

Figure 10.3 FRP shear strengthening schemes and effective depths: (*a*) four-sided; (*b*) three-sided; (*c*) two-sided.

[1] In fact, it should be noted that strain compatibility at the interface between the concrete and the FRP is not necessarily a requirement for flexural strengthening.

[2] For deep beams and walls, where the structural depth is larger, the situation is not as bad.

[3] Even though the shear force varies along the length of the member, at any section, the shear force is single-valued according to one-dimensional beam theory.

sided system. Non-fully wrapped FRP systems are susceptible to detachment failures at their ends and their strains are limited by a shear bond-reduction coefficient. In many cases, shear strengthening is performed on T-shaped sections, either produced as prefabricated T-beams or behaving as T-sections when cast monolithically with a floor slab or wall section. In this situation a four-sided shear strengthening system can still be achieved if the FRP strengthening system is applied in discrete narrow strips and slits are machined into the concrete slab or wall (which can be a time-consuming and expensive process). Slits can also be used to anchor three-sided systems more effectively in an adjoining the slab or wall. It is important to emphasize that in the case of T-sections the FRP strengthening system should never be extended around the reentrant corner unless an additional anchorage system is provided to prevent radial peeling at the corner.

Constructability issues also arise with wrapped or partially wrapped systems. The corners of the beam need to be rounded (chamfered) to prevent the sharp corner edges from causing stress concentrations and premature failure of the FRP wrap. In the case of precured systems, special discrete L-shaped elements (also called *straps*) must be manufactured, and the radius of the bend needs to be compatible with the chamfer on the beam edge (as shown in Fig. 10.4).

For laid-up FRP systems the sheets or fabrics can be applied in finite-width segments, or they can be applied continuously along the side of the member. In either case, the sheet needs to be cut to a rather short length, and due to its finite width will need to be placed in multiple sections along the beam length (as seen in Fig. 10.1). The sheet can be applied at inclined angles, α, to the beam axis, as shown in Fig 10.5. The sheet width, w_f, is measured perpendicular to the fiber principal direction, while the sheet spacing, s_f, is measured along the beam axis. The angle of inclination of the postulated

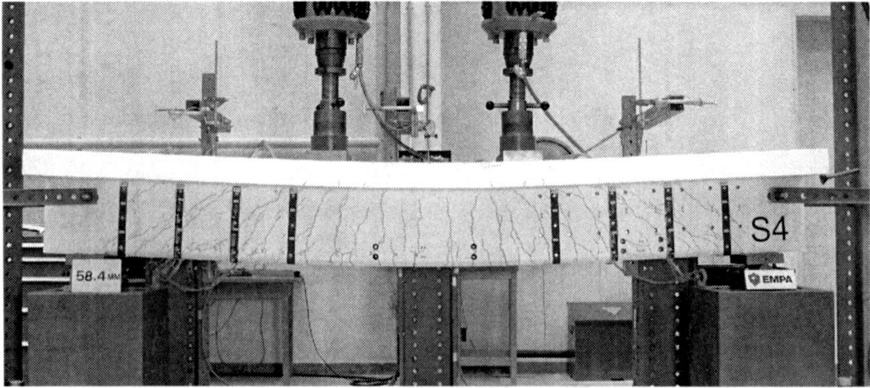

Figure 10.4 L-shaped precured FRP shear reinforcement straps. (Courtesy of Christoph Czaderski.)

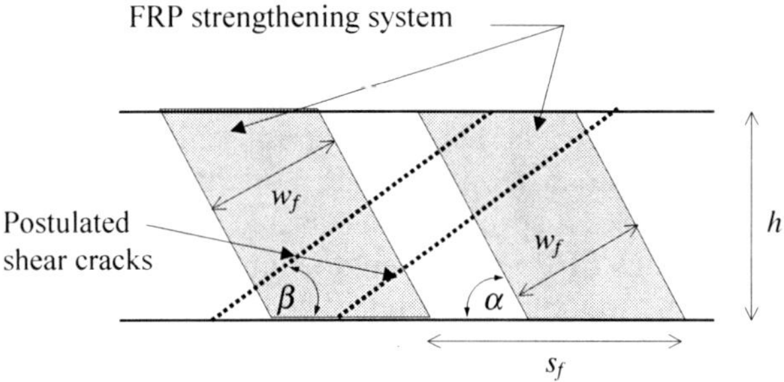

Figure 10.5 Geometric variables for FRP shear strengthening.

shear crack, typically taken as 45°, is denoted as β. If this system is wrapped, it creates a spiral along the length of the member, which can have definite advantages in terms of structural performance and may have some constructability advantages.[4] In this case, precured strips cannot be used unless they are manufactured with the appropriate angled bevel at the bottom edge.

10.3 SHEAR CAPACITY OF AN FRP-STRENGTHENED MEMBER

The nominal shear capacity of an FRP-strengthened concrete member with existing steel shear reinforcing is determined by adding the contribution of the FRP strengthening system to the existing shear capacity and is given as

$$V_n = V_c + V_s + \psi_f V_f \tag{10.1}$$

V_c is the existing shear capacity of the concrete,[5] V_s the shear capacity of the existing steel shear reinforcement, and V_f the shear capacity of the FRP strengthening system. It is assumed that the contribution of the three different contributors listed above is linearly additive and that no beneficial or harmful synergistic[6] effects occur as a result of attaching the FRP to the sides of the concrete member. Recall that the additional capacity reduction factor, ψ_f, is used on the FRP contribution. The FRP capacity reduction factor, ψ_f, is taken as 0.95 for completely wrapped sections (called *contact critical*) and as 0.85 for two- or three-sided wrapped sections (called *bond critical*).

The FRP contribution is determined using the same truss analogy as that used to develop the shear resistance provided by steel stirrups. The angle of

[4] A continuous spiral can be done only for a four-sided wrap.

[5] Due to aggregate interlock, dowel action of the main bars, and resistance of the uncracked compression zone above the shear crack.

[6] Such as an increase or decrease in V_c or V_s when the FRP is attached.

the critical shear crack with respect to the longitudinal axis of the member is taken as 45°, in accordance with the ACI 318-99 philosophy.

Two forms of the equation for the shear contribution for FRP shear strengthening are provided herein, one for intermittent strips having width (measured perpendicular to the major fiber orientation of the strengthening system) w_f and spaced at a spacing s_f (measured parallel to the member axis) and one for continuous FRP shear strengthening (i.e., full side coverage with no overlapping). Intermittent wrapping is generally preferred (although it is harder to place in the field) since it allows moisture to migrate from the concrete to the air. A continuous wrap will fully cover the concrete surface and could trap moisture under the wrap and cause degradation of the interface, substrate, or the FRP composite. In addition, a continuous wrap causes difficulties with inspection since the surface of the concrete is not visible.

For intermittent strengthening,

$$V_f = \frac{A_{fv} f_{fe} (\sin\alpha + \cos\alpha) d_f}{s_f} \qquad \text{(ACI 440.2R-02:10-3)}$$

Substituting

$$A_{fv} = 2nt_f w_f \qquad \text{(ACI 440.2R-02:10-4)}$$

gives

$$V_f = \frac{2nt_f w_f f_{fe} d_f (\sin\alpha + \cos\alpha)}{s_f} \tag{10.2}$$

with the geometric condition that

$$s_f > \frac{w_f}{\sin\alpha} \tag{10.3}$$

$f_{fe} = E_f \varepsilon_{fe}$ is the effective tensile stress in the FRP, α the inclination of the fiber in the FRP strengthening system to the longitudinal axis of the member, n the number of layers of shear reinforcement and d_f the effective depth of the FRP strengthening system (shown in Fig. 10.3).

If $s_f = w_f/\sin\alpha$, continuous strengthening (full surface coverage) occurs and

$$V_f = 2nt_f f_{fe} d_f (\sin^2\alpha + \sin\alpha\cos\alpha) \tag{10.4}$$

For a continuous inclined strip at an angle α to the beam axis that is formed out of discrete strips of width, w_f, placed side by side along the beam length, the strips must be spaced at

$$s_f = \frac{w_f}{\sin \alpha} \tag{10.5}$$

For the special case where $\alpha = 90°$, the shear contribution of the FRP strengthening systems is given as follows: For intermittent strengthening,

$$V_f = \frac{A_{fv} f_{fe} d_f}{s_f} \tag{10.6}$$

For continuous strengthening,

$$V_f = 2nt_f f_{fe} d_f \tag{10.7}$$

For the special case where $\alpha = 45°$, the shear contribution of the FRP strengthening systems is given as follows: For intermittent strengthening,

$$V_f = \frac{\sqrt{2} A_{fv} f_{fe} d_f}{s_f} \tag{10.8}$$

with the geometric condition that

$$s_f > \sqrt{2} w_f \tag{10.9}$$

For continuous strengthening,

$$V_f = 2nt_f f_{fe} d_f \tag{10.10}$$

which is the same result as for $\alpha = 90°$ for continuous strengthening.

It is interesting to note that in the case of FRP strengthening with sheets there is no apparent benefit to using inclined sheets at 45° as there is with conventional reinforcing bars, as would be implied by use of ACI 440.2R-02:10-3. This is because sheets have finite and usually quite considerable width (10 to 20 in.). In the derivation of ACI 440.2R-02:10-3[7] (which is similar to that done for steel bars) it is assumed that the shear reinforcing bars occupy a theoretically vanishing area such that they can be brought closer and closer together when inclined but not come into contact, maintaining some separation. Therefore, inclined bars are actually closer to each other than vertical bars for the same horizontal spacing, s, and intersect more inclined shear cracks, leading to the result that they are more effective as shear reinforcement. When the shear reinforcement is in the form of relatively wide sheets, such an assumption cannot be made and the sheets soon begin, the-

[7] See Wang and Salmon (2002, p. 154).

oretically, to come into contact and overlap. The apparent advantage of inclined sheets can be seen when sheets begin to resemble bars in their aspect ratios w_f/s_f (0.2 or less). In such a case, sheets become in reality more geometrically similar to near-surface-mounted (or surface-mounted) bars.

If the shear crack is not assumed to be at 45°, the equations above can be written in terms of the angle of inclination of the shear crack with respect to the longitudinal axis, β, shown in Fig. 10.5, as: For intermittent strengthening,

$$V_f = \frac{A_{fv} f_{fe}(\cot\beta \sin\alpha + \cos\alpha) d_f}{s_f} \tag{10.11}$$

for continuous strengthening,

$$V_f = 2nt_f f_{fe} d_f(\cot\beta \sin^2\alpha + \sin\alpha\cos\alpha) \tag{10.12}$$

and

$$s_f = \frac{w_f}{\sin\alpha} \tag{10.13}$$

10.4 EFFECTIVE STRAIN IN THE FRP FOR SHEAR STRENGTHENING

The effective strain in the FRP shear strengthening system is limited in ACI 440.2R-02 to prevent detachment failures and also to maintain the integrity of the concrete aggregate interlock in the concrete member. A bond-related coefficient for shear strengthening is introduced to account for the FRP shear strengthening system debonding. This is similar to the empirical approach taken for debonding of FRP systems in flexural strengthening, in which the phenomenon of debonding is accounted for by a bond-related coefficient but the capacity for each specific failure mode is not determined. Depending on the type of shear strengthening configuration used, the FRP material may fail due either to debonding or to FRP rupture. Explicit procedures to determine the failure in these two modes are discussed in Chen and Teng (2003a,b).

For completely wrapped (four-sided) FRP shear strengthening systems, the maximum effective strain in the FRP strengthening system at failure is limited to

$$\varepsilon_{fe} = 0.004 \leq 0.75\varepsilon_{fu} \qquad \text{(ACI 440.2R-02:10-6}a\text{)}$$

For three- or two-sided shear strengthening, the effective shear strain in the FRP strengthening system at failure is limited to

$$\varepsilon_{fe} = \kappa_v \varepsilon_{fu} \leq 0.004 \qquad \text{(ACI 440.2R-02:10-6}b\text{)}$$

The shear bond-reduction coefficient, κ_v, is a function of the concrete strength, the wrapping type used, and the stiffness of the FRP strengthening system and is given by ACI 440.2R-02 as

$$\kappa_v = \frac{k_1 k_2 L_e}{468\varepsilon_{fu}} \leq 0.75 \qquad \text{(ACI 440.2R-02:10-7)}$$

The equation is a function of the active bond length, L_e, over which the shear stress is transferred between the FRP and the concrete. It has been shown that it is this finite length that limits the maximum force that can be transferred between the two materials, regardless of the bonded length of the FRP strip (Maeda et al., 1997; Täljsten, 1997; Khalifa et al., 1998; Bizindavyi and Neale, 1999). In ACI 440.2R-02 the active bond length is given by an empirical equation as

$$L_e = \frac{2500}{(nt_f E_f)^{0.58}} \qquad \text{(ACI 440.2R-02:10-8)}$$

and the coefficients k_1 and k_2 are given as

$$k_1 = \left(\frac{f'_c}{4000}\right)^{2/3} \qquad \text{(ACI 440.2R-02:10-9)}$$

For three-sided shear strengthening systems,

$$k_2 = \frac{d_f - L_e}{d_f} \qquad \text{(ACI 440.2R-02:10-10)}$$

For two-sided shear strengthening systems,

$$k_2 = \frac{d_f - 2L_e}{d_f} \qquad \text{(ACI 440.2R-02:10-10)}$$

Mechanical anchorages can be used to anchor two- or three-sided wraps, in the compression zone of the web; however, design guidance is not provided by the ACI 440.2R-02 for this at the present time. Steel angles at reentrant corners, steel plates, and fiber anchors have all been proposed for this purpose. In addition, slot or slits can be drilled in T-section flanges for anchoring the FRP.

The amount of shear reinforcement in a reinforced concrete member is limited by ACI 318-99. When FRP shear strengthening is used in addition to conventional steel shear reinforcement, the reinforcement limit must hold for both types of reinforcement:

$$V_s + V_f \leq 8\sqrt{f_c'}b_w d \qquad \text{(ACI 440.2R-02:10-11)}$$

When intermittent strips are used, maximum spacing between the strips is mandated so that every shear crack will be covered by sufficient strip width. The following maximum spacing of intermittent strips is required:

$$s_f^{\max} = \frac{d_f}{4} + w_f \tag{10.14}$$

10.5 DESIGN PROCEDURE FOR SHEAR STRENGTHENING

Step 1. Determine the current and future strength requirements and check the strengthening limits. The existing shear capacity of the beam must be determined prior to design of the strengthening system.

Step 2. Choose an FRP system strengthening system and configuration. The designer needs to make some a priori choices regarding the type of strengthening system (precured or wet layup) to use and obtain manufacturer-guaranteed properties for approved systems from existing specification sheets.

Step 3. Determine the number of layers and geometry of the FRP wrap. Unlike in the flexural design, an estimating step is not required. For shear strengthening the amount of FRP material can be determined in closed form once the configuration is chosen. At first the amount of material needed for continuous coverage is determined and then the spacing of the intermittent wrap can be found to maximize the FRP used. The shear bond reduction coefficient, κ_v, is determined in this step of the design.

Step 4. Calculate the actual shear capacity of the FRP wrap. The actual FRP system used will probably be slightly different from that determined in step 3. This is because commercial sheets and precured strips are supplied in limited widths and for economic reasons the final design should attempt to be compatible with premanufactured widths.

Step 5. Calculate the factored shear capacity of the future strengthened beam. Calculate the factored shear capacity and compare with the demand.

Step 6. Check the maximum FRP reinforcement and spacing limits.

Step 7. Detail the FRP strengthening system.

Design Example 10.1: FRP Shear Strengthening of a Beam[8] A reinforced concrete T-beam with dimensions shown in Fig. 10.6 is subjected to an in-

[8] This example follows Example 14.4 in ACI 440.2R-02.

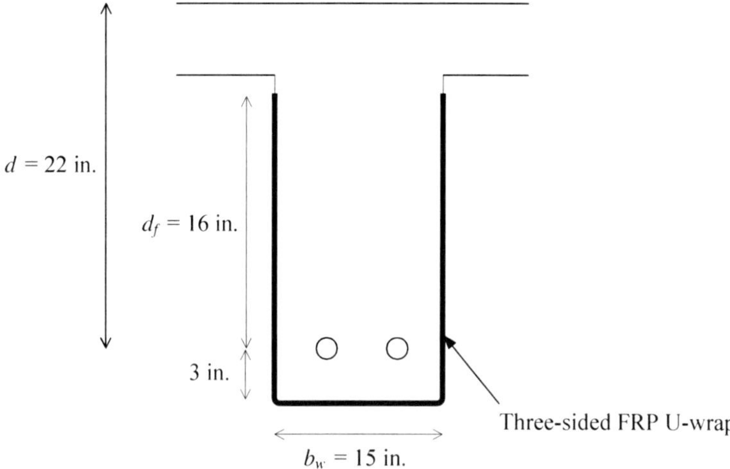

Figure 10.6 Geometry of a beam showing pertinent dimensions and an FRP wrap.

creased live load. The beam is an interior beam in a building, constructed originally using f'_c = 3000-psi concrete. Prior evaluation of the existing beam has determined that V_c = 36.4 kips and V_s = 19.6 kips. A carbon–epoxy FRP strengthening system is requested by the owner. The future total shear demand (at a distance d from the support) due to the increase in live load on the beam is V_u = 60 kips.

SOLUTION

Step 1. Determine the current and future strength requirements and check the strengthening limits. The current nominal shear capacity of the beam is given as

$$V_n = V_c + V_s = 36.4 \text{ kips} + 19.6 \text{ kips} = 56.0 \text{ kips}$$

and the current factored shear capacity is

$$\phi V_n = 0.85(56.0) = 47.6 \text{ kips}$$

Since $V_u > \phi V_n$, shear strengthening is required in the portion of the beam that has a current capacity less than that required for the future demand. It is important to recognize that the FRP does not need to cover the entire length of the beam web. For a simply supported beam this will be a region from the support a distance x along the beam. This depends on the shear force distribution (i.e., the loading) on the beam. The beam in question is loaded by a uniformly distributed load and a concentrated load. The shear force diagram for a portion of the beam near the support is shown in Fig. 10.7.

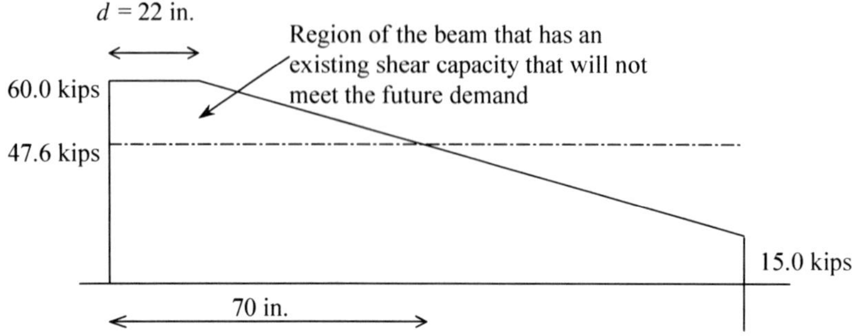

Figure 10.7 Shear force in a beam near the support.

In this example, it is assumed that the flexural capacity of the beam is adequate for the increased loads. In general, this would need to be checked. The shear strengthening limits also need to be checked. For this, the live and dead loads are needed. If these were given, the following check would be made:

$$W_{\text{limit}} = 1.2w_{\text{DL}} + 0.85w_{\text{LL}}$$

$$V_{\text{limit}} = f(W_{\text{limit}}) \leq 47.6 \text{ kips}$$

Step 2. Choose an FRP system strengthening system and configuration. A dry fiber sheet system with the following guaranteed fiber[9] properties is chosen.

$$t_f = 0.0065 \text{ in.} \qquad \varepsilon_{fu}^* = 0.017\ (1.7\%)$$

$$f_{fu}^* = 550 \text{ ksi} \qquad E_f = 33{,}000{,}000 \text{ psi}$$

Determine the design properties of the FRP system according to the environment. For this case we have interior construction with a carbon FRP and C_E = 0.95.

$$f_{fu} = 0.95(550) = 522.5 \text{ ksi}$$

$$\varepsilon_{fu} = 0.95(0.017) = 0.016\ (1.6\%)$$

$$E_f = 33{,}000{,}000 \text{ psi (unchanged)}$$

[9]Note that the fiber thickness and modulus properties are used in this example (not the FRP composite properties).

Decide on a configuration of the FRP system. Since this is a T-beam, a four-sided all-around complete wrap cannot be used (unless slits are cut through the slab). A three-sided U-wrap or a two-sided wrap can be used. It is advisable wherever possible to use the most effective level of wrapping style. In this case, a three-sided U-wrap is considered first.

Step 3. Determine the number of layers and the geometry of the FRP wrap. The shear contribution required of the FRP strengthening system can be found from

$$V_f^{\text{reqd}} = \frac{1}{\psi_f}\left(\frac{V_d}{\phi} - V_c - V_s\right) = \frac{1}{0.85}\left(\frac{60}{0.85} - 36.4 - 19.6\right) = 17.2 \text{ kips}$$

To determine the number of layers and the sizes of the FRP strengthening sheets to be used in the three-sided wrap selected several preliminary design decisions need to be made: (1) continuous or intermittent sheet; (2) inclined or 90° fiber orientation. To start it is advisable to assume a continuous 90° system to obtain an initial estimate of the number of layers. Thereafter, the intermittent system can be considered. For further optimization, an inclined FRP system can be used. However, use of an inclined system in a three- or four-sided wrap will cause constructability problems, with "spiraling" of the sheet on the opposite side of the beam.

The maximum permissible strain (effective strain) in the FRP is calculated first. This is a function of the shear bond coefficient, κ_v, found from

$$\kappa_v = \frac{k_1 k_2 L_e}{468\varepsilon_{fu}} \leq 0.75$$

Assuming one layer of the FRP will be used,

$$L_e = \frac{2500}{(nt_f E_f)^{0.58}} = \frac{2500}{[1(0.0065)(33{,}000{,}000)]^{0.58}} = 2.022 \text{ in.}$$

$$k_1 = \left(\frac{f'_c}{4000}\right)^{2/3} = \left(\frac{3000}{4000}\right)^{2/3} = 0.826$$

$$k_2 = \frac{d_f - L_e}{d_f} = \frac{16 - 2.022}{16} = 0.874 \qquad \text{for three-sided U-wraps}$$

which gives

$$\kappa_v = \frac{0.826(0.874)(2.022)}{468(0.016)} = 0.195 \leq 0.75$$

Since $\varepsilon_{fe} = \kappa_v \varepsilon_{fu} \leq 0.004$,

$$\varepsilon_{fe} = 0.195(0.016) = 0.0031 < 0.004$$

For a continuous 90° three-sided wrap, the number of layers required is

$$n = \frac{V_f^{\text{reqd}}}{2t_f f_{fe} d_f} = \frac{V_f^{\text{reqd}}}{2t_f \varepsilon_{fe} E_f d_f}$$

$$= \frac{17.2}{2(0.0065)(0.0031)(33{,}000)(16)} = 0.81 \rightarrow \text{use one layer}$$

Since the number of layers is less than 1, one continuous layer would provide too much FRP for the required design. An intermittent sheet may therefore be preferable to a continuous sheet in this case. One can obtain the sheet width-to-sheet spacing ratio, $R_{w/s} = w_f/w_s$ for an intermittent 90° wrap using

$$R_{w/s} = \frac{w_f}{s_f} = \frac{V_f^{\text{reqd}}}{2nt_f f_{fe} d_f} = \frac{17.2}{2(1)(0.0065)(102.3)(16)} = 0.81$$

(Note that the similarity between the two equations above occurs because for a continuous wrap, $w_f = s_f$.)

It is preferable to choose a sheet width that is a fraction of that of a commercially available sheet width to aid in constructability and to reduce waste. Assuming that the carbon–epoxy sheet chosen is available in 20-in. widths, a sheet width of 10 in. is chosen. For a 10-in. width, the space is calculated as

$$s_f = \frac{w_f}{R_{w/s}} = \frac{10}{0.81} = 12.35 \rightarrow \text{use 12-in. spacing}$$

Step 4. Calculate the actual shear capacity of the FRP wrap. Determine the area of the FRP in the fiber direction:

$$A_{fv} = 2nt_f w_f = 2(1)(0.0065)(10) = 0.13 \text{ in}^2$$

Determine the maximum design (effective) stress in the FRP:

$$f_{fe} = E_f \varepsilon_{fe} = 33{,}000(0.0031) = 102.3 \text{ ksi}$$

Determine the shear contribution for the 90°-oriented sheets:

$$V_f = \frac{A_f f_{fe} d_f}{s_f} = \frac{0.13(102.3)(16)}{12} = 17.7 \text{ kips} > 17.2 \text{ kips} \rightarrow \text{OK}$$

The layout schematic of the FRP strengthening system over the 70-in. critical length for this case is shown in Fig. 10.8.

Step 5. Calculate the factored shear capacity of the future strengthened beam.

$$\phi V_n = \phi(V_c + V_s + \psi_f V_f) = 0.85[36.4 + 19.6 + (0.85)(17.73)]$$
$$= 60.4 \text{ kips} > 60 \text{ kips}$$

Therefore,

$$\phi V_n > V_d$$

and the system chosen is adequate.

The percentage increase in the factored shear capacity is

$$\frac{60.4 - 47.6}{47.6}(100) = 26.9\%$$

Step 6. Check the maximum FRP reinforcement and spacing limits.

$$V_s + V_f = 19{,}600 + 17{,}700 = 37{,}300 \text{ lb} \leq 8\sqrt{f'_c}\, b_w d$$
$$= 8\sqrt{3000}(15)(16) = 105{,}162 \text{ lb} \rightarrow \text{OK}$$

$$s_f^{\max} = \frac{d_f}{4} + w_f = \frac{16}{4} + 10 = 14 \text{ in.} > 12 \text{ in.} \rightarrow \text{OK}$$

Step 7. Detail the FRP strengthening system. Extend the wrap as close as possible to the flange of the T-beam. Anchors are not required but may be

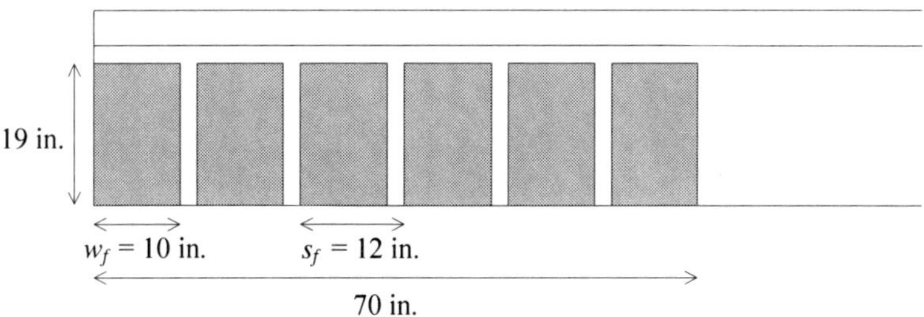

Figure 10.8 Layout of FRP shear strengthening strips.

advisable. Chamfer (round) the bottom corners of the beam to a 0.5-in. radius in the regions in which the FRP strengthening is to be applied (for a three-sided U-wrap).

Discussion of Alternative FRP Designs

A number of different FRP configurations are given next for comparison purposes.

Option a: Single layer continuous 90° three-sided FRP wrap

$$V_f = 2nt_f f_{fe} d_f = 2(0.0065)(102.3)(16) = 21.3 \text{ kips} > 17.7 \text{ kips}$$

Option b: Single layer continuous 90° two-sided FRP wrap Recalculate κ_v, ε_{fe}, and f_{fe}:

$$k_2 = \frac{d_f - 2L_e}{d_f} = \frac{16 - 2(2.022)}{16} = 0.747$$

$$\kappa_v = \frac{0.826(0.747)(2.022)}{468(0.016)} = 0.167 \leq 0.75$$

$$\varepsilon_{fe} = 0.167(0.016) = 0.0027 < 0.004$$

$$f_{fe} = E_f \varepsilon_{fe} = 33{,}000(0.0027) = 88.0 \text{ ksi}$$

Therefore,

$$V_f = 2nt_f f_{fe} d_f = 2(0.0065)(88.0)(16) = 18.3 \text{ kips} > 17.2 \text{ kips} \rightarrow \text{OK}$$

and

$$\begin{aligned}\phi V_n &= \phi(V_c + V_s + \psi_f V_f) = 0.85[36.4 + 19.6 + (0.85)(18.3)] \\ &= 60.8 \text{ kips} > 60 \text{ kips} \rightarrow \text{OK}\end{aligned}$$

This design would also work and will probably use less FRP material (this depends on the width of the particular beam since the three-sided wrap must provide material under the beam as well). In addition, it will probably be more economical to apply in the field since it will not require rounding of the corners of the beam. In this case, extend the FRP sheet to the bottom of the web of the beam (not just to the effective depth, d_f).

Option c: Single layer continuous 45° two-sided FRP wrap

$$\begin{aligned} V_f &= 2nt_f f_{fe} d_f(\sin^2\alpha + \sin\alpha\cos\alpha) \\ &= 2(1)(0.0065)(88.0)(16)(\sin^2 45 + \sin 45 \cos 45) \\ &= 2(1)(0.0065)(88.0)(16)(0.5 + 0.5) \\ &= 18.3 \text{ kips} > 17.2 \text{ kips} \end{aligned}$$

The sheets must be spaced at

$$s_f = \frac{w_f}{\sin\alpha} = \frac{10}{\sin 45} = 14.14 \text{ in.}$$

Option d: Single layer continuous 60° two-sided FRP wrap

$$\begin{aligned} V_f &= 2nt_f f_{fe} d_f(\sin^2\alpha + \sin\alpha\cos\alpha) \\ &= 2(1)(0.0065)(88.0)(16)(\sin^2 60 + \sin 60 \cos 60) \\ &= 2(1)(0.0065)(88.0)(16)(0.75 + 0.433) \\ &= 21.7 \text{ kips} > 18.3 \text{ kips} > 17.2 \text{ kips} \end{aligned}$$

$$s_f = \frac{w_f}{\sin\alpha} = \frac{10}{\sin 60} = 11.55 \text{ in.}$$

Note that the 60° continuous wrap is preferable to the 45° and the 90° wraps.

Option e: Intermittent 60° two-sided FRP wrap For $w_f = 10$ in., s_f must be greater than 11.55 in. Take $s_f = 14$ in.

$$\begin{aligned} V_f &= \frac{2nt_f w_f f_{fe} d_f(\sin\alpha + \cos\alpha)}{s_f} \\ &= \frac{2(1)(0.0065)(10)(88.0)(16)(0.866 + 0.500)}{14} \\ &= 17.9 \text{ kips} > 17.2 \text{ kips} \rightarrow \text{OK} \end{aligned}$$

Check the maximum spacing for this inclined intermittent wrap as follows:

$$s_f^{\max} = \frac{d_f}{4} + \frac{w_f}{\sin\alpha} = \frac{16}{4} + \frac{10}{\sin 60} = 4 + \frac{10}{0.866} = 15.55 > 14 \text{ in.} \rightarrow \text{OK}$$

Design Example 10.2: Shear Strengthening of a Rectangular Column[10] Given a square reinforced concrete column with a 24 × 24 in. cross section. An increase in shear strength of 60 kips is required. The column is exposed

[10]This design example follows Example 14.5 in ACI 440.2R-02.

to an exterior environment. Design an E-glass/epoxy FRP system to increase the shear capacity of the column. Consider two options: (a) a continuous wrap, and (b) an intermittent wrap.

SOLUTION

Step 1. Determine the current and future strength requirements and check the strengthening limits. Since the current details and current loads are not given, the current strength and the strengthening limits cannot be checked. All we are given for design purposes is that $V_u^{\text{future}} = V_u^{\text{current}} + 60$ kips.

Step 2. Choose an FRP system strengthening system and configuration. Choose an E-glass/epoxy system with the following FRP composite[11] properties with the following guaranteed properties:

$$t_f = 0.051 \text{ in.} \qquad \varepsilon_{fu}^* = 0.020\ (2.0\%)$$

$$f_{fu}^* = 80 \text{ ksi} \qquad E_f = 4{,}000{,}000 \text{ psi}$$

Determine the design properties of the FRP system according to the environment. For this case we have exterior construction with a glass FRP and $C_E = 0.65$.

$$f_{fu} = 0.65(80) = 52 \text{ ksi}$$

$$\varepsilon_{fu} = 0.65(0.020) = 0.013\ (1.3\%)$$

$$E_f = 4{,}000{,}000 \text{ psi (unchanged)}$$

Use a full four-sided wrap with fibers oriented perpendicular to the column axis (i.e., in the hoop direction). Place the FRP strengthening (either continuous or intermittent) over the entire height of the column from the top of the bottom story slab to the bottom of the top story beam or slab since the shear force is constant throughout the height.

Step 3. Determine the number of layers and geometry of the FRP wrap. The maximum permissible strain in the FRP for a four-sided wrap is

[11] It is important to note that in this example the thickness and modulus properties are given for the FRP composite (as a whole) and not for the fiber alone.

$$\varepsilon_{fe} = 0.004 \le 0.75\varepsilon_{fu}$$

$$0.75\varepsilon_{fu} = 0.75(0.013) = 0.0098 > 0.004$$

Therefore, $\varepsilon_{fe} = 0.004$.

Assume that all the additional shear carrying capacity is obtained from the FRP wrap (i.e., no increase in the concrete shear capacity is obtained from the confinement effect[12] or from the axial load on the column).

$$V_f^{\text{reqd}} = \frac{V_u^{\text{additional}}}{\phi\psi_f} = \frac{60}{0.85(0.95)} = 74.3 \text{ kips}$$

Determine the area of the FRP wrap required as a function of the spacing.

$$A_f^{\text{reqd}} = \frac{V_f^{\text{reqd}} s_f}{\varepsilon_{fe} E_f d_f} = \frac{74.3 s_f}{0.004(4000)(24)} = 0.1935 s_f \text{ in}^2$$

Determine the number of layers as a function of the spacing and width:

$$n = \frac{A_f^{\text{reqd}}}{2t_f w_f} = \frac{0.1935 s_f}{2(0.051) w_f} = 1.897 \frac{s_f}{w_f}$$

Option a: For a continuous wrap, $s_f = w_f$ and $n = 1.892$. Therefore, use two layers ($n = 2$).

Option b: For an intermittent wrap, assume three 5-in.-wide strips and find the required spacing. Rearranging gives us

$$n = 1.897 \frac{s_f}{w_f}$$

as

$$s_f = \frac{n w_f}{1.897} = \frac{3(5)}{1.897} = 7.9 \text{ in.} \rightarrow \text{use 7.5-in. spacing}$$

Steps 4 and 5. Calculate the shear capacity of the future strengthened beam.

Option a: For a two-layer four-sided continuous wrap,

[12] See Chapter 11 for more discussion of confinement.

$$V_f = \psi_f 2nt_f f_{fe} d_f = 0.95(2)(2)(0.051)(0.004)(4000)(24) = 74.42 \text{ kips}$$

$$\phi V_f = 0.85(74.42) = 63.26 \text{ kips} > 60 \text{ kips} \rightarrow \text{OK}$$

Option b: For a three-layer four-sided intermittent wrap,

$$V_f = \frac{\psi_f 2nt_f w_f f_{fe} d_f}{s_f} = \frac{0.95(2)(3)(0.051)(5)(0.004)(4000)(24)}{7.5}$$

$$= 74.42 \text{ kips}$$

$$\phi V_f = 0.85(74.42) = 63.26 \text{ kips} > 60 \text{ kips} \rightarrow \text{OK}$$

Step 6. Check the maximum FRP spacing limits.

Option a: Spacing limits do not apply since the wrap is continuous.
Option b: For a three-layer four-sided intermittent wrap,

$$s_f^{\max} = \frac{d_f}{4} + w_f = \frac{24}{4} + 5 = 11 \text{ in.} > 8 \text{ in.} \rightarrow \text{OK}$$

Step 7. Detail the FRP strengthening system. Cover the full height of the column with the hoop FRP wraps. If an intermittent wrap is used, ensure that the top and bottom of the column are wrapped. Apply layers continuously if possible [i.e., do not apply layers in discrete sheets (or fabrics) with overlaps; try to use a single sheet (or fabric)]. Use a minimum 6-in. overlap at the end. Chamfer (round) the bottom corners of the column to a 0.5-in. radius. In the case of continuous coverage, overlapping parallel to the fiber direction (i.e., the hoop direction) is not required. Butt the edges of the wrap and seal with epoxy.

10.6 SHEAR STRENGTHENING OF FULLY WRAPPED AXIALLY LOADED COLUMNS

Shear strengthening of axially loaded members, in particular columns, is the subject of some debate since a question arises as to whether or not to include the effects of the axial load on the column. ACI 440.2R-02 does not address the issue explicitly but does so implicitly in ACI 440.2R-02 Example 14.5, in which no account of the axial load is taken and the shear crack angle is taken as 45°. An alternative approach suggested by Priestley et al. (1996) and Seible et al. (1997) accounts for the axial effect and for the shallower angle of the shear crack that is found in axially loaded columns failing in shear that are fully wrapped (i.e., confined) by FRP systems.[13] In addition, many columns have circular cross sections, which requires an alternative formulation for the effective depth of the section for design. In this approach the nominal shear resistance[14] of an axially loaded column[15] is given as

$$V_n^a = V_c^a + V_s^a + V_p^a + V_f^a \tag{10.15}$$

where V_c^a, V_s^a, V_p^a, and V_f^a are the concrete, steel reinforcing, axial load effect, and FRP strengthening contributions, respectively, to the shear capacity of the axially loaded column, according to Priestley et al. (1996).

The V_c^a term includes the effect of the axial force on the shear strength and is given as

$$V_c^a = k\sqrt{f_c'}A_e \tag{10.16}$$

where k is a coefficient that depends on the desired column ductility (for example, $k = 1$ psi for a displacement ductility ratio of between 3 and 4) [see Chapter 11 and Priestley et al. (1996)].

and,

$$A_e = 0.8A_{\text{gross}} \tag{10.17}$$

The V_s^a term is taken as a function of the shear crack angle, β [35° is recommended by Priestley et al. (1996) instead of 45°] and a different effective depth of the section than that prescribed for circular columns by ACI 318-99 (i.e., $0.8D$) is used. For circular columns the V_s^a term is given as

$$V_s^a = \frac{\pi}{2}\frac{A_h f_{yh} D'}{s}\cot\beta \tag{10.18}$$

and for rectangular columns it is given as

$$V_s^a = \frac{A_h f_{yh} D'}{s}\cot\beta \tag{10.19}$$

where A_h and f_{yh} are the area and yield strength of the hoop reinforcement, D' the core dimension from center to center of the peripheral hoop (or tie), and β the angle of the shear crack.

[13] See Chapter 11 for a detailed discussion of FRP confining of concrete members.

[14] The a superscript denotes "axial" and indicates that these values are not the same as those without the a.

[15] See Priestley et al. (1996) for application to axially loaded beams.

The V_p^a term is a function of the angle of the vertical compression strut in the column and is given as

$$V_p^a = P \tan \alpha \tag{10.20}$$

where α is the angle[16] formed between the column axis and the compression strut, as shown in Fig. 10.9.

The V_f^a term is not premultiplied by a ψ_f factor (as in ACI 440.2R-02) and it, too, uses a shallower angle of the shear crack (35° is recommended). For circular columns,

$$V_f^a = \frac{\pi}{2} n t_f f_{fe} D \cot \beta \tag{10.21}$$

where $f_{fe} = 0.004E_f$ is the recommended effective strength of the FRP wrap of n layers with a layer thickness of t_f. For rectangular columns with continuous wraps, the shear strength provided by the FRP wrap is given as

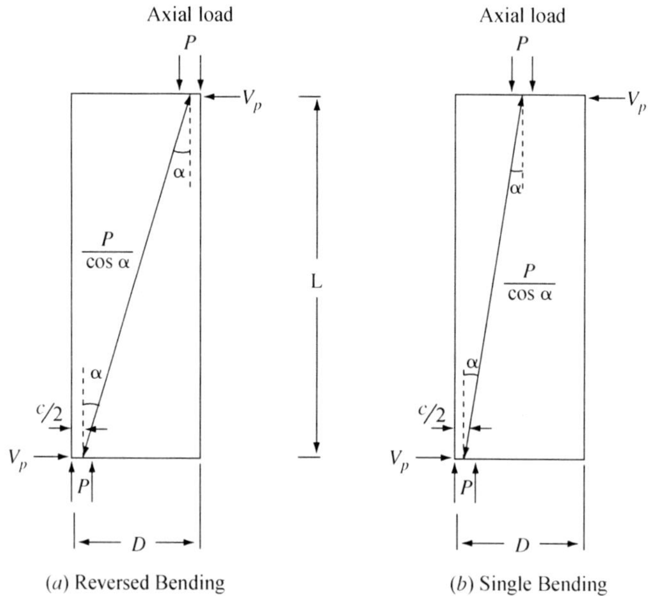

Figure 10.9 Determination of a vertical compression strut. (From Priestley et al., 1996 with permission of John Wiley & Sons.)

[16] Not to be confused with α, the angle of inclination of the shear strengthening in ACI 440.2R-02 equations.

$$V_f^a = 2nt_f f_{fe} h \cot \beta \tag{10.22}$$

where h is the overall column dimension parallel to the shear force.

Priestley's method leads to smaller quantities of FRP materials required for shear strengthening of axially loaded columns, which is important from an economical standpoint, especially where massive columns used in highways bridges and elevated roadways are strengthened. This was the case in California in the late 1990s. The reader is referred to Priestley et al. (1996) for further details. In addition to shear strengthening, the confining effect provided by a four-sided FRP wrap can also serve to increase the column axial capacity and to increase its ductility (energy absorption as evidenced by lateral deflection) to improve its overall seismic resistance. This topic is discussed further in Chapter 11.

PROBLEMS

10.1 For the FRP strengthening systems listed in Table P10.1, determine the effective active bond length, L_e, for one, two, three, and four layers of the system. Compare the effective bond lengths for the various FRP strengthening systems in a bar graph. Is a shorter or a longer effective bond length desirable for shear strengthening? Justify your answers.

10.2 For the FRP strengthening systems in Table P10.1, determine the shear bond-reduction coefficient, κ_v, for one, two, three, and four layers of the strengthening system when used in a three-sided sys-

TABLE P10.1 FRP Strengthening Systems[a]

No.	Fiber	FRP Strengthening System	f_{fu}^* (ksi)	E_f^* (Msi)	t_f (in.)	w_f (in.)
1	Carbon	SikaWrap Hex 103C (laminate property)	104.0	9.45	0.040	Varies
2	Carbon	Tyfo SCH 41 (laminate property)	121.0	11.09	0.040	Varies
3	Glass	WaboMbrace EG900 (fiber property)	220.0	10.5	0.0139	Varies
4	Carbon	Sika CarboShear L plates (laminate property)	326.0	17.4	0.055	2

[a] Properties reported in this table are design properties provided by manufacturers in their current (2006) online specification sheets and do not necessarily conform to the definition of the guaranteed property values in ACI 440.2R-02. They are assumed to be guaranteed properties for the purposes of these problems. For actual design, the user should obtain current guaranteed properties from manufacturers. These data can be compared with those obtained by the reader in Problems 8.2 and 8.3.

tem.[17] The FRP system is to be used to strengthen a 3500-psi rectangular concrete beam with a height of 24 in. and an effective depth of 21 in. The beam is in an exposed highway bridge and is subjected to water, salt, and freeze–thaw conditions (i.e., an aggressive environment). Provide a bar graph comparing the values of the bond-reduction coefficient for the various FRP systems and numbers of layers. Discuss your results and compare the effectiveness of the various materials and numbers of layers.

10.3 For the laid-up FRP strengthening systems in Table P10.1 (systems 1, 2, and 3), determine the factored FRP contribution to the shear capacity for one, two, three, and four layers when used in a four-sided (complete wrap) system. The FRP system is to be used to strengthen a 3000-psi rectangular concrete beam with a height of 18 in. and an effective depth of 15.5 in. The beam is in interior construction. Provide a bar graph comparing the values of the FRP factored shear contribution for the various FRP systems and numbers of layers. What conclusions can you draw from these data regarding the most effective FRP strengthening system for this beam?

10.4 Derive equation 440.2R-02:10.3, shown below, for the FRP contribution to the nominal shear capacity of a concrete member when intermittent FRP sheets or fabrics are used having width w_f and spacing s_f. What are the differences between this equation and the analogous equation for the shear contribution from steel reinforcing bars?

$$V_f = \frac{A_{fv} f_{fe} (\sin \alpha + \cos \alpha) d_f}{s_f}$$

10.5 Consider Design Example 10.1 and redesign the strengthening system using Sika CarboShear L plates following ACI 440.2R-02 procedures. Assume that the legs of the L plates are terminated at the top of the web. How does your design change if the L plates are anchored into slits in the flange of the T-beam?

10.6 Consider Design Example 10.1 and redesign the FRP strengthening system using the same material as in the Design Example, but use a three-sided two-layer (ply) 90° configuration. Determine the width and the spacing for the two-layer system. Compare the quantity of FRP material used per unit length with that used for the one-layer system in the design example.

10.7 Consider Design Example 10.1 and redesign the FRP strengthening system using the same material as in the design example, but use a

[17] Note that CarboShear L plates cannot actually be used in multiply layers; this is just an exercise.

two-sided two-layer (ply) configuration. Consider three options: **(a)** One +45° and one −45° layer, **(b)** one layer at 90° (vertical) and one layer at 0° (horizontal), and **(c)** two +45° layers. (Note that a 0/90 bidirectional woven fabric with equal fiber density in the 0° and 90° directions is a very commonly produced fabric style. If this is rotated through 45°, one obtains a +45°/−45° fabric.) Determine the width and spacing of each layer. Compare the quantity of FRP material used in the two options. How do you apply the bond-reduction factor when the two layers are not at the same orientation? Can you think of other advantages to using a +45° and a −45° configuration?

10.8 Determine the angle of inclination of a given intermittent FRP shear strengthening system that provides the most effective shear strength enhancement. Assume that the angle of the critical diagonal crack in the concrete is at 45°. Plot the variation of the shear strength contribution from the FRP as a function of the angle from 0° to 90°. Is this the same angle as that for a continuous FRP shear strengthening system?

10.9 Reconsider Problem 10.8 but assume that the angle of inclination of the critical diagonal shear crack is 35° rather than 45°.

10.10 An existing simply supported reinforced concrete beam in the interior of a building spans 24 ft and is constructed of 5000-psi concrete. The shear reinforcement consists of No. 3 grade 60 stirrups at 10 in. on center. The beam has a 12 × 24 in. ($b \times h$) cross section and has a 21.5-in. effective depth, d. The beam was originally designed for a dead load of 1 kip/ft (not including the self-weight) and a live load of 2 kip/ft. Due to change in use of the structure, the dead load is to be increased to 2 kips/ft (not including the self-weight) and the live load increased to 3 kips/ft. Assume that the flexural capacity of the beam is sufficient for future loads (or that it is strengthened appropriately for future loads). Consider two different FRP shear strengthening systems to carry the future loads: **(a)** A fully wrapped (four-sided) Tyfo SEH-51 glass–epoxy fabric system oriented vertically (at 90° to the beam axis), and **(b)** a two-sided VSL V-Wrap C200 carbon fiber system oriented at 60° to the beam axis. Consult manufacturers' specifications online for the guaranteed properties for design. Provide the number of layers, the width, and the spacing of the FRP strengthening system, or the number of layers if the wrap is continuous. Check the strengthening limits. For intermittent designs, use minimum widths of 12 in. Show a side view of the beam, and sketch the location and orientation of the strengthening system for each case. Determine the total area of FRP material used for each case.

10.11 Consider the shear strengthening of the reinforced concrete column described in Design Example 10.2. Redesign the strengthening system (consider both a continuous and an intermittent design) if access is available to only three sides of the column, due to an infill wall on the fourth side. Assume that the design shear force is acting parallel to the wall.

10.12 Design a carbon FRP shear strengthening system of your choice to increase the shear capacity by 100 kips in a D = 30-in.-diameter circular reinforced concrete column in an exterior environment. Assume that the column can be fully wrapped. Use only hoop direction fiber orientation in the FRP wrap. Detail the strengthening system and show the amount of overlap required. (Consider the fact that ACI 318 allows one to take the effective depth of a circular column as $0.8D$ and the width as D for the purposes of determining the concrete contribution and the steel stirrup contribution to shear capacity.)

10.13 A reinforced concrete shear wall in a building forms part of the lateral force resisting system. The wall is 24 ft long, 12 ft high, and 14 in. thick and is on the exterior perimeter of the building. The wall is reinforced with steel reinforcing bars in both directions on both faces. The horizontal wall was originally designed based on shear demand. Due to changes in building code seismic design provisions, the use of shear critical members in buildings of this type is no longer permitted. Rather, the more ductile flexural failure of the wall is desired. To increase the shear capacity of the wall, and hence make the flexural mode critical, the wall needs to be strengthened to carry an additional 300 kips of in-plane shear force. Assume that both faces of the wall can be used but that the ends are not accessible. Also assume that the flexural capacity of the wall is sufficient to carry the lateral loads. Design an FRP strengthening system to increase the shear capacity of the shear wall. Use a laid-up glass FRP strengthening system of your choice. Assume that the wall can be subjected to reverse cyclic loading, which can cause the FRP strengthening system to buckle and detach from the wall. Therefore, mechanical anchors such as fiber anchors should be considered. Sketch the layout of the strengthening system.

10.14 Reconsider the shear wall in Problem 13. Due to architectural considerations, only the exterior face of the wall is available for application of the FRP system. Can the wall still be strengthening using an FRP system? Justify your answer by citing the technical literature. Design a one-side shear wall strengthening system if you feel that it is justified. Review AC 125 (1997).

10.15 Reconsider the design of the circular reinforced concrete column described in Problem 10.12. Design the FRP shear strengthening system using the equations developed by Priestley described in Section 10.6

(Eq. 10.21). [Do not explicitly consider the shear strength enhancement due to axial loads, and do not adjust the concrete or steel shear strength contributions due to the Priestley equations.[18]]

[18] In an actual design problem where the details of the structure and materials (concrete and steel) and the design target ductility are known, these additional contributions can be incorporated.

11 FRP Confining

11.1 OVERVIEW

This chapter deals with the design of FRP systems to confine reinforced concrete members with compression regions, such as those found in columns, walls, and joint regions, with precured FRP strips and plates and laid-up FRP sheet and fabric systems. At this time, FRP confining is used primarily on columns, and in this chapter we focus on this application. FRP confining can be undertaken for two purposes: (1) to increase the axial load capacity of a column, and (2) to increase the lateral displacement capacity of a column. Only nonprestressed concrete members that are reinforced with conventional steel reinforcing bars are considered. For the purposes of this chapter, it is assumed that the reader is familiar with both the fundamentals and the details of the design of nonslender reinforced concrete columns with conventional steel reinforcing bars. The includes familiarity with construction of load–moment (P–M) interaction diagrams for nonslender columns. Some familiarity with seismic design of reinforced concrete structures is also assumed.

The design procedures presented in this chapter for axial capacity enhancement of reinforced concrete columns follow ACI 440.2R-02, which is compatible with ACI 318-99. In recent years there have been significant applications of FRP confining to increase both the load-carrying and the displacement ductility of deficient highway bridge columns in seismically active zones. ACI 440-2R.02 does not specifically address the design issues related to strengthening of columns for increased ductility. A method due to Priestley and colleagues for the design of FRP confining to increase the ductility of reinforced concrete columns, accounting for the axial loads, is presented in this chapter.

11.2 INTRODUCTION

FRP confining can be used to increase the strength and the lateral displacement capacity of reinforced concrete structural members (such as columns, walls, and beams) and reinforced concrete joints and joint regions in reinforced concrete rigid frames. The lateral displacement capacity of a column (also know as the *drift*) or the transverse displacement capacity of a beam (also known as the *deflection*) when the material response is inelastic and when energy is dissipated has been referred to historically as the *plastic ro-*

tation capacity or *ductility*. In recent years there have been attempts to extend the definition of ductility to include deformation capacity when the material behavior is elastic to failure (Bakht et al., 2000). However, in this chapter, the term *ductility* is used only to describe a state in which inelastic material response occurs and energy is dissipated, regardless of the extent of the elastic deformation that occurs or the amount of elastic (recoverable) energy that is developed in the member or the structure.

The vast majority of the work to date related to retrofitting concrete members with FRP wraps has been motivated by increasing the resistance of columns, particularly large circular or flared highway columns, to seismic forces. As discussed in Chapter 1, the first use of carbon fiber wraps to increase the ductility of reinforced concrete (RC) columns was conducted in Japan in the 1980s (Katsumata et al., 1998) and later in the United States by Fyfe and colleagues (Priestley et al., 1992). The work on FRP confining of RC columns was closely related to work on the confining of RC columns with steel jackets to increase their seismic resistance (Park et al., 1983; Chai et al., 1991). This work was in turn related to work on the effects of confinement on the concrete in a column provided by internal steel spirals (Priestley and Park, 1987). An in-depth review of the historical work and current design recommendations for seismic retrofitting of highway bridge columns is provided by Priestley et al. (1996). Only recently have studies been conducted on increasing seismic resistance of RC frames, walls, and joint regions in buildings using FRP composites (Antonopolous and Triantifillou, 2003; Paterson and Mitchell, 2003; Balsamo et al., 2005), and design recommendations for such uses of FRP materials have not been codified at this time.

The use of FRP confining to increase the seismic resistance of reinforced concrete columns is more complex than either FRP flexural strengthening or FRP shear strengthening. This is because the objective of a seismic retrofit of an RC column (or frame or wall) is usually not driven by the explicit desire to increase the strength of the member, but rather, is driven by the desire to increase the inelastic lateral displacement capacity (i.e., the ductility). Even though there may not be an explicit desire to increase strength, very often a particular strength increase is needed to increase the lateral displacement capacity. For example, it is well known that larger inelastic displacements (or deflections) in RC members (beams, columns, walls) can be obtained when a member is permitted to fail in flexure rather than in shear. This is because internal *plastic hinges* develop at locations of high bending moments, due to yielding of the internal reinforcing steel. In order, therefore, to permit these large displacements to occur, as a result of the formation of the plastic hinges, the member must often be prevented from failing in other modes before the plastic hinges can develop. These failures can occur in a number of ways. In highway columns the most common among these are shear failure (typically near midheight) and localized failures in the flexural hinge regions due to internal bar rupture or buckling or the failure of lap splices (Seible et al., 1997). In building systems, shear walls can be strengthened (and modified) to prevent brittle shear failure and cause a preferred ductile flexural failure (Paterson and Mitchell, 2003).

The visual appearance of the FRP confining system when applied to an RC member is similar in appearance to that for shear strengthening when a four-sided wrap is used. Figures 11.1 and 11.2 show circular and rectangular RC columns with an FRP confining system being applied to their exterior surfaces.

FRP confining systems are usually of the hand-laid-up type and employ fiber sheets or fabrics. However, they can also be in the form of precured shells that are bonded in staggered layers to the exterior of the column (see Fig. 1.14) or precured strips that are bonded to a column in spirals (shown in Fig. 11.3), or they can be applied by an automated fiber winding machine, as discussed in Chapter 1.

Figure 11.1 FRP system applied to a circular RC column. (Courtesy of Fyfe Company.)

Figure 11.2 FRP wrapping of a rectangular column. (Courtesy of Structural Group.)

Figure 11.3 FRP confinement with a precured spiral-wound FRP strip. (Courtesy of Urs Meier.)

As FRP shear strengthening is fundamentally different from FRP flexural strengthening from a mechanics and a constructability perspective, so, too, is FRP confining for strengthening or increasing ductility fundamentally different from both FRP flexural strengthening and FRP shear strengthening. In flexural and shear strengthening, the FRP provides an additional tensile force component to carry either the bending moment or the transverse shear force. Consequently, the FRP must be securely anchored to the RC member. In ACI 440.2R-02 a bond reduction coefficient is used to account for the susceptibility of the FRP flexural or shear strengthening system to detachment from the RC member. In addition, a partial safety factor, ψ_f, for the FRP system is used in addition to the standard member resistance factor, ϕ, which depends on the configuration of the FRP system (e.g., ψ_f is 0.85 for two- and three-sided shear systems and 0.95 for four-sided shear systems).

In the case of FRP confining, the FRP system when wrapped around the column in the hoop direction appears to perform the role of the internal steel ties or internal steel spiral reinforcement. This, however, is not really the case. Although possibly contributing to the confinement of the concrete core, steel ties and spirals are not designed for such as purpose. Their main purpose is to stabilize and restrain the longitudinal steel reinforcing bars in the compression member (and to provide shear reinforcement in beam-columns). The axial capacity of either a tied or a spirally reinforced RC column is a function of the compressive strength of the concrete, f'_c, and the yield strength, f_y, of the steel reinforcing bars. When the steel reinforcing bars yield, the RC column fails in either a more or a less ductile manner, depending on whether spirals or ties are used. Increasing the quantity of ties or spirals can lead to greater axial load-carrying capacity and ductility in an RC column (Mander et al., 1988). However, the increase in axial load capacity is not accounted for in design procedures in an explicit fashion according to ACI 318-99.[1] (It is accounted for in the different resistance factors for spiral and tied columns but is not accounted for explicitly as a function of the volume of the spiral reinforcement.)

When the FRP hoop reinforcement is added to the exterior of the column, the apparent compressive strength of the concrete is increased. This apparent increase in the concrete strength is due to the confining effect of the FRP, which encircles and wraps the column completely (and thus is often referred to as a *jacket*). This increased concrete strength, known as the *confined compressive strength* and denoted as f'_{cc}, occurs only after the concrete in the column has begun to crack and hence dilate. This typically occurs after the internal transverse reinforcing steel has yielded. By preventing the cracked concrete from displacing laterally (or radially), the FRP serves to confine the concrete and allow it to carry additional compressive stress (and hence compressive load). This well-known phenomenon also occurs when a column is

[1] ACI 318-99 Section 10.3.6.

constrained by a steel jacket (Chai et al., 1991). After the steel yields and the concrete reaches its unconfined compressive strength, f'_c, the concrete continues to increase its load-carrying capacity and displacement until it fails at its confined compressive strength, f'_{cc}.

The macroscopic stress–strain curve of an FRP confined axially loaded concrete column is shown in Fig. 11.4 together with curves for steel-confined concrete and unconfined concrete. The bilinear nature of the effective axial stress–strain curve of the FRP-confined concrete has been verified in numerous experimental investigations.[2] It is important to note that this behavior is different from that of steel jacket–confined concrete, in which the strength decreases after the steel confining jacket yields. Since the FRP jacket is linear to failure, the strength of the confined RC column continues to increase until the FRP wrap fails, due either to tensile stresses or to debonding at the overlap between layers. It has been shown that the "knee" in the bilinear stress–strain curve for the FRP-confined concrete typically occurs at an axial strain close to 0.003 and at the unconfined compression strength of the concrete, f'_c (Nanni and Bradford, 1995; Samaan et al., 1998).

Herein lies the fundamental difference between FRP flexural or shear strengthening and FRP confining. In flexural or shear strengthening the FRP serves to add an additional tensile component to that provided by the existing

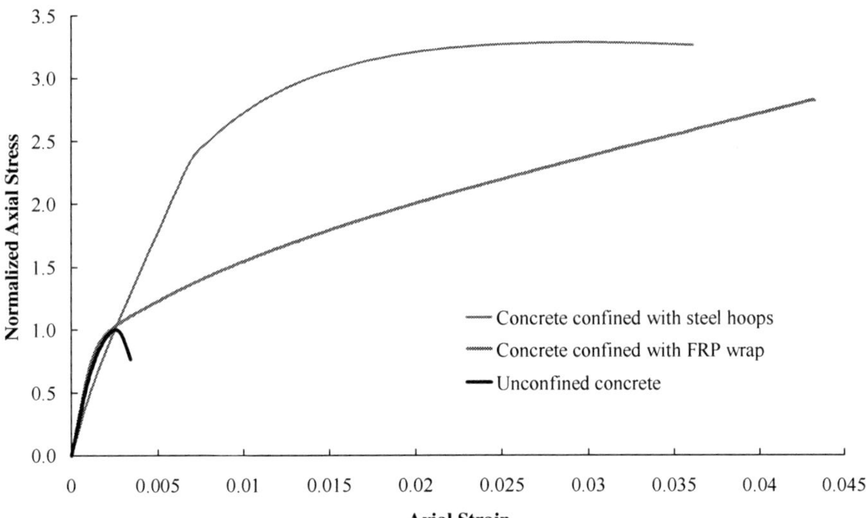

Figure 11.4 Stress–strain curves for confined and unconfined concrete. (Courtesy of A. Mirmiran.)

[2] However, many of these investigations have been conducted on small unreinforced compression test cylinders. Scale effects, and the effect of the core and cover (shell) concrete, have yet to be completely resolved for full-sized columns (Thériault et al., 2004).

steel internal reinforcing system. The properties of the existing steel and concrete are not changed by the FRP system in these cases. In the case of FRP confining, the FRP system serves to alter the effective compressive properties of the existing concrete. In this manner, FRP confining increases the contribution of the concrete in the internal force equilibrium rather than increasing the contribution of steel reinforcing, as in the case of FRP flexural or shear strengthening.

The confining pressure provided by the FRP jacket is uniform around the circumference of the column when the column is circular. A free-body diagram of a half of the cross section of a thin-walled pressure vessel[3] is shown in Fig. 11.5. The relationship between the geometric parameters of the column and the thin-walled cylinder (diameter, D, and thickness, t), the circumferential (hoop) stress, f_θ, and the radial stress due to the internal pressure, f_r, is found from equilibrium as

$$f_r = \frac{2f_\theta t}{D} \tag{11.1}$$

Substituting

$$f_\theta = E_{f\theta}\varepsilon_{f\theta} \tag{11.2}$$

and defining the cross-sectional reinforcement ratio as

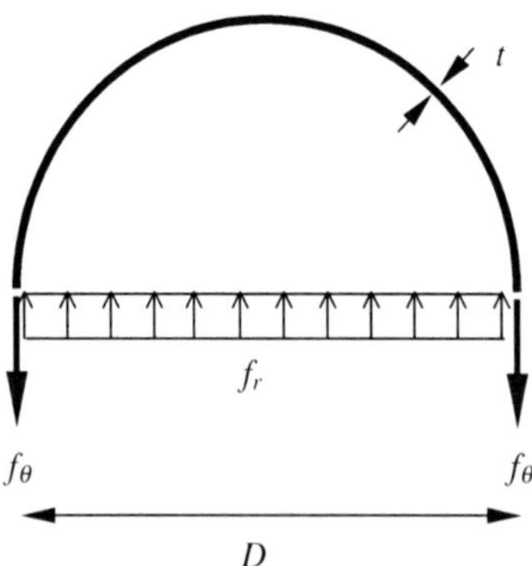

Figure 11.5 Free-body diagram of a thin-walled pressure vessel in the cross-sectional plane.

[3] Gere and Timoshenko (1997, p. 557).

$$\rho_f = \frac{A_f}{A_c} = \frac{\pi D n t_f}{\pi D^2/4} = \frac{4nt_f}{D} \tag{11.3}$$

gives the confining pressure in an FRP-wrapped circular column as

$$f_r = \frac{\rho_f E_{f\theta} \varepsilon_{f\theta}}{2} \tag{11.4}$$

where $E_{f\theta}$ is the FRP modulus in the circumferential direction, $\varepsilon_{f\theta}$ is the FRP strain in the circumferential direction, and n is the number of layers in the FRP wrap.

When wrapped around the circumference of the column, an FRP confining system and the concrete in its interior behave like a thin-walled pressure vessel, with the cracked concrete in the interior being the pressurized "liquid" in the cylinder. The amount of confining pressure is linearly related to the circumferential stress, which in the case of an FRP confining wrap is linearly related to the circumferential strain in the FRP wrap. If the FRP material has linear elastic properties in the hoop direction (typically, obtained with a hoop-only wrap), the confining stress will increase in a linear fashion until the FRP fails: hence, the bilinear stress–strain curve of FRP-confined concrete.

Another very significant difference between FRP flexural and shear strengthening and FRP confining is related to the load at which an FRP system begins to take effect. As noted previously, the confining effect takes effect only after the concrete has cracked and begins to dilate. This means that before the concrete cracks, the FRP jacket serves no structural purpose. Because it plays no part in the elastic response of the axially loaded column, it is called a *passive strengthening system.* Therefore, it is not necessary to consider the existing substrate strain in a concrete column at the time that the FRP confining wrap is attached. This is entirely different from flexural strengthening, where the FRP is active from the moment it is attached, and the prestrain on the substrate needs to be accounted for in the strengthening design. In this regard, FRP confining is somewhat similar to shear strengthening in that it begins to be effective only after shear cracking develops. For noncircular columns, the confining pressure is not uniform around the perimeter of the column and the efficiency of the FRP confining decreases significantly. Nevertheless, partial confinement may be possible to obtain with FRP wraps on noncircular columns, as discussed below.

The discussion above related to the mechanics of FRP confining has a direct bearing on the constructability of FRP jacketing systems. Since the confining pressure needs to be distributed around the entire circumference (even in the case of a noncircular cross section), the FRP wrap must fully encircle the cross section. Therefore, only complete four-sided FRP wraps can be used for FRP confining. To ensure that a four-sided wrap can maintain the confining pressure required, the FRP wrap must have sufficient strength so as not to fail in tension and must not unravel at its overlapping seams. It is not necessary that the FRP wrap actually be bonded to the surface of the concrete. In fact, it has been shown that unbonded circumferential wraps work as well as bonded ones as long as the end of the wrap does not debond from the layer directly beneath (Katsumata et al., 1988; Nanni and Bradford, 1995; Lees et al., 2002). To prevent the end from unraveling, the wrap must be detailed such that sufficient overlap length exists to prevent a debonding failure.

In what follows, FRP confining for the purpose of axial strengthening and FRP confining for the purpose of increasing ductility are discussed separately.

11.3 FRP CONFINING FOR AXIAL STRENGTHENING

Reinforced concrete structural members such as columns and beams can be confined with FRP systems to increase their axial load-carrying capacity. Axial strengthening is most suitable for circular nonslender (i.e., short) reinforced concrete columns. Combined axial and flexural strengthening of short eccentrically loaded reinforced concrete columns will increase their axial and flexural capacity. An axial load-bending moment (*P–M*) strength interaction diagram can be constructed for an FRP-strengthened reinforced concrete column in a fashion similar to that of a nonstrengthened column.

Axial strengthening is obtained by applying the FRP system oriented such that its principal fiber direction is in the circumferential (or hoop) direction of the member, perpendicular to its longitudinal axis. In addition to providing axial strengthening, hoop FRP reinforcement provides shear strengthening to the member, since it is oriented perpendicular to the member axis. As noted previously, when strengthening for only a single mode is intended, it is incumbent on the designer to determine the effects of the strengthening on the other modes and to ensure that the member has sufficient capacity in the other modes to resist the higher applied loads.

FRP axial strengthening of circular columns can be achieved using either continuous or intermittent coverage. Since the axial load is constant along the full height of the column, the FRP wrap must cover the full height of the column; however, it can be spaced intermittently, in either intermittent or spiral hoop form. It has been shown that the confining effect is reduced when intermittent hoops are used and that the confining effect depends on the spacing of the hoops (or spirals) (Saadatmanesh et al., 1994; Nanni and Bradford, 1995). Equations to estimate the confinement effectiveness of intermittent hoop strips (or straps) can be found in Saadatmanesh et al. (1994) and Mander et al. (1988). It is important to note that the ACI 440.2R-02 equations presented below apply only to continuous FRP wraps.

The theoretical concentric (nominal) axial load capacity of an FRP-strengthened nonslender nonprestressed normal-weight concrete column internally reinforced steel reinforcement is given as

$$P_0 = 0.85\psi_f f'_{cc}(A_g - A_{st}) + f_y A_{st} \qquad (11.5)$$

where A_g is the gross area of the concrete, A_{st} the area of the internal longitudinal steel, and f_y the yield stress of the internal longitudinal steel. Except for the addition of the FRP partial strength reduction factor, ψ_f, and use of the confined concrete compressive strength, f'_{cc}, instead of the conventional concrete compressive strength, f'_c, this equation is the same as that used for conventional concrete columns.

To account for unintended eccentricities, ACI 318-99 limits the concentric nominal capacity of columns, depending on whether spirals or ties are used for the hoop internal reinforcement. This is called the *maximum nominal axial load capacity.* For an FRP-strengthened nonslender nonprestressed normal-weight concrete column reinforced internally with spiral steel, this is given as

$$P_{n(\max)} = 0.85[0.85\psi_f f'_{cc}(A_g - A_{st}) + f_y A_{st}] \qquad (11.6)$$

The maximum nominal axial capacity of an FRP-strengthened nonslender nonprestressed normal-weight concrete column reinforced internally with tied steel reinforcement is given as

$$P_{n(\max)} = 0.80[0.85\psi_f f'_{cc}(A_g - A_{st}) + f_y A_{st}] \qquad (11.7)$$

where A_g is the gross area of the concrete, A_{st} the area of the internal longitudinal steel, and f_y the yield stress of the internal longitudinal steel.

According to ACI 440.2R-02, the confined compressive strength, f'_{cc} is to be taken as

$$f'_{cc} = f'_c\left(2.25\sqrt{1 + 7.9\frac{f_l}{f'_c}} - 2\frac{f_l}{f'_c} - 1.25\right) \qquad \text{(ACI 440.2R-02:11-2)}$$

which is the equation developed for confined concrete strength by Mander et al. (1988). It has been used extensively to predict the strength in columns confined by steel jackets (Chai et al., 1991). It has been shown by Spoelstra and Monti (1999) to be applicable to FRP-strengthened columns, even though the complete stress–strain curve presented by Mander et al. (1988) does not give the appropriate bilinear stress–strain curve that is generally seen in experiments with FRP-confined concrete (Mirmiran and Shahawy, 1997; Nanni and Bradford, 1995; Samaan et al., 1998). Many other equations proposed in the literature exclusively for the confined concrete strength of columns with FRP wraps (Teng et al., 2001) can be used if a full nonlinear moment curvature analysis needs to be performed. However, typically in design equations, only the maximum confined strength and strain are needed.

According to ACI 440.2R-02, the confined compressive strength, f'_{cc}, is a function of the unconfined concrete strength, f'_c, and the confining pressure provided by the FRP wrap, denoted by ACI 440.2R-02 as f_l and given as

$$f_l = \frac{\kappa_a \rho_f E_f \varepsilon_{fe}}{2} \qquad \text{(ACI 440.2R-02:11-3)}$$

which is similar to the theoretical equation (Eq. 11.4). The differences are in the notation used for the confining pressure (the l subscript has traditionally been used to denote lateral stress) and in the inclusion of κ_a, which is an efficiency factor that depends on the shape of the column. ρ_f is the FRP reinforcement ratio defined for circular columns in Eq. (11.3). For fully wrapped systems, ACI 440.2R-02 limits the effective strain in the FRP wrap to

$$\varepsilon_{fe} = 0.004 < 0.75 f_{fu} \qquad \text{(ACI 440.2R-02:11-4)}$$

At this time, ACI 440.2R-02 recommends using FRP for axial strengthening only for circular columns. In this case the efficiency factor κ_a is taken as unity ($\kappa_a = 1.0$). Although test data do show axial strength increases for rectangular columns (Teng et al., 2001), with FRP hoop wraps such strengthening is not recommended at this time by ACI 440.2R-02, and a value of the efficiency factor for axial strengthening of rectangular columns in not provided. Procedures for estimating the effect of the rectangular shape on the confining pressure and the effective area of the concrete core being confined are suggested in TR 55 (2004). Even though ACI 440.2R-02 does not permit the use of FRP wraps to increase the strength of axially loaded rectangular columns, FRP wraps can be used to increase the ductility of rectangular columns according to ACI 440.2R-02, as discussed in what follows.

It is important to note that for axial strengthening the fiber layers must all be oriented in the hoop direction around the column. If layers are also oriented in the longitudinal direction, these layers should not be considered to contribute to axial (compression) strengthening. If the FRP system is in the form of strips wrapped spirally, only the fiber stiffness in the hoop direction should be considered as contributing to the confinement. In this case, E_f is found using the transformation equations provided in Chapter 3. The effective failure strain in the hoop direction should still be taken as 0.004 in this case, since this limit is not associated with the strain capacity of the FRP itself but rather, with the shear capacity of the concrete (Priestley et al., 1996, p. 612).

11.3.1 Serviceability for FRP-Strengthened Axial Members

In the service range the concrete in the column cover (or shell) should not approach its transverse cracking strain, and the longitudinal steel should be kept below its yield strain. For this reason, ACI 440.2R-02 limits the service load stress in the concrete to $0.65f'_c$ and the service load stress in the longitudinal steel to $0.60f_y$. The stresses in the concrete and the steel at service loads are found from

$$f_{c,s} = p_s \frac{E_c}{A_c E_c + A_{st} E_s} \tag{11.8}$$

and

$$f_{s,s} = p_s \frac{E_s}{A_c E_c + A_{st} E_s} \tag{11.9}$$

where p_s is the total axial service load.

11.4 DESIGN PROCEDURE FOR FRP AXIAL STRENGTHENING OF RC CIRCULAR COLUMNS

Step 1. Determine the nominal and factored capacities of the existing column. Determine the original design dead and live loads. Determine the strengthening limit.

Step 2. Determine the amount of strengthening required. Indicate if this is to be an increase in dead load, live load, or both.

Step 3. Choose an FRP strengthening system and determine the effective design strength and strain for the hoop wraps.

Step 4. Determine the required number of layers of the strengthening system chosen. To do this, a rearranged form of the analytical equations is used.

(i) Determine the required confined concrete compressive strength f'_{cc} from

$$f'_{cc} = \frac{P_n^{\text{reqd}}/0.85 - f_y A_{st}}{\phi^h \psi_f (A_g - A_{st})} \tag{11.10}$$

($\phi^h = 0.85$ for spiral hoops, 0.80 for tied hoops. ϕ^h is the reduction factor for the maximum nominal axial capacity and must not be confused with ϕ, the strength resistance factor.)

(ii) Determine the required confining pressure, f_l, from the quadratic equation

$$f_l^2 \frac{4}{(f'_c)^2} + f_l \frac{4(f'_{cc}/f'_c + 1.25) - 40}{f'_c} + \left(\frac{f'_{cc}}{f'_c} + 1.25\right)^2 - 2.25^2 = 0 \tag{11.11}$$

(iii) Determine the number of layers of the FRP material chosen.

$$n = \frac{f_l D}{2t_f E_f \varepsilon_{fe}} \tag{11.12}$$

Step 5. Recalculate the capacity for the number of layers chosen.
Step 6. Check the service-level stresses in the concrete and the steel.

Design Example 11.1: FRP Confining for Axial Strengthening of a Concentrically Loaded Circular Column A circular spirally reinforced concentrically loaded nonslender RC column was originally designed to carry a dead load of 320 kips and a live load of 360 kips. It was designed for a concrete strength of 4000 psi and with grade 60 reinforcing steel. The column is exposed to an exterior environment.[4] Due to a change in use of the structure, the existing column is subjected to a 60% increase in live-load demand and requires axial strengthening. Two types of FRP strengthening systems are considered:

Option (*a*): a carbon fiber sheet with compatible epoxy
Option (*b*): a glass fiber fabric with compatible epoxy

Following are the details of the existing column obtained from existing plans.

Column diameter: $h = 20$ in. $A_g = 314$ in^2

Longitudinal steel: 7 No. 11 bars, $A_{st} = 10.92$ in^2

$\rho_g = 0.0347$

Spiral reinforcement: No. 3 bars at 2-in. spiral spacing

$\rho_s = 0.0115$

(Note that some design guides do not recommend FRP axial strengthening if $\rho_g > 3.0\%$, due to the confining effect of the longitudinal bars, which will prevent dilation of the concrete and activation of the FRP confinement. ACI 440.2R-02 does not impose this restriction.)

[4] The original design of this column is given by Wang and Salmon (2002, p. 484).

SOLUTION

Step 1. Determine the nominal and factored capacities of the existing column. Determine the original design dead and live loads and the strengthening limit. The maximum permitted nominal capacity, $P_{n(\max)}$, of the existing (current) column is found as

$$
\begin{aligned}
P_{n(\max)} &= 0.85[0.85(f'_c)(A_g - A_{st}) + f_y A_{st}] \\
&= 0.85[0.85(4)(314 - 10.92) + (60)(10.92)] \\
&= 1433 \text{ kips}
\end{aligned}
$$

The factored capacity (with $\phi = 0.75$ for spiral hoops per ACI 318-99) is

$$\phi P_{n(\max)} = 0.75(1433) = 1075 \text{ kips} > 1060 \text{ kips}$$

where 1060 kips is the original factored design demand $P_u = 1.4(320) + 1.7(360) = 1060$ kips. Note that the column is a little overdesigned, due to dimensions and available bar sizes, as is usually the case with a design.

Determine the strengthening limit. Since the dead load does not change, this means that the future live-load capacity is limited to

$$1.2(320) + 0.85\text{L}^{\text{future}} = 1.2(320) + 0.85\text{L}^{\text{future}} = 1075 \text{ kips}$$

$$\text{L}^{\text{future}} = 813 \text{ kips}$$

Step 2. Determine the amount of strengthening required. Indicate if this is to be an increase in dead load, live load, or both. A 60% increase in the live load is desired [i.e., the new live load is required to be 1.6(360) = 576 kips < 813 kips]. The new live load of 576 kips is less than the live-load limit of 813 kips determined in step 1. Therefore, the strengthening is permitted by ACI 440.2R-02 for normal use.

The future axial demand is

$$P_u = 1.4(320) + 1.7(576) = 1427 \text{ kips}$$

The minimum factored moment capacity, $\phi P_{n(\max)}$ must be equal to or greater than P_u. Therefore, $\phi P_{n(\max)} = 1427$ kips and the nominal capacity of the strengthened column needs to be

$$P_{n(\max)} = \frac{1427}{0.75} = 1903 \text{ kips}$$

Step 3. Choose an FRP strengthening system and determine the effective design strength and strain for the hoop wraps. For the carbon fiber sheet with a compatible epoxy, the following guaranteed properties are obtained from manufacturer specifications:

$$t_f = 0.0065 \text{ in.} \qquad \varepsilon^*_{fu} = 0.0167\ (1.67\%)$$

$$f^*_{fu} = 550 \text{ ksi} \qquad E_f = 33{,}000{,}000 \text{ psi}$$

For an exterior exposure the environmental factor for a carbon–epoxy system is 0.85, and the design properties are

$$f_{fu} = 0.85(550) = 467.5 \text{ ksi} \qquad 0.75\varepsilon_{fu} = 0.75(0.0142) = 0.0106$$

$$\varepsilon_{fu} = 0.85(0.0167) = 0.0142\ (1.42\%) \qquad \varepsilon_{fe} = 0.004 \le 0.75\varepsilon_{fu}$$

$$E_f = 33{,}000{,}000 \text{ psi (unchanged)}$$

For the glass fiber fabric with compatible epoxy, the following guaranteed properties are obtained from manufacturer specifications:

$$t_f = 0.05 \text{ in.} \qquad \varepsilon^*_{fu} = 0.022\ (2.2\%)$$

$$f^*_{fu} = 83.4 \text{ ksi} \qquad E_f = 3{,}790{,}000 \text{ psi}$$

For an exterior exposure the environmental factor for a glass–epoxy system is 0.65, and the design properties are

$$f_{fu} = 0.65(83.4) = 54.2 \text{ ksi} \qquad 0.75\varepsilon_{fu} = 0.75(0.0143) = 0.0107$$

$$\varepsilon_{fu} = 0.65(0.022) = 0.0143\ (1.43\%) \qquad \varepsilon_{fe} = 0.004 \le 0.75\varepsilon_{fu}$$

$$E_f = 3{,}790{,}000 \text{ psi (unchanged)}$$

Step 4. Determine the required number of layers of the strengthening system chosen. To do this, a rearranged form of the analytical equations is used. For the carbon fiber sheet with a compatible epoxy:

(i) Determine the required confined concrete compressive strength, f'_{cc}, from

$$f'_{cc} = \frac{(P_n^{\text{reqd}}/0.85) - f_y A_{\text{st}}}{\phi^h \psi_f (A_g - A_{\text{st}})} = \frac{1903/0.85 - (60)(10.92)}{0.85(0.95)(314 - 10.92)} = 6.471 \text{ ksi}$$

(ii) Determine the required confining pressure, f_l, from the quadratic equation:

$$f_l^2 \frac{4}{(f'_c)^2} + f_l \frac{4(f'_{cc}/f'_c + 1.25) - 40}{f'_c} + \left(\frac{f'_{cc}}{f'_c} + 1.25\right)^2 - 2.25^2 = 0$$

$$f_l^2 \frac{4}{(4.0)^2} + f_l \frac{4(6.471/4.0 + 1.25) - 40}{4.0} + \left(\frac{6.471}{4.0} + 1.25\right)^2 - 2.25^2 = 0$$

Solving and taking the appropriate root gives $f_l = 0.451$ ksi $= 451$ psi.

(iii) Determine the number of layers of the FRP material chosen.

$$n = \frac{f_l D}{2t_f E_f \varepsilon_{fe}} = \frac{451(20)}{2(0.0065)(33 \times 10^6)(0.004)} = 5.3 \qquad \therefore \text{use six layers}$$

For the glass fiber fabric with compatible epoxy:

(i) Determine the required confined concrete compressive strength, f'_{cc}, from

$$f'_{cc} = \frac{(P_n^{\text{reqd}}/0.85) - f_y A_{\text{st}}}{\phi^h \psi_f (A_g - A_{\text{st}})} = \frac{1903/0.85 - (60)(10.92)}{0.85(0.95)(314 - 10.92)} = 6.471 \text{ ksi}$$

(ii) Determine the required confining pressure, f_l, from the quadratic equation:

$$f_l^2 \frac{4}{(f'_c)^2} + f_l \frac{4(f'_{cc}/f'_c + 1.25) - 40}{f'_c} + \left(\frac{f'_{cc}}{f'_c} + 1.25\right)^2 - 2.25^2 = 0$$

$$f_l^2 \frac{4}{(4.0)^2} + f_l \frac{4(6.471/4.0 + 1.25) - 40}{4.0} + \left(\frac{6.471}{4.0} + 1.25\right)^2 - 2.25^2 = 0$$

Solving and taking the appropriate root gives $f_l = 0.451$ ksi $= 451$ psi. Note that steps 4(i) and 4(ii) do not depend on the FRP system.

(iii) Determine the number of layers of the FRP material chosen:

$$n = \frac{f_l D}{2t_f E_f \varepsilon_{fe}} = \frac{451(20)}{2(0.05)(3.79 \times 10^6)(0.004)} = 5.94 \qquad \therefore \text{use six layers}$$

Notice that the carbon and glass systems require the same number of FRP layers. The high strength of the carbon cannot be realized since the strain limit of 0.004 controls the design.

Step 5. Recalculate the capacity for the number of layers chosen. For the carbon fiber sheet with a compatible epoxy,[5] calculate the reinforcement ratio:

$$\rho_f = \frac{A_f}{A_c} = \frac{\pi D n t_f}{\pi D^2/4} = \frac{4nt_f}{D} = \frac{4(6)(0.0065)}{20} = 0.0078$$

Calculate the confining pressure:

$$f_l = \frac{\kappa_a \rho_f E_f \varepsilon_{fe}}{2} = \frac{1.0(0.0078)(33 \times 10^6)(0.004)}{2} = 514.8 \text{ psi}$$

Calculate the concrete confined compressive strength:

$$f'_{cc} = f'_c\left(2.25\sqrt{1 + 7.9\frac{f_l}{f'_c}} - 2\frac{f_l}{f'_c} - 1.25\right)$$

$$= 4{,}000\left[2.25\sqrt{1 + 7.9\left(\frac{514.8}{4000}\right)} - 2\left(\frac{514.8}{4000}\right) - 1.25\right] = 6751 \text{ psi}$$

Calculate the maximum nominal capacity:

$$P_{n(\max)} = 0.85[0.85\psi_f f'_{cc}(A_g - A_{st}) + f_y A_{st}]$$

$$= 0.85[0.85(0.95)(6.751)(314 - 10.92)] + 60(10.92) = 2127 \text{ kips}$$

Calculate the factored capacity and compare with the demand:

$$\phi P_{n(\max)} = 0.75(2127) = 1595 \text{ kips} > 1427 \text{ kips} \qquad \text{OK}$$

Step 6. Check the service-level stresses in concrete and steel. The total service load is $p_s = 320 + 576 = 896$ kips.

$$E_c = 57{,}000\sqrt{f'_c} = 57{,}000\sqrt{4000} = 3{,}604{,}000 \text{ psi} = 3604 \text{ ksi}$$

Stresses in the concrete and the steel at service loads are given by

[5] Steps 5 and 6 are shown only for the carbon–epoxy system. The glass–epoxy system follows in a similar fashion.

$$f_{c,s} = p_s \frac{E_c}{A_c E_c + A_{st} E_s} = 896\left(\frac{3604}{(314 - 10.92)(3604) + (10.92)(29{,}000)}\right)$$

$$= 2.292 \text{ ksi}$$

$$= 2292 \text{ psi} < 0.65 f'_c = 0.65(4000) = 2600 \text{ psi} \rightarrow \text{OK}$$

and

$$f_{s,s} = p_s \frac{E_s}{A_c E_c + A_{st} E_s} = 896\left(\frac{29{,}000}{(314 - 10.92)(3604) + (10.92)(29{,}000)}\right)$$

$$= 18.44 \text{ ksi}$$

$$= 18{,}440 \text{ psi} < 0.60 f_y = 0.60(60{,}000) = 36{,}000 \text{ psi} \rightarrow \text{OK}$$

11.5 FRP-STRENGTHENED ECCENTRICALLY LOADED COLUMNS

Many reinforced concrete columns are designed to carry both axial load and bending moment since they are eccentrically loaded or they are loaded simultaneously by an axial force and a bending moment. In such a case, a designer will typically construct an axial load–bending moment (*P–M*) strength interaction diagram to visualize the different combinations of axial load and bending moment that a given column will be able to carry.

To construct a *P–M* diagram for an unstrengthened column, one needs the geometric properties of the cross-section (h, d, d', b, or D, D'), the internal reinforcement properties, quantity and placement in the section (f_y, E_s, A_s, A'_s), and the concrete strength (f'_c). The maximum strain in the concrete, ε_{cu}, is taken as 0.003 according to ACI 318-99, and the Whitney stress block is used to represent the nonlinear stress–strain relationship of the concrete at failure in the compression zone. The *P–M* curve is typically constructed first by determining the balanced strain condition and obtaining the values of P_b and e_b that correspond to this balanced condition. (Recall that the balanced condition is obtained by setting the strain in the concrete to 0.003 and the strain in the tension steel to $\varepsilon_{sy} = f_y/E_s$.) Thereafter, by choosing different values of the eccentricity, e, either greater than or less than e_b points on the interaction diagram in either the tension- or compression-controlled zone are found, and the diagram is constructed. This is described in all textbooks on reinforced concrete design. These days this is accomplished readily with a simple computer program. A typical *P–M* interaction curve for the nominal capacity of an RC column is shown in Fig. 11.6.

Determination of a *P–M* interaction diagram for an FRP-strengthened column is a nontrivial matter, and limited experimental studies have been re-

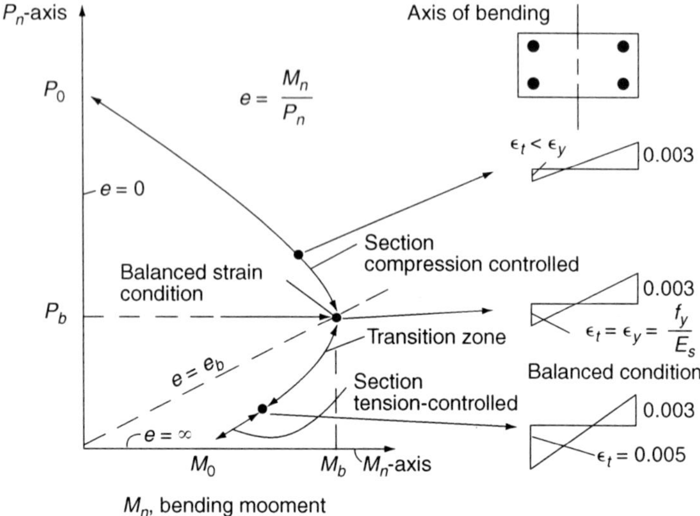

Figure 11.6 Schematic load–moment (P–M) interaction diagram for an RC column. (From Wang and Salmon, 2002 with permission from John Wiley & Sons.)

ported on tests to obtain P–M diagrams for FRP-confined columns (Nanni and Norris, 1995). Detailed experimental interaction curves for FRP-confined columns have not been obtained, mainly because these are difficult tests to do, and the demand for such applications is not high. The ACI 440.2R-02 does not directly address the issue of eccentrically loaded FRP confined columns. However, if a number of simplifying assumptions are made, theoretical P–M interaction diagrams for FRP-confined columns can be developed using the equations presented in ACI 440.2R-02.

Assumption 1. The FRP confining effect applies to the compressive strength of the concrete regardless of the extent of the compression zone. Then a P–M diagram can be constructed for an FRP-confined column in the same manner as for an unstrengthened column simply by replacing f'_c with f'_{cc} given by the Mander equation. This assumption implies that the compression zone is fully confined by the FRP wrap, even though it does not encircle the compression zone entirely. In addition, ψ_f is assumed to be 1.0 for the determination of the nominal axial capacity and nominal moment capacity for points on the P–M diagram (except for $P_{n(\max)}$. If desired a conservative value of $\psi_f = 0.85$ can be assumed.

Assumption 2. The maximum usable concrete compressive strain for an FRP-wrapped circular column given by ACI 440.2R-02 can be used to determine the neutral-axis depth in the FRP-confined concrete section. The ACI 440.2R-02 equation for the FRP-confined concrete is given as

$$\varepsilon'_{cc} = \frac{1.71(5 f'_{cc} - 4f'_c)}{E_c} \qquad \text{(ACI 440.2R-02:11-6)}$$

where E_c is the elastic modulus of the original uncracked concrete.

Based on these two assumptions and using the maximum confined concrete strain, ε'_{cc}, and the maximum confined concrete stress, f'_{cc}, defined previously, the *P–M* interaction curve for an FRP-confined concrete column can be obtained in the same fashion as for an unconfined column.

For noncircular columns it was noted previously that ACI 440.2R-02 does not permit an increase in strength due to the FRP wrap; however, ACI 440.2R-02 does permit an increase in maximum concrete strain. The maximum concrete strain in an FRP-confined rectangular FRP-confined column is given by ACI 440.2R-02:11-6, where the efficiency factor, κ_a, for rectangular columns is given as

$$\kappa_a = 1 - \frac{(b - 2r)^2 + (h - 2r)^2}{3b\,h(1 - \rho_g)} \qquad \text{(ACI 440.2R-02:11-8)}$$

where h and b are the depth and breadth of the column, r is the corner radius, and $\rho_g = (A_s + A'_s)/A_g$ is the gross reinforcement ratio. This equation is applicable only when $h/b \leq 1.5$ and when both b and h are less than 36 in., since it has been shown that the confining effect is negligible in rectangular columns with larger aspect ratios and longer sides. The reinforcement ratio for a rectangular FRP-confined column is given as

$$\rho_f = \frac{2nt_f(b + h)}{bh} \qquad \text{(ACI 440.2R-02:11-7)}$$

It is important to recognize that even though a confined compressive strength for a rectangular column is calculated as an intermediate step in calculating the confined compressive strain, this confined strength should not be used to determine any strength increase in the column. It is used only for determining the maximum confined concrete strain. Therefore, according to ACI 440.2R-02, an FRP-confined *P–M* interaction curve can be determined for a rectangular column according to the procedures outlined above by utilizing Assumption 2 only.

The procedure described above is an approximate procedure that is analogous to the approximate method used for conventional steel-reinforced columns in which the *P–M* interaction diagram is constructed using the Whitney stress block assumptions. This assumption is made whether the section is confined or unconfined (as described above). A more precise method is to construct the *P–M* diagram using the full stress–strain curve of the FRP-confined concrete (and the unconfined concrete, if necessary). In this procedure, the compressive force resultant is found by numerical integration of the nonlinear stress–strain curve for the confined concrete [using stress–strain curves for FRP-confined concrete provided in the literature (see Teng et al., 2001) and not the equation provided by Mander et al. (1988)]. If the FRP strengthening system also provides stiffness in the longitudinal direction of the member in addition to the hoop (circumferential) direction (e.g., when non-90° spiral wraps or when bidirectional fabrics are used), the *P–M* interaction curve must also account for the strengthening effect of the FRP in the flexural mode.

Analysis Example 11.2: Load–Moment Interaction Diagram for an FRP-Confined Column Consider a column with a rectangular section having a width $w = 12$ in., a depth $h = 14$ in., and a concrete strength $f'_c = 6000$ psi. It is reinforced with four No. 11 bars (two at each end) with a yield strength of $f_y = 60{,}000$ psi and has tied hoops for transverse reinforcement. The effective depth of the section is $d = 11$ in. and the depth to the compression bars is $d' = 3$. The moment is applied about the strong axis of the column. The column is in an interior unexposed environment.

The column is wrapped its full height with three layers of carbon–epoxy FRP having the following properties: $t_f = 0.0065$ in., $E_f = 33 \times 10^6$ psi, $f^*_{fu} = 550{,}000$ psi, $\varepsilon^*_{fu} = 0.017$. For the indoor environment, $C_E = 0.95$ and $f_{fe} = 522{,}500$ psi, and $\varepsilon_{fe} = 0.01615$. Prior to application of the FRP material, the corners of the concrete column are rounded with a radius of $r = 1.0$ in.

The nominal and ultimate load–moment (*P–M*) interaction diagrams are constructed for three cases:

1. The unstrengthened column[6]: $f'_c = 6000$ psi and $\varepsilon_{cu} = 0.003$.
2. The FRP-wrapped column assuming only an increase in the confined concrete strain: $f'_c = f'_{cc} = 6000$ psi and $\varepsilon_{cu} = \varepsilon'_{cc}$.
3. The FRP-wrapped column assuming an increase in the confined concrete strain and the confined concrete strength: $f'_c = f'_{cc}$ and $\varepsilon_{cu} = \varepsilon'_{cc}$. (Note that this case is *not* permitted by ACI 440.2R-02, since an increase in axial load-carrying capacity is not permitted at this time for a rectangular section. It is provided for illustrative purposes.)

SOLUTION

Step 1. Determine the confined compressive strength and strain in the FRP-wrapped column.

[6] The unstrengthened case is presented in Nawy (2003, p. 339).

$$\rho_f = \frac{2nt_f(b + h)}{bh} = \frac{2(3)(0.0065)(12 + 14)}{12(14)} = 0.00604$$

$$\rho_g = \frac{A_s + A_s'}{A_g} = \frac{3.12 + 3.12}{12(14)} = 0.0371$$

$$\kappa_a = 1 - \frac{(b - 2r)^2 + (h - 2r)^2}{3bh(1 - \rho_g)} = 1 - \frac{[12 - 2(1)]^2 + [14 - 2(1)]^2}{3(12)(14)(1 - 0.0371)}$$

$$= 0.4972$$

$$f_l = \frac{\kappa_a \rho_f E_f \varepsilon_{fe}}{2} = \frac{0.4972(0.00604)(33 \times 10^6)(0.01615)}{2} = 799.7 \text{ psi}$$

$$f'_{cc} = f'_c\left(2.25\sqrt{1 + 7.9\frac{f_l}{f'_c}} - 2\frac{f_l}{f'_c} - 1.25\right)$$

$$= 6000\left[2.25\sqrt{1 + 7.9\left(\frac{700.7}{6000}\right)} - 2\left(\frac{799.7}{6,000}\right) - 1.25\right]$$

$$= 10,243.5 \text{ psi} \approx 10,244 \text{ psi}$$

$$\varepsilon'_{cc} = \frac{1.71(5f'_{cc} - 4f'_c)}{E_c} = \frac{1.71[5(10,244) - 4(6000)]}{57,000\sqrt{6000}} = 0.0105 = 1.05\%$$

Step 2. Determine P_n and M_n for select points on the interaction diagram for the three cases considered.

(a) *P–M points A and B, axial load only* The nominal and the ACI maximum capacity for the column are calculated using the ACI formula. The nominal moment corresponding to the ACI maximum axial capacity is found by assuming a minimum eccentricity of $e_{\min} = 0.1h = 1.4$ in.

Case 1:

$$P_0 = 0.85\, f'_c(A_g - A_{st}) + f_y A_{st}$$

$$= 0.85(6000)(168 - 6.24) + 60,000(6.24)$$

$$= 1,199,376 \text{ lb} = 1199 \text{ kips}$$

$$P_{n(\max)} = 0.80P_0 = 0.8(1199) = 959 \text{ kips}$$

$$M_n = P_{n(\max)}e_{\min} = 959(1.4) = 1343 \text{ in.-kips}$$

Case 2:

$$P_0 = 0.85\psi_f f'_{cc}(A_g - A_{st}) + f_y A_{st} = 0.85(1.0)(6000)(168 - 6.24)$$
$$+ 60{,}000(6.24) = 1{,}199{,}376 \text{ lb} = 1199 \text{ kips}$$

$$P_{n(\max)} = 0.80P_0 = 0.8(1199) = 9.59 \text{ kips}$$

$$M_n = P_{n(\max)}e_{\min} = 959(1.4) = 1343 \text{ in.-kips}$$

Note: Since it is assumed that the FRP wrap does not increase concrete compressive strength, the strength reduction factor is set to unity (i.e., $\psi_f = 1.0$).

Case 3:

$$P_0 = 0.85\psi_f f'_{cc}(A_g - A_{st}) + f_y A_{st}) = 0.85(0.95)(10{,}244)(168 - 6.24)$$
$$+ 60{,}000(6.24) = 1{,}712{,}814 \text{ lb} = 1713 \text{ kips}$$

$$P_{n(\max)} = 0.80P_0 = 0.8(1713) = 1370 \text{ kips}$$

$$M_n = P_{n(\max)}e_{\min} = 1370(1.4) = 1918 \text{ in.-kips}$$

Note: $\psi_f = 0.95$ for determination of $P_{n(\max)}$ only. For all the cases above, the resistance factor for the axially loaded tied column is taken as $\phi = 0.7$ according to ACI 318-99.

(b) *P–M point C, balanced strain* The balanced strain condition defines the limit of the compression-controlled region of the *P–M* diagram. At this point, the steel yields at the same instant as the concrete fails in compression (theoretically). The Whitney stress block is used to determine the compressive force resultant in the concrete.

Case 1:

$$c = \frac{\varepsilon_{cu}d}{\varepsilon_{cu} + \varepsilon_{sy}} = \frac{0.003(11)}{0.0003 + 0.00207} = 6.509 \text{ in.}$$

$$a = \beta_1 c = 0.75(6.509) = 4.882 \text{ in.}$$

$$\varepsilon'_s = \varepsilon_{cu}\frac{c - d'}{c} = 0.003\left(\frac{6.509 - 3}{6.509}\right)$$
$$= 0.00162 < 0.00207 \qquad \therefore \text{ compression steel has not yielded}$$

$$f'_s = (29 \times 10^6)(0.00162) = 46{,}980 \text{ psi}$$

$$C_c = 0.85f'_c ba = 0.85(6000)(12)(4.882) = 298{,}778 \text{ lb}$$

$$C_s = f'_s A'_s = 46{,}980(3.12) = 146{,}578 \text{ lb}$$

$$T_s = f_y A_s = 60{,}000(3.12) = 187{,}200 \text{ lb}$$

$$P_n = C_c + C_s - T_s = 298{,}778 + 146{,}578 - 187{,}200 = 258{,}156 \text{ lb}$$

$$= 258 \text{ kips}$$

$$M_n = C_c\left(\frac{h}{2} - \frac{a}{2}\right) + C_s\left(\frac{h}{2} - d'\right) + T_s\left(d - \frac{h}{2}\right)$$

$$= 298{,}778\left(\frac{14}{2} - \frac{4.882}{2}\right) + 146{,}578\left(\frac{14}{2} - 3\right) + 187{,}200\left(11 - \frac{14}{2}\right)$$

$$= 2{,}697{,}241 \text{ in.-lb} = 2697 \text{ in.-kips}$$

$$e = \frac{M_n}{P_n} = \frac{2697}{258} = 10.45 \text{ in.}$$

Case 2:

$$c = \frac{\varepsilon'_{cc} d}{\varepsilon'_{cc} + \varepsilon_{sy}} = \frac{0.0105(11)}{0.0105 + 0.00207} = 9.189 \text{ in.}$$

$$a = \beta_1 c = 0.75(9.189) = 6.861 \text{ in.}$$

$$\varepsilon'_s = \varepsilon'_{cc}\,\frac{c - d'}{c} = 0.0105\left(\frac{9.189 - 3}{9.189}\right)$$

$$= 0.00707 > 0.00207 \qquad \therefore \text{ compression steel has yielded}$$

$$f'_s = f_y = 60{,}000 \text{ psi}$$

$$C_c = 0.85f'_c ba = 0.85(6000)(12)(6.861) = 419{,}893 \text{ lb}$$

$$C_s = f'_s A'_s = 60{,}000(3.12) = 187{,}200 \text{ lb}$$

$$T_s = f_y A_s = 60{,}000(3.12) = 187{,}200 \text{ lb}$$

$$P_n = C_c + C_s - T_s = 419{,}893 + 187{,}200 - 187{,}200 = 419{,}893 \text{ lb}$$

$$= 420 \text{ kips}$$

$$M_n = C_c\left(\frac{h}{2} - \frac{a}{2}\right) + C_s\left(\frac{h}{2} - d'\right) + T_s\left(d - \frac{h}{2}\right)$$

$$= 419{,}893\left(\frac{14}{2} - \frac{6.861}{2}\right) + 187{,}200\left(\frac{14}{2} - 3\right) + 187{,}200\left(11 - \frac{14}{2}\right)$$

$$= 2{,}996{,}408 \text{ in.-lb} = 2996 \text{ in.-kips}$$

$$e = \frac{M_n}{P_n} = \frac{2996}{420} = 7.13 \text{ in.}$$

Case 3:

$$c = \frac{\varepsilon'_{cc} d}{\varepsilon'_{cc} + \varepsilon_{sy}} = \frac{0.0105(11)}{0.0105 + 0.00207} = 9.189 \text{ in.}$$

$$a = \beta_1 c = 0.65(9.189) = 5.973 \text{ in. } (\beta_1 = 0.65 \text{ since } f'_{cc} > 8{,}000 \text{ psi})$$

$$\varepsilon'_s = \varepsilon'_{cc}\frac{c - d'}{c} = 0.0105\left(\frac{9.189 - 3}{9.198}\right)$$

$$= 0.00707 > 0.00207 \qquad \therefore \text{ compression steel has yielded}$$

$$f'_s = f_y = 60{,}000 \text{ psi}$$

$$C_c = 0.85 f'_{cc} ba = 0.85(10{,}244)(12)(5.973) = 624{,}112 \text{ lb}$$

$$C_s = f'_s A'_s = 60{,}000(3.12) = 187{,}200 \text{ lb}$$

$$T_s = f_y A_s = 60{,}000(3.12) = 187{,}200 \text{ lb}$$

$$P_n = C_c + C_s - T_s = 624{,}112 + 187{,}200 - 187{,}200 = 624{,}112 \text{ lb}$$

$$= 624 \text{ kips}$$

$$M_n = C_c\left(\frac{h}{2} - \frac{a}{2}\right) + C_s\left(\frac{h}{2} - d'\right) + T_s\left(d - \frac{h}{2}\right)$$

$$= 624{,}112\left(\frac{14}{2} - \frac{5.973}{2}\right) + 187{,}200\left(\frac{14}{2} - 3\right) + 187{,}200\left(11 - \frac{14}{2}\right)$$

$$= 4{,}002{,}474 \text{ in.-lb} = 4002 \text{ in.-kips}$$

$$e = \frac{M_n}{P_n} = \frac{4{,}002}{624} = 6.41 \text{ in.}$$

Note that for all other points on the interaction curve:

If $e > e_b$, the section is tension controlled.
If $e < e_b$, the section is compression controlled.

The resistance factor is equal to $\phi = 0.70$ for all cases for both M_n and P_n since the section is still regarded as being compression controlled.

(c) *P–M point D, tension-controlled transition point* The tensile strain in the steel at the tension-controlled transition point is set to 0.005 to ensure a ductile flexural failure of the section according to the ACI code.

Case 1:

$$c = \frac{\varepsilon_{cu} d}{\varepsilon_{cu} + \varepsilon_s} = \frac{0.003(11)}{0.003 + 0.005} = 4.125 \text{ in.}$$

$$a = \beta_1 c = 0.75(4.125) = 3.094 \text{ in.}$$

$$\varepsilon_s' = \varepsilon_{cu} \frac{c - d'}{c} = 0.003\left(\frac{4.125 - 3}{4.125}\right)$$

$$= 0.00082 < 0.00207 \qquad \therefore \text{ compression steel has not yielded}$$

$$f_s' = (29 \times 10^6)(0.00082) = 23{,}727 \text{ psi}$$

$$C_c = 0.85 f_c' ba = 0.85(6000)(12)(3.094) = 189{,}353 \text{ lb}$$

$$C_s = f_s' A_s' = 23{,}727(3.12) = 74{,}028 \text{ lb}$$

$$T_s = f_y A_s = 60{,}000(3.12) = 187{,}200 \text{ lb}$$

$$P_n = C_c + C_s - T_s = 189{,}353 + 74{,}028 - 187{,}200 = 76{,}181 \text{ lb}$$

$$= 76 \text{ kips}$$

$$M_n = C_c\left(\frac{h}{2} - \frac{a}{2}\right) + C_s\left(\frac{h}{2} - d'\right) + T_s\left(d - \frac{h}{2}\right)$$

$$= 187{,}353\left(\frac{14}{2} - \frac{3.094}{2}\right) + 74{,}028\left(\frac{14}{2} - 3\right) + 187{,}200\left(11 - \frac{14}{2}\right)$$

$$= 2{,}066{,}548 \text{ in.-lb} = 2{,}067 \text{ in.-kips}$$

$$e = \frac{M_n}{P_n} = \frac{2067}{76} = 27.20 \text{ in.}$$

Case 2:

$$c = \frac{\varepsilon'_{cc} d}{\varepsilon'_{cc} + \varepsilon_s} = \frac{0.0105(11)}{0.0105 + 0.005} = 7.452 \text{ in.}$$

$$a = \beta_1 c = 0.75(7.452) = 5.589 \text{ in.}$$

$$\varepsilon'_s = \varepsilon'_{cc} \frac{c - d'}{c} = 0.0105\left(\frac{7.452 - 3}{7.452}\right)$$

$$= 0.00627 > 0.00207 \qquad \therefore \text{ compression steel has yielded}$$

$$f'_s = f_y = 60{,}000 \text{ psi}$$

$$C_c = 0.85 f'_c ba = 0.85(6000)(12)(5.589) = 342{,}046 \text{ lb}$$

$$C_s = f'_s A'_s = 60{,}000(3.12) = 187{,}200 \text{ lb}$$

$$T_s = f_y A_s = 60{,}000(3.12) = 187{,}200 \text{ lb}$$

$$P_n = C_c + C_s - T_s = 342{,}046 + 187{,}200 - 187{,}200 = 342{,}046 \text{ lb}$$

$$= 342 \text{ kips}$$

$$M_n = C_c\left(\frac{h}{2} - \frac{a}{2}\right) + C_s\left(\frac{h}{2} - d'\right) + T_s\left(d - \frac{h}{2}\right)$$

$$= 342{,}046\left(\frac{14}{2} - \frac{5.589}{2}\right) + 187{,}200\left(\frac{14}{2} - 3\right) + 187{,}200\left(11 - \frac{14}{2}\right)$$

$$= 2{,}936{,}074 \text{ in.-lb} = 2936 \text{ in.-kips}$$

$$e = \frac{M_n}{P_n} = \frac{2936}{342} = 8.58 \text{ in.}$$

Case 3:

$$c = \frac{\varepsilon'_{cc} d}{\varepsilon'_{cc} + \varepsilon_s} = \frac{0.0105(11)}{(0.0105 + 0.005)} = 7.452 \text{ in.}$$

$$a = \beta_1 c = 0.65(7.452) = 4.844 \text{ in.}$$

$$\varepsilon_s' = \varepsilon_{cc}' \frac{c - d'}{c} = 0.0105\left(\frac{7.452 - 3}{7.452}\right)$$

$$= 0.00627 > 0.00207 \qquad \therefore \text{ compression steel has yielded}$$

$$f_s' = f_y = 60{,}000 \text{ psi}$$

$$C_c = 0.85 f_{cc}' ba = 0.85(10{,}244)(12)(4.844) = 506{,}144 \text{ lb}$$

$$C_s = f_s' A_s' = 60{,}000(3.12) = 187{,}200 \text{ lb}$$

$$T_s = f_y A_s = 60{,}000(3.12) = 187{,}200 \text{ lb}$$

$$P_n = C_c + C_s - T_s = 506{,}144 + 187{,}200 - 187{,}200 = 506{,}144 \text{ lb}$$

$$= 506 \text{ kips}$$

$$M_n = C_c\left(\frac{h}{2} - \frac{a}{2}\right) + C_s\left(\frac{h}{2} - d'\right) + T_s\left(d - \frac{h}{2}\right)$$

$$= 506{,}144\left(\frac{14}{2} - \frac{4.844}{2}\right) + 187{,}200\left(\frac{14}{2} - 3\right) + 187{,}200\left(11 - \frac{14}{2}\right)$$

$$= 3{,}814{,}727 \text{ in.-lb} = 3815 \text{ in.-kips}$$

$$e = \frac{M_n}{P_n} = \frac{3815}{506} = 7.54 \text{ in.}$$

The resistance factor is equal to $\phi = 0.90$ for all cases for both M_n and P_n since the section is tension controlled. The region where $0.00207 < \varepsilon_s < 0.005$ is known as the *transition region* and the resistance factor ϕ varies between 0.70 and 0.90 (according to ACI 318-99), depending on the tensile strain in the internal steel reinforcing bars.

(d) *P–M point E, pure bending* The flexural capacity of the section is calculated assuming that $A_s' = 0$. At this point the column is, in fact, a beam, and a traditional flexural analysis is performed.

Case 1:

$$a = \frac{A_s f_y}{0.85 f_c' b} = \frac{3.12(60{,}000)}{0.85(6{,}000)(12)} = 3.059 \text{ in.}$$

$$c = \frac{a}{\beta_1} = \frac{3.059}{0.75} = 4.078 \text{ in.}$$

$$\varepsilon_s = \frac{d - c}{c}\varepsilon_{cu} = \frac{11 - 4.078}{4.078}(0.003 = 0.0051 > 0.005 \qquad \therefore \phi = 0.90$$

$$M_n = A_s f_y\left(d - \frac{a}{2}\right) = 3.12(60{,}000)\left(11 - \frac{3.059}{2}\right)$$
$$= 1{,}772{,}878 \text{ in.-lb} = 1773 \text{ in.-kips}$$

Case 2:

$$a = \frac{A_s f_y}{0.85 f'_c b} = \frac{3.12(60{,}000)}{0.85(6{,}000)(12)} = 3.059 \text{ in.}$$

$$c = \frac{a}{\beta_1} = \frac{3.059}{0.75} = 4.078 \text{ in.}$$

$$\varepsilon_s = \frac{d - c}{c}\varepsilon'_{cc} = \frac{11 - 4.078}{4.078}(0.0105) = 0.0178 >> 0.005 \qquad \therefore \phi = 0.90$$

$$M_n = A_s f_y\left(d - \frac{a}{2}\right) = 3.12(60{,}000)\left(11 - \frac{3.059}{2}\right)$$
$$= 1.772{,}878 \text{ in.-lb} = 1773 \text{ in.-kips}$$

Case 3:

$$a = \frac{A_s f_y}{0.85 f'_{cc} b} = \frac{3.12(60{,}000)}{0.85(10{,}244)(12)} = 1.792 \text{ in.}$$

$$c = \frac{a}{\beta_1} = \frac{1.792}{0.65} = 2.756 \text{ in.}$$

$$\varepsilon_s = \frac{d - c}{c}\varepsilon'_{cc} = \frac{11 - 2.756}{2.756}(0.0105) = 0.0314 >> 0.005 \qquad \therefore \phi = 0.90$$

$$M_n = A_s f_y\left(d - \frac{a}{2}\right) = 3.12(60{,}000)\left(11 - \frac{1.792}{2}\right)$$
$$= 1.891{,}469 \text{ in.-lb} = 1892 \text{ in.-kips}$$

Table 11.1 shows the results obtained above for the three cases considered. The table also includes data for a number of additional points on the *P–M*

TABLE 11.1 ***P–M*** **Data for Three Cases**

Case	Point	ε_s	c (in.)	P_n (kips)	M_n (in.-kips)	e (in.)	ϕ
1	A	$>-\varepsilon_{sy}$	∞	1199	0	0	0.70
	B	$>-\varepsilon_{sy}$	—	960	1344	1.4	0.70
		0	11.0	692	2200	3.17	0.70
		0.0003	10.0	619	2349	3.79	0.70
		0.0011	8.0	435	2555	5.87	0.70
	C	0.00207	6.60	261	2699	10.36	0.70
	D	0.0050	4.13	76	2067	27.2	0.90
	E	0.0051	4.08	0	1773	∞	0.90
2	A	$>-\varepsilon_{sy}$	∞	1199	0	0	0.70
	B	$>-\varepsilon_{sy}$	—	960	1344	1.4	0.70
		0	11.0	692	2200	3.17	0.70
		0.0011	10.0	551	2260	4.75	0.70
	C	0.00207	9.19	421	2997	7.11	0.70
	D	0.0050	7.45	342	2963	8.58	0.90
	E	0.0107	4.08	0	1773	∞	0.90
3	A	$>-\varepsilon_{sy}$	∞	1712	0	0	0.70
	B	$>-\varepsilon_{sy}$	—	1506	2108	1.4	0.70
		0	11.0	934	3307	3.54	0.70
		0.0011	10.0	771	3675	4.68	0.70
	C	0.00207	9.19	624	4002	6.41	0.70
	D	0.0050	7.45	506	3814	7.54	0.90
	E	0.0314	2.76	0	1891	∞	0.90

curves whose calculation is not shown above but follows in an identical manner.

The schematic nominal load–moment interaction diagram is shown in Fig. 11.7. It can be seen that the effect of the FRP confinement for the rectangular column is small when only the confining effect on the compressive strain in the concrete is considered (case 2). If the confining effect on the concrete compressive strength is considered (which is not permitted by ACI 440.2R-02 at this time), the *P–M* diagram is expanded significantly. In both cases 2 and 3, the transition region (C to D) from the steel strain of 0.005 to the yield strain (0.00207) is much smaller than for the unconfined concrete (case 1).

It is also important to recall that this is a schematic of the load–moment interaction diagram, where the discrete points (obtained by using the Whitney approximation) that were evaluated have been joined by straight lines. The true load–moment interaction diagram obtained by using the nonlinear stress strain relations for the concrete will have curved lines (as seen in Fig. 11.6). In addition, the horizontal portion of the interaction diagram, which represents the ACI maximum axial load cutoff, intersects the current diagram at a point defined by the minimum eccentricity. Using a nonlinear stress–strain relation for the concrete would enable one to find this point precisely.

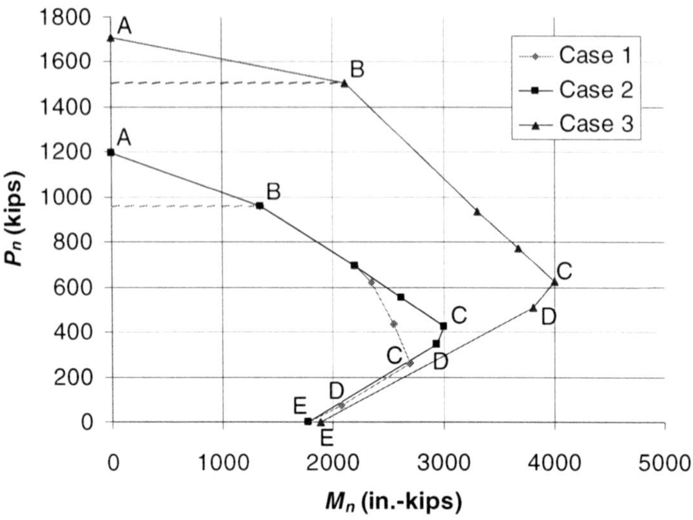

Figure 11.7 Nominal load–moment interaction diagram.

The ultimate load–moment interaction diagram is shown in Fig. 11.8 (without the ACI maximum axial load cutoff, for the sake of clarity). It is plotted on the same axes as Fig. 11.6 so that the reduction in the extent of the diagram can be seen. In addition to the reduced size, it is interesting to note the shape of the diagram in the tension transition region up to the balanced strain point of the FRP confined column (cases 2 and 3). This dramatic discontinuity is not seen in the unconfined column (case 1). This is due to the increased

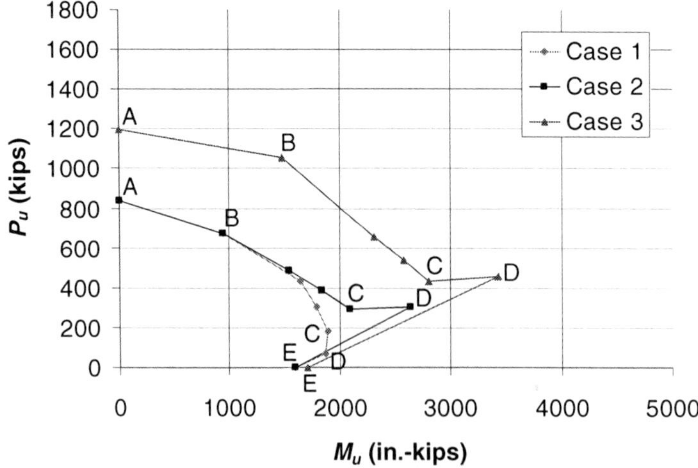

Figure 11.8 Ultimate load–moment interaction diagram.

compressive strain in the confined concrete. The design of FRP-confined columns in the transition region should be avoided. In addition, the FRP reduction factor, ψ_f, has not been included in the ultimate (and nominal) diagrams. This will tend to decrease the size of these diagrams for the FRP strengthened columns.

11.6 FRP CONFINING FOR INCREASED DUCTILITY

The equations presented above for the maximum confined concrete strain for both circular and rectangular columns can be used to design a column to increase its lateral displacement capacity, which is of utmost importance in the seismic design of reinforced concrete structures (Paulay and Priestley, 1992). At this time, ACI 440.2R-02 does not provide detailed instructions on how to use the confined concrete strength to design an FRP retrofit system for increasing the ductility of a column. Since this topic is of increasing importance to structural engineers wishing to use FRP materials to retrofit concrete structures for increased seismic resistance it is reviewed briefly in this section.

At this time the only existing codified equations for the design of FRP systems to increase the lateral displacement capacity of RC columns have been published in Japanese guidelines (JSCE, 2001). The Japanese code equations are not presented in this book as they refer back to symbols and derivations in the Japanese code that cannot be fully explained here; however, the reader is encouraged to consult this source for further code-based procedures on using FRP to increase lateral ductility in columns. Design equations have also been developed by Priestley et al. (1996) and Seible et al. (1997) for confining RC columns (primarily, highway bridge columns) using FRP jackets, based on pioneering work by these authors and colleagues on retrofitting of RC columns with steel and FRP jackets. In what follows, the Priestley approach is outlined.

11.6.1 Lateral Displacement Ductility

Lateral displacement ductility of a structural system depends on the lateral displacement ductility of individual components of a structure, one of these being the member's lateral displacement ductility (Priestley et al., 1996). In what follows, only the member's (in particular, a column's) lateral displacement ductility is addressed. For the relationship between the member's ductility and the structural ductility, the reader is referred to Paulay and Priestley (1992) and Priestley et al. (1996).

The lateral displacement, or drift, of a column when subjected to horizontal forces is due primarily to flexure of the column and is defined as the relative horizontal displacement between the bottom and the top of the column. The total or ultimate lateral displacement of a reinforced concrete column at failure is denoted as Δ_u. This ultimate lateral displacement is made up of an elastic part and an inelastic (or plastic) part. The lateral displacement at the end of the elastic response is known as the *yield displacement,* Δ_y. The difference between the ultimate displacement and the yield displacement is known as the *plastic displacement,* $\Delta_p = \Delta_u - \Delta_y$. The member lateral ductility displacement factor is defined as the ratio of the yield displacement to the ultimate displacement:

$$\mu_\Delta = \frac{\Delta_u}{\Delta_y} = 1 + \frac{\Delta_p}{\Delta_y} \tag{11.13}$$

Displacement ductility factors of 6 to 8 are generally recommended for current seismic design (Priestley et al., 1996). Many older reinforced concrete columns have displacement ductility factors of 2 to 3, making them extremely susceptible to failure under seismically induced lateral loads. To increase the displacement ductility factor of a column, the plastic displacement ductility must be increased. This is achieved by confining the concrete in the column to increase its plastic flexural hinge capacity while simultaneously preventing the column from failing prematurely in shear or debonding of the internal reinforcing bars, due to less than required splice overlap lengths. These debonding failures may be either at the column ends, where bars are lap-spliced to foundation "starter" bars, or at the interior of the column, at bar flexural cutoff points. It is very important to note that the contribution of the confining FRP wrap to the shear capacity, the flexural hinge capacity, and the bar lap splice capacity are interrelated. The design of an FRP wrap for increasing the lateral displacement ductility of a column implies that all three of these modes are addressed. The Japanese code (JSCE, 1997) equation for lateral ductility enhancement includes the shear capacity explicitly. Others (Seible et al., 1997) provide separate equations for shear (discussed in Chapter 10), flexural hinge, and lap splice FRP confinement design.

11.6.2 Flexural Hinge Confinement

The confining of the flexural hinge region serves to increase the failure strain of the concrete. The plastic flexural hinge capacity can also be increased by providing additional longitudinal FRP strengthening in the plastic hinge region; however, this requires very careful detailing to ensure that the tensile loads in the FRP can be transferred to the foundation at the column ends at the location of the largest moments.

Corresponding to the yield and ultimate displacements mentioned previously, yield moments and curvatures, M_y and ϕ_y, and the ultimate moments and curvatures, M_u and ϕ_u, are defined.[7] According to Paulay and Priestley

[7] Determination of the yield and ultimate moments and curvatures requires a full non-linear moment-curvature analysis of the section which is beyond the scope of this book. Therefore, no example problems are presented in this section and the reader is referred to Priestley et al. (1996) and Seible et al. (1997) for example calculations.

(1992), the yield moment may be taken as the moment at the "knee" in an approximated bilinear moment–curvature curve for the column. This occurs when all the tensile steel is not at the same depth from the neutral axis and does not yield at the same time (e.g., in a circular column with axial reinforcement distributed equally around the circumference). In this case the yield moment is typically greater than the moment at first yielding of the tensile steel in the column and can be approximated as (Paulay and Priestley, 1992, p. 136)

$$M_y = 1.33 M_{yi} \tag{11.16}$$

where M_{yi} is the moment corresponding to the first (or initial) yield in the section.

The *yield curvature,* ϕ_y, is taken from a linearization of the moment–curvature graph for the section. According to Paulay and Priestley (1992, p. 138), ϕ_y can be approximated as (except when very high axial reinforcement ratios or high axial forces are present)

$$\phi_y = 1.33\phi_{yi} \tag{11.17}$$

where ϕ_{yi} is the curvature at initial yielding of the section, given as

$$\phi_{yi} = \frac{\varepsilon_{sy}}{d - c_{yi}} \tag{11.18}$$

where ε_{sy} is the yield strain in the tension steel and c_{yi} is the depth of the neutral axis at the time of first steel yielding in the section, which is obtained from a moment-curvature analysis of the section.

When high axial forces (or high axial reinforcement ratios) exist the concrete will "yield" before the steel and ϕ_{yi} is taken as

$$\phi_{yi} = \frac{\varepsilon_c}{c_y} \tag{11.19}$$

with ε_c taken as 0.0015.

The *ultimate moment,* M_u, is defined as the bending moment at ultimate capacity and includes effects of "overstrength" as well as the effects of confinement on the concrete maximum compressive strain and strength (if applicable) (see Paulay and Priestley, 1992). The *ultimate curvature* can be determined from the concrete strains in the ultimate state as

$$\phi_u = \frac{\varepsilon_{cu}}{c_u} \tag{11.20}$$

where ε_{cu} is the maximum compression strain at the extreme fiber in the concrete and c_u is the depth of the section neutral axis at the ultimate moment, which is obtained from a full moment-curvature analysis of the section.

It is important to remember that in all of the columns considered in this section, the column is subjected to both axial and flexural loads and that the depth of the neutral axis depends on both the value of the axial force and the value of the bending moment. In general, the presence of the axial load will increase ϕ_y and decrease ϕ_u (Paulay and Priestley (1992) p. 138).

When the yield moment is reached at either the bottom or the top of the column (or both), local yielding will occur in this finite-length region, which is known as the *flexural plastic hinge region.* According to Priestley et al. (1996), the length of the plastic hinge region in a column confined by internal spirals or ties is

$$L_p = 0.08L + 0.15 f_y d_b \geq 0.3 f_y d_b \tag{11.21}$$

where L is the distance from the point of maximum moment (in the plastic hinge region) to the point of zero moment in the column, f_y is the yield strength (in ksi) of the longitudinal reinforcement in the plastic hinge region, and d_b is the diameter of a longitudinal reinforcing bar.

When the column is confined by an external FRP jacket, the length of the plastic hinge region can be taken as the lower bound of equation (11.21):

$$L_p = 0.3 f_y d_b + g \tag{11.22}$$

where g is the radial gap between the jacket and the concrete column. For FRP wraps, g can usually be taken as zero since the FRP wrap is applied directly to the surface of the column. The curvature ductility is then given as

$$\mu_\phi = \frac{\phi_u}{\phi_y} \tag{11.23}$$

and the member displacement ductility factor is given as (Priestley et al., 1996, p. 310)

$$\mu_\Delta = \frac{M_u}{M_y} + 3(\mu_\phi - 1)\frac{L_p}{L}\left(1 - 0.5\frac{L_p}{L}\right) \tag{11.24}$$

If it is assumed that the confinement effect on the bending moment capacity of the column is small (as demonstrated in Example 11.2), and the additional flexural overstrength is ignored, $M_u = M_y$ and equation (11.24) becomes (see Paulay and Priestley, 1992)

$$\mu_\Delta = 1 + 3\left(\frac{\phi_u}{\phi_y} - 1\right)\frac{L_p}{L}\left(1 - 0.5\frac{L_p}{L}\right) \tag{11.25}$$

For the FRP confined section, it is further assumed that the ultimate curvature can be determined from the confined concrete strain

$$\phi_u = \frac{\varepsilon'_{cc}}{c_u} \tag{11.26}$$

11.7 DESIGN PROCEDURE FOR FLEXURAL HINGE CONFINEMENT

For a required ductility, the confined concrete compressive strain and the depth of the neutral axis at yield and at ultimate need to be determined for a given axial load. For a trial number of FRP layers of a selected FRP material system, the confined compressive stress can be found, and thereafter the confined compressive strain can be found (both using the ACI 440.2R-02 formula). With the known value of the confined compressive strain, the depth of the neutral axis at ultimate can be found and the ductility factor can be determined. This requires a trial-and-error procedure using a full moment-curvature analysis.

An alternative empirical procedure is presented by Seible et al. (1997) to obtain the confined compressive strain in the FRP-wrapped concrete (denoted as ε'_{cu} to differentiate it from both ε'_{cc} and ε_{cu}). In their approach, the confined compressive strength, f'_{cc}, is taken as 1.5 times the unconfined compressive strength, and the confined compressive strain in a circular column is given as

$$\varepsilon'_{cu} = 0.004 + \frac{2.8\rho_f f_{fe}\varepsilon_{fe}}{f'_{cc}} = 0.004 + \frac{2.8\rho_f f_{fe}\varepsilon_{fe}}{1.5 f'_c} \tag{11.27}$$

where f_{fe} and ε_{fe} are the effective FRP strength and maximum strain in the hoop direction, respectively. For rectangular columns, Seible et al. (1997) recommend modifying equation (11.27) by assuming that the FRP wrap is half as effective, to give

$$\varepsilon'_{cu} = 0.004 + \frac{1.4\rho_f f_{fe}\varepsilon_{fe}}{f'_{cc}} = 0.004 + \frac{1.4\rho_f f_{fe}\varepsilon_{fe}}{1.5 f'_c} \tag{11.28}$$

In the Seible et al. (1997) approach, the curvature ductility is taken as

$$\mu_{\phi} = \frac{\varepsilon'_{cu}}{c_u \, \phi_y} \tag{11.29}$$

where c_u and ϕ_y are obtained from a full moment–curvature analysis of the section.

Following ACI 440.2R-02 procedures, a flexural strength resistance factor of $\phi = 0.9$ and an FRP strength reduction factor of $\psi_f = 0.95$ should be used in determine the required number of layers of FRP for confinement of the flexural hinge region. For circular columns with aspect ratios defined by $M/VD > 4$, additional checks should be performed to ensure that bar buckling does not occur in the flexural hinge region (Priestley et al., 1996, p. 314).

11.8 LAP SPLICE REGION CONFINEMENT

FRP wraps in the location of the lap splice in the internal reinforcing bars can be used to ensure that column failure will not occur prematurely due to bond failure of internal reinforcing bars. In experimental studies it has been found that lap splice failures in concrete columns occur when the circumferential strains in the concrete are between 0.001 and 0.002. The lower limit of 0.001 is recommended for FRP wraps (Seible et al., 1997). In Seible et al. (1997) the required confining pressure in the lap splice region (also known as the *clamping pressure*) is given as

$$f_l = \frac{A_b f_y}{[(p/2n_b) + 2(d_b + c_c)]L_s} \tag{11.30}$$

where A_b, and d_b are the area and diameter of a single steel internal bar, n_b is the number of spliced bars along the perimeter, p, and c_c is the clear concrete cover to the main bars (requiring clamping). L_s is the length of the lap splice for the internal bars and f_y is the yield strength of the internal bars.

Ignoring the confining pressure provided by existing spirals or ties, the lateral pressure can then be written in terms of the FRP reinforcement ratio using the limiting design strain in the FRP of $\varepsilon_{fe} = 0.001$, to give

$$f_l = \frac{\kappa_a \rho_f E_f \varepsilon_{fe}}{2} = 0.0005\kappa_a \rho_f E_f \tag{11.31}$$

Rearranging gives the required FRP reinforcement ratio as

$$\rho_f = \frac{2000 f_l}{\kappa_a E_f} \tag{11.32}$$

For a circular column with diameter D and $\kappa_a = 1.0$, the number of layers of the FRP wrap chosen is therefore

$$n = \frac{500Df_l}{E_f t_f} \tag{11.33}$$

The FRP wrap should extend over the length of the bar lap splice, L_s. If lap splice lengths are too short, confinement of the lap splice region may not be possible (Priestley et al., 1996).

11.9 PLASTIC SHEAR OVERSTRENGTH DEMAND

Equations for designing a four-sided FRP wrap for shear strengthening of a column according to ACI 440.2R-02 were presented in Chapter 10. Alternative procedures following Priestley et al. (1996) that include the effects of axial load on shear strength were also presented in Chapter 10. As noted previously, to increase the lateral displacement ductility of a column, its shear strength must be adequate to resist the shear forces in both the flexural hinge region, where large plastic rotations occur, and in the interior of the column. To ensure sufficient shear strength to enable the full inelastic flexural capacity of the section to be developed, the shear demand is increased for columns that are confined to increase their lateral ductility. This is known as the *maximum plastic shear demand* or *shear overstrength demand* (Paulay and Priestley, 1992) and is denoted as

$$V_0 = \lambda_0 \frac{M_y}{L} \tag{11.34}$$

where M_y is the yield moment capacity of the column and L is the height of the column for a column in single curvature (cantilever) and half the height for a column in double curvature (fixed–fixed). For FRP confinement the shear overstrength parameter, λ_0, is recommended to be 1.5 (Seible et al., 1997).

Since different amounts of FRP may be needed for lap-splice confinement, flexural hinge confinement, and shear strength enhancement of a single column, three different wrap reinforcement ratios (and types of reinforcement) could be used simultaneously. The lengths, number of layers, and types of FRP used in the various regions must be clearly marked out on construction plans. It is also important to recognize that if the column requires only flexural hinge confinement to increase its displacement ductility (assuming that the shear strength of the original column is sufficient and that lap splice confinement is not needed), only a portion of the column (near the joint) may need to be wrapped. This is the reason that many photographs of highway bridge retrofits (see Fig. 1.12, for example) show FRP wraps only near the bottom of columns, not over their entire height.

PROBLEMS

11.1 For the FRP strengthening systems listed in Table P11.1, determine the confining pressure, f_l, provided by one, two, three, and four layers of the system when used on interior circular reinforced concrete columns with diameters, D, of 16, 24, and 36 in. Provide bar graphs comparing the confining pressures for the various numbers of layers and various column diameters for each FRP system.

11.2 For the FRP strengthening systems listed in Table P11.1, determine the confined concrete compressive strength, f'_{cc}, when three layers of the system are used to confine an exterior circular reinforced concrete column with a diameter, D, of 30 in. Consider unconfined concrete compressive strengths, f'_c, of 3000, 5000, and 8000 psi. Provide bar graphs comparing the confined concrete compressive strength for the various unconfined concrete strengths for each FRP system.

11.3 For the FRP strengthening systems listed in Table P11.1, determine the maximum confined concrete compressive strain, ε'_{cc}, when three layers of the system are used to confine an exterior circular reinforced concrete column with a diameter, D, of 30 in. Consider unconfined concrete compressive strengths, f'_c, of 3000, 5000, and 8000 psi. Provide bar graphs comparing the maximum confined concrete compressive strain for the various unconfined concrete strengths for each FRP system.

11.4 For the FRP strengthening systems listed in Table P11.1, determine the maximum confined concrete compressive strain, ε'_{cc}, when three layers of the system are used to confine an interior rectangular reinforced concrete column with a depth, h, of 30 in. and a breadth, b, of 20 in. Assume a longitudinal steel reinforcement ratio, ρ_g, of 3%

TABLE P11.1 FRP Strengthening Systems[a]

No.	Fiber	FRP Strengthening System	f^*_{fu} (ksi)	E^*_f (Msi)	t_f (in.)
1	Glass	SikaWrap Hex 107G (laminate property)	86.0	3.57	0.040
2	Aramid	Tyfo SAH 41 (laminate property)	88.0	4.64	0.050
3	Carbon	VSL V-Wrap C-100 (laminate property)	105.0	8.20	0.020
4	Carbon	WaboMBrace C530 (fiber properties)	510.0	54.0	0.0065

[a] Properties reported in this table are design properties provided by manufacturers in their current (2006) online specification sheets and do not necessarily conform to the definition of the guaranteed property values in ACI 440.2R-02. They are assumed to be guaranteed properties for the purposes of these problems. For actual design, the user should obtain current guaranteed properties from manufacturers. These data can be compared with those obtained by the reader in Problems 8.2 and 8.3.

and that the corners of the column are rounded to a radius, r, of 0.75 in. Consider unconfined concrete compressive strengths, f'_c, of 3000, 5000, and 8000 psi. Provide bar graphs comparing the maximum confined concrete compressive strain for the different unconfined concrete strengths for each FRP system.

11.5 A 18-in.-diameter short exterior concrete column needs to be strengthened to resist an additional 100 kips of concentric axial load (50% dead load and 50% live load). The column has 5% longitudinal grade 60 steel, has tied hoops and has f'_c = 4000-psi-design strength concrete. Design and detail a SikaWrap Hex 107G strengthening system to increase the axial load-carrying capacity of the column. Check the strengthening limits (assuming a dead load/live load ratio of 1:1 in the original column) and the limiting strains in the FRP and steel in the strengthened column.

11.6 A 16-in.-diameter short exterior concrete column needs to be strengthened to resist an additional 50 kips of concentric axial load (50% dead load and 50% live load). The column has 2% longitudinal grade 60 steel and has tied hoops and has f'_c = 4000-psi-design strength concrete. Design and detail a strengthening system to increase the axial load-carrying capacity of a VSL V-Wrap C-100 strengthening system. Check the strengthening limits (assuming a dead load/live load ratio of 1:1 in the original column) and the limiting strains in the FRP and steel in the strengthened column.

11.7 To strengthen a circular reinforced concrete column, it is proposed to use a precured Sika Carbodur 2-in.-wide FRP strip wound spirally around the column at an angle θ to the horizontal (see Fig. 11.3 for an example). Since the confining effect is due to the hoop reinforcement, the effective modulus, $E_f(\theta)$, in the hoop (horizontal) direction is needed to do the design (assuming that the strength of the system is governed by the 0.004 strain limit). Plot the modulus of the Carbodur system in the hoop direction as a function of the wind angle of the strip relative to the horizontal. At what spiral angle does the system start to become ineffective for confinement? (*Hint:* You need to use the stiffness transformation equations in Chapter 3. To do this you need the four in-plane orthotropic elastic constants of the strip. Since the manufacturer does not supply these, you will need to make assumptions and estimates using micromechanics. Justify all your assumptions.) Also note that to use this type of system for confinement, equations for intermittent strips provided in the literature must be used.

11.8 Consider the 20-in.-diameter circular column in Design Example 11.1. Determine the maximum confined concrete strain that can be obtained when the column is wrapped with six layers of the carbon sheet system (option a) and with six layers of the glass fabric system (option b).

11.9 Construct a schematic *P–M* interaction diagram for the circular column in Design Example 11.1 (option b, glass fiber fabric) in the same manner as that done for the rectangular column in Analysis Example 11.2. Use an appropriate formula to transform the circular column in to an idealized equivalent rectangular column (e.g., Nawy, 2003, p. 331).

11.10 Consider the rectangular column in the *P–M* interaction diagram in Analysis Example 11.2. Construct nominal and ultimate *P–M* diagrams for the column wrapped with five layers of a Sikawrap Hex 107G FRP strengthening system. Compare the *P–M* diagram with that obtained in Analysis Example 11.2.

SUGGESTED PROJECT ACTIVITY

11.1 *Laboratory assignment.*[8] The objectives of this assignment are (1) to wrap a concrete "column" (actually, a standard 6 × 12 in. cylinder) with various FRP strengthening systems, (2) to test the wrapped columns and to determine the strengthened axial capacity experimentally, and (3) to use ACI analytical models to predict the axial capacity enhancement of the wrapped columns and to compare the prediction with the test results.

Directions:

1. Obtain samples of FRP strengthening systems.[9] Wrap and test two strengthened cylinders and two unstrengthened cylinders.
2. Use a single layer of the FRP wrap in the hoop direction only.
3. Overlap one-fourth of the circumference.
4. Leave about $\frac{1}{4}$ in. at the top and bottom of the cylinder uncovered so that the FRP does not contact the compression test machine's load head.

[8] *Note to instructor:* This is an easy FRP strengthening project to assign to students. It does not take much in the way of materials or resources. It is best to cast a batch of cylinders at the beginning of the semester for both the strengthened and unstrengthened cylinder test. If the class is large, different materials and numbers of layers can be assigned to different groups and the results compared.

[9] If samples of proprietary FRP strengthening systems cannot be obtained from manufacturers, purchase glass fiber woven fabric (about 9 to 12 oz/yd^2) and epoxy resin from any marine supply store. This is for laboratory purposes only and should not be used for actual FRP strengthening projects.

5. Allow the FRP to cure, then test the cylinders. Test in displacement control (if possible) to a displacement beyond the peak load. Observe the behavior and the failure modes of the strengthened and unstrengthened cylinders.
6. Plot the stress–strain curves for the strengthened and unstrengthened cylinders on the same graph.
7. Determine the experimental maximum axial capacity of each specimen.
8. Use the ACI 440.2R-02 equations to calculated the maximum confined strength, the axial capacity, and the maximum confined axial strain in the concrete.
9. Compare the experimental and theoretical values for strength and strain at failure.

SUGGESTED FRP STRENGTHENING STRUCTURAL DESIGN PROJECTS

The following design projects are suggested for students in a composites for construction design class. The types of projects selected are those in which FRP strengthening systems are known to be beneficial. The design project should preferably be done in groups of two or three students. Students should be given 4 to 6 weeks to compete the design project (which therefore needs to be assigned early in the semester). The final deliverable should be a design proposal[10] that includes the problem statement, scope, codes and specifications, loads, materials, design calculations, design drawings, and a cost analysis. A presentation of the design proposal should be given in class. A discussion of the current condition of the existing structure and any problems is recommended. Invite owners and designers to form part of the project jury to obtain feedback from industry on the designs presented.

11.1 *FRP strengthening of parking garage T-beams.* Design an FRP shear strengthening system for the webs of typical double-T precast and prestressed concrete beams. Scout your local area and identify a precast parking garage built in recent years using a system of this type. Use the dimensions of one of the typical double-T beams in the structure. Obtain plans, if possible, or take approximate measurements from the structure. (Typical double-Ts have a top flange 12 ft wide by 4 in. deep, a tapered web 30 in. by approximately 5 in., and spans of 60 to 90 ft.) Assume that the shear capacity of the webs needs to be increased to carry an additional superimposed dead load of 20 lb/ft^2 on the simply supported double-Ts.

[10] Items listed are at the discretion of the instructor.

11.2 *Strengthening a reinforced concrete slab.* Design an FRP flexural and shear strengthening system to increase the load-carrying capacity of a reinforced concrete pedestrian bridge. Scout your local area and identify a reinforced concrete bridge or elevated walkway (such as a footbridge over a road or an elevated walkway between buildings) that has a concrete slab floor (or deck). Obtain plans for the structure, if possible. Design a flexural strengthening system to increase the live-load capacity on the slab by 50%. If the slab is supported on concrete beams and stringers, design the strengthening system for these elements as well. If the walkway is supported on columns (without drop panels), consider the punching shear capacity of the slab.

11.3 *Strengthening a floor slab to accommodate a cutout.* Design a FRP strengthening system to strengthen a reinforced concrete floor slab in a university building to accommodate a 3 × 5 ft cutout (penetration) for new mechanical equipment to be installed. Place the cutout at the center of the slab. Obtain plans for a typical floor beam and column layout. Consult with local experts on how this would be accomplished with conventional engineering materials and propose an FRP strengthening alternative.

11.4 *Increasing the capacity of a concrete column.* Design an FRP wrap to increase the axial load capacity of a reinforced concrete column. Scout your local area and identify a free-standing exposed exterior reinforced concrete column (either as part of a building frame or a support for a walkway or small bridge). Design an FRP strengthening system to increase the axial load capacity on the column by 40%. If the column is rectangular, suggest alternatives design guides to ACI 440.2R-02 to achieve this objective.

12 Design Basis for FRP Profiles

12.1 OVERVIEW

In this chapter we provide an introduction to the design of structures with fiber-reinforced polymer (FRP) profiles produced by the pultrusion method. In this and the following chapters we focus on commercially manufactured thin-walled pultruded glass FRP profiles of I-, box-, channel-, and angle-shaped sections. However, there is no standard for the dimensions or properties of these profiles. Guidance is provided on where to obtain these data. There is also no approved code or guide for the design of pultruded structures, and therefore the design basis for pultruded structural design needs to be defined by the designer (as opposed to simply being chosen by the designer of FRP rebars and strengthening systems as in previous chapters). Guidance is provided on how to develop both an allowable stress design (ASD) basis and a load and resistance factor design (LRFD) basis for pultruded structures. Specific forms of an ASD and an LRFD basis are recommended for use in design. These are used in later chapters to design pultruded structural members, such as tension members, beams, columns, and connections.

Thin-walled FRP components can also be produced using other FRP manufacturing processes, such as filament winding, resin-transfer molding, and centrifugal casting. Such specialty FRP components are generally not intended for use by structural engineers in truss or framed structures, are not identified as structural sections by their manufacturers, and are therefore not covered by the design guidelines presented in what follows.

The reader is assumed to have some familiarly with the properties of orthotropic materials and the mechanics of laminated plates, discussed in Chapter 3. A review of that chapter is highly recommended prior to studying this section of the book. Background and familiarity in the behavior of thin-walled steel sections are required. The reader should know how to design steel truss and frame structures, including the design of chords, ties, struts, beams, columns, and connections and should also have knowledge of the failure modes of thin-walled steel members and structures. The reader is expected to have some familiarity with both allowable stress design (ASD) and load and resistance factor design (LRFD) procedures, for steel structures in particular (AISC, 2005). In addition, reference is made to the European limits states design (LSD) basis. Some knowledge of the design of structures following the European partial factor approach is recommended.

12.2 INTRODUCTION

Truss structures and braced framed structures have been designed and constructed with FRP pultruded members for over 30 years. Pultruded structural profiles[1] (referred to as *pultrusions,* or *pultruded shapes,* or *pultruded members*) have been used in a significant number of structures to date, including pedestrian bridges, vehicular bridges, building frames, stair towers, cooling towers, and walkways and platforms, as described and illustrated in Chapter 1. Pultruded profiles are often the structural members of choice where significant corrosion and chemical resistance is required (such as in food and chemical processing plants, cooling towers, and offshore platforms), where electromagnetic transparency is required (such as in electronics manufacturing plants), or where accessibility is limited and lightweight skeletal structures are assembled on site (such as for pedestrian bridges in parklands).

Pultrusion is a continuous and highly cost-effective manufacturing technology for producing constant-cross-section fiber-reinforced-plastic[2] (FRP) structural profiles.[3] Pultruded profiles are made of pultruded materials. Pultruded materials consist of fiber reinforcements (typically, glass fiber or carbon fiber) and thermosetting resins (typically, polyester, vinylester, and epoxy polymers). The fiber architecture within a thin panel or plate (such as a web or a flange) in a pultruded profile typically consists of longitudinal continuous fiber bundles (called *rovings* or *tows*) and *continuous filament mats* (CFMs).[4] In specialized custom pultruded profiles, bidirectional stitched fabrics are also used to improve mechanical or physical properties. The volume fraction of the fiber reinforcement in a pultruded profile is typically between 30 and 50%. In addition to the base polymer resin, pultruded materials typically contain inorganic fillers, chemical catalysts and promoters, release agents, ultraviolet retardants, fire retardants, pigments, and surfacing veils. Commercially produced pultruded profiles made with glass fibers cost \$2 to \$4 per pound. Custom pultruded profiles utilizing high-performance carbon fibers, sophisticated fabrics, and high-performance resin systems may be considerably more expensive.

Pultruded profiles are produced for use in commercial products (such as ladders and window frames) and for use in building and bridge structures (such as beams, columns, and truss members). The former are often referred to as nonstructural applications; the latter are referred to by civil and structural

[1] In what follows, the term *pultruded profile* is used when referring to the part itself. *Pultruded member* (or *beam* or *column*) is used when referring to the structural design.

[2] The term *fiber-reinforced plastic* as opposed to *polymer* is used most commonly in the industry when referring to these FRP profiles.

[3] The reader is referred to Chapter 2 for a more in-depth discussion of pultruded materials and pultrusion processing.

[4] More commonly known in the United States as *continuous strand mats.*

engineers as structural applications.[5] Most commercial product profiles are custom designed and have unique shapes that are compatible with their end markets. Most structural product profiles are produced in conventional profile shapes similar in geometry to those of metallic materials (e.g., steel and aluminum) such as the I-, wide flange (WF), angle, tube, and channel profiles (Fig. 12.1).

Pultruded profiles for commercial and structural applications are produced by many pultrusion manufacturers around the world. Most manufacturers produce custom profiles for commercial applications. Only the larger manufacturers produce the conventional structural profiles discussed in this book. These include Strongwell, Creative Pultrusions, and Bedford Reinforced Plastics in the United States, Fiberline in Denmark, TopGlass in Italy, and Pacific Composites in Australia, Asia, and the UK. A number of industry groups represent and loosely coordinate the activities of pultrusion manufacturers. Leading groups are the Pultrusion Industry Council of the American Composites Manufacturers Association (ACMA) and the European Pultrusion Technology Association (EPTA).

There is a similarity in the geometry and properties of the pultruded profiles produced by different manufacturers; however, no standard geometries and standard mechanical and physical properties for pultruded profiles are used by all (or even some) manufacturers at this time (2006). Nevertheless, these conventional profiles are commonly referred to as *standard pultruded structural profiles*. In this book they are referred to as *conventional profiles*. Company-specific literature provides details of the geometries and mechanical properties of their conventional profiles (e.g., Fiberline, 2003; Creative Pultrusions, 2004; Strongwell, 2002; Bedford, 2005). Commonly produced conventional pultruded profiles that are usually available from stock inventories include I, wide-flange, square tube, rectangular tube, channel, and angle profiles, as well as plate materials. Standard profiles range from 2 in. (50 mm)

Figure 12.1 Conventional pultruded profiles: I, wide flange, angle, tube, and channel. (Courtesy of Racquel Hagen.)

[5] This is not meant to imply that nonstructural pultruded profiles may not have a structural purpose—rather, that they are not used in structural engineering as load-carrying structural members in buildings and bridges.

in height and width to approximately 12 in. (300 mm) in height and width and have pultruded material thicknesses of $\frac{1}{4}$ to $\frac{1}{2}$ in. (6 to 13 mm).

At this time there are no consensus guidelines for the design of framed structures using either conventional or custom pultruded structural profiles as there are for concrete structures reinforced with FRP rebars or strengthened with FRP strengthening systems. However, two design manuals that have been developed by consensus procedures are available for structural engineers: the *Structural Plastics Design Manual* (ASCE, 1984) and the *Eurocomp Design Code and Handbook* (Eurocomp, 1996). Guidance provided by these two consensus manuals is presented below.

Since the early 1980s, many laboratory studies have been conducted to investigate the behavior of pultruded structural members (mostly on wide-flange and tubular profiles used in beams and columns). Based on these studies, there is sufficient evidence to confirm that the analytical equations presented in the consensus manuals and in the technical literature for structural members are suitable for use when designing framed structures with conventional pultruded profiles. Analytical equations for designing connections between pultruded members are less well developed or validated, and as discussed below, are still mostly empirically based (which is similar to connections in steel structures). Although the suitability of the specific analytical equations provided in the manuals to predict the behavior of pultruded members has been demonstrated, there is significantly less consensus on what safety factors (for allowable stress design) and resistance factors (for load and resistance factor design) to use in structural design.

In addition to the two consensus manuals noted above, a number of companies have produced their own design manuals over the years. The most comprehensive manuals of this type at present are those by Creative Pultrusions (2004), Strongwell (2002), and Fiberline (2003). According to current company literature, the first Extren engineering manual was published by Strongwell (then known as Morrison Molded Fiberglass Company) in 1979, and the first Pultex design manual was published by Creative Pultrusions in 1973. Fiberline published their first design manual in 1995. Recently, Bedford Reinforced Plastics published their first manual (Bedford, 2005). Equations and load tables provided for design in these manuals are based on company testing and are intended to be used only with the profiles produced by the company whose manual is being consulted. These company design manuals and other useful property and design data are available at the Internet addresses listed in the reference section.

There are many excellent classic texts on the mechanics and to some extent the design of composite materials for aerospace and mechanical engineering structures. These include classics by Tsai and Hahn (1980), Daniel and Ishai (1994), and Agarwal and Broutman (1990). Texts by Hollaway (1993), Kim (1995), Barbero (1999), and Kollár and Springer (2003) bridge the gap between the mechanics of composite materials and composite structures and have more emphasis on applications to civil engineering but do not explicitly cover the design of structures with pultruded profiles.

12.3 PROPERTIES OF PULTRUDED PROFILES

Conventional pultruded profiles are constructed of flanges and webs of orthotropic thin plates in which the principal axes of orthotropy of the discrete plates (referred to as *walls* in what follows) in the profile are aligned with global axes of the profile as illustrated in Fig. 12.2. This type of orthotropic plate, known as a *specially orthotropic plate,* was discussed in Chapter 3. The internal architecture of the plate elements, which may consist of roving bundles, fabrics, and mats, is such that the laminate is assumed to be both balanced and symmetric, in order to prevent extensional-shear coupling and extensional-flexural coupling (Tsai and Hahn, 1980, p. 217) in the plate and in the profile.

The in-plane constitutive relation for a specially orthotropic plate made of a laminated material is given as (see Chapter 3)

$$\begin{Bmatrix} \varepsilon_1^0 \\ \varepsilon_2^0 \\ \varepsilon_6^0 \end{Bmatrix} = \begin{bmatrix} \dfrac{1}{E_1^0} & -\dfrac{\nu_{12}^0}{E_2^0} & 0 \\ -\dfrac{\nu_{21}^0}{E_1^0} & \dfrac{1}{E_2^0} & 0 \\ 0 & 0 & \dfrac{1}{E_6^0} \end{bmatrix} \begin{Bmatrix} \overline{\sigma}_1 \\ \overline{\sigma}_2 \\ \overline{\sigma}_6 \end{Bmatrix} \tag{12.1}$$

where ε_i^0 are the in-plane strains in the plate and $\overline{\sigma}_i$ are the (average) in-plane stresses in the plate in the longitudinal (1), transverse (2), and shear (6) directions, respectively. The in-plane engineering constants are E_1^0, longitudinal

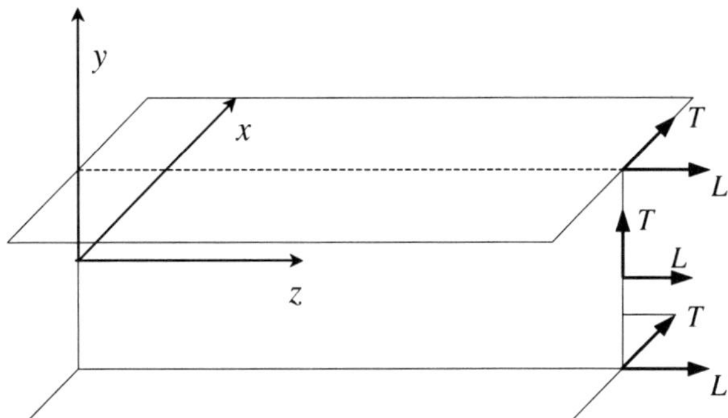

Figure 12.2 Relationship between local and global coordinates in a pultruded profile.

modulus; E_2^0, transverse modulus; ν_{12}^0, major Poisson ratio; ν_{21}^0, minor Poisson ratio; and E_6^0, in-plane shear modulus. As noted in Chapter 3, only four of these five properties are independent.

In structural engineering, a different notation is used to identify the directions and hence the constants and the stresses and strains in a plate of a pultruded profile. The in-plane directions are identified as the longitudinal (L), transverse (T), and shear (LT) directions. The out-of-plane direction is referred to as the *through-the-thickness direction* (TT). The in-plane stiffness constants are defined as E_L, longitudinal modulus; E_T, transverse modulus; ν_L, major Poisson ratio[6]; ν_T, minor Poisson ratio; and G_{LT}, in-plane shear modulus. The in-plane stresses are identified as σ_L, σ_T, and τ; the in-plane strains are identified as ε_L, ε_T, and γ. [Note, however, that in manufacturers' design guides, the longitudinal direction is often referred to as the *machine* or *pultrusion direction* and called the *lengthwise* (LW) *direction,* and the transverse direction is called the *crosswise* (CW) *direction.*] In analysis it is often assumed that the longitudinal and transverse moduli are the same in tension and compression. This is often not the case. Manufacturers generally report tensile and compressive moduli separately. In this case, the superscript t is used for tension, and the superscript c is used for compression.

The in-plane strengths of an orthotropic plate in a pultruded profile are characterized with five independent constants. In Chapter 3 these five strength constants were identified as X_t, longitudinal tensile strength; X_c, longitudinal compressive strength; Y_t, transverse tensile strength; Y_c, transverse compressive strength, and S, in-plane shear strength. In structural engineering these are identified as, $\sigma_{L,t}$, $\sigma_{L,c}$, $\sigma_{T,t}$, $\sigma_{T,c}$, and τ_{LT}. The out-of-plane (through-the-thickness) properties of an orthotropic plate of pultruded material are generally not measured and are not used in design. The exception to this is the interlaminar shear strength (ILSS), also known as the short beam shear (SBS) strength, which is measured routinely. It is a measure of the through-the-thickness (TT) shear strength of the laminate and is identified as τ_{TT}. In addition, for design purposes a quasistructural pin-bearing strength in either the longitudinal or transverse direction is also measured and identified as, $\sigma_{L,b}$ or $\sigma_{T,b}$.

Since a plate of pultruded material is not homogeneous through its thickness [i.e., it consists of layers or laminae (or plies) of fiber-reinforced materials that have different in-plane properties], the in-plane flexural properties are not the same as the in-plane extensional properties. However, as noted in Chapter 3, laminates are typically thin and loaded in-plane in structural profiles. Consequently, in design formulas for pultruded profiles in structural engineering, the in-plane effective engineering properties of the laminate are

[6]A shorthand form is used for Poisson ratios in this book for the sake of brevity and to avoid confusion with the order of the subscripts for the major and minor Poisson ratios, which depends on the mechanics definition used (i.e., ν_{LT} or ν_{TL}).

used in design calculations. In thin-shell structures where laminates are loaded out of plane, the flexural properties are used.

The in-plane stiffness properties may be calculated using classical lamination theory (CLT), in which the pultruded plate is characterized by its in-plane extensional stiffness coefficients, A_{ij}, as discussed in Chapter 3. Alternatively, the in-plane engineering stiffness properties may be obtained from standard tests on coupons extracted from the pultruded profile. In this approach, the laminate is assumed to be homogeneous. Standard ASTM tests that are recommended for determining the properties of pultruded materials are listed in Table 3.1.

Since the orthotropic plates in the pultruded profile are assumed to be homogeneous on a macromechanics level, the plate flexural properties can be calculated from their in-plane extensional engineering properties (obtained either from test data or from the in-plane extensional matrix). This assumption is made routinely in the analysis of pultruded structures and is appropriate given the geometrical properties of pultruded profiles. The orthotropic plate flexural rigidities (the equivalents of *EI* per unit width for a beam) are given as

$$D_L = \frac{E_L t_p^3}{12(1 - \nu_L \nu_T)} \tag{12.2a}$$

$$D_T = \frac{E_T t_p^3}{12(1 - \nu_L \nu_T)} \tag{12.2b}$$

$$D_{LT} = \frac{\nu_T E_L t_p^3}{12(1 - \nu_L \nu_T)} = \frac{\nu_L E_T t_p^3}{12(1 - \nu_L \nu_T)} \tag{12.2c}$$

$$D_S = \frac{G_{LT} t_p^3}{12} \tag{12.2d}$$

where D_L, D_T, D_{LT}, and D_S are the longitudinal, transverse, coupling, and shear flexural rigidities and t_p is the plate thickness. The flexural rigidities relate the plate bending moments (per unit length) to the plate curvatures and are measured in units of force × length. In classical texts on plate behavior, the equations of state are written in terms of plate flexural rigidities (e.g., Timoshenko and Woinowsky-Krieger, 1959), and this notation is often used in analytical equations for pultruded profiles.

The in-plane strength properties may be obtained from theoretical calculations or from testing of coupons taken from the laminate. Where theoretical predictions are used, the first ply failure (FPF) is assumed to represent the strength of the laminate. Coupon testing is recommended for obtaining the strength properties for structural design, as discussed in Chapter 3.

The European Standard EN 13706 (CEN 2002b) is currently the only published standards document that specifies minimum properties for various grades of pultruded materials. The two grades specified for pultruded profiles by EN 13706 are E23 grade and E17 grade. The grade number is taken from the minimum longitudinal tensile modulus of the grade in GPa. The minimum properties for the two grades and the tests required for obtaining them are listed in Table 12.1.[7] Also listed in Table 12.1 are additional mechanical and physical properties that EN 13706 defines as being optionally reported properties; no limits are specified for these.

Property data provided in pultrusion manufacturer design guides are obtained from tests on coupons of pultruded materials taken from pultruded profiles. Data reported are typically applicable to a broad range of profile sizes and types. The specific fiber architectures (volume fractions of roving, mats, and fabrics) in the different profile sizes and types varies. Fiber architecture may also vary within the profile (e.g., web reinforcement architecture may be different from flange reinforcement architecture). The properties given by manufacturers can be assumed to be lower bounds for the class of profiles indicated in the manuals. No data are provided to determine the statistics of these properties for use in probability-based design. In addition, the property data are not related to the capacity of members subjected to specific loading conditions (e.g., axial load, flexure). Published data are based on coupon testing. Representative mechanical property data for pultruded materials used in commercially available pultruded profiles, flat sheets, and rods are given in Table 1.4. The properties of a pultruded carbon–epoxy precured strip for FRP strengthening is shown for comparison purposes. Selected physical and electrical property data for pultruded materials reported by manufacturers is shown in Table 1.5.

For design purposes, it is generally assumed that pultruded materials behave in a linear elastic manner in tension, compression, and shear in both the longitudinal and transverse directions and that failure is brittle. These assumptions are reasonable in the service range (~20% ultimate) but are not reasonable, especially in the shear and transverse directions, at higher strains, where the stress–strain behavior is found to be highly nonlinear and the failure may be progressive. Since the current design of the majority of pultruded profiles is controlled by serviceability criteria (e.g., deflections) and large factors of safety[8] are currently used in the ultimate limit state, service stress levels around 20% are reasonable.

Conventional pultruded profiles are often assumed to be homogeneous on the profile level (i.e., the profile consists of plates all having the same prop-

[7] These test methods are discussed in greater detail in Chapter 3.

[8] Currently, the factors of safety used in design are larger than for conventional materials. However, they are expected to decrease as our understanding of the properties and design methods for pultruded structures increases.

TABLE 12.1 Properties of Pultruded Materials

Property	Test Method	Units[a]	Minimum Properties E23 Grade	Minimum Properties E17 Grade
Minimum Properties That Are Required for Each Grade				
Full-section bending modulus	Annex D, EN 13706-2	GPa (Msi)	23 (3.3)	17 (2.5)
Tensile modulus				
Longitudinal[b]	EN ISO 527-4	GPa (Msi)	23 (3.3)	17 (2.5)
Transverse	EN ISO 527-4	GPa (Msi)	7 (1.0)	5 (0.7)
Tensile strength				
Longitudinal	EN ISO 527-4	MPa (ksi)	240 (34.8)	170 (24.7)
Transverse	EN ISO 527-4	MPa (ksi)	50 (7.3)	30 (4.4)
Pin-bearing strength				
Longitudinal	Annex E, EN 13706-2	MPa (ksi)	150 (21.8)	90 (12.1)
Transverse	Annex E, EN 13706-2	MPa (ksi)	70 (10.2)	50 (7.3)
Flexural strength				
Longitudinal	EN ISO 14125	MPa (ksi)	240 (34.8)	170 (24.7)
Transverse	EN ISO 14125	MPa (ksi)	100 (14.5)	70 (10.2)
Interlaminar shear strength, longitudinal	EN ISO 14130	MPa (ksi)	25 (3.6)	15 (2.2)
Material Properties That May Be Reported[c]				
Compression strength				
Longitudinal	EN ISO 14126	MPa		
Transverse	EN ISO 14126	MPa		
Fiber content by weight	ISO 1183	%		
Density	ISO 1183	kg/m^3		
Poisson ratio				
Longitudinal	EN ISO 527-4			
Transverse	EN ISO 527-4			
Thermal expansion				
Longitudinal	ISO 11359-2	$10^{-6}\ °C^{-1}$		
Transverse	ISO 11359-2	$10^{-6}\ °C^{-1}$		
In-plane shear modulus	ISO 15310	GPa		

Source: EN 13706.

[a] Original minimum values are in SI units. U.S. Customary values are converted and are rounded off.

[b] EN 13706 refers to the longitudinal direction as the axial direction.

[c] No minimum (or maximum) properties are stipulated for either grade material in EN 13706.

erties). However, it should be noted that some manufacturers (e.g., Creative Pultrusions) have modified their profiles and provide properties separately for the webs and the flanges of many of their profiles. For the purposes of analysis, it is assumed that the properties of the junctions between the plates (e.g., the web–flange junctions in an I-profile) are the same as those of the plates themselves. Testing, shows this not to be the case, however. Determination of these very local properties is exceedingly difficult (Turvey and Zhang, 2005).

The ratio of the longitudinal modulus to the shear modulus, E_L/G_{LT}, for pultruded orthotropic plates can be larger than that for isotropic plates. For typical glass-reinforced pultruded profiles it is approximately 6 (as opposed to 2.6 for isotropic materials). Because of this fact and the fact that pultruded profiles are generally more stocky[9] than metal profiles, shear deformation can play a significant role in the analysis of thin-walled pultruded profiles. The effects of shear deformation should be accounted for in deflection calculations and may need to be considered in stability calculations, as discussed below.

The properties of pultruded profiles are affected by the environment in which the profiles are used. These environmental-use effects can include time, temperature, radiation, solvents, fire, impact, abrasion, and fatigue. The way in which these effects are accounted for in design is not yet clear. For some conditions (e.g., temperature and creep) some guidance is available. The safety factors recommended by pultrusion companies for use in design are intended to account for some of these effects.

In the case of elevated-temperature service, pultrusion companies specifically recommend reductions in strength and stiffness properties as a function of temperature for various resin systems (usually, isophthalic polyester or vinylester). Table 12.2 gives retention data for strength and stiffness properties of pultruded profiles recommended by Strongwell for their proprietary pultruded materials. Long-term changes in stiffness and strength can be accounted for using analytical methods and test data. For creep deflections the models due to Findley proposed by the *Structural Plastics Design Manual* (ASCE, 1984) can be used to determine effective viscoelastic full-section moduli for use in predicting long-term deflections of pultruded structures.

Protection against long-term degradation of pultruded profiles due to corrosive environments is typically determined by use of corrosion-resistance guides, which provide recommendations for use of pultruded materials having specific resins in a variety of chemical environments and temperatures. These guides are based on in-house coupon testing according to the standard practice in ASTM C 581 (see the list of standards in Chapter 3). Research studies to develop models to predict the long-term degradation of FRP composites in various service environments are ongoing but cannot yet be used for reliable lifetime prediction.

[9] That is, they are deeper than metal profiles when they span the same length.

TABLE 12.2 Property Retention of Conventional Pultruded Profiles as a Function of Ambient Temperature

Property	Temperature [°F (°C)]	Property Retained (%)	
		Glass Polyester Pultruded Material	Glass Vinylester Pultruded Material
Strength	100 (37.8)	85	90
	125 (51.7)	70	80
	150 (65.6)	50	80
	175 (79.4)	do not use	75
	200 (93.3)	do not use	50
Modulus	100 (37.8)	100	100
	125 (51.7)	90	95
	150 (65.6)	85	90
	175 (79.4)	do not use	88
	200 (93.3)	do not use	85

Source: Strongwell, 2002.

12.4 DESIGN BASIS FOR FRP PULTRUDED STRUCTURES

In the two previous topics covered in this book—FRP rebars (Chapters 4 to 7) and FRP strengthening systems (Chapters 8 to 11)—the limit states design basis of the American Concrete Institute design guides, ACI 440.1R-06 and ACI 440.2R-02, were adopted. As such, a design basis did not need to be developed for the design of structures using FRP composites for these applications. In the case of design with FRP pultruded profiles, a consensus-based design basis on the same level as those developed by the ACI (or other code bodies related to FRP rebars and strengthening systems) does not yet exist. Design bases are provided in the *Structural Plastics Design Manual* (ASCE, 1984) and the *Eurocomp Design Code and Handbook* (Eurocomp, 1996), but these two documents do not have the model code–like status of the ACI (or other national model code–like) documents developed for FRP rebars and strengthening systems.

It is a professional requirement for a structural engineer to declare the basis for design of a structure clearly and unambiguously. The basis for design is so well established for conventional materials that structural engineers routinely note in construction documents and plans the appropriate building codes used for the design. [For example, in the United States this includes building codes IBC (2003) or NFPA (2003) and the model design codes and specifications ASCE 7-02 (ASCE, 2002), ANSI/AISC 360-5 (AISC, 2005) and ACI 318-05 (ACI, 2005) for steel and concrete structures.[10]] For the case

[10]Or applicable national building codes and specifications, depending on the location of the structure being built and the governing codes in that location.

of design of structures with FRP profiles, this step in the design process is not routine, and the designer must take care to provide details of the design basis used and provide additional written special provisions in the construction documents to define aspects of the design basis that may be ambiguous in referenced documents. It is also important that the client (or owner) be aware of and accept the design basis proposed before developing the detailed design of an FRP structure.

In what follows two different design bases are suggested for use in design of structures using FRP pultruded profiles: (1) allowable stress design (ASD) and (2) load and resistance factor design (LRFD). A performance-based design (PBD) basis is mentioned briefly, but at this time a procedure for its use is not suggested for use with FRP profiles. Performance-based design is currently (2005) being developed for materials and structures in structural engineering and may have advantages in the future for structural design with FRP composites (Inokuma, 2002).

12.4.1 Allowable Stress Design

Allowable stress design (ASD) is based on the philosophy that the safety of a structure is obtained by ensuring that no structural member reaches its ultimate strength under nominal service loads. This is accomplished by ensuring that the design stress[11] in every member, or the required stress, σ_{reqd}, obtained from calculations using elastic theory and using the nominal service loads on the structure, be less than the ultimate strength, σ_{ult}, of the material used in the structural member, divided by a factor of safety, SF. The ultimate strength divided by the safety factor is termed the allowable stress, σ_{allow}, and the fundamental ASD equations are given as

$$\sigma_{\text{reqd}} \leq \sigma_{\text{allow}} \tag{12.3}$$

$$\sigma_{\text{allow}} = \frac{\sigma_{\text{ult}}}{SF} \tag{12.4}$$

$$\sigma_{\text{reqd}} = f(\text{loads, geometry, elastic analysis}) \tag{12.5}$$

With respect to serviceability, ASD philosophy is to assign different factors (analogous to strength safety factors) to the material moduli to determine deflections and displacements under the nominal service loads. The nominal service loads used in design are obtained from building codes or model load codes (such as ASCE 7-02).

Allowable stress design is recommended by most U.S. pultrusion companies in their design manuals. The safety factors recommended by U.S. pul-

[11] σ is used symbolically to represent any design stress (e.g., shear stress, tensile stress, bearing stress).

trusion companies, and generally accepted by the structural engineering community in the United States at this time, are given in Table 12.3.

These factors are applied to the material properties provided in manufacturers' design guides and are intended for use with analytical equations provided in these design guides. Many of these are empirical design equations that differ from manufacturer to manufacturer. Material properties for use in design (ultimate strengths and moduli) that are obtained using designated tests (either ASTM standard test methods or in-house test methods) are provided in the design guides for a variety of pultruded materials (according to different resin systems) and profiles (according to shape).

Allowable stress design is also recommended by AASHTO (2001) for designing FRP luminaire and sign support structures. The analytical equations provided in the AASHTO specifications are based on theoretical and empirical equations for orthotropic plates, beams, and columns of linear elastic materials made of glass-reinforced polyester (the only fiber and resin types permitted) and apply to members manufactured by pultrusion, centrifugal casting, or filament winding. The AASHTO specification recommends the minimum safety factors shown in Table 12.4 that are to be used with material properties obtained from specified ASTM tests (stipulated in AASHTO, 2001).

Designs are based on loads provided in the AASHTO (2001) specification, according to which, when the analytical allowable stress design method is used, "design calculations provided shall be verified by documented test results on similar structures" (AASHTO, 2001, p. 8-5). As an alternative to the analytical design approach based on the allowable stress design basis, full-scale testing according to ASTM 4923 (Standard Specification for Reinforced Plastic Thermosetting Poles) can be used in a performance-based design philosophy to design an FRP luminaire.

Allowable stress design is also commonly used by structural designers in conjunction with design equations based on theoretical equations for ortho-

TABLE 12.3 Safety Factors for Allowable Stress Design Provided by Pultrusion Manufacturers

Strength or Modulus	Typical Member	Design Objective	Safety Factor
Flexural strength	Beam	Strength	2.5
Flexural modulus	Beam	Serviceability	1.0
Shear strength	Beam	Strength	3.0
Shear modulus	Beam	Serviceability	1.0
Compressive strength	Column or truss member	Strength	3.0
Compression modulus	Column or truss member	Serviceability	1.0
Tensile strength	Tie or truss member	Strength	not provided
Tensile modulus	Tie or truss member	Serviceability	not provided
Bearing strength	Joint or connection	Strength	4.0[a]
Bearing modulus	Joint or connection	Serviceability	not applicable

[a] The SF for connections should be applied to all stresses in a connection; however, this assumes that connections in pultruded structures will be dimensioned to fail in bearing (see Chapter 15).

TABLE 12.4 Safety Factors for Allowable Stress Design

Strength or Modulus	Typical Member	Design Objective	Safety Factor[a]
Flexural strength	Beam	Strength	2.5
Flexural modulus	Beam	Serviceability	not provided
Shear strength	Beam	Strength	3.0
Shear modulus	Beam	Serviceability	not provided
Compressive strength	Column or truss member	Strength	3.0
Compression modulus	Column or truss member	Serviceability	not provided
Tensile strength	Tie or truss member	Strength	2.0
Tensile modulus	Tie or truss member	Serviceability	not provided
Bearing strength	Joint or connection	Strength	not provided
Bearing modulus	Joint or connection	Serviceability	not applicable

Source: AASHTO (2001).

[a]The SF for connections should be applied to all stresses in a connection; however, most connections in pultruded structures are dimensioned to fail in bearing (see Chapter 15).

tropic plates, beams, and columns provided in the *Structural Plastics Design Manual* (ASCE, 1984) or the *Eurocomp Design Code and Handbook* (Eurocomp, 1996). Material properties are often obtained from independent testing using standard ASTM tests or are taken from manufacturers' specifications. Safety factors recommended by pultrusion companies or by AASHTO (2001) are generally used in this approach. Designs are based on nominal service loads taken from building codes or model load codes (e.g., ASCE 7-02).

The allowable stress design basis is not considered to provide a known degree of structural reliability. That is, even with a known safety factor, it is not known with a quantifiable degree of confidence what the probability is that the structure will be able to carry the design loads in the various limit states. This is due to the fact that (1) the safety factors that are used in the ASD basis have been developed by pultrusion manufacturers from simple tests on pultruded members and accepted conventions for safety factors in civil engineering structures that range from 2 to 3; (2) the material property data provided by manufacturers that are used in the ASD basis are provided without statistical data (mean, variation, number of tests, etc.) and therefore the variability in the properties and the confidence limits on the values cannot be established in a quantifiable manner; and (3) the stresses determined in developing the safety factors from test data depend on the complexity (or naivety) of the analytical and empirical equations used in their determination.

However, although no statistical data are provided for the pultruded material properties given in manufacturers' design guides, it is well known that the values provided are conservative and represent the lower ranges of the properties measured. In addition, the ASD basis is generally used with loads obtained from probabilistically based load codes such as ASCE 7 (even though probabilistically based load factors are not used). This implies that there is some statistically and probabilistically based element to the way in which the ASD basis is used today. Nevertheless, since the material data and the safety factors are not statistically and probabilistically derived to yield a defined structural reliability, the ASD basis is regarded as providing an ad hoc level of safety and is not regarded as a modern scientific design basis (Ellingwood, 2003). However, as discussed in Section 12.4.2, a probability-based design basis for FRP pultruded structures is not yet fully developed, and therefore the ASD basis is still used in the design of the vast majority of pultruded structures in the United States.

In what follows a version of the ASD basis is presented as a basis for the design of FRP structures. The elements of the ASD basis used in this book are:

1. Analytical equations presented in the *Structural Plastics Design Manual* (ASCE, 1984) and the *Eurocomp Design Code and Handbook* (Eurocomp, 1996) and supplemented with equations from the technical literature (detailed in what follows)
2. Design properties of pultruded materials taken from manufacturers' design guides
3. Safety factors taken from AASHTO (2001)
4. Nominal service loads and service load combinations taken from ASCE 7-02

12.4.2 Load and Resistance Factor Design

The load and resistance factor design (LRFD) basis used in the United States and the closely related limit states design (LSD) basis used in Europe have at their heart the fundamental philosophy that the level of safety or serviceability provided by a design must produce a quantifiable level of structural reliability. In the LRFD basis, (1) the nominal loads, P, are factored by probabilistically derived load factors, γ, which depend on load type (dead load, live load, etc.) and load combinations, and (2) the material or structural nominal resistances (e.g., the material strengths or member load-carrying capacities), R_n, are factored by resistance factors, ϕ, that depend on the material variability, the type of failure mode predicted (brittle or ductile), and the type of resistance required (e.g., shear, flexural). The required member resistance, R_{reqd}, is determined using elastic or inelastic analytical methods as a function of the factored loads. This is represented in a general form as

$$R_{\text{reqd}} \leq \phi R_n \tag{12.6}$$

$$R_{\text{reqd}} = f(\gamma_i P_i) \tag{12.7}$$

A reliability index, β, is used to quantify the structural reliability in probability-based design (Nowak and Collins, 2000). For conventional structural materials such as steel and concrete, reliability indices between $\beta = 2.0$ and 4.0 are common (Ellingwood, 2003). A reliability index of $\beta = 2.5$ corresponds to a probability of failure of $P_f = 0.005$. When ASCE 7 load factors are used, target reliability indices of $\beta = 2.5$ to 4.5 are generally accepted for structural engineering design with conventional materials (Ellingwood, 2003; Szerszen and Nowak, 2003). For members (beams, columns, etc.) where analytical methods are well established to predict deformation and failure, the lower values of β in the range noted above are generally accepted, whereas for connections and joints, where analytical methods to predict deformation and failure are less developed, higher values of β are generally accepted. For materials that fail in a brittle fashion and whose long-term behavior is not well known, higher values of β are also used. For this reason target reliability indexes for FRP pultruded structures are expected to be in the β range 3.0 to 4.5 until more data become available. The reliability index is not used explicitly by structural engineers to design a structure. It is used by code-writing bodies and researchers to determine the appropriate resistance factors, ϕ, for use in the LRFD procedure. Structural engineers then use the probability-based LRFD procedure in a conventional deterministic manner.

In the LRFD basis (as in the ASD basis), two primary limit states are considered in design: the ultimate limit state and the serviceability limit state. The *ultimate limit state* is associated with life safety and is controlled by collapse, instability, and failure of the structure or structural members. The *serviceability limit state* is associated with short- and long-term serviceability of the structure and is controlled by deflection, vibration, and degradation of the materials, the members, and the connections in the structure (Eurocomp, 1996).

A probability-based LRFD specification or design procedure for FRP pultruded structures does not exist at this time. However, use of the LRFD approach for the design of pultruded FRP structures is quite possible, and a prestandard outline has been prepared by ASCE (Chambers, 1997). Fundamental issues associated with developing an LRFD design procedure for pultruded FRP structural profiles, with measurable reliability indexes, are discussed in Ellingwood (2003).

The difficulties associated with developing an LRFD basis for FRP pultruded structures are associated primarily with the determination of the statistical material design properties for common pultruded profiles and the appropriate analytical models to use for predicting the member resistances. Since a standard material specification is not followed by pultrusion manufacturers at this time, the statistical properties of pultruded materials are not available. In the future it is anticipated that manufacturers will follow a standard material specification such as European Standard EN 13706 (2002) or will develop one based on proposed model specifications such as that of Bank et al. (2003).

An LRFD procedure for the design of doubly symmetric and singly symmetric pultruded columns that accounts for global flexural buckling, global torsional buckling, flexural–torsional buckling, and material compression has been presented by Zureick and Scott (1997), Zureick (1998), and Zureick and Steffen (2000). In these studies, resistance factors were determined based on detailed tests of glass–vinylester pultruded materials (from a single manufacturer) and on full-scale buckling experiments conducted on pultruded columns having wide-flange, square-tube (box), and equal-leg-angle profiles. Analytical equations based on orthotropic plate theory were used to predict the buckling loads. The resistance factors to be used with ASCE 7 load factors, determined in these studies, are given in Table 12.5. Unfortunately, these resistance factors are only applicable to the types of profiles and pultruded materials tested in these studies.

For doubly symmetric profiles, Zureick and Scott (1997) did not conduct a rigorous reliability analysis, and the resistance factors were obtained from lower bounds of comparisons between experimental data and theoretical predictions. Reliability indices are therefore not provided for these resistance factors. For the singly symmetric angles, Zureick and Steffen (2000) chose target reliability indices of $\beta = 3$, 3, and 4 for the flexural buckling, flexural–torsional buckling, and material compressive failure limit states. Design property values for the pultruded materials in this study were based on the 95% lower confidence limit on the 5th percentile of the test data. A three-parameter Weibull distribution was used to characterize the test data statistically. Based on further work Zureick and co-workers (Alqam et al., 2002) have recommended a two-parameter Weibull distribution to characterize pultruded materials statistically.

European Limits States Design Basis In the LSD basis used predominantly in Europe, instead of using resistance factors to account for the variability in the material, safety factors or material partial factors (coefficients) are used (Hollaway and Head, 2001). The characteristic material property value,[12] X_k,

TABLE 12.5 LRFD Resistance Factors for Glass–Vinylester Pultruded Columns

Profile	Failure Mode	Limit State	Resistance Factor, ϕ	Reliability Index, β
WF or tube	Flexural (Euler) buckling	Ultimate	0.85	not available
	Axial shortening	Serviceability	0.80	not available
Single-angle	Flexural (Euler) buckling	Ultimate	0.65	3.0
	Flexural–torsional buckling	Ultimate	0.85	4.0
	Material compression	Ultimate	0.50	4.0

[12] Typically defined as the 5th percentile of the test data.

is divided by the material partial coefficient as opposed to being multiplied by the resistance factor in the LRFD approach in the following general form:

$$R_{\text{reqd}} \leq R_d \tag{12.8}$$

$$R_d = f\left(\frac{X_k}{\gamma_m}, a_0\right) \tag{12.9}$$

where R_d is the design resistance and a_0 is the effect on resistance of geometric variation of the section. The LSD safety factor approach is somewhat different from the LRFD approach in that the resistance factor in the LRFD approach is selected based on the member type (flexural, compression, shear) and failure mode, whereas in the LSD method the safety factors are selected primarily based on material properties. However, analytical factors similar to the geometric factors can be introduced to account for failure modes. The development of limit states design procedures for FRP structures for structural applications was initiated in Europe by Head and Templeman (1986). As noted previously, the *Structural Plastic Design Manual,* developed in the United States in the late 1970s, also follows a limit states design procedure, but explicit factors for design use are not provided in this document (ASCE, 1984).

A difficulty exists with the use of the LSD basis for FRP pultruded structures similar to that with the LRFD basis (Hollaway and Head, 2001): that is, determination of the material partial factors with known probability so that a structural reliability analysis can be conducted. In Europe, the load factors and load combinations are those prescribed by Eurocode 1 (CEN, 2002b).

An approach to obtaining the material partial factors for member design[13] is provided in the *Eurocomp Design Code and Handbook* (Eurocomp, 1996), following the Eurocode approach, in which the material partial factor, γ_m, is obtained as the product of coefficients that are intended to account for separate and quantifiable influences on the material properties as follows:

$$\gamma_m = \gamma_{m,1}\gamma_{m,2}\gamma_{m,3} \tag{12.10}$$

where, $\gamma_{m,1}$ accounts for the method in which the material property data were obtained, $\gamma_{m,2}$ accounts for the material manufacturing process,[14] and $\gamma_{m,3}$ accounts for the effects of environment and the duration of loading on the material properties. According to the Eurocomp design code, the material

[13] For connection and joint design a similar approach is followed, however, additional safety factors are used to account for local connection effects and for the method of analysis used (Eurocomp, 1996, Sec. 5).

[14] The Eurocomp design code permits any manufacturing process and is not applicable only to pultruded profiles.

partial factor may not be less than 1.5 for ultimate limit states (nor need be greater than 10) and not less than 1.3 for serviceability limit states. However, in the opinion of many designers, the 1.3 material partial factor for the serviceability limit state is regarded as too conservative if the material characteristic properties (see Table 12.7) are obtained from manufacturer-published data. A material partial factor of 1.0 for the serviceability limit state is recommended if manufacturer material data are used in design. Table 12.6 lists key values of the material partial factors for the ultimate limit state (ULS) for commercially produced pultruded materials, as listed in the *Eurocomp Design Code and Handbook.*[15]

According to the Eurocomp design code, the material manufacturer is required to provide proof that the material is fully cured in order for the designer to use the fully cured partial safety coefficients. Although the Eurocomp design code is written from an LSD or LRFD basis, detailed data are not provided as to how the material partial safety coefficients were determined, and therefore the structural reliability of a structure designed with these coefficients cannot be ascertained with certainty at this time (Hollaway and Head, 2001). However, the Eurocomp design code does provide tables of characteristic values for pultruded materials as a function of the fiber volume fraction and ratio of the longitudinal fiber to a continuous strand mat. The characteristic properties that are stated have been determined from the mean of test data minus 1.64 standard deviations; however, the number of tests and a detailed description of the materials are not provided to verify this claim. Characteristic property data for three typical types of glass–FRP pultruded materials specified by the Eurocomp design code are listed in Table 12.7. The data provided for pultruded material 2 are very similar to those presented

TABLE 12.6 ULS Material Partial Factors

Source of Factor	Material Partial Factor
Properties of the pultruded plate are derived from test	$\gamma_{m,1} = 1.15$
Properties of the pultruded plate are derived from laminate theory	$\gamma_{m,1} = 1.5$
Fully cured pultruded material	$\gamma_{m,2} = 1.1$
Non-fully cured pultruded material	$\gamma_{m,2} = 1.7$
Short-term loading for $25 < T < 50$°C and $T_g{}^a > 90$°C	$\gamma_{m,3} = 1.0$
Short-term loading for $0 < T < 25$°C and $T_g > 80$°C	$\gamma_{m,3} = 1.0$
Long-term loading for $25 < T < 50$°C and $T_g > 90$°C	$\gamma_{m,3} = 2.5$
Long-term loading for $0 < T < 25$°C and $T_g > 80$°C	$\gamma_{m,3} = 2.5$

Source: Eurocomp Design Code and Handbook (1996).

[a] T_g is the glass transition temperature of the FRP pultruded material. For commercially manufactured, fully cured pultruded profiles, it should not be less than 80°C.

[15] Consult the Eurocomp design code for a full list of safety factors.

TABLE 12.7 Characteristic Properties of Pultruded Materials for Design (Eurocomp)

	Pultruded Material 1	Pultruded Material 2	Pultruded Material 3
Fiber volume (typical)	48%	38%	65%
Fiber architecture	Roving and CFM (ratio 4:1)	Roving and CFM (ratio 1:1)	Roving only
Resin system	Polyester, vinylester	Polyester	Polyester
Typical part	Special nonstructural pultruded profile (e.g., grating I-bar, ladder rail)	Conventional structural pultruded profile (I, WF, L, □, [)	Round solid pultruded rod
Strength properties ($\times 10^3$ psi)			
Tensile			
Longitudinal	59.5	30.0	100.1
Transverse	6.4	7.0	not provided
Compressive			
Longitudinal	39.2	30.0	60.0
Transverse	not provided	14.9	not provided
Shear			
In-plane	2.2	4.5	5.5
Out-of-plane	not provided	not provided	not provided
Flexural			
Longitudinal	58.0	30.0	100.1
Transverse	16.7	10.0	not provided
Bearing			
Longitudinal	not provided	not provided	not provided
Transverse	not provided	not provided	not provided

Stiffness properties ($\times 10^6$ psi)			
Tensile			
Longitudinal	3.9	2.5	5.9
Transverse	0.5	0.8	not provided
Compressive			
Longitudinal	3.5	2.5	not provided
Transverse	0.7	1.0	not provided
Shear, in-plane	0.6	0.4	not provided
Flexural			
Longitudinal	2.0	2.0	5.9
Transverse	1.2	0.8	not provided
Poisson Ratio			
Longitudinal	0.2	0.33	not provided
Transverse	0.1	0.11	not provided

historically in U.S. manufacturers' design manuals for conventional glass–FRP pultruded shapes and (e.g., Strongwell, 2002) and would appear to have "embedded" knock-down factors included. It is unlikely that they are truly characteristic values for conventional pultruded materials obtained (as can be seen by comparison with test data in Zureick and Scott, 1997, for example).

In what follows a version of the LRFD basis is presented as an acceptable design basis for FRP structures. The elements of the LRFD basis used in this book are:

1. Analytical equations presented in the *Structural Plastics Design Manual* (ASCE, 1984) and *Eurocomp Design Code and Handbook* (Eurocomp, 1996), supplemented with equations from the technical literature
2. Characteristic design properties of pultruded materials taken from European Standard EN 13706 (CEN, 2002a) or from U.S. manufacturers' design guides
3. Resistance factors taken as the inverse of the material partial factors given in the *Eurocomp Design Code and Handbook* (Eurocomp, 1996)
4. Nominal loads, load factors, and factored load combinations taken from ASCE 7-02

However, it must be noted that the LRFD basis as described above and presented in what follows in the design examples does not have a fundamental probability-based foundation and does not yield known reliability indices. Nevertheless, based on comparisons with the ASD procedure, it is expected that this approach will yield reliability indexes in the range $\beta = 2.5$ to 3.0 (Ellingwood, 2003).

12.5 PERFORMANCE-BASED DESIGN

The structural design of an FRP pultruded structure or a portion of the structure can be based on a performance specification (ICC, 2003). In this approach, the entire structure or a portion of the structure is required to meet specific structural performance objectives, which are evaluated in a predetermined manner according to predetermined performance criteria. The performance objectives are typically applied to both local and global deformations and capacities. Full-scale testing of the structure or a portion of the structure is usually required to meet the performance specification. Both proof testing of the actual structure and failure testing of a full-sized mock-up of the structure (or parts thereof) may be used. A factor of safety or a structural reliability index must be defined and agreed upon by designer and client as part of the performance specification. The performance specification is usually appended to the construction documents and plans and is part of the construction contract. Performance specifications are typically developed on a case-by-case basis. They are not uncommon in the FRP industry. In highway construction documents in the United States, they are often appended to the construction documents as "special provisions."

In some situations, a combined prescriptive and performance approach is used. A prescriptive set of material specifications is mandated along with a number of structural-level performance specifications. The material specification may stipulate the manufacturing method, the limiting mechanical and physical properties of the materials produced, and the requirements for quality assurance testing.

PROBLEMS

12.1 Visit the following pultrusion manufacturers' Web sites[16] and request or download copies of their most recent pultrusion design guides and their pultruded profile geometry and property specifications (typically, included in the design guides). These will be needed for solving design problems in the chapters to follow.

Strongwell: www.strongwell.com

Creative Pultrusions: www.pultrude.com

Bedford Reinforced Plastics: www.bedfordplastics.com

Fiberline: www.fiberline.com

TopGlass: www.topglass.it

12.2 From U.S. manufacturers' specification sheets (Strongwell, Creative Pultrusions, and Bedford Reinforced Plastics), extract and list the manufacturer-reported geometric and mechanical properties (in-plane stiffness and strength properties for the web and the flange) for the following pultruded sections[17] commonly used in truss and light-frame pultruded structures:

Glass–polyester $3 \times \frac{1}{4}$ $(b \times t)$ in. square tube

Glass–polyester $2 \times \frac{1}{4}$ $(b \times t)$ in. square tube

Glass–polyester $6 \times \frac{1}{4} \times 1\frac{5}{8} \times \frac{1}{4}$ $(d \times t_w \times b_f \times t_f)$ in. channel

Glass–polyester $3 \times \frac{1}{4}$ $(b \times t)$ in. equal leg angle

12.3 From U.S. manufacturers' specification sheets (Strongwell, Creative Pultrusions, and Bedford Reinforced Plastics) extract and list the manufacturers' reported geometric and mechanical properties (in-plane stiffness and strength properties for the web and the flange) for the

[16]No endorsement of the manufacturers listed is implied. The reader is free to choose any manufacturer's products to use in the examples that follow, and is encouraged to collect similar data for locally available FRP pultruded shapes.

[17]Not all manufacturers may produce all shapes listed in Problems 12.2 to 12.4.

following pultruded sections commonly used in heavy-frame pultruded structures:

Glass–polyester $8 \times 8 \times \frac{3}{8}$ ($d \times b \times t$) in. wide-flange I-shaped section

Glass–polyester $10 \times 5 \times \frac{1}{2}$ ($d \times b \times t$) in. narrow-flange I-shaped section

Glass–vinylester $6 \times 6 \times \frac{1}{4}$ ($d \times b \times t$) in. wide-flange I-shaped section

Glass–vinylester $6 \times \frac{3}{8}$ ($b \times t$) in. square tube

Glass–vinylester $6 \times \frac{1}{2}$ ($b \times t$) in. equal leg angle

Glass–polyester $10 \times \frac{1}{2} \times 2\frac{3}{4} \times \frac{1}{2}$ ($d \times t_w \times b_f \times t_f$) in. channel

12.4 From European manufacturers' specification sheets (Fiberline or TopGlass), extract and list the manufacturer-reported geometric and mechanical properties (in-plane stiffness and strength properties for the web and the flange) for the following pultruded sections[18] commonly used in heavy-frame pultruded structures. Provide your answers in both metric and U.S. units to allow for comparison with properties of U.S.-produced profiles.

Glass–vinylester/polyester[19] $240 \times 120 \times 12$ ($d \times b \times t$) mm narrow-flange I-shaped section

Glass–vinylester/polyester $300 \times 8 \times 150 \times 12$ ($d \times t_w \times b_f \times t_f$) mm narrow-flange I-shaped section

Glass–vinylester/polyester 200×10 ($b \times t$) mm square tube

Glass–vinylester/polyester 150×8 ($b \times t$) inch equal leg angle

Glass–vinylester/polyester $240 \times 7 \times 72 \times 10$ ($d \times t_w \times b_f \times t_f$) mm channel

12.5 Contact pultrusion companies or local distributors of pultruded profiles and obtain current pricing (per unit length, ft or m) of the pultruded beams listed in Problems 12.2 to 12.4.

12.6 Read and write a one-page report on the following technical papers that discuss design bases for pultruded structures:

Head, P., and Templeman, R. B. (1986), The application of limit states design principles to fibre reinforced plastics, *Proceedings of the British Plastics Federation Congress '86,* Nottingham, England, September 17–19, pp. 69–77.

Chambers, R. E. (1997), ASCE design standard for pultruded fiber-reinforced plastic (FRP) structures, *Journal of Composites for Construction,* Vol. 1, pp. 26–38.

[18] If the exact size is not manufactured by the manufacturer, choose one close to the size listed.

[19] European manufacturers generally do not provide different properties for polyester and vinylester profiles.

Ellingwood, B. R. (2003), Toward load and resistance factor design for fiber-reinforced polymer composite structures, *Journal of Structural Engineering*, Vol. 129, pp. 449–458.

SUGGESTED ACTIVITY

12.1 Contact a pultrusion manufacturer[20] and ask them to send you some samples of pultruded profiles (1-ft-long lengths of an I-shape and a tube are good). The simple qualitative experiments listed below will enable you to understand more about the properties of the material and its workability. Write short (e.g., two-page) reports on the materials, methods, and results of your examinations. Consider the following:

(a) Weigh and measure the sample and determine the mass density of the pultruded material. Compare measured densities to the manufacturers' reported densities.

(b) Mark and drill holes of $\frac{1}{4}$- and 1-in. diameter in the pultruded material. Experiment with various ways of marking the material (pencil, pen, scoring, etc.) and various drill bits (solid, hole, etc.)

(c) If possible, cut a coupon 1 in. wide $\times$ 3 in. long from the material in both the longitudinal and transverse directions. Use a masonry blade or diamond-tipped blade and a power saw (table, circular, or radial arm saw) or a hand hacksaw or box saw (cross-cut saw).

(d) Expose the coupon to a constant flame in a Bunsen burner (in a fully ventilated hood with an exterior exhaust system). Observe what happens. Examine the residue and record your observations.

(e) Sand and polish (to 200 grit, if possible) the cut ends of the coupon and examine the cut ends in the longitudinal and transverse directions under an optical microscope. Try to distinguish between roving and mat layers. Look for voids in the material.

[20] It is preferable for the instructor to contact the manufacturer on behalf of the class early in the semester to ensure that the materials are available when needed.

13 Pultruded Flexural Members

13.1 OVERVIEW

In this chapter, design procedures are presented for flexural members made of pultruded FRP profiles. Axially loaded members in tension and compression or combined axial and flexural members are discussed in Chapter 14. Connections and joints for pultruded profiles are discussed in Chapter 15. Theoretical mechanics-based equations are presented here for analyzing strength, stability, and deformation limit states of flexurally loaded pultruded structural members.

Design procedures for the allowable stress design (ASD) and the load and resistance factor design (LRFD) bases described in Chapter 12 are presented. Since pultruded profiles behave essentially as linear elastic (or linear viscoelastic) structural members, the analytical equations developed for predicting stresses and deformations in these members for the ultimate and the serviceability limit states are valid for either of the design bases selected. Therefore, it is only the safety factors for ASD or the load and resistance factors for the LRFD that differentiate the design procedures for the two design bases.

13.2 INTRODUCTION

The design of pultruded profiles that are symmetric with respect to the plane of loading and consist of horizontal plates (flanges) and vertical plates (webs) is discussed in this section. This includes conventionally manufactured I-shaped,[1] square tube, rectangular tube, round tube, and multicelled tubular profiles. It also includes unsymmetric or singly symmetric sections, such as equal or unequal leg angles, or channel sections, or built-up sections, if they are loaded in their plane of symmetry through the shear center or are used back to back to create a symmetric composite profile. Typical member cross sections and orientations for loading in the vertical plane (i.e., the plane of the web) are shown in Fig. 13.1.

In vertical braced frame or horizontal unbraced grid structures, wide-flange and tubular sections are most frequently used as beams and girders since they are readily connected using clip angles or gusset plates and traditional me-

[1] The term *I-shaped profiles* is used to describe both narrow- and wide-flange I-shaped profiles.

Figure 13.1 Cross sections and orientations of typical flexural members.

chanical connectors, such as nuts and bolts (e.g., see Figs. 1.15 and 1.18). In vertical braced frames the beams may need to be designed as combined load members (beam-columns or tension-bending members), which are discussed in Chapter 14. The choice of profile used as a beam often depends on the connection detailing required, as discussed further in Chapter 15.

In trusses or stick-built frames, back-to-back angles and small channels, as well as small square tubular elements, are most commonly used (see e.g., Figs. 1.17 and 1.19) as the primary structural elements. In this case, the elements are usually treated as continuous beams (such as chords, girts, purlins, or rafters) or as axially loaded members (such as columns, struts, or ties) or combined load members (beam-columns or tension-bending members). Axially loaded and combined load members are discussed in Chapter 14.

13.3 STRESSES IN FLEXURAL MEMBERS

The flexural member discussed in this chapter is designed to resist stress resultants (or internal forces) in the global x,y,z coordinate system that are caused by the applied transverse distributed loads, $w(z)$, or concentrated loads, P, in the y–z plane, as shown in Fig. 13.2. These stress resultants are the

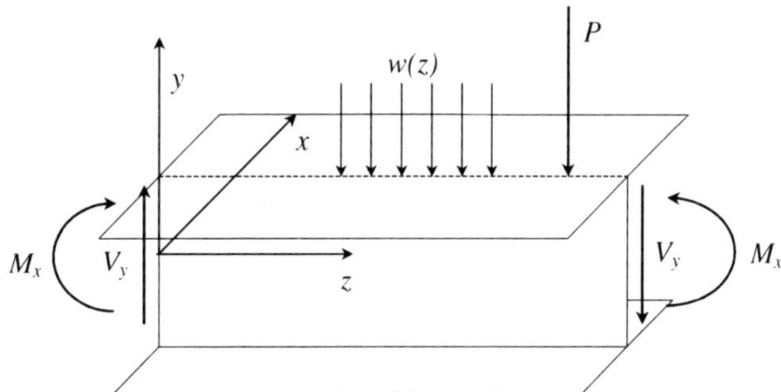

Figure 13.2 Transversely loaded beam.

bending moment, M_x, and the transverse shear force, V_y. According to one-dimensional beam theory for beams of homogeneous material,[2] the bending moment, M_x, causes the member to develop a linearly varying axial (flexural) stress, σ_x, through the depth of the cross section and along its length that is given by the flexural stress equation

$$\sigma_z = -\frac{M_x y}{I_x} \tag{13.1}$$

where I_x is the second moment of area of the cross section about the x-axis and y is the vertical distance from the neutral axis. For the (positive) moment shown in Fig. 13.2, a compressive (negative) stress is obtained in the top part of the cross section above the neutral axis, and a tensile (positive) stress is obtained in the bottom part of the cross section below the neutral axis.

For a beam with nonhomogeneous properties through the depth of the cross section, such as a pultruded profile with different longitudinal elastic moduli in its webs and flanges, as may be the case in some pultruded profiles, the relationship between the bending moment and the axial (flexural) stress is not closed formed but is written in general form as

$$M_x = -\int_A \sigma_z y \, dA \tag{13.2}$$

where the integral is taken over the cross-sectional area of the beam. Since the pultruded material is assumed to be linear elastic in this one-dimensional beam theory, the strain through the depth of the flexural member, ε_z, varies linearly, and the stress at any height of the section, σ_z, can be found directly using the longitudinal modulus, E_L, at the position of interest. Beams of this type have traditionally been known as *composite beams* since they are composed of different materials[3] and are treated extensively in many texts (e.g., Gere and Timoshenko, 1997). Stress analysis of beams of this type is commonly performed using the transformed section method or the composite mechanics method. These methods were used extensively in previous chapters dealing with FRP-reinforced and FRP-strengthened concrete flexural members (Chapters 5 and 9).

The shear force resultant, V_y, causes a homogeneous material beam to develop a nonuniformly varying transverse shear stress,[4] τ, along its length

[2] This implies that the longitudinal stiffness does not vary through the depth in the cross section. This assumption will be revisited when pultruded profiles having differing properties in their webs and flanges are discussed.

[3] Although the historically used term *composite* used to describe these beams may cause some initial confusion, it is actually perfectly appropriate for the FRP beams being considered.

[4] No subscript is used for the shear stress in one-dimensional thin-walled beam theory, since it is used to represent either τ_{yz} (in the web) or τ_{xz} (in the flange).

and throughout its cross section that is given by the familiar thin-walled section shear stress equation as

$$\tau = \frac{V_y Q_x}{I_x t} \tag{13.3}$$

where Q_x is the first moment of area of the cross section calculated at the point of interest, I_x the second moment of area of the cross section, and t the thickness wall (perpendicular to the contour) at the point of interest. In a nonhomogeneous beam, the properties Q_x and I_x of the transformed section must be used in equation (13.3). Equation (13.3) can only be used to determine the in-plane shear stress in the direction of the shear flow in the thin-walled section. It cannot be used to determine the interlaminar shear stress through the thickness of the flange. Since the shear stress causes shear distortion (as opposed to extension or contraction in the case of the normal stress due to the bending moment) in the individual orthotropic plates (or walls) of the thin-walled section, the sign of the shear stress is not significant.[5] Therefore, the stress given by equation (13.3) is always taken as positive regardless of the sign of V_y.

In addition to the stresses described above, which are developed as a result of the global bending moment and transverse shear force and are obtained using one-dimensional beam theory, additional local stresses at points of concentrated loads must be accounted for in the design of pultruded members. These local, primarily bearing-type stresses are not obtained from beam theory but are obtained using commonly employed engineering approximations for the "length of bearing" similar to those used for steel profiles.

Due to concentrated loads along the beam and reactions at the supports, the profile is assumed to develop a uniform compressive bearing stress in the vertical plates (web or webs) of the section at the points of load application or at the support given by

$$\sigma_y = \frac{F}{A_{\text{eff}}} \tag{13.4}$$

where F is the transverse force and A_{eff} is the effective area over which the force is applied. For a web with outstanding flanges on both sides (e.g., an I-shaped profile) the effective area is given as

[5] It should be noted that the strength of a highly orthotropic unidirectional laminate can depend on the sign of the shear stress when certain multiaxial failure criteria are used to predict failure (Agarwal and Broutman, 1990, p. 181). However, no distinction is made between positive and negative shear stress for the purpose of design of pultruded profiles where only uniaxial failure criteria are used in design.

$$A_{\text{eff}} = (t_w + 2t_f + 2t_{\text{bp}})L_{\text{eff}} \tag{13.5}$$

For a web with an outstanding flange on one side only (e.g., a corner of a square tube profile) the effective area is given as

$$A_{\text{eff}} = (t_w + t_f + t_{\text{bp}})L_{\text{eff}} \tag{13.6}$$

where L_{eff} is the effective bearing length along the beam in the z-direction (taken as the width of the support or the length over which the concentrated load is applied) and t_w, t_f, and t_{bp} are the web thickness, the flange thickness, and the thickness of a bearing plate under the flange (if applicable), respectively, as shown in Fig. 13.3. For a section with multiple webs, the contributions to the effective area from the webs having two outstanding flanges and a single outstanding flange are added as appropriate. For example, for a single-cell tube section (also known as a box section) having two webs, each with one outstanding flange, the effective area is given as twice the area obtained from equation (13.6) (assuming that both webs have a thickness of t_w).

The stresses (or force resultants) calculated must be less than the material strengths and critical buckling stresses (or the critical resistances or buckling capacities) according to the design basis selected.

13.4 DEFORMATIONS IN FLEXURAL MEMBERS

Due to the stress resultants, M_x and V_y, the member will deflect in its plane of loading. The elastic curve that describes the deflected shape of the beam is a function of the flexural rigidity, EI, and the transverse shear rigidity, KAG, of the member. The deflection can be obtained using Timoshenko beam theory (also known as shear deformation beam theory) (Timoshenko, 1921). In Timoshenko beam theory, the elastic curve is defined by two independent var-

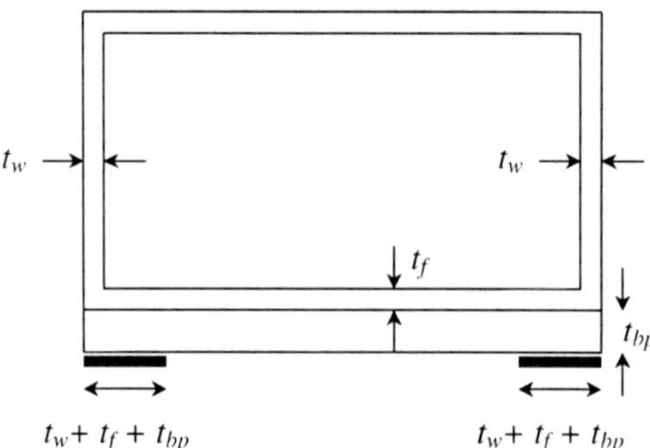

Figure 13.3 Effective bearing width for beam webs (shown as a thick dark line).

iables, the transverse deflection, $\delta(z)$, due to bending and shearing, and the slope of the elastic curve, $\phi(z)$, due to bending only.[6] They are obtained by simultaneous solution of the differential equations

$$\frac{d\phi}{dz} = \frac{M_x}{EI} \tag{13.7}$$

$$\frac{dy}{dz} - \phi = -\frac{V_y}{KAG} \tag{13.8}$$

and boundary conditions for the deflection and bending slope at fixed, pinned, and free ends as in Euler–Bernoulli beam theory. The solution to equations (13.7) and (13.8) with appropriate boundary conditions leads to the general expression for the total beam deflection as a sum of the deflection due to bending deformation and the deflection due to shear deformation:

$$y(z) = y_b(z) + y_s(z) = \frac{f_1(z)}{EI} + \frac{f_2(z)}{KAG} \tag{13.9}$$

where $f_1(z)$ and $f_2(z)$ are functions that depend on loading and boundary conditions. For determinate beams with common loads and support conditions, the functions are given in Table 13.1. It can be seen that the deflection function, $f_1(z)$, that represents the bending contribution to the deflection is the same as that obtained by traditional Euler–Bernoulli beam theory. In addition, it is important to recognize that for statically indeterminate beams (and frames) the member internal forces, M_x and V_y (and N_z in frames), will also be a function of the shear rigidity of the beam and will not be the same as those for the same indeterminate beam when Euler–Bernoulli beam theory is used.[7] For determinate beams the internal forces are not functions of the deflections; therefore, the internal forces will be the same regardless of the beam theory used.

For the serviceability limit state, the calculated maximum deflections must be less than the building code–stipulated deflections. The *Structural Plastics Design Manual* (ASCE, 1984) gives typical deflection limits as $L/180$ for visual appearance to $L/400$ for vibration sensitivity. The *Eurocomp Design Code and Handbook* (Eurocomp, 1996) recommends limiting instantaneous deflections to $L/300$ and long-term deflection to $L/250$ for frame structures.

[6] In Timoshenko beam theory the bending slope, $\phi(z)$, is an independent variable and is not equal to dy/dz as in Euler–Bernoulli beam theory. In Timoshenko beam theory the total slope is the sum of the bending slope and the shear slope (i.e., $dy/dz = \phi + \gamma$).

[7] Indeterminate beams can by solved by hand using Timoshenko beam theory and the force method. In computerized analysis of building frames, a shear-deformable beam element should always be used for frames of pultruded members. These elements are available in most commercial structural analysis codes.

TABLE 13.1 Deflection Functions for Common Determinate Beams[a]

Beam	$f_1(z)$	$f_2(z)$	$f_1\ (\delta_{max})$	$f_2\ (\delta_{max})$	$z\ (\delta_{max})$
Simply supported beam					
Uniformly distributed load (w)	$-\frac{w}{24}(z^4 - 2\ell z^3 + \ell^3 z)$	$-\frac{w}{2}(\ell z - z^2)$	$-\frac{5w\ell^4}{384}$	$-\frac{w\ell^2}{8}$	$\frac{\ell}{2}$
Concentrated load (P) at midspan ($0 < z < \ell/2$)	$-\frac{P}{48}(3\ell z^2 - 4z^3)$	$-\frac{P}{2}(z)$	$-\frac{P\ell^3}{48}$	$-\frac{P\ell}{4}$	$\frac{\ell}{2}$
Cantilever beam					
Uniformly distributed load (w)	$-\frac{w}{24}(z^4 - 4\ell z^3 + 6\ell^2 z^2)$	$-\frac{w}{2}(2\ell z - z^2)$	$-\frac{w\ell^4}{8}$	$-\frac{w\ell^2}{2}$	ℓ
Concentrated load (P) at tip	$-\frac{P}{6}(3\ell z^2 - z^3)$	$-P(z)$	$-\frac{P\ell^3}{3}$	$-P\ell$	ℓ

[a] Loads are defined as positive as shown in Fig. 13.2.

Ultimately, local building code regulations should be followed regarding allowable service load deflections. It is generally accepted that deflections should not exceed $L/240$ when brittle nonstructural elements such as floors, partition walls, ducts, and piping are attached to the structure.

Analysis Example 13.1: Timoshenko Beam Theory Obtain the equation of the elastic curve for the uniformly loaded cantilever beam shown in the Fig. 13.4 using Timoshenko beam theory.

SOLUTION The free-body diagram of a portion of the beam at the fixed end is shown in Fig. 13.5. The bending moment and shear force along the length of the beam are expressed as

$$M_x(z) = w\ell z - \frac{w\ell^2}{2} - \frac{wz^2}{2}$$

and

$$V_y(z) = w\ell - wz$$

The mathematical boundary conditions for the physical conditions of no transverse deflection and no bending slope at the fixed end are

$$y(0) = 0 \qquad \phi(0) = 0$$

The first differential equation of Timoshenko beam theory is written as

$$\frac{d\phi}{dz} = \frac{1}{EI}\left(w\ell z - \frac{w\ell^2}{2} - \frac{wz^2}{2}\right)$$

Integrating gives

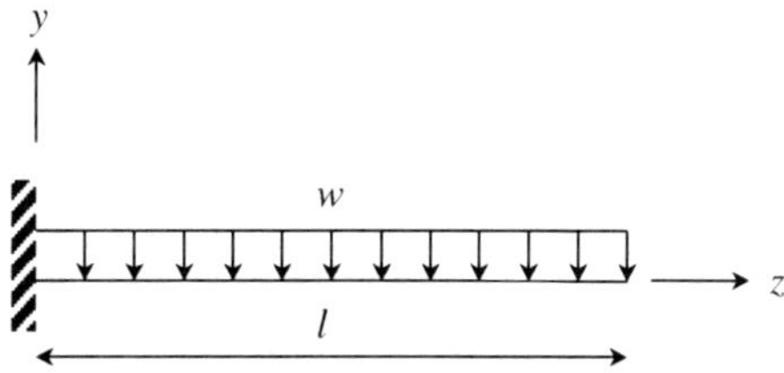

Figure 13.4 Uniformly loaded cantilever beam.

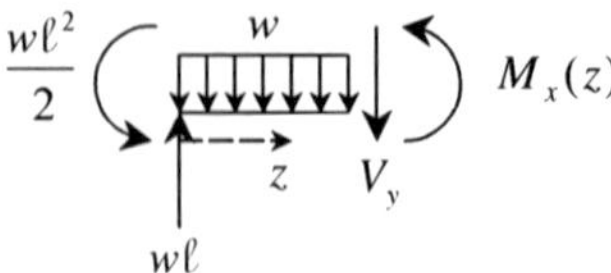

Figure 13.5 Free-body diagram of the fixed end of a cantilever beam.

$$\phi(z) = \frac{1}{EI}\left(\frac{w\ell z^2}{2} - \frac{w\ell^2 z}{2} - \frac{wz^3}{6}\right) + C_1$$

Substituting the boundary condition $\phi(0) = 0$ gives $C_1 = 0$. The second differential equation of the Timoshenko beam theory is then written as

$$\frac{dy}{dz} = -\frac{1}{KAG}(w\ell - wz) + \frac{1}{EI}\left(\frac{w\ell z^2}{2} - \frac{w\ell^2 z}{2} - \frac{wz^3}{6}\right)$$

and integrating gives

$$y = \frac{1}{KAG}\left(-w\ell z + \frac{wz^2}{2}\right) + \frac{1}{EI}\left(\frac{w\ell z^3}{6} - \frac{w\ell^2 z^2}{4} - \frac{wz^4}{24}\right) + C_2$$

Substituting the boundary condition $y(0) = 0$ gives $C_2 = 0$. Finally, the solution is rearranged and written in its usual form, $y(z) = f_1(z)/EI + f_2(z)/KAG$, as

$$y(z) = \frac{1}{EI}\left(\frac{w\ell z^3}{6} - \frac{w\ell^2 z^2}{4} - \frac{wz^4}{24}\right) + \frac{1}{KAG}\left(-w\ell z + \frac{wz^2}{2}\right)$$

$$= \frac{-w}{24EI}(z^4 + 4\ell z^3 - 6\ell^2 z^2) - \frac{w}{2KAG}(-z^2 + 2\ell z)$$

which can be found in the third row of Table 13.1.

13.5 DETERMINATION OF DEFLECTIONS AND STRESSES FOR SERVICEABILITY AND ULTIMATE LIMIT STATES

For flexural members the deflections are determined for the serviceability limit state (SLS) and stresses (or stress resultants) are determined for the ultimate limit state (ULS). Since pultruded profiles have low stiffness-to-strength ratios relative to conventional structural materials, the serviceability limit states of flexural members are checked first (i.e., the deflection calculation is the first step in the design procedure). Thereafter, the ultimate strength limit states are checked. These are usually governed by critical buckling stresses. The axial and shear material strengths very seldom control the design of flexural members. The order of the limit states presented follows this order. Notice that this is typically not the order in which limit states are considered for design with profiles of metallic materials that are braced against global lateral–torsional buckling. In such cases the material strength criteria are checked first and thereafter, shear stresses and transverse deflections are checked. This is because most standard I-shaped steel sections (profiles) used as flexural members are classified as *compact sections,* which means that they will reach their full plastic moment capacity prior to local buckling of their flanges. If pultruded profiles are produced with the same cross-sectional geometries as steel compact sections, they will not be compact, due to their different mechanical properties. In such steel-like geometry pultruded profiles, local buckling of flanges (and occasionally, webs) becomes a critical limit state.[8]

13.6 SERVICEABILITY LIMITS STATES

13.6.1 Deformation Limit State: Transverse Deflection

Timoshenko shear deformation beam theory is used to determine the deflection of a pultruded beam. This is the procedure recommended by the *Structural Plastic Design Manual* (1984) and the *Eurocomp Design Code and Handbook* (1996). The use of shear deformation beam theory is especially important in pultruded beams because of the relatively low longitudinal modulus (leading to beams with short spans) and the relatively high E_L/G_{LT} ratios, due to the low shear modulus of pultruded materials (Bank, 1989a,b). To calculate the deflection, appropriate choices of the flexural rigidity, EI, and the transverse shear rigidity, KAG, are required.

For homogeneous pultruded beams having the same properties in the flanges and webs of the profile, the flexural rigidity is obtained as

$$EI = E_L I_x \tag{13.10}$$

where I_x is the second moment of the cross section and E_L is the longitudinal modulus of the pultruded material in the profile. For nonhomogeneous pultruded profiles having different longitudinal moduli in the parts of the cross section, the flexural rigidity can be found using the composite section method[9] or the transformed-section method. The longitudinal modulus, especially that

[8] Due to the high longitudinal strength and low longitudinal stiffness of glass-reinforced pultruded profiles, it is difficult to design a compact type of section in which compressive failure would occur before local flange buckling, as we discuss further below.

[9] As discussed previously.

of the flanges, is the dominant material property in the flexural rigidity. It has been shown that flexural rigidity is insensitive to the transverse modulus of the pultruded material.

For homogeneous pultruded beams, having the same properties in the flanges and webs of the profile, the transverse shear rigidity is obtained from

$$KAG = k_{\text{tim}} A G_{LT} \approx A_{\text{web}} G_{LT(\text{web})} \tag{13.11}$$

where, k_{tim} is the Timoshenko shear coefficient, A the area of the entire cross section, G_{LT} the in-plane shear modulus of the webs and the flanges (assumed to be equal), A_{web} the total area of the web or webs of the section, and $G_{LT(\text{web})}$ the in-plane shear modulus of the web or webs. The transverse shear rigidity is a function of the shear flow in the flanges and the webs of open, single-cell, and multicell thin-walled sections (Cowper, 1966; Bank, 1987; Bank and Melehan, 1989). Equations for k_{tim} for I- and single-cell box-shaped sections are given in Table 13.2 following the procedure described in Bank (1987) for profiles for which the in-plane properties of the pultruded material are the same in the webs and the flanges. For profiles in which the properties of the pultruded material in the webs and the flanges are not the same, a modified form of the Timoshenko beam theory is required (Bank, 1987).

For conventionally manufactured glass-reinforced pultruded I- and box-shaped profiles it has been shown that the transverse shear rigidity can be approximated by either the full-section shear rigidity defined as AG_b or by the area of the web multiplied by the in-plane shear modulus of the web, $A_{\text{web}} G_{LT(\text{web})}$ (Bank 1989a; Nagaraj and GangaRao, 1997). The full-section shear modulus is obtained from tests on full-section pultruded beams (Bank, 1989b; Roberts and Al-Ubaidi, 2002; Giroux and Shao, 2003).

Analysis Example 13.2: Determination of the Flexural and Shear Rigidities for Pultruded Profiles Determine the value of the flexural rigidity, EI, and the shear rigidity, KAG, for the homogeneous FRP beams shown in Table 13.3. Compare the predictions of the exact theory (i.e., $KAG = K_{\text{tim}} A G_{LT}$) with those of the approximate theory (i.e., $KAG = A_{\text{web}} G_{LT(\text{web})}$).

SOLUTION The values calculated for the profiles considered are displayed in Table 13.4. It can be seen from the table that the difference between the exact and approximate methods of determining the transverse shear rigidity for I-shaped profiles is between 10 and 20%, with the approximate method giving a lower value. (For box-shaped sections, the approximate method will generally give a higher value than that of the exact method.) A lower value of transverse shear stiffness translates into a higher value of predicted shear deflection, which is conservative from a design perspective. The effect of shear deformation increases as the ratio of the flexural stiffness to the transverse shear stiffness increases. It can be seen from the last column in Table 13.4 that the effect of shear deformation will be larger in sections with wider

TABLE 13.2 Timoshenko Shear Coefficient for Homogeneous I and Box Profiles

Profile	Homogeneous Profile	
	Symmetric Geometry (E_L, G_{LT}, ν_L constant and $t_w = t_f$, $b = h$)	Unsymmetric Geometry (E_L, G_{LT}, ν_L constant, $n = b/h$ and $m = t_f b/t_w h$)
I-shaped section	$\dfrac{80}{192 + (\nu_L G_{LT}/E_L)(33)}$	$\dfrac{20(1 + 3m)^2}{(180m^3 + 300m^2 + 144m + 60m^2n^2 + 60mn^2 + 24) + (\nu_L G_{LT}/E_L)(30m^2 + 40mn^2 + 60m^2n^2 + 6m - 4)}$
Box-shaped section	$\dfrac{80}{192 + (\nu_L G_{LT}/E_L)(-12)}$	$\dfrac{20(1 + 3m)^2}{(180m^3 + 300m^2 + 144m + 60m^2n^2 + 60mn^2 + 24) + (\nu_L G_{LT}/E_L)(30m^2 - 50mn^2 - 30m^2n^2 + 6m - 4)}$

Source: Bank (1987).

TABLE 13.3 Geometric and Mechanical Properties of Pultruded Profiles for Example 13.2

Profile[a]	b (in.)	h (in.)	t_f (in.)	t_w (in.)	I_x (in^4)	A (in^2)	A_{web} (in^2)	E_L (Msi)	$G_{LT} = G_{LT(web)}$ (Msi)	ν_{LT}
GFRP WF section $8 \times 8 \times \frac{3}{8}$	8	8	0.375	0.375	99.18	8.72	3.0	2.6	0.425	0.33
GFRP I section $4 \times 8 \times \frac{3}{8}$	4	8	0.375	0.375	55.45	5.72	3.0	2.6	0.425	0.33
GFRP I section $4 \times \frac{1}{2} \times 8 \times \frac{3}{8}$	4	8	0.500	0.375	67.05	6.63	3.0	2.6	0.425	0.33
GFRP box section $4 \times \frac{1}{2} \times 8 \times \frac{3}{8}$	4	8	0.500	0.375	77.80	9.26	6.0	2.6	0.425	0.33
CFRP I section $4 \times 8 \times \frac{3}{8}$	4	8	0.375	0.375	55.45	5.72	3.0	13.7	0.450	0.31

[a] GFRP, properties reported by manufacturers of glass–vinylester conventional pultruded profiles; CFRP, calculated properties of custom pultrusion: three-ply glass CFM (3 × 0.025 in.) + two-ply 50% unidirectional carbon–vinylester (2 × 0.15 in.). Only the first two profiles are conventional off-the-shelf profiles.

TABLE 13.4 Flexural and Transverse Shear Rigidities of Pultruded Beams in Example 13.2

Profile	$E_L I_x$ (Mlb-in^2)	k_{tim}	$k_{\text{tim}} A G_{LT}$ (Mlb)	$A_{\text{web}} G_{LT(\text{web})}$ (Mlb)	Percent diff.	$\frac{E_L I_x}{k_{\text{tim}} A G_{LT}}$
GFRP WF $8 \times 8 \times \frac{3}{8}$	257.9	0.413	1.530	1.257	−17.8	169
GFRP I $4 \times 8 \times \frac{3}{8}$	144.2	0.608	1.478	1.257	−14.9	98
GFRP I $4 \times \frac{1}{2} \times 8 \times \frac{3}{8}$	174.3	0.554	1.561	1.257	−19.5	112
GFRP box $4 \times \frac{1}{2} \times 8 \times \frac{3}{8}$	200.4	0.558	2.196	2.514	+14.5	91
CFRP $4 \times 8 \times \frac{3}{8}$	759.7	0.610	1.570	1.350	−14.0	484

flanges and in profiles made from pultruded materials having higher in-plane anisotropy ratios (i.e., E_L/G_{LT} ratios), such as a carbon fiber–reinforced beam. However, the total deflection in the beam will be significantly less, due to its significantly larger flexural rigidity, EI.

13.6.2 Long-Term Deflection in Pultruded Beams

The moduli of a pultruded profile will decrease over time if the profile is subjected to sustained loads. This is because fiber-reinforced polymeric materials are viscoelastic and creep under sustained loads. Although reinforced polymers do not creep as much as unreinforced polymers, they are nevertheless susceptible to increases in deflection over time without a commensurate increase in load. Reinforced polymers with oriented continuous fibers creep less than those with random reinforcement, as do those with higher volume fractions of fibers (Scott et al., 1995). As shown in Table 1.5, most conventional pultruded profiles have approximately 40% fiber reinforcement by volume, of which about 60 to 70% consists of continuous longitudinal reinforcement. Therefore, pultruded profiles are considered "reasonably" creep resistant, particularly when the sustained loads are small, such as in walkways, platforms, and short bridges. However, where pultruded structures are subjected to a significant amount of sustained load, long-term creep deformation must be considered in design.

To predict the long-term deflection, time-dependent versions of longitudinal modulus and in-plane shear modulus are used in deflection equations. They are referred to as *viscoelastic moduli* and are used instead of the conventional *instantaneous moduli* to predict long-term deflection using the Timoshenko beam theory described previously. The viscoeleastic moduli are identified with a superscript v to distinguish them from the instantaneous moduli that are typically reported for pultruded materials and measured using standard coupon tests. The longitudinal viscoelastic modulus, $E_L^v(t)$, and the in-plane viscoelastic shear modulus, $G_{LT}^v(t)$, are defined as

$$E_L^v(t) = \frac{E_L E_L^t}{E_L^t + E_L t^{n_e}} = \frac{E_L}{1 + (E_L/E_L^t)\, t^{n_e}} \tag{13.12}$$

and

$$G_{LT}^v(t) = \frac{G_{LT} G_{LT}^t}{G_{LT}^t + G_{LT}\, t^{n_g}} = \frac{G_{LT}}{1 + (G_{LT}/G_{LT}^t)\, t^{n_g}} \tag{13.13}$$

where t is time in hours and E_L and G_{LT} are the instantaneous (i.e., the usual) moduli. E_L^t and G_{LT}^t are known as creep moduli and n_e and n_g are the creep rate exponents. The creep moduli and creep rate exponents are constants for a given material type and are obtained from long-term creep tests using the linearized version of Findley's theory (ASCE, 1984). The theory has been applied to determine the creep constants for conventional pultruded sections by a number of authors (Bank and Mosallam, 1992; Mottram, 1993; Scott and Zureick, 1998; Shao and Shanmugam, 2003). The values of the creep moduli and creep rate exponents recommended for design are given in Table 13.5. They were obtained by Mosallam and Bank (1991) for flexural members, and by Scott and Zureick (1998) for compression members (discussed later).

In typical pultruded structures, the long-term sustained stress is usually between 10 and 20% of the ultimate strength of the material. The constants above are intended to be used at these load levels. At higher sustained stress levels the creep moduli show a time dependency and the constant reported in Table 13.5 cannot be used. The creep rate constants appear to be relatively insensitive to stress level (Scott and Zureick, 1998; Shao and Shanmugam, 2003). The reason for the difference in the creep constants for pultruded members in flexure and in pure compression has not been fully explained but is felt to be due to structural effects in flexural members, such as complex stress states at load and support points and the nonuniformity of the stress along the member length, which influences long-term tests. Greater creep

TABLE 13.5 Creep Constants for Use in Design of Pultruded Structural Members

Sustained Loading Type	E_L^t (psi)	n_e	G_{LT}^t (psi)	n_g
Flexure	180×10^6	0.30	27×10^6	0.30
Compression	216×10^6	0.25	NA	NA

deformation is obtained in a pultruded structural member as both the creep rate exponent and the ratio of the instantaneous modulus to the creep modulus increase.

Using instantaneous longitudinal moduli $E_L^0 = 2.6 \times 10^6$ psi and $G_{LT}^0 = 0.425 \times 10^6$ psi for glass–vinylester pultruded profiles, the following expressions are obtained for viscoelastic moduli for flexural members for use in design calculations:

$$E_L^v = \frac{468 \times 10^6}{180 + 2.6t^{0.3}} \quad \text{psi} \tag{13.14}$$

$$G_{LT}^v = \frac{11.475 \times 10^6}{.27 + 0.425t^{0.3}} \quad \text{psi} \tag{13.15}$$

Using these expressions, Table 13.6 shows the values calculated for the viscoleastic longitudinal and shear moduli for these pultruded members under sustained loading durations of 1, 10, 30, and 50 years.

Analysis Example 13.3: Instantaneous and Long-Term Deflection of a Pultruded Beam Under Sustained Load A simply supported WF 8 × 8 × 3/8 glass–vinylester pultruded beam spans 12 ft. Its properties are given in the first row of Table 13.3. The beam has a nominal self-weight of 6.97 lb/ft. It is loaded with a uniformly distributed sustained dead load of 150 lb/ft and a live load of 200 lb/ft. Determine the instantaneous midspan deflection and the midspan deflection after 30 years of sustained load on the beam under self-weight, sustained dead load, and live load. Assume that none of the live load is sustained.

SOLUTION Using the data from Tables 13.4 and 13.5 the following stiffness properties are obtained for the beam for the instantaneous and 30-year long-term deflection calculations.

$$E_L I_x = 257.9 \times 10^6 \text{ lb-in}^2$$

$$k_{\text{tim}} A G_{LT} = 1.530 \times 10^6 \text{ lb}$$

TABLE 13.6 Longitudinal and Shear Moduli of Glass–Vinylester Pultruded Profiles for Various Sustained Load Durations

	Initial	1 Year	10 Years	30 Years	50 Years
E_L^v ($\times 10^6$ psi)	2.60	2.13	1.81	1.62	1.52
G_{LT}^v ($\times 10^6$ psi)	0.425	0.343	0.287	0.255	0.239

$$E_L^v I_x = (1.62 \times 10^6)(99.18) = 160.67 \times 10^6 \text{ lb-in}^2$$

$$k_{\text{tim}} A G_{LT}^v = 0.413(8.72)(0.255 \times 10^6) = 0.918 \times 10^6 \text{ lb}$$

The maximum midspan deflection is given as

$$\delta_{\max} = -\frac{5w\ell^4}{384(EI)} - \frac{w\ell^2}{8(KAG)}$$

The deflections due to the sustained loads and live loads are calculated separately and added. For the live load deflection, the instantaneous moduli are used at all times (according to the Boltzmann superposition principle; ASCE, 1984). Substituting in the deflection equation, the instantaneous deflection is

$$\delta_{\max(sw+\text{DL})} = -\frac{5(156.97/12)(144)^4}{384(257.9 \times 10^6)} - \frac{(156.97/12)(144)^2}{8(1.530 \times 10^6)}$$
$$= -0.284 - 0.0222 = -0.3060 \text{ in.}$$

$$\delta_{\max(\text{LL})} = -\frac{5(200/12)(144)^4}{384(257.9 \times 10^6)} - \frac{(200/12)(144)^2}{8(1.530 \times 10^6)}$$
$$= -0.3618 - 0.0282 = -0.3900 \text{ in.}$$

The total instantaneous midspan deflection is therefore

$$\delta_{\max(\text{instantaneous})} = -0.3060 - 0.3900 = -0.6962 \text{ in.} \qquad \text{which is } \ell/206$$

The 30-year (long-term) deflection is

$$\delta_{\max(sw+\text{DL})} = -\frac{5(156.97/12)(144)^4}{384(160.67 \times 10^6)} - \frac{(156.97/12)(144)^2}{8(0.918 \times 10^6)}$$
$$= -0.4558 - 0.0369 = -0.4927 \text{ in.}$$

The live load deflection calculation is unchanged:

$$\delta_{\max(\text{LL})} = -\frac{5(200/12)(144)^4}{384(257.9 \times 10^6)} - \frac{(200/12)(144)^2}{8(1.530 \times 10^6)}$$
$$= -0.3618 - 0.0282 = -0.3900 \text{ in.}$$

The total long-term (30-year) midspan deflection is therefore

$$\delta_{\max(30\text{ year})} = -0.4927 - 0.3900 = -0.8827 \text{ in.} \qquad \text{which is } \ell/163$$

This is a 26.8% increase in the total deflection and a 61.0% increase in the sustained load deflection over the 30-year period chosen. It can also be observed that the shear deflection for both instantaneous and long-term deflection is approximately 8% of the total deflection in a beam that has the mechanical and geometric properties stipulated.

13.7 ULTIMATE LIMIT STATES

13.7.1 Lateral–Torsional Buckling

The term *lateral–torsional* buckling is used to describe a specific type of physical instability that commonly occurs in open-section (typically, I-shaped) flexural members loaded by transverse loads. When the critical lateral–torsional buckling load is reached, the flanges displace laterally (or sideways) relative to the transverse load direction, and the web twists, causing the entire beam to move out of its vertical plane. Figure 13.6 shows lateral–torsional buckling of a pultruded wide-flange I-shaped profile.

Lateral–torsional buckling of doubly symmetric pultruded profiles has been studied experimentally by a number of researchers (Mottram, 1992; Turvey, 1996; Davalos and Qiao, 1997). It is generally accepted that the well-known equation that is used for isotropic beam sections (see, e.g., Salmon and Johnson, 1996) can be used for conventional pultruded I-shaped profiles provided that the appropriate values of E and G are used in the equations. Inclusion of the effects of shear deformation on lateral–torsional buckling of conven-

Figure 13.6 Lateral–torsional buckling of an I-shaped pultruded profile. (Copyright G. J. Turvey and Y.-S. Zhang, published with permission.)

tional pultruded profiles is small and can generally be neglected, as shown by Roberts (2002).

The critical lateral–torsional buckling stress for homogeneous doubly symmetric open profiles is given as

$$\sigma_{cr}^{lat} = \frac{C_b}{S_x}\sqrt{\frac{\pi^2 E_L I_y G_{LT} J}{(k_f L_b)^2} + \frac{\pi^4 E_L^2 I_y C_\omega}{(k_f L_b)^2 (k_\omega L_b)^2}} \tag{13.16}$$

and the critical lateral–torsional bending moment (for a homogeneous section[10]) is given as

$$M_{cr}^{lat} = \sigma_{cr}^{lat} S_x \tag{13.17}$$

where C_b is a coefficient that accounts for moment variation along the length of the beam, $S_x = I_x/c$ is the section modulus about the strong axis, J is the torsional constant, I_y the second moment about the weak (i.e., vertical) axis, C_ω the warping constant, k_f the effective length coefficient for flexural buckling about the weak axis (i.e., $k_f = 1.0$ for simple supports that permit rotation about the y-axis at the beam ends), k_ω an effective length coefficient for torsional buckling of the section (i.e., $k_\omega = 1.0$ for supports that prevent twisting about the neutral axis but permit the flanges to warp and move in the z-direction at the beam ends), and L_b the unbraced length of the member (or a portion thereof).

The lateral–torsional buckling load depends on the height of the load relative to the neutral axis of the beam (Timoshenko and Gere, 1961; Mottram, 1992). Equation (13.16) assumes that the transverse load acts at the neutral axis of the section in its undeformed configuration. When the load acts on the top flange (above the neutral axis), the beam is more susceptible to lateral–torsional buckling; when the load acts on the bottom flange (below the neutral axis), the beam is less susceptible to lateral–torsional buckling.

For I-shaped profiles, $C_\omega = I_y(d^2/4)$, where d is the depth of the profile. For other common thin-walled sections, values of the warping constant, identified as C_ω, can be found in texts such as Timoshenko and Gere (1961) and Salmon and Johnson (1996). The coefficient C_b can also be found in the aforementioned texts. For the common case of a simply supported uniformly loaded beam with $k_f = k_\omega = 1.0$, $C_b = 1.13$.

The values of k_f and k_ω depend on the connection detailing. k_f is usually assumed to be equal to k_ω in pultruded structures as it is in metallic structures. In typical shear-plate (clip-angle) connections in pultruded structures, little warping restraint of the flanges can be expected and $k_\omega = 1.0$ for simply supported (z-axis) beams. Some degree of rotational restraint may be possible

[10] For a nonhomogeneous beam composite section, properties in the buckling equation and the critical bending moment are found by integrating the axial stress as described previously.

about the y-axis; however, this is also likely to be small in pultruded connections, and $k_f = 1.0$ is recommended for simply supported (z-axis) beams unless rotational restraint can be demonstrated by testing.

A flexural member in a pultruded structure will typically be designed to be braced along its length to prevent lateral–torsional buckling. Often, this bracing is provided continuously along the length of the compression flange of the member by a continuous diaphragm (e.g., a floor plate attached to the top flange). Where continuous support in not provided to the top flange, equation (13.16), should be used to calculate the distance required between lateral braces, L_b, to prevent lateral–torsional buckling.

The torsional resistance of a closed cross section such as a square tube is much larger than that of an open section, and lateral–torsional buckling is not likely to be a critical limit state. To determine the critical lateral–torsional buckling stress for a closed section, the first term only in equation (13.16) can be used (ASCE, 1984). For singly symmetric sections such as channels and angles few experimental data are available for pultruded profiles. Use of appropriate equations for isotropic metallic members is recommended at this time, with substitution of the orthotropic material properties in the isotropic equations (Razzaq et al., 1996).

If the flexural member is subjected to long-term sustained loading, the instantaneous longitudinal and shear moduli in the lateral buckling equation should be replaced by the appropriate viscoelastic moduli. This will reduce the lateral–torsional buckling load for the member and will decrease the design unbraced length. Although such an approach is recommended, no test data exist that have documented the occurrence of creep buckling of pultruded profiles under sustained load.

13.7.2 Local Buckling of Walls Due to In-Plane Compression

Conventional pultruded GFRP profiles are especially susceptible to local buckling under transverse loads due to the low in-plane moduli and the slenderness (width-to-thickness ratio) of the plate elements (known as *walls*) that make up the thin-walled profile. Local buckling in compression flanges of beams has been demonstrated in numerous tests (e.g., Barbero et al., 1991; Bank et al., 1994b; 1996a). Figure 13.7 shows local compression flange buckling in the constant moment region of a transversely loaded wide-flange pultruded profile. If the load is increased beyond the elastic buckling load, the profile will typically fail as a result of separation of the flange from the web due to high transverse tensile stresses (Bank and Yin, 1999), which is followed immediately by in-plane buckling of the then-unsupported web, as shown in Fig. 13.8.

The critical buckling load (or stress) in a wall (or panel) of a profile is a function of the boundary conditions on the longitudinal edges of the wall. In wide-flange I-shaped profiles the flange is particularly susceptible to buckling since the one edge is free while the other edge is elastically restrained at the

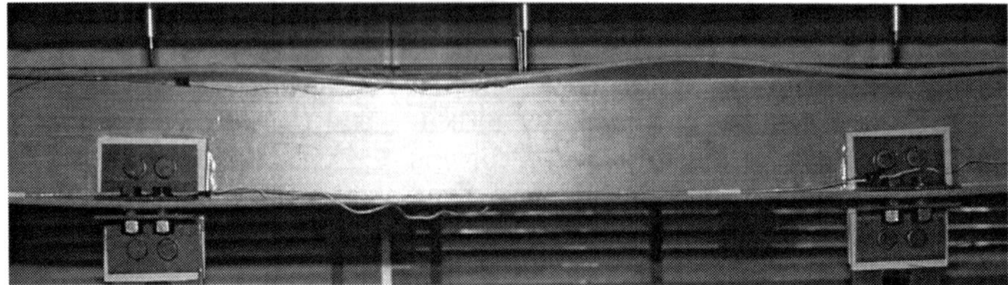

Figure 13.7 Local compression flange buckling in a wide-flange pultruded profile.

web–flange junction. There is no exact closed-form solution for the case of a free and rotationally restrained orthotropic plate. The characteristic transcendental equations required to obtain the buckling load for the plate in terms of the coefficient of edge restraint of the plate are given in Bank and Yin (1996).

For I-shaped profiles, both the *Structural Plastics Design Manual* (ASCE, 1984) and the *Eurocomp Design Code and Handbook* (Eurocomp, 1996) recommend that the elastically restrained edge be assumed to be simply supported. This is known to be an overly conservative assumption. Experimental tests on both pultruded beams and short columns (where local flange buckling also is critical) clearly show that the local buckling stress is significantly higher than that predicted by the free/simply supported plate assumption (Bank et al., 1995a; Qiao et al., 2001; Mottram, 2004). However, it is also well known that the flange buckling load is lower than that predicted by assuming that the restraining web provides a fixed edge condition (Bank et al., 1995a).

An approximate method to obtain closed-form equations for the buckling load for free and rotationally restrained orthotropic plates has been proposed by Kollár (2002). Kollár subsequently extended this work to give closed-form equations for buckling of many different thin-walled sections with orthotropic walls (Kollár, 2003), using his equations together with existing equations in the literature for other boundary conditions. Kollár's results show very good

Figure 13.8 Ultimate failure following local flange buckling.

agreement with finite element analyses and tests of conventional pultruded profiles and reduce to well-known solutions in the literature for plates with special properties. Kollár's equations are presented below for the design of pultruded profiles. Kollár's method has the ability to distinguish between flange buckling in beams subjected to transverse loads (flexure) and flange buckling in columns subjected to axial loads (pure compression) since the web restraint is given as a function of the end conditions of the restraining plate following the procedure presented by Bleich (1952). None of the other existing empirical equations have this ability (Mottram, 2004). Kollár's method also allows the user to determine a priori whether flange buckling or web buckling will control the design based on the slenderness ratios of the flange and web of the section. Kollár has presented results for I-, box-, single-leg-angle-, channel-, and Z-shaped sections. Only the design equations for doubly symmetric I- and box-shaped profiles are presented in this book, due to the fact that at present these profiles are most commonly used as beams and columns in pultruded structures. This includes built-up sections that are I- and box-shaped. The reader is referred to Kollár's work for singly symmetric sections. The cross-sectional notation for I- and box-shaped profiles is shown in Fig. 13.9. Note that the wall widths and depths are measured from the centerline of the walls (shown by the dashed lines in Fig. 13.9).

Since local buckling of the flange and web of a section due to normal loads are interrelated, flange and web buckling are treated together in this section. Local buckling due to shear (tangential) loads is treated separately. It should be noted, however, that only local flange buckling has been observed experimentally in I-shaped profiles since conventional pultruded profiles have aspect ratios that lead to flange buckling occurring before web buckling. Neither has in-plane shear buckling been observed experimentally in beam tests.

To use Kollár's method, first the buckling stresses (or loads) of the walls are found, assuming that they are simply supported at their restrained edges. These buckling stresses are then used to determine which wall buckles first and the coefficient of edge restraint for the critical wall. The final solution

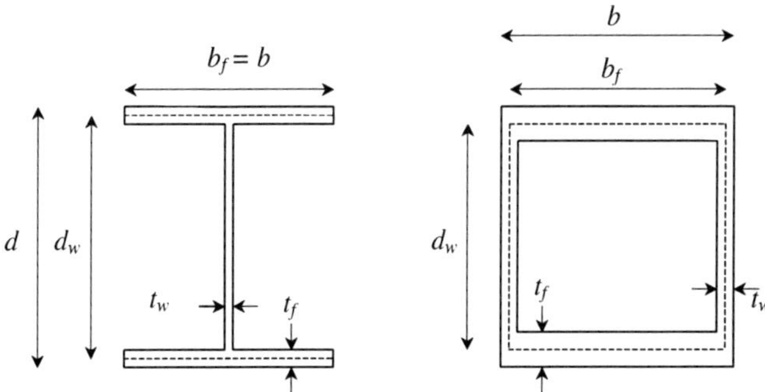

Figure 13.9 Cross-sectional notation for I- and box-shaped profiles.

for the buckling stress for the restrained wall is then given by a closed-form equation that includes the coefficient of edge restraint, ζ, of the critical wall. Three types of simply supported walls[11] are needed for the first part of the solution.

1. *A wall that is free and simply supported on its edges under uniform compressive stress.* This solution is used to determine the critical stress in the flange of an I-shaped profile.[12] It is given in terms of the flexural rigidities[13] of the flange as

$$(\sigma_{\text{free}}^{\text{ss}})_f = \frac{\pi^2}{t_f(b_f/2)^2}\left[D_L\left(\frac{b_f/2}{a}\right)^2 + \frac{12}{\pi^2}D_S\right] \qquad (13.18)$$

   where, t_f the flange thickness, b_f the profile width (twice the nominal width of the flange), and a the length of the flange.[14] For long flanges, as is typically the case, the first term in brackets is negligible and the critical stress can be found from

$$(\sigma_{\text{free}}^{\text{ss}})_f = \frac{4t_f^2}{b_f^2}G_{LT} \qquad (13.19)$$

   In this mode the plate buckles with one buckle half-wavelength, a, equal to the plate length (like a simply supported Euler column).
2. *A wall that is simply supported along both edges under uniform compressive stress.* This solution is used to determine the critical stress in the flange of a box section. It is given in terms of the flexural rigidities of the flange as

$$(\sigma_{\text{ss}}^{\text{ss}})_f = \frac{2\pi^2}{t_f b_f^2}(\sqrt{D_L D_T} + D_{LT} + 2D_S) \qquad (13.20)$$

   The length of the buckle half-wavelength for this mode is given by

[11] For all cases it is assumed that the loaded edges are simply supported. The super- and subscripts are used to identify the longitudinal unloaded edges of the wall.

[12] A different notation from that used in Kollár (2003) is used in what follows, to make the equations more familiar to structural engineers unfamiliar with laminated plate and composite material notation. Equations are given for plate buckling stresses and not plate buckling loads, for the same reason.

[13] See Chapter 12 for a definition of plate flexural rigidities.

[14] Note that in many texts the parameter b is used to denote the width of the flange only and is equal to one-half of b_f.

$$(a_{ss}^{ss})_f = b_f \sqrt[4]{\frac{E_L}{E_T}} \tag{13.21}$$

3. *A wall that is simply supported along both edges under linearly varying compressive stress.* This solution is used to determine the critical stress in the web of an I or box section. It is given in terms of the flexural rigidities of the web as

$$(\sigma_{ss}^{ss})_w = \frac{\pi^2}{t_w d_w^2}(13.9\sqrt{D_L D_T} + 11.1D_{LT} + 22.2D_S) \tag{13.22}$$

where t_w is the thickness of the web and d_w is the depth of the web. The length of the buckle half-wavelength for this mode is given by

$$(a_{ss}^{ss})_w = 0.707d_w \sqrt[4]{\frac{E_L}{E_T}} \tag{13.23}$$

Coefficient of Restraint A nondimensional coefficient known as the coefficient of restraint (Bleich, 1952), ζ, is defined to account for the combined effect of the rotational stiffness of the junction itself and the plate geometric and mechanical properties on the critical buckling stress.[15] It is defined as

$$\zeta = \frac{D_T}{kL_T} \tag{13.24}$$

where k is the rotational spring constant of the junction between the walls and L_T and D_T are the width and flexural rigidity of the plate perpendicular to the edge being restrained (i.e., in the transverse direction of the orthotropic plate, hence the subscript T).

Local Buckling Equations for an I-Shaped Profile in Bending About Its Major Axis If $(\sigma_{free}^{ss})_f/(E_L)_f < (\sigma_{ss}^{ss})_w/(E_L)_w$, the flange will buckle before the web. The flange will therefore be "restrained" by the web. Except in very unusual circumstances, this will be the local buckling mode for conventional pultruded GFRP profiles. The spring constant is given as

[15] That is, even if the junction has a relatively high rotational spring stiffness but the plate is very flexible and wide (small D_T and large L_T), the restraining effect of the junction on the buckling load will not be very significant.

$$k_{\text{I-flange}} = \frac{2(D_T)_w}{d_w}\left[1 - \frac{(\sigma_{\text{free}}^{\text{ss}})_f(E_L)_w}{(\sigma_{SS}^{SS})_w(E_L)_f}\right] \tag{13.25}$$

and the local buckling stress for the rotationally restrained flange is given in terms of the properties of the flange as[16]

$$\sigma_{\text{cr}}^{\text{local,I-flange}} = \frac{1}{(b_f/2)^2 t_f}\left(7\sqrt{\frac{D_L D_T}{1 + 4.12\zeta_{\text{I-flange}}}} + 12D_S\right) \tag{13.26}$$

where

$$\zeta_{\text{I-flange}} = \frac{(D_T)_f}{k_{\text{I-flange}}(b_f/2)} = \frac{(E_T)_f t_f^3}{6k_{\text{I-flange}} b_f[1 - (\nu_T)_f(\nu_L)_f]} \tag{13.27}$$

The critical bending moment for local flange buckling for a homogeneous I-shaped section for bending about the major axis is given as

$$M_{\text{cr}}^{\text{local,I-flange}} = \sigma_{\text{cr}}^{\text{local,I-flange}} S_x \tag{13.28}$$

The buckle half-wavelength, a, for this case is given in terms of the properties of the flange as (Kollar and Springer, 2003, Table 4.10)

$$a_{\text{I-flange}} = 1.675\frac{b_f}{2}\sqrt[4]{\frac{D_L}{D_T}(1 + 4.12\zeta_{\text{I-flange}})} \tag{13.29}$$

The length of the buckle half-wavelength may be used to determine the required spacing of bracing members to prevent local flange buckling in I-shaped members. The length of buckle half-wavelength is typically between 1.0 to 2.0 times the flange width, b_f, for conventional FRP pultruded profiles. This implies that to brace the section against local buckling, the compression flange needs to be supported (or restrained) continuously by a flooring or wall system. The spacing of the fasteners used to attach the flooring system to the compression flange must be less than the buckle half-wavelength, and the flooring system must be stiff enough, and attached firmly enough, to prevent the local instability of the flange.

If $(\sigma_{\text{free}}^{\text{ss}})_f/(E_L)_f > (\sigma_{\text{ss}}^{\text{ss}})_w/(E_L)_w$, the web will buckle before the flange. A closed-form expression to predict the buckling stress of the web when re-

[16]Kollár (2003) provides an alternative, more detailed equation for greater accuracy for highly orthotropic walls. The equations presented in this book are suitable for the design of conventional pultruded profiles.

strained by the flanges and loaded with a linearly varying axial stress is not currently available. Equation (13.22), for a wall simply supported on its two edges and loaded with a linearly varying axial stress, can be used as a reasonable approximation, (i.e., $(\sigma_{ss}^{ss})_w = \sigma_{cr}^{local,I\text{-}web}$) as the flanges will not provide significant rotational restraint to the web.

The critical bending moment for local web buckling for a homogeneous I-shaped section for bending about the major axis is given as

$$M_{cr}^{local,I\text{-}web} = \sigma_{cr}^{local,I\text{-}web} \frac{2I_x}{d - 2t_f} \tag{13.30}$$

Bending of an I-shaped profile about its minor axis is not treated in this book since a conventional pultruded I-shaped profile will generally be used in bending about its major axis because of serviceability deflection requirements. When used as a beam-column (discussed in Chapter 14), it will be assumed that the beam is also bent about its major axis and braced about its minor axis. Local buckling of the flanges of an I-shaped profile in bending about its minor axis can be treated in a fashion similar to local buckling of the leg of a single-angle section (presented in Kollár, 2003).

Local Buckling Equations for a Rectangular Box Profile in Bending If $(\sigma_{ss}^{ss})_f/(E_L)_f < (\sigma_{ss}^{ss})_w/(E_L)_w$, the compression flange will buckle before one of the webs. The flange will therefore be restrained by both webs, one on either longitudinal edge of the flange. The spring constant is then given as

$$k_{\text{box-flange}} = \frac{4(D_T)_w}{d_w}\left[1 - \frac{(\sigma_{ss}^{ss})_f(E_L)_w}{(\sigma_{ss}^{ss})_w(E_L)_f}\right] \tag{13.31}$$

and the local buckling stress for the rotationally restrained flange is given in terms of the properties of the flange as

$$\sigma_{cr}^{local,box\text{-}flange} = \frac{\pi^2}{b_f^2 t_f}\left[2\sqrt{(D_L D_T)}(1 + 4.139\xi) + (D_{LT} + 2D_S)(2 + 0.62\xi_{\text{box-flange}}{}^2)\right] \tag{13.32}$$

where

$$\xi_{\text{box-flange}} = \frac{1}{1 + 10\zeta_{\text{box-flange}}} = \frac{1}{1 + 10\,[(D_T)_f/k_{\text{box-flange}} b_f]} \tag{13.33}$$

The critical bending moment for local flange buckling for a homogeneous box-shaped section for bending about the major axis is given as

$$M_{\text{cr}}^{\text{local,box-flange}} = \sigma_{\text{cr}}^{\text{local,box-flange}} \frac{2I_x}{d} \tag{13.34}$$

The buckle half-wavelength, a, for this case is given in terms of the properties of the flange as (Kollár and Springer, 2003, Table 4.10)

$$a_{\text{box-flange}} = b_f \sqrt[4]{\frac{D_L}{D_T}\left(\frac{1}{1 + 4.139\xi_{\text{box-flange}}}\right)} \tag{13.35}$$

If $(\sigma_{\text{ss}}^{\text{ss}})_f/(E_L)_f > (\sigma_{\text{ss}}^{\text{ss}})_w/(E_L)_w$, the webs will buckle before the flange. A closed-form expression to predict the buckling stress of the web when restrained by the top and bottom flanges and loaded with a linearly varying axial stress is not currently available. Equation (13.22), for a web simply supported on its two edges, can be used as a conservative approximation (i.e. $(\sigma_{\text{ss}})_w = \sigma_{\text{cr}}^{\text{local,box-web}}$). For conventional pultruded GFRP square-box beams, the flange will buckle before the web. Only in the case of rectangular box profiles with large depth-to-width ratios (>2) will the web possibly buckle before the flange.

The critical bending moment for local web buckling for a homogeneous box-shaped section for bending about the major axis is given as

$$M_{\text{cr}}^{\text{local,box-web}} = \sigma_{\text{cr}}^{\text{local,box-web}} \frac{2I_x}{d - 2t_f} \tag{13.36}$$

Approximate Procedure for Local Buckling of I-Shaped Profiles To use the equations due to Kollár (2003) presented previously, all of the in-plane properties of the orthotropic walls of the section are required to predict the buckling stress. However, the in-plane shear modulus is often not reported by manufacturers and is not required to be reported by the European Standard EN 13706 (CEN, 2002a). In this situation, the simplified form suggested in the *Structural Plastics Design Manual* (ASCE, 1984, p. 676) for obtaining the critical flange buckling stress for the free and rotationally restrained flange in an I-profile can be used as an approximation (Mottram, 2004):

$$\sigma_{\text{cr,approx}}^{\text{local,I-flange}} = \frac{\pi^2 t_f^2}{(b_f/2)^2}\left[\left(0.45 + \frac{b_f^2}{4a^2}\right)\frac{\sqrt{E_L E_T}}{12(1 - \nu_L \nu_T)}\right] \tag{13.37}$$

where the mechanical properties are those of the flange. This expression requires the length of the buckle half-wavelength, a, which is not known a priori. To use this equation, the buckle half-wavelength can be taken conservatively as $2b_f$ for wide-flange and $3b_f$ for narrow-flange conventional pultruded I-shaped profiles. Alternatively, the first term in brackets can be taken even more conservatively as 0.45.

The approximate critical bending moment for the local flange buckling of a homogeneous I-shaped profile loaded in flexure is therefore

$$M_{cr,approx}^{\text{local,I-flange}} = \sigma_{\text{cr,approx}}^{\text{local,I-flange}} \frac{2I_x}{d} \qquad (13.38)$$

Further Discussion of Flange Buckling in I-Sections In many designs of I-shaped profiles, particularly those with wide flanges, the local buckling stress in the flange will control the design and the ultimate compressive or tensile strength of the material will not be realized such that the pultruded material will be underutilized (from a strength and therefore economic perspective). In standard steel profiles, flange width-to-thickness slenderness ratios are used to ensure that local flange buckling does not occur prior to the material in the flange reaching its yield strength (or its fully plastic moment capacity, known as compact sections in steel ASD terminology) (Salmon and Johnson, 1996). This convenient method is not used in the design of conventional pultruded profiles for the following primary reason: Conventional pultruded profiles are manufactured in far, far fewer (by two orders of magnitude) sizes than are steel profiles. The typical dimensions of these profiles in the United States were established in the late 1970s and early 1980s and were controlled by the flexural stiffness, *EI*, of the section and the need to prevent excessive deflections. For this reason, wide-flange profiles with thin walls were designed. The designs were also a function of the pultrusion processing technology of that time, which made pultruding shapes with variable cross-sectional thickness difficult. For this reason, large-radius fillets at the web and flange junctions were not produced. Although many of these processing obstacles have been overcome in recent years, the manufacturers have a large inventory of pultrusion dies and continue to produce shapes with these existing dies. Some manufactures have attempted to improve the properties of their shapes by using different layups in the flanges and webs of their sections (e.g., SuperStructural sections from Creative Pultrusions). However, in the United States the geometric shapes have not changed much since the 1980s. I-shaped profiles currently produced by Fiberline Composites (Denmark) have narrow flanges which in some cases are thicker than the webs, in order to increase the local buckling resistance of the profiles. Nevertheless, designers should check all I-shaped sections for local buckling. For this reason, designers of pultruded structures must contend with local buckling and cannot simply choose compact shapes from conventional sections.

In addition to the above, it is not really desirable to develop a profile that fails before local buckling of the walls, for the following reasons:

1. Pultruded materials are linear elastic in the roving direction and neither a yield (plastic) moment or a fully plastic moment exist for pultruded profiles. This means that postbuckling capacity cannot be developed. It

has been demonstrated that when elastic local instability occurs, the profile is very close to its ultimate capacity (Bank et al., 1994b). Therefore, a compact section designation is not applicable. A partially compact designation may be possible (Salmon and Johnson, 1996).

2. The material strength (yield for steel, ultimate for pultruded materials)-to-stiffness ratio for typical GFRP pultruded materials is much higher than for steel profiles (30/2,600 versus 50/29,000), meaning that the flanges would have to be either very thick or very narrow to prevent local buckling prior to ultimate material failure. This would make the profiles difficult to manufacture.

To increase the local flange buckling capacity of an I-shaped profile, several details can be used:

1. The flange stiffness can be increased by bonding a higher stiffness plate to the profile. This is the same philosophy as welding a steel cover plate to a steel section (or FRP strengthening of a concrete beam). Care must be taken to ensure that the cover plate does not delaminate from the flange of the profile when subjected to compressive stress (leading to delamination buckling, as it is known). Composite section mechanics can be used to obtain the properties of the built-up section (and built-up web) in this case.
2. The rotational stiffness of the junction can be increased by bonding a small single-leg angle into the fillet region of the profile, as shown in Fig. 13.10. In this case, the mechanics analysis is more complicated and testing is recommended.

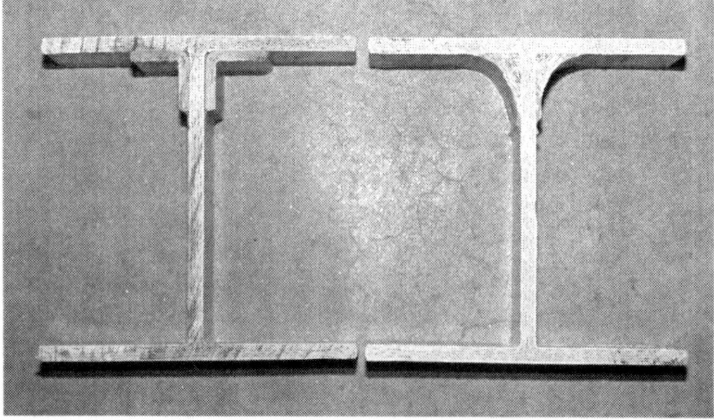

Figure 13.10 Details to increase the rotational stiffness of a web–flange junction.

13.7.3 Local Buckling of Walls Due to In-Plane Shear

The vertical web of an I- or box-shaped profile can buckle in shear at locations of high shear forces that typically occur at concentrated load points or near support locations. The equation for the buckling of an orthotropic plate in pure shear is a function of the restraint provided by the flanges, the aspect ratio, and the orthotropy ratio (E_L/E_T). Experimental evidence of this buckling mode in conventional pultruded GFRP beams has not been reported in the literature. The critical shear stress for local shear buckling of an orthotropic web is given as

$$\tau_{\text{cr}}^{\text{local}} = \frac{4\, k_{LT}\sqrt[4]{D_L D_T^3}}{t_w d_w^2} \tag{13.39}$$

The critical shear force for local web buckling of a homogeneous beam,

$$V_{\text{cr}}^{\text{local}} = \tau_{\text{cr}}^{\text{local}} \frac{I_x t_w}{Q_{\text{max}}} \tag{13.40}$$

may be approximated for an I-shaped profile as

$$V_{\text{cr}}^{\text{local}} \approx \tau_{\text{cr}}^{\text{local}} A_{\text{web}} = \tau_{\text{cr}}^{\text{local}}\, t_w d \tag{13.41}$$

where d_w is the depth of the web, d the nominal depth of the profile, t_w the thickness of the web, and k_{LT} a shear buckling coefficient given in charts in the *Structural Plastics Design Manual* (ASCE, 1984, p. 682) or in an approximate closed-form by Kollár (2003) as

$$k_{LT} = 8.125 + 5.045K \qquad \text{for } K \leq 1 \tag{13.42}$$

where

$$K = \frac{2D_S + D_{LT}}{\sqrt{D_L D_T}} = \frac{2G_{LT}(1 - \nu_L \nu_T) + \nu_T E_L}{\sqrt{E_L E_T}} \tag{13.43}$$

Kollár also presents equations for $K > 1$; however, for the orthotropic materials in conventional GFRP pultruded profiles, $K < 1$, and these equations are not presented here. For isotropic materials, $K = 1$.

When the beam is subjected simultaneously to high shear forces and high bending moments,[17] the web is subjected to combined in-plane shear stress, τ, and in-plane axial compressive (flexural) stress, σ_z, the critical web buck-

[17] Such as near the interior support of a continuous beam.

ling stress may be reduced. The *Structural Plastics Design Manual* (ASCE, 1984, p. 667) recommends using a second-order interaction equation taken from isotropic plate theory to determine the critical transverse load on the beam, given in terms of stresses as

$$\left(\frac{\sigma_z}{\sigma_{cr}^{local}}\right)^2 + \left(\frac{\tau}{\tau_{cr}^{local}}\right)^2 \leq 1.0 \tag{13.44}$$

or in terms of forces as

$$\left(\frac{M_x}{M_{cr}^{local}}\right)^2 + \left(\frac{V_y}{V_{cr}^{local}}\right)^2 \leq 1.0 \tag{13.45}$$

13.7.4 Web Crushing and Web Buckling in the Transverse Direction

Webs of pultruded beams are particularly susceptible to local failure at concentrated loads or reactions, due to the relatively low transverse compressive strength and stiffness of the web. This is often referred to as *web resistance to transverse forces.* Bearing plates at load and reaction points should always be placed directly over (or under) the web of a section and not on outstanding cantilevered flanges. Due to processing, pultruded sections tend to have slightly concave flanges (i.e., "curled" inward) such that load is transferred directly to the web or webs in a concentrated fashion (Mottram, 1991). Local failure and deformation at a roller support of a wide-flange I-section is shown in Fig. 13.11.

To design for web crushing[18] failure, the critical crushing stress is assumed to be equal to the transverse compressive strength of the material in the web,

$$(\sigma_y)_{cr}^{crush} = \sigma_{T,c} \tag{13.46}$$

and the critical crushing force is

$$F_{cr}^{crush} = (\sigma_y)_{cr}^{crush} A_{eff} \tag{13.47}$$

In addition to the web being susceptible to crushing at locations of local concentrated forces, the web may also buckle in the vertical plane as if it were a wide but slender column. In this case it is assumed that the web acts as a plate simply supported on its unloaded edges (parallel to the load direction) in the transverse direction with the load applied over an effective width, b_{eff}. The equation to predict the critical local buckling stress is the same as

[18]This is sometimes referred to as *crippling failure.* Since the web does not yield prior to the failure in local compression, the term *crushing* is preferred.

Figure 13.11 Local bearing failure at a simple support.

that used for the wall simply supported on its two longitudinal edges presented in Eq. (13.20); however, in this case the plate is oriented in the transverse direction of the web. In terms of the effective width (along the beam length), it is written as

$$(\sigma_y)_{cr}^{local} = \frac{2\pi^2}{t_w b_{eff}^2}(\sqrt{D_L D_T} + D_{LT} + 2D_S) \tag{13.48}$$

and the critical transverse buckling load is

$$F_{cr}^{local} = (\sigma_y)_{cr}^{local} A_{eff} \tag{13.49}$$

In the equations above, b_{eff} is taken as the web depth, d_w, or the distance between the vertical web stiffeners, for point loads. At simply supported beam ends, b_{eff} is taken as equal to the length of bearing plus half the web depth, $L_{eff} + d_w/2$ (ASCE, 1984, p. 724).

13.7.5 Additional Factors Affecting Local Buckling in Pultruded Profiles

Since the pultruded material is viscoelastic and will creep, local buckling of the slender walls of a pultruded profile can be affected by reduction in the moduli of the thin walls due to sustained axial (flexural) and shear stresses.

Since the flanges are the most highly stressed portions of the profile under flexural loads, where high levels of sustained stresses occur, viscoelastic moduli can be used to determine the critical buckling loads under sustained stresses as a function of time. Since the sustained stress is primarily in the longitudinal direction of the flange, the *Structural Plastics Design Manual* (ASCE, 1984), following Bleich (1952, p. 343), suggests[19] multiplying by $\sqrt{E_L^v/E_L}$ the critical buckling stress obtained using the instantaneous elastic constants. For example, considering the properties in Table 13.6, after 30 years of sustained load on the flange the critical buckling stress would be 79% of the instantaneous buckling stress calculated (i.e., $\sqrt{1.62/2.6} = 0.79$).

Local buckling may also be influenced by shear deformation effects in the same manner as shear deformation affects global deformation. A higher-order shear-deformable plate theory is required to evaluate such phenomena; however, they are expected to be small in conventional GFRP pultruded profiles since the effect of shear deformation on global buckling of pultruded profiles has been shown to be small (less than 5% in most cases) (Roberts, 2002).

Shear lag in very wide flanges of I-shaped and rectangular box profiles may reduce the axial stress and increase the in-plane shear stress carried by the flanges of the profile in bending. This is not expected to be a significant source of error in the design of conventional pultruded GFRP profiles. In experiments on beams, very little change in the axial strain across the width of the flanges has been observed.

In some situations, coupling between local wall buckling and global lateral–torsional buckling of I-shaped profiles, referred to as *distortional buckling,* may occur (Davalos and Qiao, 1999) in a similar fashion to that of steel I-shaped sections (Hancock, 1978). Design guidance is not provided for this coupled buckling mode, as most conventional pultruded beams will be braced against lateral–torsional buckling, and therefore local buckling criteria will govern.

Web Stiffeners Vertical web stiffeners can be used to increase the critical buckling stress in a web for in-plane shear stresses, in-plane flexural (compressive) stresses, and transverse compression stresses at supports and load points. Based on isotropic material assumptions, the *Structural Plastics Design Manual* (ASCE, 1984) recommends that the minimum flexural rigidity, $(EI)_{\text{stiffener}}$ of a vertical web stiffener about the plane of the web be taken as

$$(EI)_{\text{stiffener}} = 0.34\,\frac{(E_L)_s d_w^4 t_w^3}{b_s^3} \tag{13.50}$$

[19]Note that Eq. 6.73 in the SPDM is not correct, as it will lead not to smaller buckling loads, as would be expected, but to larger loads, due to viscoelastic effects.

where b_s is the spacing of the stiffeners along the length of the beam and $(E_L)_s$ is the longitudinal modulus of the stiffener in the vertical direction, which must be equal to or larger than the longitudinal modulus of the pultruded material in the web, $(E_L)_w$. Single-leg angles and small square tubes are commonly used as web stiffeners. They should be bonded or bolted to the web and extended to the compression flange of the section wherever possible (see Chapter 15 for a detailed discussion of connection issues).

Bearing Stiffeners At the location of concentrated loads and vertical supports, bearing stiffeners should be used to stabilize the section and distribute the loads into the profile at the location of the loads. Without such stiffeners, severe local deformation can occur at locations of concentrated loads (see the support point in Fig. 13.6). These stiffeners should bear on the inside surface of the loaded flange and on the inside surface of the opposite flange. If possible, they should also be attached to the web of the section. In I-shaped profiles, bearing stiffeners should be placed symmetrically on both sides of the web. Figure 13.12 shows vertical bearing stiffeners of this type at a simple support. Vertical bearing stiffeners can be designed as pinned–pinned columns with height d_w, subjected to one-half the applied load at the point, according to the procedures detailed in Chapter 14.

13.7.6 Flange and Web Longitudinal Material Failure

For pultruded profiles where the slenderness ratio of the flange, $0.5b_f/t_f$, and the slenderness ratio of the web, d_w/t_w, are small and local buckling does not

Figure 13.12 Vertical bearing stiffeners at a simple support.

occur (or is prevented by multiple longitudinal and transverse stiffeners), the flexural member may fail due to compressive crushing or tensile rupture of the pultruded material due to flexural stresses. To design for flange or web material compressive failure, the critical compressive stress is taken as the longitudinal ultimate compressive strength of the pultruded material in the web or the flange,

$$\sigma_{\text{cr}}^{\text{comp}} = \sigma_{L,c} \tag{13.51}$$

The critical bending moment for a homogeneous profile due to compressive material failure is

$$M_{\text{cr}}^{\text{comp}} = \sigma_{\text{cr}}^{\text{comp}} \frac{I_x}{y} \tag{13.52}$$

where y is the distance to the point of interest from the neutral axis.

To design for flange or web material tensile failure, the critical tensile stress is taken as the longitudinal ultimate tensile strength of the pultruded material in the web or the flange,

$$\sigma_{\text{cr}}^{\text{ten}} = \sigma_{L,t} \tag{13.53}$$

The critical bending moment for a homogeneous profile due tensile material failure is

$$M_{\text{cr}}^{\text{ten}} = \sigma_{\text{cr}}^{\text{ten}} \frac{I_x}{y} \tag{13.54}$$

where y is the distance to the point of interest from the neutral axis.

13.7.7 Flange and Web Material Shear Failure

To design for flange or web material in-plane shear failure, the critical shear stress in the flanges and webs of a pultruded profile is taken as the ultimate in-plane shear strength of the pultruded material:

$$\tau_{\text{cr}}^{\text{shear}} = \tau_{LT} \tag{13.55}$$

The critical shear force for a homogeneous profile due to web material failure is

$$V_{cr}^{local} = \tau_{cr}^{local} \frac{I_x t}{Q} \approx \tau_{cr}^{local} A_{web} \tag{13.56}$$

where t is the thickness of the flange or web at the point of interest and Q is the first moment of area at the point of interest.

Manufacturers often do not report the in-plane shear strength of pultruded profiles. However, they usually report the out-of-plane (or interlaminar) shear strength, τ_{TT}. If necessary, the out-of-plane shear strength can be used in place of the in-plane shear strength, although this will lead to overly conservative designs for conventional pultruded profiles.

Unlike in the case of buckling stresses, interaction between in-plane normal stresses and shear stresses is not generally considered for pultruded profiles. Since the maximum stress failure criterion[20] is typically used to determine material failure in pultruded materials, no interaction is assumed between the effects of normal and shear stress on material failure, and each condition is considered separately. More sophisticated interactive failure criteria, such as the Tsai-Wu and Tsai-Halpin criteria, are seldom, if ever, used in the design of pultruded profiles. In addition, the principal stresses are not calculated. Since the material is orthotropic, the value of the principal normal and maximum shear stresses at a point are not particularly relevant. In orthotropic materials, the stresses parallel and perpendicular to the major axes of orthotropy are those that control in design.

13.8 DESIGN PROCEDURE FOR FLEXURAL MEMBERS

The design procedure presented here for flexural members permits design by the allowable stress design (ASD) basis or the load and resistance factor design (LRFD) basis, as discussed in Chapter 12. Both of these bases use the analytical, mechanics-based equations presented in preceding sections. They do not use the empirical equations presented in the U.S. pultrusion manufacturers' design guides mentioned in Chapter 12. Consequently, the load tables presented in manufacturers' design guides are not used in this book to size conventional pultruded profiles. However, the empirical equations presented in the guides, and hence the load tables, can be used to estimate the size or the profile, or its load-carrying capacity, if necessary for the purposes of a preliminary design.

Step 1. Determine the design loads, choose the design basis, and determine the basis factors. The distributed and concentrated loads on flexural members and their types are determined from the structural geometry and load-

[20] Discussed in Chapter 3.

ing requirements. The support conditions for the beam are established. The design basis for the flexural member is chosen and the appropriate safety factors (ASD) and resistance factors (LRFD) are determined. Load factors for the LRFD method are listed. Maximum short- and long-term deflection criteria are taken from codes or established by the project specifications.

Step 2. Select a profile type and select a trial size for serviceability. Using the maximum allowable deflection criteria, select a profile and size to meet this criterion (for the short and long terms). Neglect shear deformation for the first trial. Check the chosen trial profile and size with the shear deformation included. Service loads are used for both the ASD and LRFD procedures.

Step 3. Determine the maximum design stresses or forces. The maximum flexural, shear, and bearing stresses on a member subjected to the nominal loads are calculated for the ASD procedure. The maximum bending moment, shear force, and reactions on the member subjected to the factored loads are calculated for the LRFD procedure.

Step 4. Determine the critical stresses or forces. For the profile chosen, determine (or calculate) the lateral–torsional buckling, local buckling (normal and shear), web buckling, web crippling, compressive strength, tensile strength, and shear strength critical stresses (ASD) or forces (LRFD).

Step 5. Determine the factored critical stresses or forces. For ASD, divide the critical stresses determined in step 4 by the appropriate safety factors for flexural, compressive, or tensile stresses. For LRFD, multiply the critical forces by the appropriate resistance factors for flexural members.

Step 6. Check the ultimate strength or capacity of the profile selected. For ASD, check that the design stresses are less than the allowable stresses. For LRFD, check that the factored design forces are less than the factored resistances. Return to step 3 if the trail size does not work.

Step 7. Dimension the ancillary bracing, stiffeners, and bearing plates. Dimension (or design if quantitative procedures are available) braces, stiffeners, and bearing plates for the flexural member. Provide a sketch of the beam showing all ancillary elements, dimensions, and important construction notes. (If the beam is connected to other framing members, the connection should be dimensioned and detailed following the design procedure described in Chapter 15.)

Design Example 13.4: Pultruded Beam Design A conventional commercially produced pultruded E-glass/vinylester wide-flange I-shaped profile is to be used as a simply supported interior floor beam of a building structure with a typical repeating floor plan as shown in Fig. 13.13. Pultruded gratings attached with grating clips are to be used as the flooring system. Assume that the flooring system does not provide any restraint against global lateral–torsional buckling or local compression flange buckling. The floor dead load is 20 psf and the floor live load is 40 psf. Live load reduction is not permitted.

Figure 13.13 Typical repeating floor plan.

TABLE 13.7 Pultruded Material Properties in Flange and Web

Symbol	Description	Value[a]
E_L^c	Longitudinal compressive modulus	2.6×10^6 psi
E_L^t	Longitudinal tensile modulus	2.6×10^6 psi
E_T^c	Transverse compressive modulus	1.0×10^6 psi
E_T^t	Transverse tensile modulus	0.8×10^6 psi
G_{LT}	In-plane shear modulus	NR
ν_L	Major (longitudinal) Poisson ratio	0.33
$\sigma_{L,c}$	Longitudinal compressive strength	30,000 psi
$\sigma_{L,t}$	Longitudinal tensile strength	30,000 psi
$\sigma_{T,c}$	Transverse compressive strength	16,000 psi
$\sigma_{T,t}$	Transverse tensile strength	7,000 psi
τ_{TT}	Interlaminar shear strength	4,500 psi
τ_{LT}	In-plane shear strength	NR
E_b	Full-section flexural modulus	2.5×10^6 psi
G_b	Full-section shear modulus	0.425×10^6 psi

[a] NR, not reported.

The beams are to be designed for a maximum short-term deflection of $L/240$. The beams are loaded on their top flanges and are connected to the columns with web clip angles[21] (simple shear connections). The material properties for a conventional pultruded glass FRP profile taken from a manufacturer design guide (e.g., Strongwell, Bedford) are listed in Table 13.7.

SOLUTION

Step 1. Determine the design loads, choose the design basis, and determine the basis factors.

Design loads and internal forces For the given tributary width of a typical beam of 6 ft, the beam loads are calculated as follows. For ASD and LRFD serviceability limit state calculations:

$$\text{Dead load:} \quad w_{\text{DL}} = (20 \text{ psf} \times 6 \text{ ft}) + 10 \text{ lb/ft} = 130 \text{ lb/ft}$$

(assuming a beam self-weight of 10 lb/ft)

$$\text{Live load:} \quad w_{\text{LL}} = (40 \text{ psf} \times 6 \text{ ft}) = 240 \text{ lb/ft}$$

$$\text{Total service load:} \quad 130 + 240 = 370 \text{ lb/ft}$$

For LRFD design the ASCE 7 load factors for dead and live loads are 1.2 and 1.6, respectively, and the factored loads are

$$\text{Dead load:} \quad W_{\text{D}} = 1.2(130) = 156 \text{ lb/ft}$$

$$\text{Live load:} \quad W_{\text{L}} = 1.6(240) = 384 \text{ lb/ft}$$

$$\text{Total factored load:} \quad W_u = 156 + 384 = 540 \text{ lb/ft}$$

ASD and LRFD factors The ASD safety factors to be used are taken from Chapter 12:

$$\text{flexure} = 2.5$$

$$\text{compression} = 3.0$$

$$\text{shear} = 3.0$$

The LRFD resistance factors are determined using the Eurocomp material factors (Chapter 12), as follows:

[21] The design of the columns and clip angle connections are treated in Chapters 14 and 15, respectively.

$\gamma_{m,1}$ = 1.15 (material properties obtained from test data)

$\gamma_{m,2}$ = 1.1 (fully cured pultruded material)

$\gamma_{m,3}$ = 1.0 (T_g > 80° and room-temperature service conditions)

No other environmental or long-term reduction factors are assumed for this problem.

Therefore,

$$\gamma_m = \gamma_{m,1}\gamma_{m,s}\gamma_{m,3} = 1.15(1.1)(1.0) = 1.265 \leq 1.5$$

Therefore, γ_m = 1.5 (minimum value permitted by Eurocomp for ultimate limit state).

The LRFD resistance factor is taken as the inverse of the Eurocomp material factor:

$$\phi = \frac{1}{\gamma_m} = \frac{1}{1.5} = 0.67$$

For the serviceability limit state, the material factor is taken as 1.0 and the resistance factor is therefore also 1.0.

Step 2. Select a profile type and select a trial size for serviceability. The maximum allowable midspan deflection of the beam is

$$\delta_{\max} = \frac{L}{240} = \frac{15 \times 12}{240} = 0.75 \text{ in.}$$

The minimum required second moment of the section to limit the deflection to this is

$$I_{\text{reqd}} = \frac{5}{384}\frac{wL^4}{E\delta_{\max}} = \frac{5}{384}\frac{(370/12)(15 \times 12)^4}{(2.6 \times 10^6)(0.75)} = 216.13 \text{ in}^4$$

Using U.S. manufacturers' conventional wide-flange profiles, choose a W 10 × 10 × $\frac{1}{2}$ profile with the geometric properties given in Table 13.8.

TABLE 13.8 Geometric Properties of Conventional Glass FRP W 10 × 10 × $\frac{1}{2}$ Profile

Symbol	Description	Value
I_x	Second moment about major axis	256.20 in^4
I_y	Second moment about minor axis	83.42 in^4
S_x	Major axis section modulus	51.20 in^3
A_z	Area	14.55 in^2
J	Torsional constant	1.208 in^4
t_w	Web thickness	0.5 in
t_f	Flange thickness	0.5 in
d	Nominal section depth	10 in
d_w	Web depth	9.5 in
b	Nominal section width (breadth)	10 in
b_f	Flange width (breadth)	10 in
w	Nominal weight	11.64 lb/ft

Recalculate design loads with actual beam weight. For ASD and LRFD serviceability limit state calculations,

$$\text{Dead load:} \quad w_{\text{DL}} = (20 \text{ psf} \times 6 \text{ ft}) + 11.64 \text{ lb/ft} = 131.64 \text{ lb/ft} \cong 132 \text{ lb/ft}$$

$$\text{Live load:} \quad w_{\text{LL}} = (40 \text{ psf} \times 6 \text{ ft}) = 240 \text{ lb/ft}$$

$$\text{Total service load:} \quad 140 + 240 = 372 \text{ lb/ft}$$

For LRFD design the ASCE 7 load factors for dead and live loads are 1.2 and 1.6, respectively, and the factored loads are

$$\text{Dead load:} \quad W_{\text{D}} = 1.2(131.64) = 157.97 \text{ lb/ft} \simeq 158 \text{ lb/ft}$$

$$\text{Live load:} \quad W_{\text{L}} = 1.6(240) = 384 \text{ lb/ft}$$

$$\text{Total factored load:} \quad W_u = 158 + 384 = 542 \text{ lb/ft}$$

Determine the actual deflection for the profile chosen using shear deformation beam theory.

$$KAG = k_{\text{tim}} A G_{LT} = 0.413(14.55)(0.425 \times 10^6) = 2.55 \times 10^6 \text{ lb}$$

Note that the value of G_{LT} is taken as the full-section shear modulus G_b since the in-plane shear modulus is not reported by the manufacturer.

$$EI = E_L I_x = (2.6 \times 10^6)(256.2) = 666.12 \times 10^6 \text{ lb–in}^2$$

Note that the longitudinal modulus of the flange, E_L, is used in this calculation. If the full-section flexural modulus, E_b, is used, a more conservative (i.e., larger deflection) prediction is obtained. The full-section flexural modulus is a value that may already include the effects of shear deformation, depending on how the test was conducted (not reported by the manufacturer).

$$\delta_{\max} = \frac{5wL^4}{384EI} + \frac{wL^2}{8KAG} = \frac{5(372/12)(15 \times 12)^4}{384(666.12 \times 10^6)} + \frac{(372/12)(15 \times 12)^2}{8(2.55 \times 10^6)}$$

$$= 0.64 + 0.05 = 0.69 \text{ in.} < 0.75 \text{ in.} \qquad \text{OK}$$

(Note that the shear deflection is only 7% of the total deflection for this relatively long pultruded beam. For shorter beams, the shear deflection will be a higher percentage of the total deflection.)

Step 3. Determine the maximum design stresses or forces. The maximum design service load moment, shear force, and stresses are

$$m_{\max} = \frac{wL^2}{8} = \frac{372(15)^2}{8} = 10{,}462.5 \text{ ft-lb}$$

$$\sigma_{\max} = \frac{m_{\max}}{S_x} = \frac{10{,}462.5(12)}{51.20} = 2452 \text{ psi}$$

$$\sigma_y = \frac{w}{t_w} = \frac{372/12}{0.5}$$

$$= 62.0 \text{ psi (transverse stress in the web due to the distributed load)}$$

$$v_{\max} = \frac{wL}{2} = \frac{372(15)}{2} = 2790 \text{ lb}$$

$$\tau_{\max} = \frac{v_{\max}}{A_w} = \frac{2790}{(10)(0.5)} = 558 \text{ psi}$$

The maximum design ultimate bending moment, shear force, and transverse force are

$$M_{\max} = \frac{W_u L^2}{8} = \frac{542(15)^2}{8} = 15{,}243.8 \text{ ft-lb}$$

$$V_{\max} = \frac{W_u L}{2} = \frac{542(15)}{2} = 4065 \text{ lb}$$

$$F_{y,\max} = W_u d = 542(10/12)$$

$$= 451.7 \text{ lb (transverse force on web acting over } L_{\text{eff}} = d)$$

Step 4. Determine the critical stresses and forces.

Global lateral–torsional buckling For the simply supported uniformly loaded beam the moment coefficient $C_b = 1.13$ and the end-restraint coefficients k_f and $k_\omega = 1.0$. The warping constant is

$$C_\omega = \frac{I_y d^2}{4} = \frac{83.4(10)^2}{4} = 2086 \text{ in}^6$$

and the critical lateral torsional buckling stress is

$$\sigma_{\text{cr}}^{\text{lat}} = \frac{C_b}{S_x}\sqrt{\frac{\pi^2 E_L I_y G_{LT} J}{(k_f L_b)^2} + \frac{\pi^4 E_L^2 I_y C_\omega}{(k_f L_b)^2 (k_\omega L_b)^2}}$$

$$= \frac{1.13}{51.2}\sqrt{\frac{\pi^2(2.6 \times 10^6)(83.42)(0.425 \times 10^6)(1.208)}{(1.0 \times 15 \times 12)^2} + \frac{\pi^4(2.6 \times 10^6)^2(83.42)(2{,}085)}{(1.0 \times 15 \times 12)^2(1.0 \times 15 \times 12)^2}}$$

$$= 8348 \text{ psi}$$

The critical lateral–torsional bending moment is

$$M_{\text{cr}}^{\text{lat}} = \sigma_{\text{cr}}^{\text{lat}} S_x = 8348(51.2) = 427{,}418 \text{ lb-in.} = 35{,}618 \text{ lb-ft}$$

Local buckling Calculate the plate minor (transverse) Poisson ratio:

$$\nu_T = \frac{E_T^c}{E_L^c}\nu_L = \frac{1.0}{2.6}(0.33) = 0.13$$

Calculate the plate flexural rigidities:

$$D_L = \frac{E_L^c t_p^3}{12(1 - \nu_L \nu_T)} = \frac{(2.6 \times 10^6)(0.5)^3}{12(1 - (0.33)(0.13))} = 28{,}297 \text{ lb-in.}$$

$$D_T = \frac{E_T^c}{E_L^c} D_L = \frac{1.0}{2.6}(28{,}297) = 10{,}883 \text{ lb-in.}$$

$$D_{LT} = \nu_T D_L = 0.13(28{,}297) = 3679 \text{ lb-in.}$$

$$D_S = \frac{G_{LT} t_p^3}{12} = \frac{(0.425 \times 10^6)(0.5)^3}{12} = 4227 \text{ lb-in.}$$

Calculate the buckling stresses for the simply supported flange and web to determine which wall buckles first:

$$(\sigma_{\text{free}}^{\text{ss}})_f = \frac{4t_f^2}{b_f^2} G_{LT} = \frac{4(0.5)^2(0.425 \times 10^6)}{(10)^2} = 4250 \text{ psi}$$

$$\begin{aligned}
(\sigma_{\text{ss}}^{\text{ss}})_w &= \frac{\pi^2}{t_w d_w^2}(13.9\sqrt{D_L D_T} + 11.1 D_{LT} + 22.2 D_S) \\
&= \frac{\pi^2}{0.5(9.5)^2}[13.9\sqrt{(28{,}297)(10{,}883)} + 1.11(3679) + 22.2(4427)] \\
&= 83{,}778 \text{ psi}
\end{aligned}$$

Since the longitudinal compressive modulus in the flange and the web are the same, the flange buckles first, since $(\sigma_{\text{free}}^{\text{ss}})_f/(E_L)_f < (\sigma_{\text{ss}}^{\text{ss}})_w/(E_L)_w$, or $4250 < 83{,}655$ psi. Now calculate the local buckling stress for the flange. First calculate the junction stiffness, k, and then the coefficient of restraint, ζ.

$$\begin{aligned}
k_{\text{I-flange}} &= \frac{2(D_T)_w}{d_w}\left[1 - \frac{(\sigma_{\text{free}}^{\text{ss}})_f (E_L)_w}{(\sigma_{\text{ss}}^{\text{ss}})_w (E_L)_f}\right] = \frac{2(10{,}883)}{9.5}\left(1 - \frac{4250}{83{,}655}\right) \\
&= 2174.9 \text{ lb}
\end{aligned}$$

$$\begin{aligned}
\zeta_{\text{I-flange}} &= \frac{D_T}{k_{\text{I-flange}} L_T} = \frac{10{,}883}{2174.9(10/2)} \\
&= 1.001 \text{ (nondimensional coefficient of restraint)}
\end{aligned}$$

and

$$\sigma_{\text{cr}}^{\text{local,I-flange}} = \frac{1}{(b_f/2)^2 t_f}\left(7\sqrt{\frac{D_L D_T}{1 + 4.12\zeta_{\text{I-flange}}}} + 12D_S\right)$$

$$= \frac{1}{(10/2)^2(0.5)}\left[7\sqrt{\frac{28{,}297(10{,}883)}{1 + 4.12(1.001)}} + 12(4427)\right]$$

$$= 8591.3 \text{ psi}$$

The length of the buckle half-wavelength (for information purposes) is

$$a_{\text{I-flange}} = 1.675\frac{b_f}{2}\sqrt[4]{\frac{D_L}{D_T}(1 + 4.12\zeta_{\text{I-flange}})}$$

$$= 1.675\left(\frac{10}{2}\right)\sqrt[4]{\frac{28{,}297}{10{,}883}[1 + 4.12(1.001)]} = 16.0 \text{ in.}$$

The critical bending moment at which local flange buckling occurs is

$$M_{\text{cr}}^{\text{local,I-flange}} = \sigma_{\text{cr}}^{\text{local,I-flange}} S_x = \frac{8591.3(51.2)}{12} = 36{,}656.2 \text{ lb-ft}$$

Local web shear buckling. To calculate the critical shear stress for local in-plane shear buckling of the web, the shear buckling coefficient, k_{LT}, is first found as a function of the shear anisotropy ratio, K.

$$K = \frac{2D_S + D_{LT}}{\sqrt{D_L D_T}} = \frac{2(4427) + 3679}{\sqrt{(28{,}297)(10{,}883)}} = 0.714 < 1.0 \qquad \text{OK}$$

$$k_{LT} = 8.125 + 5.045K = 8.125 + 5.045(0.714) = 11.727$$

$$\tau_{cr}^{\text{local}} = \frac{4\, k_{LT}\sqrt[4]{D_L D_T^3}}{t_w d_w^2} = \frac{4(11.728)\sqrt[4]{28{,}297(10{,}883)^3}}{0.5(9.5)^2} = 14{,}365.7 \text{ psi}$$

The critical in-plane web shear force is

$$V_{\text{cr}}^{\text{local}} \approx \tau_{\text{cr}}^{\text{local}} A_{\text{web}} = \tau_{\text{cr}}^{\text{local}} t_w d = 14{,}365.7(0.5)(10) = 71{,}829 \text{ lb}$$

Local transverse web buckling The effective web width, b_{eff}, that resists the distributed load as a "column" is equal to the depth of the web, d_w, for a distributed load applied to the top flange of the section. The critical transverse buckling stress is

$$(\sigma_y)_{cr}^{local} = \frac{2\pi^2}{t_w b_{eff}^2} (\sqrt{D_L D_T} + D_{LT} + 2D_S)$$

$$= \frac{2\pi^2}{0.5(9.5)^2} [\sqrt{28{,}297(10{,}883)} + 3679 + 2(4427)] = 13{,}159 \text{ psi}$$

The critical transverse buckling load is

$$F_{cr}^{local} = (\sigma_y)_{cr}^{local} A_{eff} = 13{,}159(9.5)(0.5) = 62{,}505 \text{ lb}$$

Tensile, compressive, and shear material failure Using the material tensile, compressive, and shear strengths, the critical bending, shear, and transverse resistances are obtained as

$$M_{cr}^{comp} = \sigma_{cr}^{comp} \frac{I_x}{y} = (30{,}000) \frac{256.2}{5} = 1{,}537{,}200 \text{ in.-lb} = 128{,}100 \text{ ft-lb}$$

$$M_{cr}^{tens} = \sigma_{cr}^{tens} \frac{I_x}{y} = (30{,}000) \frac{256.2}{5} = 1{,}537{,}200 \text{ in.-lb} = 128{,}100 \text{ ft-lb}$$

$$V_{cr}^{local} = \tau_{cr}^{local} A_{web} = 4500(0.5)(10) = 22{,}500 \text{ lb}$$

$$F_{cr}^{crush} = (\sigma_y)_{cr}^{crush} A_{eff} = 16000(0.5)(9.5) = 76{,}000 \text{ lb}$$

Step 5. Determine the factored critical stresses or forces. The critical stresses and loads are summarized in Table 13.9. Using the ASD safety factors and the LRFD resistance determined in step 1, the allowable stresses for ASD and the member resistances for LRFD are tabulated in Table 13.9.

Step 6. Check the ultimate strength or capacity of the profile selected. For the ultimate limit state, the design stress and forces are compared to the allowable stresses and member resistances in Table 13.10. For the serviceability limit state, the maximum midspan deflection is compared with the prescribed allowable deflection limit of $L/240$ for both the ASD and LRFD bases. Based on the comparisons shown, it is seen that the W $10 \times 10 \times \frac{1}{2}$ conventional pultruded profile meets the design demands. The controlling critical ultimate limit state is lateral-torsional buckling for this beam. However, the design is actually controlled by the serviceability limit state. This can be seen by comparing the ratio of design deflection to the allowable deflection limit (0.684/0.75 = 91%) to that of the flexural stress to the critical buckling stresses ratio (2452/3339 = 73%). However, it is also observed that these two efficiency ratios are reasonable high (~75% or more) and that the section is reasonably well optimized for both the ULS and the SLS.

TABLE 13.9 Critical Stress, ASD Design Stresses and Loads, and LRFD Design Resistances

Mode	σ_{cr}	SF	σ_{allow}	R_{cr}	ϕ^{a}	ϕR_{cr}
Lateral–torsional buckling	8,348 psi	2.5	3,339 psi	35,618 ft-lb	0.67	23,864 ft-lb
Local buckling (flange controls)	8,591 psi	2.5	3,436 psi	36,656 ft-lb	0.67	24,560 ft-lb
Flange tensile failure	30,000 psi	2.5	12,000 psi	128,100 ft-lb	0.67	85,827 ft-lb
Flange compressive failure	30,000 psi	2.5	12,000 psi	128,100 ft-lb	0.67	85,827 ft-lb
Web transverse buckling	13,159 psi	3.0^{b}	4,380 psi	62,505 lb	0.67	41,878 lb
Web transverse crushing	16,000 psi	3.0^{b}	5,333 psi	76,000 lb	0.67	50,920 lb
Web shear buckling	14,366 psi	3.0	4,772 psi	71,829 lb	0.67	48,125 lb
Web shear failure	4,500 psi	3.0	1,500 psi	22,500 lb	0.67	15,075 lb

[a] According to the Eurocomp partial factor method, different ultimate modes of failure (e.g., flexure, compression) are not treated independently as in a traditional LRFD or ASD approach.
[b] The web is considered to be a slender column and the SF for compressive stress is used.

PROBLEMS

13.1 Using the Timoshenko beam theory, derive the expressions for the bending and shear deflections [$f_1(z)$ and $f_2(z)$] for a simply supported beam with a span, ℓ, having a concentrated load, P, at its midspan. Provide the expression for the midspan deflection. Check your result with the expressions in Table 13.1, line 2.

13.2 Using the Timoshenko beam theory, derive the expressions for the bending and shear deflections [$f_1(z)$ and $f_2(z)$] for a simply supported beam with a span, ℓ, and having concentrated loads, P, at its quarter points. Provide the expression for the midspan deflection and for the deflection under one of the load points. Compare your result with those published in the literature for this case.

13.3 Using the Timoshenko beam theory, determine the reactions and draw the shear force and bending moment diagrams for a uniformly loaded,

TABLE 13.10 Comparison of Design Demands with Permitted Values for ASD and LRFD

Mode	σ_{design}	vs.	σ_{allow}	R_{design}	vs.	ϕR_{cr}
Ultimate limit state						
Lateral torsional buckling	2,452 psi	<	3,339 psi	15,244 ft-lb	<	23,864 ft-lb
Local buckling (flange)	2,452 psi	<	3,436 psi	15,244 ft-lb	<	24,560 ft-lb
Flange tensile failure	2,452 psi	<	12,000 psi	15,244 ft-lb	<	85,827 ft-lb
Flange compressive failure	2,452 psi	<	12,000 psi	15,244 ft-lb	<	85,827 ft-lb
Web transverse buckling	62 psi	<	4,380 psi	452 lb	<	41,878 lb
Web transverse crushing	62 psi	<	5,333 psi	452 lb	<	50,920 lb
Web shear buckling	558 psi	<	4,772 psi	4,065 lb	<	48,125 lb
Web shear failure	558 psi	<	1,500 psi	4,065 lb	<	15,075 lb
Serviceability limit state						
Midspan deflection	0.684 in.	<	0.75 in.	0.684 in.	<	0.75 in.

w, continuous beam on three simple (pinned) supports having two equal spans of length, ℓ. Determine an expression for the maximum deflection of the beam. (To aid you in your solution, you can assume that the maximum deflection occurs at approximately the same location in the span as in an Euler–Bernoulli beam.) Compare the maximum negative moment obtained at the middle support with that of an Euler–Bernoulli beam for E/G ratios of 2.6 (isotropic material), 5, 10, 15, and 20. Assume that $K = 1$ and $A = 0.1I$. [*Hint:* Use the force method (method of consistent deformations).]

13.4 Analyze the two-span continuous Timoshenko beam in Problem 13.3 using a commercial structural analysis computer program using beam elements that permit the inclusion of shear deformation effects. Choose appropriate values of I, A, w, ℓ, E and G for your numerical computations in order to compare your results with your closed-form solutions developed in Problem 13.3.

13.5 For profiles a to e from U.S. manufacturers (Strongwell, Creative Pultrusions, and Bedford Reinforced Plastics) and profiles f to h from

European manufacturers (Fiberline or TopGlass) listed in Table P13.5,[22] determine the Timosheko shear coefficient, k_{tim}, using the equations listed in Table 13.2. For nonhomogeneous profiles from Creative Pultrusions, use average values of the mechanical properties in the webs and flanges of the sections. For I sections, assume bending about the major axis.

13.6 For profiles *a* to *e* from U.S. manufacturers (Strongwell, Creative Pultrusions, and Bedford Reinforced Plastics) and profiles *f* to *h* from European manufacturers (Fiberline or TopGlass) listed in Table P13.5, determine the flexural rigidity, *EI*, and the shear rigidity, *KAG*, for bending about the major axis. Calculate the shear rigidity using the $k_{\mathrm{tim}}AG_{LT}$ and the $A_{\mathrm{web}}G_{LT(\mathrm{web})}$ methods and compare the results. For nonhomogeneous profiles, use the transformed section method or the composite mechanics method to determine *EI*. For nonhomogeneous profiles, use average values of the properties in the webs and flanges of the sections to calculate the shear rigidity *KAG* when using the Timoshenko shear coefficient. [For a more rigorous treatment of the problem of shear deflection in nonhomogeneous profiles using the Timosheko beam theory, see Bank (1987).]

13.7 A $10 \times 5 \times \frac{1}{2}$ glass–polyester narrow-flange I section (profile *b* in Table P13.5) produced by Strongwell is used as a simply supported beam and spans 9 ft. The beam has a nominal self-weight of 4.61 lb/ft. It is loaded with a uniformly distributed sustained dead load of 100 lb/ft and a live load of 200 lb/ft. Determine the instantaneous midspan deflection and the midspan deflection after 25 years of sustained load on the beam under the self-weight, sustained dead load, and the live load.

TABLE P13.5 Pultruded Profiles

a	Glass–polyester $8 \times 8 \times \frac{3}{8}$ $(d \times b \times t)$ in. wide-flange I-shaped section
b	Glass–polyester $10 \times 5 \times \frac{1}{2}$ $(d \times b \times t)$ in. narrow-flange I-shaped section
c	Glass–vinylester $6 \times 6 \times \frac{1}{4}$ $(d \times b \times t)$ in. wide-flange I-shaped section
d	Glass–vinylester $6 \times \frac{3}{8}$ $(b \times t)$ in. square tube
e	Glass-polyester $3 \times \frac{3}{8}$ $(b \times t)$ in. square tube
f	Glass–vinylester $200 \times 100 \times 10$ $(d \times b \times t)$ mm narrow-flange I-shaped section
g	Glass–vinylester $300 \times 8 \times 150 \times 12$ $(d \times t_w \times b_f \times t_f)$ mm narrow-flange I-shaped section
h	Glass–vinylester 120×8 $(b \times t)$ mm square tube

[22] Instructors may assign only selected beams or may choose other singly symmetric sections from manufacturer design guides.

13.8 Solve Problem 13.7 using the Creative Pultrusions $10 \times 5 \times \frac{1}{2}$ nonhomogeneous section. Compare the results with those obtained in Problem 13.7.

13.9 Determine the critical lateral torsional buckling stress for the I-shaped pultruded profiles *a*, *b*, and *c* (U.S. manufacturers' properties[23]) and profiles *f* and *g* (European manufacturers' properties[24]) in Table P13.5 when loaded by a uniformly distributed load over a 12-ft unbraced simply supported span. What is the maximum nominal (no ASD or LRFD factors) uniformly distributed load that can be placed on the beam based on the critical lateral torsional buckling stress of the chosen beam?

13.10 Determine the critical local buckling stress, critical bending moment, and length of the buckle half-wavelength, *a*, for the I-shaped pultruded profiles *a*, *b*, and *c* (U.S. manufacturers' properties[25]) and profiles *f* and *g* (European manufacturers' properties) in Table P13.5 when loaded in flexure.

13.11 Calculate the critical local buckling stress for the I-shaped beams in Problem 13.10 using the approximate equations [equation (13.37)] suggested by the *Structural Plastics Design Manual.* Compare these predictions with those obtained in Problem 13.10 using the exact method.

13.12 Determine the critical local buckling stress, critical bending moment, and length of the buckle half-wavelength, *a*, for square-tube profiles *d* and *e* (U.S. manufacturer properties) and profile *h* (European manufacturers' properties) in Table P13.5 when loaded in flexure.

13.13 Determine the critical local buckling stress, critical bending moment, and length of the buckle half-wavelength, *a*, for a Strongwell $9 \times 6 \times \frac{5}{16}$ $(d \times b \times t)$ in. rectangular-tube beam when loaded in bending about **(a)** its major axis and **(b)** its minor axis.

13.14 Determine the critical web in-plane shear buckling stress and the critical shear force for profiles *a* and *c* (U.S. manufacturers) and profiles *g* and *h* (European manufacturers) in Table P13.5.

13.15 Consider a $24 \times 3 \times 0.25$ $(d \times b_f \times t)$ in. glass–vinylester channel section with material properties given in Table 13.7. Determine **(a)** the critical web in-plane shear buckling stress and **(b)** the critical local in-plane buckling load when the beam is bent about its major axis and loaded through its shear center. To determine the local flange

[23] Use nonhomogeneous materials properties for Creative Pultrusions' beams when appropriate.
[24] If available.
[25] Use nonhomogeneous materials properties for Creative Pultrusions' beams when appropriate.

buckling stress for the channel, the formula for an I-shaped section is used with two modifications: (1) the channel flange width, b_f (3 in. for this case), is used in place of $b_f/2$ in the buckling stress equation for the I-shaped section [equation (13.26)], and (2) the expression for k [equation (13.25)] is multiplied by 2 since the web restrains only one flange (Kollár, 2003).

13.16 Consider a built-up section of two of the channels in Problem 13.15 used back to back to form a deep I-section with dimensions of $12 \times \frac{1}{2} \times 6 \times \frac{1}{4}$ ($d \times t_w \times b \times t_f$) inches. (Assume that the channel webs are fully bonded to each other back to back and act as a single $\frac{1}{2}$-in.-thick wall.) The beam is to be used as a simply supported beam loaded by a concentrated load at midspan (applied at the beam centroidal axis) over a 14-ft-long span. Determine the **(a)** critical lateral torsional buckling stress, **(b)** critical in-plane buckling stress, and **(c)** critical web shear buckling stress for this built-up section. Assume that the section behaves as an I-shaped section for the purposes of calculating in-plane buckling of the web and flanges and shear buckling of the web.

13.17 Determine the maximum concentrated load, P, that can be applied to the top flange (directly above the web) of a $12 \times 6 \times \frac{1}{2}$ Strongwell glass–vinylester pultruded I-shaped section when the load is controlled by **(a)** local web compression buckling or **(b)** local web compression crushing. The concentrated load is applied over a bearing length (lengthwise along the beam axis) of 3 in. Vertical web plate stiffeners are placed 6 in. apart on either side of the load point on both sides of the web.

13.18 Determine the maximum reaction (or shear force at the beam end), R, that can be carried by a creative pultrusions $6 \times \frac{3}{8}$ glass–vinylester square tube when used as a simply supported beam. The beam is seated on a support having a 4-in. width. A pultruded material bearing plate $\frac{3}{8}$ in. thick is placed between the support and the beam.

13.19 Redesign the beam in Design Example 13.4 using a vinylester wide-flange I-shaped profile from the Creative Pultrusions SuperStructural series. Design the beam using **(a)** the ASD basis and **(b)** the LRFD basis. Compare the section selected to that published in Creative Pultrusions' design guide, "Allowable Uniform Load Tables," for the loading given.

13.20 Consider a $12 \times 6 \times \frac{1}{2}$ in. Strongwell glass–vinylester narrow-flange I-shaped section for the beam in Design Example 13.4 and determine if it will be an adequate section to carry the design loads.

13.21 Consider the floor plan shown in Fig. 13.13 and assume that this represents the plan for a single-story structure. To decrease deflections, 30-ft-long pultruded beams that are continuous over the columns spaced at 15 ft on center are suggested in place of the 15-ft-long beams considered in Design Example 13.4. Design the 30-ft-long continuous pultruded beams using the material properties given in Table 13.7. Design using **(a)** the ASD basis and **(b)** the LRFD basis. [*Note:* Include effects of shear deformation in the calculation of the beam deflection (see Problem 13.3), use appropriate values of C_b in the lateral torsional buckling equations that account for the moment gradient, and examine local buckling at both positive and negative moment locations. Justify all design assumptions in writing.]

13.22 Design a laterally braced simply supported wide–flange glass-vinylester pultruded section to carry a uniformly distributed load of 95 lb/ft (not including beam self-weight) on a 15-ft span. Assume that a maximum short-term deflection of $L/180$ is permitted. Do not consider long-term deflection. Determine the spacing of the lateral braces to prevent lateral torsional buckling of the beam selected. Consider pultruded materials from three different U.S. manufacturers and use the ASD and LRFD design bases. Compare your final design to that given in the allowable load tables of the manufacturer selected.

13.23 Design **(a)** a conventional steel I-beam, **(b)** a cold-formed steel I-beam, **(c)** an aluminum I-beam, and **(d)** a structural grade lumber beam to meet the requirements of Design Example 13.4. Use appropriate design bases and design specifications for the conventional material beams. Choose the lowest-possible strength and stiffness grades of the conventional materials for the sizes chosen. Compare the cross-sectional sizes, unit weights, and costs of these conventional material beams to that of the pultruded beam designed in Design Example 13.4.

14 Pultruded Axial Members

14.1 OVERVIEW

In this chapter, design procedures are presented for axial members made of pultruded FRP profiles. This includes axially loaded compression members called *columns*, axially loaded tension members, and combined axial and flexural members (called *beam-columns* when the axial load is compressive). Flexural members are discussed in Chapter 13. Connections and joints for pultruded profiles are discussed in Chapter 15. Theoretical mechanics-based equations are presented here for determining strength, stability, and deformation limit states of axially loaded pultruded structural members.

Design procedures for the allowable stress design (ASD) and the load and resistance factor design (LRFD) bases described in Chapter 12 are presented. Since pultruded profiles behave essentially as linear elastic (or linear viscoelastic) structural members, the analytical equations developed for predicting stresses and deformations in these members for the ultimate and serviceability limit states are valid for either of the design bases selected. Therefore, it is only the safety factors for ASD or the load and resistance factors for the LRFD that differentiate the design procedures for the two design bases.

14.2 INTRODUCTION

The design of pultruded profiles that are doubly symmetric with respect to their principal coordinate axes and consist of horizontal plates (flanges) and vertical plates (webs) are discussed in this section. This includes the conventionally manufactured I-shaped,[1] square tube, rectangular tube, round tube, and multicelled tubular profiles. Typical member cross sections and orientations are shown in Fig. 14.1.

In braced frame structures, wide-flange I-sections and tubular sections are most frequently used as columns and bracing members, since they are readily connected using profiles or gusset plates and traditional mechanical connectors such as nuts and bolts (see, e.g., Figs. 1.15 and 1.18). The choice of

[1] The term *I-shaped profiles* is used to describe both narrow- and wide-flange I-shaped profiles.

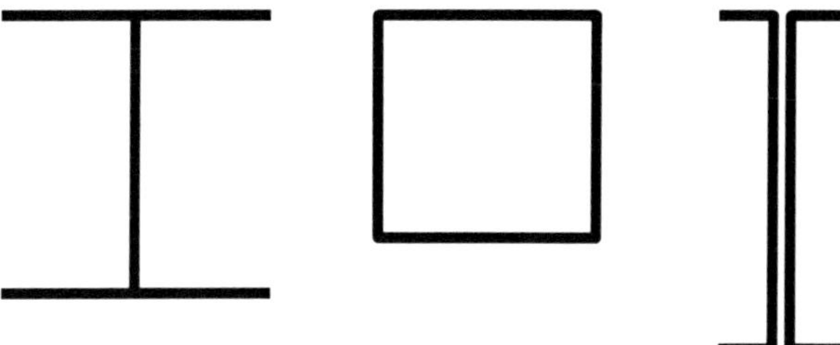

Figure 14.1 Cross sections and orientations of typical axial members.

profile used as a column or brace often depends on the connection detailing required, as discussed further in Chapter 15.

In trusses or stick-built frames, back-to-back angles and small channels, as well as small square tubular elements, are most commonly used (see, e.g., Figs. 1.17 and 1.19) as the primary structural elements. In this case, the elements are usually treated as continuous beams (such as chords, girts, purlins, or rafters), as axially loaded members (such as columns, struts, or ties), or as combined load members (beam-columns or tension-bending members). Flexural members are discussed in Chapter 13.

The design of axially loaded members is divided here into three subsections: (1) concentrically loaded compression members, (2) concentrically loaded tension members, and (3) combined load members.

14.3 CONCENTRICALLY LOADED COMPRESSION MEMBERS

The compression member discussed in this section is designed to resist stress resultants (or internal forces) in the global x,y,z coordinate system that are caused by the concentric axial load, P_z, applied at the centroid of the cross section in the y–z plane, as shown in Fig. 14.2. The single stress resultant

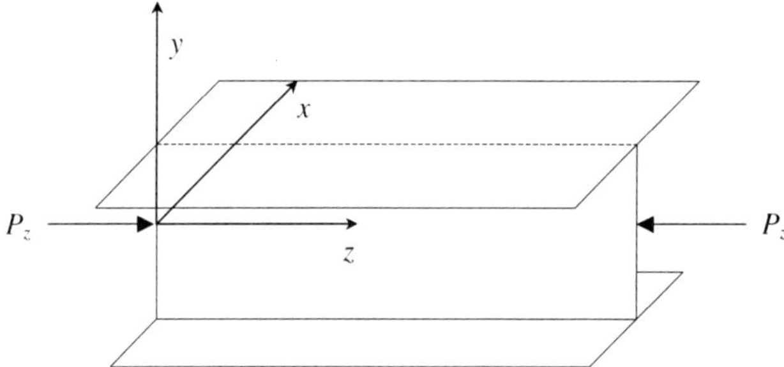

Figure 14.2 Concentrically loaded compression member.

that the concentrically loaded compression member is designed for is the axial force, P_z. According to one-dimensional bar theory, if the member is homogeneous, this concentric axial force causes the member to develop a uniform axial (compressive) stress, σ_z, throughout its cross section along its length, given by the normal stress equation

$$\sigma_z = \frac{P_z}{A_z} \tag{14.1}$$

where A_z is the cross-sectional area of the member.

For a beam with nonhomogeneous properties through the depth of the cross section, such as a pultruded profile with different longitudinal elastic moduli in its webs and flanges, the relationship between the axial force and the axial stress is not closed form but is written in general form as

$$P_z = \int_A \sigma_z \, dA \tag{14.2}$$

where the integral is taken over the cross-sectional area of the beam. Since the pultruded material is assumed to be linear elastic in this one-dimensional bar theory, the strain through the depth of the axial member, ε_z, is constant and the stress at point on the cross section, σ_z, can be found directly using the longitudinal modulus, E_L, at the point of interest. One-dimensional bar theories for bars having material of different moduli in the axial direction are covered in all elementary texts on mechanics of materials texts (e.g., Gere and Timoshenko, 1997). In what follows the properties of the inhomogeneous section are referred to as composite section properties, as noted in Chapter 13. In a typical pultruded profile consisting of webs and flanges having different longitudinal moduli, the axial force is found by summation as

$$P_z = \sum_A \sigma_{z(\text{flanges})} A_{z(\text{flanges})} + \sigma_{z(\text{webs})} A_{z(\text{webs})} \tag{14.3}$$

As shown in Fig. 14.2, the axial force will have a negative sign, and therefore compressive stresses will have a negative sign; however, for concentrically loaded members, the negative sign will be dropped here, and all equations for compression members will be written with a positive sign, as is customary in structural engineering.

The axial stress (or axial force) calculated must be less than the material strength and critical buckling stress (or the critical resistance or buckling capacity) according to the design basis selected.

14.4 DEFORMATIONS IN CONCENTRICALLY LOADED COMPRESSION MEMBERS

Due to the concentric axial force, P_z, a homogeneous member will shorten by the amount

$$\delta_z = \frac{P_z L}{A_z E} \tag{14.4}$$

where $A_z E$ is the axial rigidity of the member in compression (defined below.) If the member is inhomogeneous, its axial rigidity is found using composite section mechanics.

14.5 DETERMINATION OF DEFLECTIONS AND STRESSES FOR SERVICEABILITY AND ULTIMATE LIMIT STATES

For compression members the deflections are determined for the serviceability limit state and stresses (or stress-resultants) are determined for the ultimate limit state. As in the case of pultruded beam design discussed in Chapter 13, the low longitudinal modulus of conventional pultruded profiles will lead to appreciable axial deflections in column and truss members that should not be ignored in design. However, these deflections are nevertheless not usually the controlling limit state for compression members as they are for beams, due to the shorter lengths of compression members in most pultruded structures and the larger cross sections needed for axial members to resist global and local buckling. Consequently, the design procedure for compression members is slightly different from that for beams. Nevertheless, the serviceability limit states are first discussed.

14.6 SERVICEABILITY LIMIT STATES: AXIAL SHORTENING

Uniaxial bar theory is used to determine the deflection of pultruded compression members. This is the procedure recommended by the *Structural Plastic Design Manual* (1984) and the *Eurocomp Design Code and Handbook* (Eurocomp, 1996). For the calculation of axial deformation, the axial rigidity is taken as

$$A_z E = A_z E_L \tag{14.5}$$

where E_L is the longitudinal modulus of the pultruded material in compression. It is generally assumed that the tensile and compressive longitudinal moduli of the FRP material are the same, and most manufacturers report the same value for the tensile and compressive moduli or do not differentiate between the two. Some manufacturers report different values for the tensile and compressive moduli. If compressive moduli are reported, they should be used in deformation and buckling calculations for compression members. In the case of a nonhomogenous profile having different moduli in different walls of the profile, the composite axial rigidity, $\Sigma_i E_i A_i$, is used (provided that the property variation is symmetric with respect to the centroidal axes). To determine long-term axial shortening, the viscoelastic longitudinal modulus, E_L^v, with the creep parameters for compression given in Table 13.5 should be used.

The axial shortening of compression members must be less than code-stipulated maxima for axially loaded members. A limit of $L/1500$ is suggested for design of pultruded compression members; however, in a typical section, local buckling will probably occur well before this limit is reached. Often, these deformation requirements will be enforced via global deflection or lateral displacement limits on the entire structure.

14.7 ULTIMATE LIMIT STATES

14.7.1 Global Flexural Buckling

The term *global flexural buckling* is used to describe the commonly observed overall physical instability that can occur in compression members loaded by axial loads. It is commonly referred to as *Euler buckling.* When the critical global load is reached, the entire profile displaces laterally about one of the centroidal axis planes (usually, the one with the larger slenderness ratio, kL/r), causing the entire beam to move out of its vertical plane. Figure 14.3 shows global buckling of an I-shaped profile about its weak (minor) axis.

Global flexural buckling of conventional pultruded profiles has been studied in a number of experimental investigations (Barbero and Tomlin, 1993; Zureick and Scott, 1997; Zureick and Steffen, 2000; Mottram et al., 2003a). It has been shown that the classical Euler equation can be used to describe the global flexural buckling of conventional pultruded profiles. However, since pultruded profiles can be used as columns with low story heights (8 to 10 ft) and due to a higher E_L/G_{LT} ratio than the E/G ratio for isotropic columns, the effects of shear deformation should be considered, although in many cases the influence will be small ($< 5\%$) (Zureick and Scott, 1997; Roberts, 2002).

The critical flexural buckling load for a homogeneous member, including the effects of shear deformation, is given as (Zureick, 1998),

$$P_{cr}^{flex} = \frac{P_{euler}}{1 + P_{euler}/k_{tim}A_z G_{LT}} \tag{14.6}$$

where

Figure 14.3 Global buckling of an I-shaped profile. (Courtesy of Pizhong Qiao and Julio Davalos.)

$$P_{\text{euler}} = \frac{\pi^2 E_L I}{(kL)^2} \tag{14.7}$$

As discussed in Chapter 13, the transverse shear rigidity, KAG, can be approximated by

$$k_{\text{tim}} A_z G_{LT} \approx A_{\text{web}} G_{LT(\text{web})}$$

The critical flexural buckling stress, including the effects of shear deformation, is given as

$$\sigma_{cr}^{flex} = \frac{\pi^2 E_L}{(kL/r)_{max}^2} \left[\frac{1}{1 + (1/k_{tim} G_{LT})(\pi^2 E_L/(kL/r)_{max}^2)} \right] \tag{14.8}$$

where $(kL/r)_{max}$ is the maximum slenderness ratio of the profile, k the end restraint coefficient, L the unbraced length (often called the *height, h*, in the case of a column), and r the radius of gyration for the axis under consideration. k_{tim} is the Timoshenko shear coefficient (defined in Chapter 13) and A_z is the cross-sectional area of the section. To neglect the effects of shear deformation, the term in brackets in equation (14.8) is set to unity.

For nonhomogeneous members, composite section properties are used in the equations above. The Timoshenko shear coefficient and the transverse shear rigidity, KAG, for a nonhomogeneous member can be found following procedures described in Bank (1987).

14.7.2 Global Torsional Buckling

Open-section compression members can buckle in a pure torsional mode. For doubly symmetric sections such as I-shaped profiles and tubular box sections, the torsional rigidity is high and torsional buckling is seldom a limiting state. However, for open sections in which the walls all meet at the centroid (such as a cruciform section), torsional buckling can be critical. Neglecting the influence of shear deformation, the critical torsional buckling stress in a homogeneous member is

$$\sigma_{cr}^{tor} = \frac{1}{I_p} \left[\frac{\pi^2 E_L C_\omega}{(k_\omega L)^2} + G_{LT} J \right] \tag{14.9}$$

where A_z is the cross-sectional area, I_p the polar second moment of area, J the torsional constant, C_ω the warping constant, k_ω an end restraint coefficient for torsional buckling, and L the unbraced length of the member. G_{LT} is the previously defined in-plane shear modulus of the pultruded material. The $G_{LT}J$ term is a conservative approximation for the Saint-Venant torsional rigidity, GJ, of a pultruded profile. Its value can be obtained from full-section uniform torsion tests (Roberts and Al-Ubaidi, 2002).

The critical torsional buckling load for homogeneous members is

$$P_{cr}^{tor} = \sigma_{cr}^{tor} A_z \tag{14.10}$$

An approximate equation for the critical torsional buckling load that includes the effect of shear deformation (Roberts, 2002) for profiles with $J/I_p << 1$ is

$$P_{cr}^{tor+sd} = \frac{P_{cr}^{tor}}{1 + P_{cr}^{tor}/KAG} \tag{14.11}$$

The torsional shear rigidity, KAG, can be approximated conservatively as the transverse shear stiffness, $k_{tim}A_z G_{LT}$, or can be obtained by full-section torsional tests (Roberts and Ul-Ubaidi, 2002). For a nonhomogeneous member, the composite section properties need to be used in the equations above. The Timoshenko shear coefficient and the transverse shear rigidity, KAG, for a nonhomogeneous member can be found following procedures described in Bank (1987). The composite warping stiffness, $E_L C_\omega$, and the composite Saint-Venant/torsional stiffness, $G_{LT}J$, are not easily obtained, and approximate values of these values may need to be used for design. Guidance can be found in texts dealing with Vlasov thin-walled bar theory (Gjelsvik, 1981). Alternatively, full-section torsion tests can be performed to obtain the values for a specific section as described by Roberts and Ul-Ubaidi (2002).

Singly symmetric sections such as angle profiles are not covered in this book; however, the reader should note that they are susceptible to global buckling in a flexural–torsional mode consisting of Euler buckling about the major axis of the cross section and torsional buckling about the longitudinal axis of the member. Equations for this special case are presented by Zureick and Steffen (2000).

14.7.3 Local Buckling Due to Axial Loads

Conventional pultruded GRFP profiles are especially susceptible to local buckling when subjected to axial loads, due to the low in-plane moduli and the slenderness (width-to-thickness ratio) of the plate elements (known as *walls*) that make up the thin-walled profile. Local buckling in compression flanges of columns has been demonstrated in numerous tests (e.g., Yoon et al., 1992; Tomblin and Barbero, 1994; Pecce and Cosenza, 2000; Mottram, 2004). Figure 14.4 shows local compression flange buckling in a concentrically loaded wide-flange pultruded profile. The local buckling seen in concentrially loaded columns is physically very similar in appearance to that in the compression flanges of the flexurally loaded beams discussed in Chapter 13 and shown in Fig. 13.7. However, in axially loaded columns, both flanges buckle at the same time, whereas in a beam, only the compression flange buckles. Therefore, the restraint provided by the web on the buckling of the flange in the column is less than that provided by the web of a beam, and the local buckling equations are different from those for a beam.

As discussed for beams in Chapter 13, the critical buckling load (or stress) in a wall (or panel) of a profile is a function of the boundary conditions on the longitudinal edges of the wall. In I-shaped profiles[2] the flange is partic-

[2] The term *I-shaped profiles* is used to describe both narrow- and wide-flange I-shaped profiles.

Figure 14.4 Local compression flange buckling in both flanges of a wide-flange pultruded profile subjected to axial load. (Courtesy of J. T. Mottram.)

ularly susceptible to buckling, since one edge is free while the other edge is elastically restrained at the web–flange junction. As in the case of the transversely loaded beam, the method due to Kollár (2003) is presented for local buckling of the walls of a concentrically loaded column. The reader is referred to the discussion in Chapter 13 related to the choice of Kollár's method.

Only the design equations for doubly symmetric I- and box-shaped profiles are presented in this book, due to the fact that at this time, these profiles are most commonly used as beams and columns in pultruded structures. This includes built-up sections that are of I or box shape. The reader is referred to Kollár's work for singly symmetric sections. The cross-sectional notation for I- and box-shaped profiles are shown in Fig. 13.9. Since local buckling of the flange and the web of a section due to in-plane normal loads are interrelated, flange and web buckling are treated together in this section.

To use Kollár's method, first the buckling stresses (or loads) of the walls, assuming they are simply supported at their restrained edges, are found. These stresses are then used to determine which wall buckles first and the coefficient of edge restraint for the critical wall. The final solution for the buckling stress for a restrained wall is then given by a closed-form equation that includes the coefficient of edge restraint, ζ, of the critical wall. Two types of simply supported walls[3] are needed for the first part of the solution[4]:

1. *A wall that is free and simply supported on its edges under uniform axial stress.* This solution is used to determine the critical stress in the

[3] For all cases it is assumed that the loaded edges are simply supported. The descriptors are used to identify the longitudinal unloaded edges of the wall.

[4] Many of the equations presented in this section are identical to those presented for local buckling in Chapter 13. They are repeated in this chapter to aid the reader and to make the section on axial members self-contained.

flange of an I-shaped profile.[5] It is given in terms of the flexural rigidities[6] of the flange as

$$(\sigma_{\text{free}}^{\text{ss}})_f = \frac{\pi^2}{t_f(b_f/2)^2}\left[D_L\left(\frac{b_f/2}{a}\right)^2 + \frac{12}{\pi^2}D_S\right] \quad (14.12)$$

where t_f is the flange thickness, b_f the profile width (twice the nominal width of the flange), and a the buckle half-wavelength, which is equal to the length of the flange in this special case.[7] For long flanges, as is typically the case, the first term in brackets in equation (14.12) is negligible and the critical stress can be found from

$$(\sigma_{\text{free}}^{\text{ss}})_f = \frac{4t_f^2}{b_f^2}G_{LT} \quad (14.13)$$

In this mode the plate buckles with one buckle half-wavelength, equal to the plate length (like a simply supported Euler column).

2. *A wall that is simply supported along both edges under uniform axial stress.* This solution is used to determine the critical stress in the web of an I-shape and the web and flange of a box section. It is given in terms of the flexural rigidities of the wall, i, as

$$(\sigma_{\text{ss}}^{\text{ss}})_i = \frac{2\pi^2}{t_t b_i^2}(\sqrt{D_L D_T} + D_{LT} + 2D_S) \quad (14.14)$$

where the subscript i implies w for the web and f for the flange and the symbol b implies b for the flange and d for the web. The length of the buckle half-wavelength for this mode is given by

$$(a_{SS}^{\text{ss}})_i = b_i\sqrt[4]{\frac{E_L}{E_T}} \quad (14.15)$$

Local Buckling Equations for an I-Shaped Profile in Axial Compression
If $(\sigma_{\text{free}}^{\text{ss}})_f/(E_L)_f < (\sigma_{\text{ss}}^{\text{ss}})_w/(E_L)_w$, the flange will buckle before the web. The flange will therefore be restrained by the web. Except in very unusual cir-

[5] A different notation from that used in Kollár (2003) is used in what follows to make the equations more familiar to structural engineers unfamiliar with laminated plate and composite material notation. For the same reason, equations are given for plate buckling stresses but not for plate buckling loads.

[6] See Chapter 12 for a definition of plate flexural rigidities.

[7] Note that in many texts the parameter b is used to denote the width of the flange only and is equal to one-half of b_f.

cumstances, this will be the buckling mode that will control for conventional pultruded GFRP profiles. The spring constant is then given as

$$k_{\text{I-flange}} = \frac{(D_T)_w}{d_w}\left[1 - \frac{(\sigma^{ss}_{\text{free}})_f(E_L)_w}{(\sigma^{ss}_{ss})_w(E_L)_f}\right] \tag{14.16}$$

and the buckling stress for the rotationally restrained flange is given in terms of the properties of the flange as[8]

$$\sigma_{cr}^{\text{local,I-flange}} = \frac{1}{(b_f/2)^2 t_f}\left(7\sqrt{\frac{D_L D_T}{1 + 4.12\zeta_{\text{I-flange}}}} + 12D_S\right) \tag{14.17}$$

where

$$\zeta_{\text{I-flange}} = \frac{(D_T)_f}{k_{\text{I-flange}}\,(b_f/2)} = \frac{(E_T)_f t_f^3}{6k_{\text{I-flange}}\; b_f\,[1 - (\nu_T)_f(\nu_L)_f]} \tag{14.18}$$

The buckle half-wavelength, a, for this case is given in terms of the properties of the flange as (Kollár and Springer, 2003, Table 4.10)

$$a_{\text{I-flange}} = 1.675\,\frac{b_f}{2}\sqrt[4]{\frac{D_L}{D_T}(1 + 4.12\zeta_{\text{I-flange}})} \tag{14.19}$$

The critical axial load for the local flange buckling of a homogeneous I-shaped profile loaded in axial compression is therefore

$$P_{cr}^{\text{local,I-flange}} = \sigma_{cr}^{\text{local,I-flange}}\,A_z \tag{14.20}$$

In the case of a nonhomogeneous section, the critical axial load is found by integration of the axial stresses over the various property walls in the section. The strain corresponding to the critical stress in the flange is first found and then the stresses in all the walls of the section are determined.

If $(\sigma^{ss}_{\text{free}})_f/(E_L)_f > (\sigma^{ss}_{ss})_w/(E_L)_w$, the web will buckle before the flange. The spring constant is then obtained by assuming that the outstanding flanges act as longitudinal stiffeners at the web edges (Kollár, 2003), and the equivalent spring constant is given as

[8] Kollár (2003) provides an alternative equation for this case that provides greater accuracy if needed.

$$k_{\text{I-web}} = \frac{4(D_T)_f}{b_f}\left[1 - \frac{(\sigma^{\text{ss}}_{\text{ss}})_w(E_L)_f}{(\sigma^{\text{ss}}_{\text{free}})_f(E_L)_w}\right] \tag{14.21}$$

and the buckling stress for the rotationally restrained web is given in terms of the properties of the web as

$$\begin{aligned}\sigma^{\text{local,I-web}}_{\text{cr}} = \frac{\pi^2}{d_w^2 t_w}&[2\sqrt{(D_L D_T)}(1 + 4.139\xi)\\ &+ (D_{LT} + 2D_S)(2 + 0.62\xi^2_{\text{I-web}})]\end{aligned} \tag{14.22}$$

where

$$\xi_{\text{I-web}} = \frac{1}{1 + 10\zeta_{\text{I-web}}} = \frac{1}{1 + 10\,[(D_T)_w/k_{\text{I-web}}d_w]} \tag{14.23}$$

The buckle half-wavelength, a, for this case is given in terms of the properties of the web as (Kollár and Springer, 2003, Table 4.10)

$$a_{\text{I-web}} = b_f \sqrt[4]{\frac{D_L}{D_T}\left(\frac{1}{1 + 4.139\ \xi_{\text{I-web}}}\right)} \tag{14.24}$$

The critical axial load for the local web buckling of a homogeneous I-shaped profile loaded in axial compression is therefore

$$P^{\text{local,I-web}}_{\text{cr}} = \sigma^{\text{local,I-web}}_{\text{cr}} A_z \tag{14.25}$$

For a nonhomogeneous section, the critical axial load is found by integration of the axial stresses over the various property walls in the section. The strain corresponding to the critical stress in the web is first found, and then the stresses in all the walls of the section are determined.

Local Buckling Equations for a Rectangular Box Profile in Axial Compression If $(\sigma^{\text{ss}}_{\text{ss}})_f/(E_L)_f < (\sigma^{\text{ss}}_{\text{ss}})_w/(E_L)_w$, the flanges will buckle before the webs. The flanges will therefore be restrained by both of the webs, one on either longitudinal edge of the flange. The spring constant is then given as

$$k_{\text{box-flange}} = \frac{2(D_T)_w}{d_w}\left[1 - \frac{(\sigma^{\text{ss}}_{\text{ss}})_f(E_L)_w}{(\sigma^{\text{ss}}_{\text{ss}})_w(E_L)_f}\right] \tag{14.26}$$

and the buckling stress for the rotationally restrained flange is given in terms of the properties of the flange as

$$\sigma_{cr}^{\text{local,box-flange}} = \frac{\pi^2}{b_f^2 t_f} [2\sqrt{(D_L D_T)}(1 + 4.139\xi) + (D_{LT} + 2D_S)(2 + 0.62\xi_{\text{box-flange}}^2)] \tag{14.27}$$

where

$$\xi_{\text{box-flange}} = \frac{1}{1 + 10\zeta_{\text{box-flange}}} = \frac{1}{1 + 10\ [(D_T)_f / k_{\text{box-flange}} b_f]} \tag{14.28}$$

The buckle half-wavelength, a, for this case is given in terms of the properties of the flange as (Kollár and Springer, 2003, Table 4.10)

$$a_{\text{box-flange}} = b_f \sqrt[4]{\frac{D_L}{D_T}\left(\frac{1}{1 + 4.139\ \xi_{\text{box-flange}}}\right)} \tag{14.29}$$

The critical axial load for the local flange buckling of a homogeneous box-shaped profile loaded in axial compression is therefore,

$$P_{cr}^{\text{local,box-flange}} = \sigma_{cr}^{\text{local,box-flange}} A_z \tag{14.30}$$

For a nonhomogeneous section, the critical axial load is found by integration of the axial stresses over the different property walls in the section. The strain corresponding to the critical stress in the flange is found first, and then the stresses in all the walls of the section are determined.

If $(\sigma_{ss}^{ss})_f/(E_L)_f > (\sigma_{ss}^{ss})_w/(E_L)_w$, the webs will buckle before the flanges. The webs will therefore be restrained by both of the flanges, one on either longitudinal edge of each web. The spring constant is then given as

$$k_{\text{box-web}} = \frac{2(D_T)_f}{b_f}\left[1 - \frac{(\sigma_{ss}^{ss})_w (E_L)_f}{(\sigma_{ss}^{ss})_f (E_L)_w}\right] \tag{14.31}$$

and the buckling stress for the rotationally restrained web is given in terms of the properties of the web as

$$\sigma_{cr}^{\text{local,box-web}} = \frac{\pi^2}{d_w^2 t_w} [2\sqrt{(D_L D_T)}(1 + 4.139\xi) + (D_{LT} + 2D_S)(2 + 0.62\xi_{\text{box-web}}^2)] \tag{14.32}$$

where

$$\xi_{\text{box-web}} = \frac{1}{1 + 10\zeta_{\text{box-web}}} = \frac{1}{1 + 10\,[(D_T)_w / k_{\text{box-web}} d_w]} \tag{14.33}$$

The buckle half-wavelength, a, for this case is given in terms of the properties of the web as (Kollár and Springer, 2003, Table 4.10)

$$a_{\text{box-web}} = d_w \sqrt[4]{\frac{D_L}{D_T}\left(\frac{1}{1 + 4.139\,\xi_{\text{box-web}}}\right)} \tag{14.34}$$

The critical axial load for the local web buckling of a homogeneous box-shaped profile loaded in axial compression is therefore

$$P_{cr}^{\text{local,box-web}} = \sigma_{cr}^{\text{local,box-web}} A_z \tag{14.35}$$

For a nonhomogeneous section, the critical axial load is found by integration of the axial stresses over the various property walls in the section. The strain corresponding to the critical stress in the web is first found and then the stresses in all the walls of the section are determined. It can be observed that the buckling equations for the webs and flanges of the rectangular box beams are of the same form but have the web and flange properties interchanged.

Approximate Procedure for I-Profiles To use the equations presented previously due to Kollár (2003), all of the in-plane properties of the orthotropic walls of the section are required to predict the buckling stress. However, the in-plane shear modulus is often not reported by manufacturers and is not required to be reported by the European Standard EN 13706 (CEN, 2002a). In this situation a simplified form suggested in the *Structural Plastics Design Manual* (ASCE, 1984, p. 676) for obtaining the critical flange buckling stress for the free and rotationally restrained flange in an I-shaped profile can be used (Mottram, 2004):

$$\sigma_{cr,approx}^{\text{local,I-flange}} = \frac{\pi^2 t_f^2}{(b_f/2)^2}\left[\left(0.45 + \frac{b_f^2}{4a^2}\right)\frac{\sqrt{E_L E_T}}{12(1 - \nu_L \nu_T)}\right] \tag{14.36}$$

where the mechanical properties are those of the flange. This expression requires the length of the buckle half-wavelength, a, which is not known a priori. To use this equation the buckle half-wavelength can be taken conservatively as $2b_f$ for wide-flange and $3b_f$ for narrow-flange conventional pultruded I-shaped profiles. Alternatively, the first term in brackets can be taken even more conservatively as 0.45. Note that Equation (14.36) is identical to Equation (13.37) as the simplified method does not distinguish between flange buckling in a beam and a column. The approximate critical axial load for the local flange buckling of a homogeneous I-shaped profile loaded in axial compression is therefore

$$P_{\text{cr,approx}}^{\text{local,I-flange}} = \sigma_{\text{cr,approx}}^{\text{local,I-flange}} A_z \tag{14.37}$$

Additional Issues Related to Buckling of Pultruded Profiles A comparison between the equations presented in this chapter for predicting local buckling of webs and flanges in axially loaded compression members with those presented in Chapter 13 for predicting similar phenomena reveals that it is only the coefficients of restraint at the web flange junction that leads to different, and expected, results for these members.

For further discussion on local buckling-related phenomena in pultruded profiles, the reader should refer to the discussions in Section 13.7.2 under the heading "Further Discussion of Flange Buckling in I-Sections" related to slenderness ratios of thin walls in pultruded profiles, and in Section 13.7.5 related to the influence of sustained load and creep deformation on buckling of pultruded profiles. These discussions are also applicable to pultruded compression members.

14.7.4 Interaction Between Local and Global Buckling Modes in Intermediate-Length Compression Members

Whether a pultruded compression member fails due to global flexural buckling, local buckling, or material compressive failure depends on the slenderness or the column, usually represented by the nondimensional slenderness ratio, kL/r. Failure due to material crushing will seldom occur in conventional pultruded I-shaped sections and larger box sections (b, $h > 4$ in.). Rather, short columns ($kL/r < kL_c/r$) will fail due to local buckling instability of the flanges or webs, whereas long columns ($kL/r > kL_c/r$) will fail due to global flexural buckling. kL_c/r is the critical slenderness ratio of a column for which bifuraction theory predicts that local and global buckling will occur at the same time.

An interesting and important phenomenon occurs in the neighborhood of the critical length (or critical slenderness ratio), where the buckling mode changes from the local buckling mode to the global buckling mode. In this transition region, the local and global buckling modes interact due to imperfection sensitivity and reduce the overall buckling load of the section (Barbero and Tomblin, 1994; Barbero et al., 2000; Lane and Mottram, 2002). Columns in this region are often referred to as *intermediate-length columns.* In addition, the failure that occurs when the column fails in this combined coupled buckling mode[9] is unstable and catastrophic, whereas when the column fails due

[9]This is often referred to as the *tertiary buckling mode.*

to buckling in either of the two other uncoupled modes, it is stable and occurs with significant lateral or local deformations. The instability in the coupled mode typically starts as global Euler buckling, but then a local buckle at the location of maximum deformation occurs, which leads rapidly to a catastrophic failure and a separation of the flange from the web of the section (Mottram et al., 2003b).

Since the coupled buckling failure is unstable and catastrophic, it must be evaluated with care by designers of structures containing pultruded compression members. This phenomenon has an analogy in steel structures; however, in steel profiles the transition region occurs between the material yielding failure mode and the global buckling mode since local buckling is not a critical failure mode for most rolled steel sections.

The extent of the transition region and the reduction in the load-carrying capacity in the transition region is not precisely known. An empirical equation to predict the critical column buckling load, P_{cr}, that includes local, global, and coupled mode buckling in the transition region for I-shaped profiles has been proposed by Barbero and Tomblin (1994) in the form

$$P_{cr}^{int} = k_i P_L \tag{14.38}$$

and

$$\sigma_{cr}^{int} = \frac{P_{cr}^{int}}{A_z} \tag{14.39}$$

where P_L is the column buckling load when it is controlled by the local buckling mode (i.e., the column is short). k_i is a factor that accounts for local and global imperfections and is given as

$$k_i = k_\lambda - \sqrt{k_\lambda^2 - \frac{1}{c\lambda^2}} \tag{14.40}$$

with

$$k_\lambda = \frac{1 + (1/\lambda^2)}{2c} \tag{14.41}$$

where c is an empirical curve-fitting constant and λ is a nondimensional slenderness ratio, defined as

$$\lambda = \frac{kL}{\pi}\sqrt{\frac{P_L}{(EI)_{min}}} = \sqrt{\frac{P_L}{P_E}} \tag{14.42}$$

where $(EI)_{\min}$ is the flexural rigidity of the section about the minor axis. Figure 14.5 is a plot of the nondimensional failure load, k_i, versus the nondimensional column slenderness ratio, λ, and comparisons with published experimental data.

The critical length at which the local and global curves intersect, can be found from Eq. (14.7) as

$$kL = \sqrt{\frac{\pi^2(EI)_{\min}}{P_L}} \tag{14.43}$$

It should also be noted that in this formulation the critical load flexural buckling is the Euler load and does not include the effects of shear deformation. It therefore predicts slightly higher critical loads than if shear deformation were included. [To include shear deformation, P_{cr}^{flex} given in Eq. (14.6) can be used in place of P_E in Eq. (14.42)]

Based on a combination of testing and theoretical predictions of material properties and individual mode buckling loads on 4 × 4 × ¼ and 6 × 6 × ¼ I-shaped profiles, Barbero and Tomblin (1994) initially recommended a value of $c = 0.84$ for I-shaped conventional pultruded profiles. Using a more extensive database of test results, Barbero and DeVivo (1999) later proposed

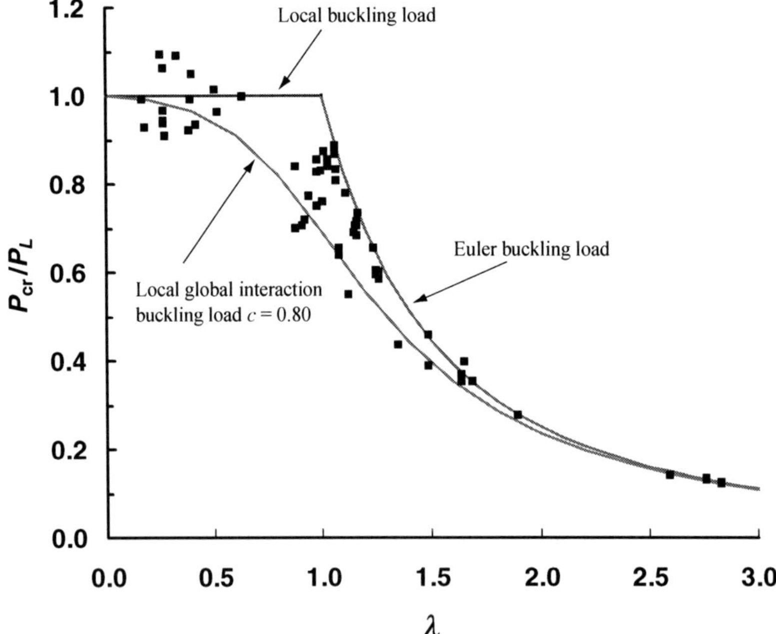

Figure 14.5 Local–global interaction equation compared with test data. (Courtesy of J. T. Mottram.)

a value of $c = 0.65$. However, recent test data by Lane and Mottram (2002) on W $8 \times 8 \times 3/8$ conventional pultruded profiles appear to indicate less imperfection sensitivity than that reported by Barbero and DeVivo (1999) and that the 0.65 value for c suggested by Barbero and co-authors may be unduly conservative.

Based on the discussion above, the following is recommended for the design of intermediate-length pultruded columns. If $0.5 \leq \lambda \leq 1.5$, local–global interaction should be considered, and the Barbero and Tomblin equation should be used with a value of $c = 0.80$. This value should be used with the local buckling loads calculated using the Kollár equations presented previously, together with manufacturer-supplied material properties. As discussed previously, these material property data are typically lower than the measured material property data reported in experimental investigations conducted by Barbero and co-workers and by Mottram and co-workers, for which comparisons, and hence c values, with column test data were obtained. Therefore, at this time, the Barbero interaction equation with $c = 0.80$ is felt to be a conservative approach. If $\lambda < 0.5$, it can be assumed that local buckling will control, and if $\lambda > 1.5$, it can be assumed that global flexural buckling (with shear deformation) will control.

14.7.5 Flange and Web Longitudinal Material Failure

For pultruded profiles where the slenderness ratio of the flange, $0.5b_f/t_f$, and the slenderness ratio of the web, d_w/t_w, are small and local buckling does not occur (or is prevented by multiple longitudinal and transverse stiffeners), the compression member may fail, due to crushing of the pultruded material due to compressive stresses. To design for flange or web material compressive failure, the ultimate compressive strength is taken as the longitudinal compressive strength of the pultruded material in the web or the flange,

$$\sigma_{\text{cr}}^{\text{comp}} = \sigma_{L,c} \tag{14.44}$$

The critical axial load in a homogeneous member due to compressive material failure is

$$P_{\text{cr}}^{\text{comp}} = \sigma_{\text{cr}}^{\text{comp}} A_z \tag{14.45}$$

14.8 DESIGN PROCEDURE FOR CONCENTRICALLY LOADED COMPRESSION MEMBERS

The design procedure presented in what follows for compression members (or columns) permits design by the allowable stress design (ASD) basis or the load and resistance factor design (LRFD) basis as discussed in Chapter 12. Both of these bases use the analytical, mechanics-based equations presented in preceding sections. They do not use the empirical equations presented in the U.S. pultrusion manufacturers' design guides mentioned in Chapter 12. Consequently, the load tables presented in manufacturers' guides are not used in this book to size conventional pultruded profiles. However, the empirical equations presented in the manufacturers' guides, and hence the load tables, can be used to estimate the size or the profile, or its load-carrying capacity, if necessary, for the purposes of a preliminary design. It should be kept in mind that none of the manufacturers' design guides currently accounts for global and local coupling in intermediate-length columns.

Step 1. Determine the design loads, choose the design basis, and determine the basis factors. The concentric axial load on the compression member is determined from the structural geometry and loading requirements. End conditions used to establish the effective length of the column, which are compatible with the end connection details, are established. The design basis for the compression member is chosen and the appropriate safety factors (ASD) and resistance factors (LRFD) are determined. Load factors for the LRFD method are listed. Maximum short- and long-term axial shortening criteria are taken from codes or established by the project specifications.

Step 2. Select a profile type and trial size. Select a trial profile. The trial size can be chosen assuming a maximum axial shortening of the column of $L/1500$ and finding the area of the cross section required to achieve this shortening. Alternatively, manufacturers' design guides can be used as a first approximation.

Step 3. Determine the maximum design stresses or forces. The axial stress on the member subjected to the nominal loads are calculated for the ASD procedure. The axial load on the member subjected to the factored loads is calculated for the LRFD procedure.

Step 4. Determine the critical stresses and forces. Determine (or calculate) global flexural buckling, torsional buckling, local buckling, and compressive strength critical stresses (ASD) or axial forces (LRFD). If the column is in the intermediate-length range, calculate the critical stress or axial force for the intermediate column in addition to the above.

Step 5. Determine the factored critical stresses or forces. For ASD divide the critical stresses determined in step 4 by the appropriate safety factor for compressive stress. For LRFD, multiply the critical forces by the appropriate resistance factors for compression members.

Step 6. Check the ultimate strength or capacity of the selected profile. For ASD, check that the design stresses are less than the allowable stresses. For LRFD, check that the factored design forces are less than the factored resistances. Return to step 2 if the trail size does not work.

Step 7. Check the profile for serviceability. Check that the axial shortening in the chosen trial profile does not exceed the code-stipulated maximum. Service loads are used for both the ASD and LRFD procedures.

Step 8. Dimension the ancillary bracing, stiffeners, and bearing plates. Dimension (or design if quantitative procedures are available) braces, stiffeners, and bearing plates for the compression member. Provide a sketch of the compression member showing all ancillary elements, dimensions, and important construction notes. (If the column is connected to other framing members, dimension and detail the connection following the design procedure given in Chapter 15.)

Design Example 14.1: Pultruded Concentrically Loaded Column Design A conventional commercially produced pultruded E-glass/vinylester wide-flange I-shaped profile is to be used as a column to support the floor beams of a three-story building structure with a typical repeating floor plan as shown in Fig. 13.13. The nominal story height is 9 ft. The floor dead load is 20 psf and the floor live load is 40 psf. Assume that the roof load is the same as the typical floor load. Live load reduction is not permitted. The beams are loaded on their top flanges and are connected to the columns with web clip angles (simple shear connections). The column is to be designed for a maximum axial shortening per floor of $\delta_{\max} = L/1500$. The short-term material properties taken from the manufacturer's design guide (Creative Pultrusions' SuperStructural specifications) for the pultruded glass FRP profile[10] series selected for the design are listed in Table 14.1.

TABLE 14.1 Pultruded Material Properties in Flange and Web

Symbol	Description	Flange Value[a]	Web Value[a]
E_L^c	Longitudinal compressive modulus	3.85×10^6 psi	2.8×10^6 psi
E_L^t	Longitudinal tensile modulus	4.16×10^6 psi	3.1×10^6 psi
E_T^c	Transverse compressive modulus	1.9×10^6 psi	1.9×10^6 psi
E_T^t	Transverse tensile modulus	NR	1.4×10^6 psi
G_{LT}	In-plane shear modulus	NR	NR
ν_L	Major (longitudinal) Poisson ratio	0.35	0.35
$\sigma_{L,c}$	Longitudinal compressive strength	52,500 psi	43,125 psi
$\sigma_{L,t}$	Longitudinal tensile strength	46,000 psi	35,000 psi
$\sigma_{T,c}$	Transverse compressive strength	20,400 psi	16,330 psi
$\sigma_{T,t}$	Transverse tensile strength	NR	12,000 psi
τ_{TT}	Interlaminar shear strength	4,500 psi	3,900 psi
τ_{LT}	In-plane shear strength	NR	7,000 psi
E_b	Full-section flexural modulus	3.9×10^6 psi	
G_b	Full-section shear modulus	0.5×10^6 psi	

[a]NR, not reported.

[10]Note that this is referred to as a conventional pultruded profile since it has a conventional U.S. pultruded shape. However, the properties are not the same in the web and flange of the profile.

SOLUTION

Step 1. Determine the design loads, choose the design basis, and determine the basis factors.

Design loads and internal forces For the given tributary area of 6 ft × 15 ft = 90 ft², beam weights of 10 × (15 + 6) = 210 lb per floor, and an assumed column weight of 10 lb/ft, the column loads are calculated as follows. For ASD and LRFD serviceability limit state calculations:

Dead load:

$$p_{\mathrm{DL}} = ((20 \text{ psf} \times 90 \text{ ft}^2) + 210 \text{ lb} + [(10 \times 9) \times 3 \text{ floors}] = 6300 \text{ lb}$$

Live load: $p_{\mathrm{LL}} = (40 \text{ psf} \times 90 \text{ ft}^2) \times 3 \text{ floors} = 10{,}800 \text{ lb}$

Total service load: $p_S = 6300 + 10{,}800 = 17{,}100 \text{ lb}$

For LRFD design the ASCE 7 load factors for dead and live loads are 1.2 and 1.6, respectively, and the factored loads are

Dead load: $P_{\mathrm{D}} = 1.2(6300) = 7560 \text{ lb}$

Live load: $P_{\mathrm{L}} = 1.6(10{,}800) = 17{,}280 \text{ lb}$

Total factored load: $P_u = 7560 + 17{,}280 = 24{,}840 \text{ lb}$

ASD and LRFD factors The ASD safety factors to be used are taken from Chapter 12: Compression = 3.0. The LRFD resistance factors are determined using the Eurocomp material factors (Chapter 12) as follows:

$$\gamma_{m,1} = 1.15 \text{ (material properties obtained from test data)}$$

$$\gamma_{m,2} = 1.1 \text{ (fully cured pultruded material)}$$

$$\gamma_{m,3} = 1.0\ (T_g > 80^\circ \text{ and room-temperature service conditions})$$

No other environmental or long-term reduction factors are assumed for this problem. Therefore,

$$\gamma_m = \gamma_{m,1}\gamma_{m,s}\gamma_{m,3} = 1.15(1.1)(1.0) = 1.265 \leq 1.5$$

Therefore, $\gamma_m = 1.5$ (minimum value permitted by Eurocomp for the ultimate limit state) The LRFD resistance factor is taken as the inverse of the Eurocomp material factor:

$$\phi = \frac{1}{\gamma_m} = \frac{1}{1.5} = 0.67$$

For the serviceability limit state the material factor is taken as 1.0 and the resistance factor is therefore also 1.0.

Step 2. Select a profile type and trial size. Assuming a trail size based on an axial shortening limit of $L/1500$ gives EA_{reqd} as follows:

$$EA_{\text{reqd}} = 1{,}500P_{\text{service}} = 1500(17{,}100) = 2.56 \times 10^7 \text{ lb}$$

For a wide-flange I-section with different compressive moduli in the flanges and web, the average compressive modulus is approximated for the composite section by assuming that the area of the flanges is double that of the web area (taken as unity):

$$\begin{aligned} E_{\text{ave}} &= \frac{E_{L(\text{flange})}A_{\text{flanges}} + E_{L(\text{web})}A_{\text{web}}}{A_{\text{flanges}} + A_{\text{web}}} \\ &\approx \frac{(3.85 \times 10^6)(2.0) + (2.8 \times 10^6)(1.0)}{2.0 + 1.0} = 3.50 \times 10^6 \text{ psi} \end{aligned}$$

The approximate required area of the profile is then

$$A_{\text{reqd}} = \frac{EA_{\text{reqd}}}{E_{\text{ave}}} = \frac{2.56 \times 10^7}{3.50 \times 10^6} = 7.31 \text{ in}^2$$

A W8 $\times$ 8 $\times$ $\frac{3}{8}$ profile from the manufacturer's conventional shapes in the W series[11] has an area of 8.82 in^2, which is selected since it has an area closest to (but larger than) the required area of 7.31 in^2. The geometric properties of the profile are given in Table 14.2.

It can be noted that for a required axial load of 17,100 lb, Creative Pultrusions' design guide suggests the same profile, gives its axial capacity as 26,942 lb, and indicates that the capacity is controlled by Euler buckling about the weak axis (referred to as the *long-column mode*).

Step 3. Determine the maximum design stresses or forces. Since the longitudinal moduli of the flanges and the web are different, the loads and stresses carried in the flanges and web are calculated separately. The area of the web is calculated (assuming that the small fillet region is part of the web) as

[11] W or WF is typically used to identify a pultruded wide-flange profile. I is used to identify the narrow-flange profile.

TABLE 14.2 Geometric Properties of Conventional Glass FRP W 8 × 8 × $\frac{3}{8}$ Profile

Symbol	Description	Value
I_x	Second moment about major axis	100.35 in^4
I_y	Second moment about minor axis	31.65 in^4
A_z	Gross area	8.82 in^2
J	Torsional constant	0.422 in^4
t_w	Web thickness	0.375 in.
t_f	Flange thickness	0.375 in.
d	Nominal section depth	8 in.
d_w	Web depth	7.625 in.
b	Nominal section width (breadth)	8 in.
b_f	Flange width (breadth)	8 in.
w	Nominal weight	6.61 lb/ft

$$A_{\text{web}} = A_g - A_{\text{flanges}} = 8.82 - 2.0(8.0)(0.375) = 2.82 \text{ in}^2$$

The nominal axial load will be carried as follows:

$$P_{\text{flanges}} = P_s \frac{E_{\text{flange}} A_{\text{flanges}}}{E_{\text{flange}} A_{\text{flanges}} + E_{\text{web}} A_{\text{web}}} = 17{,}100\left[\frac{3.85(6.0)}{3.85(6.0) + 2.8(2.82)}\right]$$

$$= 12{,}744 \text{ lb}$$

$$P_{\text{web}} = P_s - P_{\text{flanges}} = 17{,}100 - 12{,}744 = 4356 \text{ lb}$$

That is, the flanges and the web carry 74.5 and 25.5% of the nominal service load, respectively.

The ASD design stresses in the flanges and webs are therefore

$$\sigma_{\text{flange}} = \frac{P_{\text{flanges}}}{A_{\text{flanges}}} = \frac{12{,}744}{6.0} = 2124 \text{ psi}$$

$$\sigma_{\text{web}} = \frac{P_{\text{web}}}{A_{\text{web}}} = \frac{4356}{2.82} = 1545 \text{ psi}$$

For LRFD the axial design load is

$$P_u = 24{,}840 \text{ lb}$$

For the profile chosen, the axial shortening under the service load for the first-floor column is

$$\delta_{\max} = \frac{PL}{\sum AE} = \frac{17{,}100(108)}{6.0(3.85 \times 10^6) + 2.82(2.8 \times 10^6)}$$

$$= 0.06 \text{ in.} \rightarrow \frac{L}{1800} < \frac{L}{1500}$$

Step 4. Determine the critical stresses and forces.

Global flexural buckling Determine the composite flexural stiffness of the profile about the major and minor axes, respectively.

$$EI_{\text{major}} = 2(3.85 \times 10^6)\left[\frac{8(0.375)^3}{12} + (8 \times 0.375)\left(4 - \frac{0.375}{2}\right)^2\right]$$

$$+ (2.8 \times 10^6)\,\frac{0.375(7.25)^3}{12}$$

$$= 3.69 \times 10^8 \text{ lb-in}^2$$

Note that the second moment of a flange about its own centroidal axis is usually neglected because it is very small. It is included here for completeness.

$$EI_{\text{minor}} = 2(3.85 \times 10^6)\left[\frac{0.375(8)^3}{12}\right] + (2.8 \times 10^6)\,\frac{7.25(0.375)^3}{12}$$

$$= 1.23 \times 10^8 \text{ lb-in}^2$$

Determine the Euler buckling load for the minor axis buckling:

$$P_{\text{euler}} = \frac{\pi^2 E_L I}{(kL)^2} = \frac{\pi^2 (EI)_{\min}}{(kL)^2} = \frac{\pi^2(1.23 \times 10^8)}{(1.0 \times 108)^2} = 104{,}078 \text{ lb}$$

Determine the shear correction using the full-section shear modulus since the in-plane shear modulus is not given. Also, the shear coefficient, k_{tim}, for a homogeneous section bending about its major axis is used as an approximation for the shear coefficient for the inhomogeneous section bending about its minor axis. This will not have a significant effect since the profile does not have large (i.e., order of magnitude) property differences between the web and the flanges, and it will be seen that the shear deformation contribution is quite small (~6%).

$$k_{\text{tim}} A_z G_{LT} = k_{\text{tim}} A_z G_b = 0.413(8.82)(0.5 \times 10^6) = 1.82 \times 10^6 \text{ lb}$$

(or alternatively use $A_{\text{flanges}} G_b = (6.0)(0.5 \times 10^6) = 3 \times 10^6$ lb for the minor axis *KAG* term)

Determine the critical flexural buckling load:

$$P_{cr}^{flex} = \frac{P_{euler}}{1 + P_{euler}/k_{tim}A_zG_{LT}} = \frac{104{,}078}{1 + 104{,}078/(1.82 \times 10^6)} = 98{,}448 \text{ lb}$$

Note that the shear deformation reduces the critical flexural buckling load by approximately 6%.

The critical flexural buckling stresses in the web and flange are found based on the percentage of load carried by flange (74.5%) and web (25.5%) since the member is linear elastic to failure. Also, since the distribution of the load to the flanges and the web for the global ultimate failure modes is in the same proportion to the design loads, this means that the flange and the web will fail at the same load; therefore, only one of them needs to be checked in what follows (Tables 14.3 and 14.4). The flange and web stresses for global flexural buckling critical limits state are

$$\sigma_{cr(flange)}^{flex} = \frac{P_{cr}^{flex}(0.745)}{A_{z(flange)}} = \frac{98{,}448(0.745)}{6.0} = 12{,}224 \text{ psi}$$

$$\sigma_{cr(web)}^{flex} = \frac{P_{cr}^{flex}(0.255)}{A_{z(web)}} = \frac{98{,}448(0.255)}{2.82} = 8902 \text{ psi}$$

Global torsional buckling The polar second moment is

TABLE 14.3 Critical Stress, ASD Design Stresses and Loads, and LRFD Design Resistances

Mode	σ_{cr}	SF	σ_{allow}	R_{cr}	ϕ^a	ϕR_{cr}
Global flexural buckling[b]	12,224 psi	3.0	4,075 psi	98,448 lb	0.67	65,960 lb
Global torsional buckling[b]	16,543 psi	3.0	5,514 psi	133,236 lb	0.67	89,268 lb
Local buckling (flange controls)	8,959 psi	3.0	2,986 psi	72,128 lb	0.67	48,326 lb
Global–local interaction buckling[b]	7,171 psi	3.0	2,390 psi	57,751 lb	0.67	38,693 lb
Flange compressive failure	52,500 psi	3.0	17,500 psi	422,672 lb	0.67	283,190 lb
Web compressive failure	43,125 psi	3.0	14,375 psi	477,398 lb	0.67	319,857 lb

[a] According to the Eurocomp partial factor method, different ultimate modes of failure (e.g., flexure, compression) are not treated independently as in a traditional LRFD or ASD approach.

[b] The flange values are listed for the ASD. As noted, flanges and web critical stresses for these modes are proportional to the design stress; therefore, either one can be listed.

TABLE 14.4 Comparison of Design Demands with Values Permitted for ASD and LRFD

Mode	σ_{design}	vs.	σ_{allow}	R_{design}	vs.	ϕR_{cr}
Ultimate limit state						
Global flexural buckling	2,124 psi	<	4,075 psi	24,840 lb	<	65,960 lb
Global torsional buckling	2,124 psi	<	5,514 psi	24,840 lb	<	89,268 lb
Local buckling (flange controls)	2,124 psi	<	2,986 psi	24,840 lb	<	48,326 lb
Global–local interaction buckling	2,124 psi	<	2,390 psi	24,840 lb	<	38,693 lb
Flange compressive failure	2,124 psi	<	17,500 psi	24,840 lb	<	283,190 lb
Web compressive failure	1,545 psi	<	14,375 psi	24,840 lb	<	319,857 lb
Serviceability limit state						
Axial shortening	0.060 in.	<	0.072 in.	0.060 in.	<	0.072 in.

$$I_p = \sqrt{I_x^2 + I_y^2} = \sqrt{(100.35)^2 + (31.65)^2} = 105.22 \text{ in}^4$$

The longitudinal modulus of the composite section under axial load is

$$E_L = \frac{E_{L(\text{flanges})}A_{\text{flanges}} + E_{L(\text{web})}A_{\text{web}}}{A_{\text{flanges}} + A_{\text{web}}}$$

$$= \frac{(3.85 \times 10^6)(2.0)(3.0) + (2.8 \times 10^6)(1.0)(2.82)}{2.0(3.0) + 1.0(2.82)} = 3.51 \times 10^6 \text{ psi}$$

$$C_\omega = \frac{I_y d^2}{4} = \frac{31.65(8)^2}{4} = 506.4 \text{ in}^6$$

Note that this is an approximation of the composite warping stiffness, $\Sigma\, E_L C_\omega$. For the Saint-Venant torsional stiffness, the full-section shear modulus is used, also as an approximation.

The critical torsional buckling stress[12] (not including shear deformation effects) is

[12] This can be thought of as the effective critical buckling stress for the section as a whole.

$$\sigma_{cr}^{tor} = \frac{1}{I_p}\left[\frac{\pi^2 E_L C_\omega}{(k_\omega L)^2} + G_{LT}J\right] = \frac{1}{105.22}\left[\frac{\pi^2(3.51 \times 10^6)(506.4)}{(1.0 \times 108)^2}\right.$$

$$\left. + (0.5 \times 10^6)(0.422)\right] = 16{,}299 \text{ psi}$$

The critical torsional buckling load is

$$P_{cr}^{tor} = \sigma_{cr}^{tor} A_z = 16{,}299(8.82) = 143{,}760 \text{ lb}$$

Including the shear correction gives

$$P_{cr}^{tor+sd} = \frac{P_{cr}^{tor}}{1 + P_{cr}^{tor}/KAG} = \frac{143{,}760}{1 + 143{,}760/(1.82 \times 10^6)} = 133{,}236 \text{ lb}$$

Note that the shear deformation reduces the critical torsional buckling load by approximately 8%. The critical torsional buckling stresses in the flanges and the web are

$$\sigma_{cr(flange)}^{tor+sd} = \frac{P_{cr}^{tor+sd}(0.745)}{A_{z(flange)}} = \frac{133{,}236(0.745)}{6.0} = 16{,}543 \text{ psi}$$

$$\sigma_{cr(web)}^{tor+sd} = \frac{P_{cr}^{tor+sd}(0.255)}{A_{z(web)}} = \frac{133{,}236(0.255)}{2.82} = 12{,}048 \text{ psi}$$

Local buckling Since the material properties of the flanges and the webs are different for this profile, the plate rigidities are calculated separately for the flange properties and for the web properties. Care should be taken to use the correct properties in the local buckling equations that follow.

For the flange rigidities, calculate the flange minor (transverse) Poisson ratio:

$$\nu_T = \frac{E_T^c}{E_L^c}\nu_L = \frac{(1.9)}{(3.85)}0.35 = 0.17$$

Calculate the flange flexural rigidities:

$$D_L = \frac{E_L^c t_f^3}{12(1 - \nu_L \nu_T)} = \frac{(3.85 \times 10^6)(0.375)^3}{12[1 - 0.35(0.17)]} = 17{,}989 \text{ lb-in.}$$

$$D_T = \frac{E_T^c}{E_L^c} D_L = \frac{1.9}{3.85}(17{,}989) = 8878 \text{ lb-in.}$$

$$D_{LT} = \nu_T D_L = 0.17(17{,}989) = 3058 \text{ lb-in.}$$

$$D_S = \frac{G_{LT}t_f^3}{12} = \frac{(0.5 \times 10^6)(0.375)^3}{12} = 2197 \text{ lb-in.}$$

For the web rigidities, calculate the web minor (transverse) Poisson ratio:

$$\nu_T = \frac{E_T^c}{E_L^c}\nu_L = \frac{1.9}{2.8}(0.35) = 0.24$$

Calculate the web flexural rigidities:

$$D_L = \frac{E_L^c t_w^3}{12(1 - \nu_L\nu_T)} = \frac{(2.8 \times 10^6)(0.375)^3}{12(1 - (0.35)(0.24)} = 13{,}433 \text{ lb-in.}$$

$$D_T = \frac{E_T^c}{E_L^c}D_L = \frac{1.9}{2.8}(13{,}433) = 9115 \text{ lb-in.}$$

$$D_{LT} = \nu_T D_L = 0.24(13{,}433) = 3244 \text{ lb-in.}$$

$$D_S = \frac{G_{LT}t_w^3}{12} = \frac{(0.5 \times 10^6)(0.375)^3}{12} = 2197 \text{ lb-in.}$$

Note that the flange is more anisotropic than the web. This is a function of the fiber architecture of the flanges and indicates that the flanges have more unidirectional rovings than the web.

Calculate the buckling stresses for the simply supported flange and web to determine which wall buckles first:

$$(\sigma_{\text{free}}^{\text{ss}})_f = \frac{4t_f^2}{b_f^2}G_{LT} = \frac{4(0.375)^2(0.5 \times 10^6)}{(8)^2} = 4395 \text{ psi}$$

$$(\sigma_{\text{ss}}^{\text{ss}})_w = \frac{2\pi^2}{t_w d_w^2}(\sqrt{D_L D_T} + D_{LT} + 2D_S)$$

$$= \frac{2\pi^2}{0.375(7.625)^2}[\sqrt{13{,}433(9115)} + 3244 + 2(2197)]$$

$$= 16{,}915 \text{ psi}$$

Determine which wall buckles first by checking:

$$\frac{(\sigma_{\text{free}}^{\text{ss}})_f}{(E_L)_f} \text{ vs. } \frac{(\sigma_{\text{ss}}^{\text{ss}})_w}{(E_L)_w} \rightarrow \frac{4395}{3.85 \times 10^6} \text{ vs. } \frac{16{,}915}{2.8 \times 10^6}$$

$$\rightarrow 1.14 \times 10^{-3} < 6.04 \times 10^{-3}$$

Therefore, the flange buckles first. Now calculate the local buckling stress for the flange. First, calculate the junction stiffness, k, and then the coefficient of restraint, ζ.

$$k_{\text{I-flange}} = \frac{(D_T)_w}{d_w}\left[1 - \frac{(\sigma^{\text{ss}}_{\text{free}})_f (E_L)_w}{(\sigma^{\text{ss}}_{\text{ss}})_w (E_L)_f}\right]$$

$$= \frac{9115}{7.625}\left[1 - \frac{(4395)(2.8 \times 10^6)}{(16{,}915)(3.85 \times 10^6)}\right] = 969.8 \text{ lb}$$

$$\zeta_{\text{I-flange}} = \frac{D_T}{k_{\text{I-flange}}\, L_T} = \frac{9{,}115}{(969.8)(8/2)}$$

$$= 2.289 \text{ (nondimensional coefficient of restraint)}$$

and

$$\sigma_{\text{cr}}^{\text{local,I-flange}} = \frac{1}{(b_f/2)^2 t_f}\left(7\sqrt{\frac{D_L D_T}{1 + 4.12\zeta_{\text{I-flange}}}} + 12 D_S\right)$$

$$= \frac{1}{(8/2)^2(0.375)}\left[7\sqrt{\frac{17{,}989(8878)}{1 + (4.12)(2.289)}} + 12(2{,}179)\right]$$

$$= 8959 \text{ psi}$$

The length of the buckle half-wavelength (for information purposes) is

$$a_{\text{I-flange}} = 1.675\,\frac{b_f}{2}\sqrt[4]{\frac{D_L}{D_T}(1 + 4.12\zeta_{\text{I-flange}})}$$

$$= 1.675\,\frac{8}{2}\sqrt[4]{\frac{17{,}989}{8{,}878}[1 + 4.12(2.289)]} = 14.37 \text{ in.}$$

Since the section is inhomogeneous the critical axial load at which local buckling occurs is,

$$P_{\text{cr}}^{\text{local,I-flange}} = \sigma_{\text{cr}}^{\text{local,I-flange}} A_{z(\text{flange})} + \sigma_{\text{cr}}^{\text{local,I-flange}} \frac{E_{L(\text{web})}}{E_{L(\text{flange})}} A_{z(\text{web})}$$

$$= 8959(6.0) + (8959)\frac{2.8}{3.85}(2.82) = 72{,}128 \text{ lb}$$

Local–global buckling interaction To determine if the column falls into the intermediate-length region where local–global buckling interaction can occur, the nondimensional buckling ratio, λ, is calculated:

$$\lambda = \frac{kL}{\pi}\sqrt{\frac{P_L}{(EI)_{\min}}} = \sqrt{\frac{P_L}{P_E}} = \sqrt{\frac{72{,}128}{104{,}078}} = 0.832$$

Since $0.5 \leq \lambda \leq 1.5$, local–global interaction should be considered and the column falls into the intermediate-length range. It can also be observed that the global flexural buckling load, predicted not including shear deformation effects, is 104,078 lb and the critical local buckling load is 72,128 lb, which are reasonably close (within 30%), and therefore the column should fall into the intermediate-length transition region.

The critical length for the profile can be determined as

$$L = \frac{1}{k}\sqrt{\frac{\pi^2(EI)_{\min}}{P_L}} = \frac{1}{1.0}\sqrt{\frac{\pi^2(1.23 \times 10^8)}{72{,}128}}$$

$$= 129.7 \text{ in.} \quad \text{or} \quad 10 \text{ ft } 9.7 \text{ in.}$$

The reduced critical buckling stress, accounting for the local–global interaction, is now calculated from

$$k_\lambda = \frac{1 + (1/\lambda^2)}{2c} = \frac{1 + (1/0.832^2)}{2(0.80)} = 1.528$$

$$k_i = k_\lambda - \sqrt{k_\lambda^2 - \frac{1}{c\lambda^2}} = 1.528 - \sqrt{(1.528)^2 - \frac{1}{0.8(0.832)^2}} = 0.801$$

and the critical buckling load (with interaction) is

$$P_{\text{cr}}^{\text{int}} = k_i P_L = 0.801(72{,}128) = 57{,}751 \text{ lb}$$

The critical buckling stresses in the flanges and web (with interaction) are

$$\sigma_{\text{cr(flange)}}^{\text{int}} = \frac{P_{\text{cr}}^{\text{int}}(0.745)}{A_{z\text{(flange)}}} = \frac{57{,}751(0.745)}{6.0} = 7171 \text{ psi}$$

$$\sigma_{\text{cr(web)}}^{\text{int}} = \frac{P_{\text{cr}}^{\text{int}}(0.255)}{A_{z\text{(web)}}} = \frac{57{,}751(0.255)}{2.82} = 5222 \text{ psi}$$

Compressive material failure in the web and flange The critical axial loads based on the flange- and web-based material strengths are

$$P_{cr}^{comp,flange} = \sigma_{cr}^{comp,flange} A_{z(flange)} + \sigma_{cr}^{comp,flange} \frac{E_{(web)}}{E_{L(flange)}} A_{z(web)}$$

$$= 52{,}500(6.0) + (52{,}500)\frac{2.8}{3.85}(2.82) = 422{,}672 \text{ lb}$$

$$P_{cr}^{comp,web} = \sigma_{cr}^{comp,web} A_{z(web)} + \sigma_{cr}^{comp,web} \frac{E_{L(flange)}}{E_{L(web)}} A_{z(flange)}$$

$$= 43{,}125(2.82) + (43{,}125)\frac{3.85}{2.8}(6.0) = 477{,}398 \text{ lb}$$

Therefore, the flange critical load will control in material failure. This can also be seen from the following: Since the ratio of the flange-to-web longitudinal modulus (3.85/2.80 = 1.375) is larger than the ratio of the flange-to-web longitudinal strength (52,500/43,125 = 1.217), the flanges will be more highly stressed than the web but do not have a commensurate strength advantage over the web. However, also note that these values are an order of magnitude higher than those for the buckling modes and that they will very seldom control the capacity of a large (b, $h > 6$ in.) pultruded member.

Step 5. Determine the factored critical stresses or forces. The critical stresses and loads are summarized in Table 14.3. Using the ASD safety factors and the LRFD resistance factors determined in step 1, the allowable stresses calculated for ASD and the member resistances calculated for LRFD are tabulated in Table 14.3.

Step 6. Check the ultimate strength or capacity of the profile selected. For the ultimate limit state, the design stresses and member forces are compared to the allowable stresses and member resistances in Table 14.4. For the serviceability limit state, the maximum axial shortening is compared with the prescribed limit of $L/1500$ for both the ASD and LRFD bases.

Based on the comparisons shown in Table 14.4, it is seen that the W8 × 8 × $\frac{3}{8}$ SuperStructural pultruded profile meets the design demands according to both the ASD and LRFD design bases. It is seen that the local–global buckling interaction determines the critical load on column. According to the ASD basis, the maximum concentric axial design load is 57,751/3 = 19,250 lb, while the maximum design load according to the LRFD basis is 57,751 × 0.67 = 38,693 lb. Accounting for the LRFD load factors (a weighted average of 1.46 for this example), the LRFD gives a larger load-carrying capacity than the ASD basis (26,502 lb versus 19,250 lb). As noted previously, the manufacturer reports a maximum concentric design load for this profile of 26,942 lb.

14.9 CONCENTRICALLY LOADED TENSION MEMBERS

The tension member discussed in this section is designed to resist stress resultants (or internal forces) in the global x,y,z coordinate system that are caused by the concentric axial load, P_z, applied at the centroid of the cross section in the y–z plane as shown in Fig. 14.6.

The single stress resultant that the concentrically loaded tension member is designed for is the axial force, P_z (note that the letter T is often used for *tension*). According to homogeneous one-dimensional bar theory, this concentric axial force causes the member to develop uniform axial (tensile) stress, σ_z, throughout its cross section along its length and that is given by the normal stress equation

$$\sigma_z = \frac{P_z}{A_z} \tag{14.46}$$

where A_z is the gross cross-sectional area of the member. For nonhomogenous members the composite section properties are used as previously described for axially loaded compression members. As shown in Fig. 14.6, the sign of both the axial force and that of the stress will be positive, indicating tensile stress.

Since buckling does not occur in a tension member and therefore the longitudinal material strength limit state will control the capacity of the member, the stress on the net cross-sectional area, $\sigma_{z,\text{net}}$, that exists at locations of holes and cutouts at any point along the member length must also be calculated:

$$\sigma_{z,\text{net}} = \frac{P_z}{A_{z,\text{net}}} \tag{14.47}$$

where $A_{z,\text{net}}$ is the net cross-sectional area of the member. The net cross-sectional area is determined from

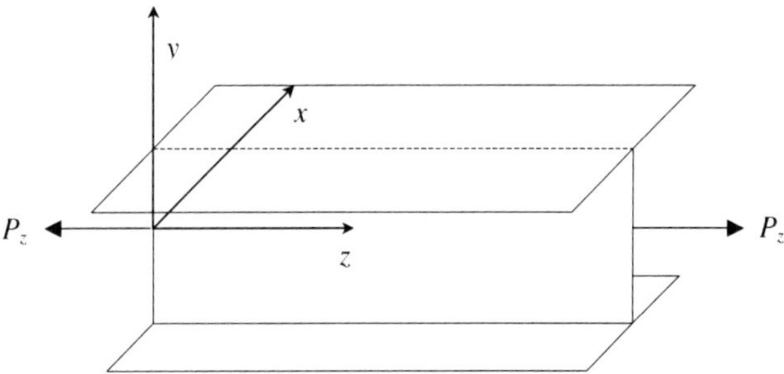

Figure 14.6 Concentrically loaded tension member.

$$A_{z,\text{net}} = A_z - \sum_i d_i t_i \tag{14.48}$$

where d_i is the diameter of a hole or the width of the cutout in the transverse direction of the member and t_i is the wall thickness at the location of the hole or cutout.

The axial stress (or axial force) and net axial stress (or net axial force) calculated must be less than the longitudinal tensile material strength, accounting for strength reduction due to holes and cutouts according to the design basis selected.

14.9.1 Deformations in Concentrically Loaded Tension Members

Due to the concentric axial force, P_z, the member will elongate by the amount

$$\delta_z = \frac{P_z L}{A_z E} \tag{14.49}$$

where $A_z E$ is the axial rigidity of the member in tension.

14.10 DETERMINATION OF DEFLECTIONS FOR SERVICEABILITY: AXIAL ELONGATION

Uniaxial bar theory is used to determine the elongation of a pultruded tension members. This is the procedure recommended by the *Structural Plastic Design Manual* (1984) and the *Eurocomp Design Code and Handbook* (Eurocomp, 1996). For the calculation of axial deformation, the axial rigidity is taken as

$$A_z E = A_z E_L \tag{14.50}$$

where E_L is the longitudinal modulus of the pultruded material in tension. It is generally assumed that the tensile and compressive longitudinal moduli of the FRP material are the same, and most manufacturers report the same value for the tensile and compressive moduli or do not differentiate between the two. Some manufacturers report different values for the tensile and compressive moduli. If tensile moduli are reported they should be used in deformation calculations for tension members. In the case of a nonhomogenous profile having different moduli in different walls of the profile, the composite axial rigidity, $\Sigma_i E_i A_i$, is used (provided that the property variation is symmetric with respect to the centroidal axes). To determine long-term axial elongation, the viscoelastic longitudinal modulus, E_L^v, with creep parameters for flexure given in Table 13.5 should be used.

The axial elongation of tension members must be less than code-stipulated amounts for axially loaded members. A limit of $L/800$ is suggested for design of pultruded tension members. Often, these deformation requirements will be enforced via global deflection or lateral displacement limits on the entire structure.

14.11 ULTIMATE LIMIT STATES

14.11.1 Longitudinal Material Failure on the Gross Area

To design for material tensile failure (in the flange or the web) at the gross cross-sectional area of the profile subjected to a uniform axial stress, the critical tensile strength is taken as the longitudinal tensile strength of the pultruded material,

$$\sigma_{\text{cr}}^{\text{tens}} = \sigma_{L,t} \tag{14.51}$$

The critical axial force for a homogeneous member due to tensile material failure on the gross cross section is

$$P_{\text{cr}}^{\text{tens}} = \sigma_{\text{cr}}^{\text{tens}} A_z \tag{14.52}$$

14.11.2 Longitudinal Material Failure on the Net Area

To design for material tensile failure (in the flange or the web) at the net cross-sectional area of the profile, the critical strength is taken as the ultimate longitudinal tensile strength of the pultruded material multiplied by a strength reduction factor (or ratio), k_F, to account for the reduction in strength at the holes (Daniel and Ishai, 1994, p. 360),

$$\sigma_{\text{cr}}^{\text{net-tens}} = k_F \sigma_{L,t} \tag{14.53}$$

The critical axial force for a homogeneous member due to tensile material failure on the net cross section is

$$P_{\text{cr}}^{\text{net-tens}} = \sigma_{\text{cr}}^{\text{net-tens}} A_{z,\text{net}} \tag{14.54}$$

The *Eurocomp Design Code and Handbook* (Eurocomp, 1996) suggests a value of $k_F = 0.90$ for conventional pultruded GFRP materials. The strength reduction factor can be obtained experimentally using the ASTM D 5766 open-hole strength test. The strength reduction factor depends on the fiber layup, fiber and resin properties, and hole diameter-to-member (or wall) width ratio, d_h/w. This recommendation should only be used for d_h/w ratios below 0.2 (or w/d_h ratios greater than 5) and only when considering the axial stress in the roving (or pultrusion) direction of the member.

This high value of k_F would suggest that a pultruded material's longitudinal strength at the net section is not greatly affected by the presence of the holes (i.e., is not very notch sensitive). This appears to be counterintuitive given that FRP composite materials are generally described as linear elastic brittle materials. However, conventional pultruded materials, which have 20 to 30% of their fiber volume consisting of continuous filament mats, are capable of redistributing stresses at holes and other discontinuities due to material fragmentation and can exhibit pseudo-elastoplastic behavior (Rizzo et al., 2005). However, this also depends on the stiffness and strength of the longitudinal roving. Experiments on pultruded materials taken from plate stock have indicated significantly lower values (as low as $k_F = 0.4$) for strength reduction for holes in pultruded materials (Turvey and Wang, 2003; Saha et al., 2004).

It is also important to recognize that failure at the net section will typically occur where the tension member is connected to other members with bolts. In these cases the hole (or holes) will be pin-loaded and will not be an open hole. The preferred[13] design failure mode of bolted joints and connections in pultruded structures is local bearing failure (as opposed to net tension, shear out, cleavage, or block shear failure). This is discussed further in Chapter 15. When a conventional pultruded material fails in bearing, local material fragmentation occurs in the vicinity of the bolt, and the maximum load transferred into the member is limited by the bearing strength of the material.

14.12 DESIGN PROCEDURE FOR CONCENTRICALLY LOADED TENSION MEMBERS

The design procedure presented in what follows for tension members (or ties) permits design by the allowable stress design (ASD) basis or the load and resistance factor design (LRFD) basis, as discussed in Chapter 12. Both of these bases use the analytical mechanics-based equations presented in preceding sections.

Step 1. Determine the design loads, choose the design basis, and determine the basis factors. The concentric axial load on the tension member is determined from the structural geometry and loading requirements. The design basis for the tension member is chosen and the appropriate safety factors (ASD) and resistance factors (LRFD) are determined. Load factors for the LRFD method are listed. Maximum short- and long-term axial elongation criteria are taken from codes or established by the project specifications.

Step 2. Determine the maximum design stresses or forces. The axial tensile stress on the gross and net cross sections of a member subjected to nominal loads are calculated for the ASD procedure. The axial load on the gross

[13] As discussed further in Chapter 15, this is preferred but may not be possible to achieve or even predict and should not necessarily be assumed.

and net cross sections of a member subjected to factored loads is calculated for the LRFD procedure.

Step 3. Select a profile type and trial size and determine the critical stresses or forces. For the profile chosen, determine (or calculate) the critical stresses (ASD) or loads (LRFD) on the gross and net cross sections.

Step 4. Determine the factored critical stresses or forces. For ASD, divide the critical stresses determined in step 3 by the appropriate safety factor for tensile stress. For LRFD, multiply the critical forces by the appropriate resistance factors for tension members.

Step 5. Check the ultimate strength or capacity of the profile selected. For ASD, check that the design stresses are less than the allowable stresses. For LRFD, check that the factored design forces are less than the factored resistances. Return to step 3 if the trail size does not work.

Step 6. Check the profile selected for serviceability. Check that the axial elongation in the trial profile chosen does not exceed the code-stipulated maximum. Service loads are used for both the ASD and LRFD procedures.

Step 7. Dimension the ancillary bracing, stiffeners, and bearing plates. Dimension (or design if quantitative procedures are available) braces, stiffeners, and bearing plates for the tension member. Provide a sketch of the tension member showing all ancillary elements, dimensions, and important construction notes. (If the tension member is connected to other framing members, dimension and detail the connection following the design procedure in Chapter 15.)

14.13 COMBINED LOAD MEMBERS

14.13.1 Members Subjected to Combined Flexure and Compression (Beam-Columns)

The design of thin-walled steel sections subjected to combined bending moments, M_x and M_y, and axial compression load, P, is usually addressed with interaction equations of the form (Galambos, 1998)

$$f\left(\frac{P}{P_u}, \frac{M_x}{M_{ux}}, \frac{M_y}{M_{uy}}\right) \leq 1.0 \qquad (14.55)$$

where P_u, M_{ux}, and M_{uy} are the ultimate resistances of the profile under axial load and bending about the x and y centroidal axes, respectively, when each load is considered to act independently. The interaction equation gives the ultimate capacity of the member when subjected to the combined action of the loads when they act simultaneously and also gives the member capacity when only one of the loads acts individually. Consequently, interaction equation formula are desirable from a design perspective.[14]

The nominal capacity of a thin-walled steel section subjected to axial load and bending about one axis only is usually given by a linear interaction equation in the form (Salmon and Johnson, 1996)

$$\frac{P}{P_n} + \frac{M}{M_n}\left(\frac{C_m}{1 - P/P_E}\right) \leq 1.0 \tag{14.56}$$

where C_m is a moment magnifier that depends on the loading and end conditions and is given in standard tables (Salmon and Johnson, 1996, p. 709) and P_E is the Euler buckling load for a pinned–pinned column,

$$P_E = \frac{\pi^2 EI}{L^2} \tag{14.57}$$

P_n is the nominal axial load-carrying capacity and is a function of the slenderness ratio (i.e., for long columns it approaches the Euler load-carrying capacity, and for short columns it approaches the material compressive strength or local buckling strength capacity). M_n is the nominal bending moment capacity based on the material tensile or compressive strength or the member lateral torsional buckling strength. If the beam-column is part of an unbraced frame, additional factors are introduced to account for the additional moment due to the differential lateral displacement between the ends of the column.

Only in the case of the perfectly axially loaded column can sudden global instability occur due to buckling. In the case of a beam-column, the bending moment will cause the initially perfect column to have a deflected shape. Application of the axial load to the already deflected column will cause it to increase its lateral deflection (for the case of bending in single curvature) until it fails due to local material failure or exceeds a deflection limit. This is identical to the case of a column subjected to an eccentric axial load or a column with an initial curvature or an assumed imperfect shape (Timoshenko and Gere, 1961, p. 31). Sudden lateral deflection does not occur in a beam-column.[15] Rather, at the critical axial load, the lateral deflection and bending moments in a beam-column increase indefinitely.

Few test data are available for pultruded profiles subjected to combined flexure and axial compression (Barbero and Turk, 2000; Mottram et al.,

[14] It can be recalled that in the case of a reinforced concrete column discussed in Chapter 11, a continuous interaction equation is not available, and the P–M interaction diagram must be constructed numerically.

[15] In reality, neither does it occur for a perfect column. It occurs mathematically only for a perfect column.

2003b). Both studies consider only axial load and single-axis bending. Barbero and Turk (2000) tested conventional pultruded columns under eccentric axial loads with bending about the minor axis and with small eccentricities (much smaller than the depth of the profile). Mottram et al. (2003b) tested conventional pultruded columns under eccentric axial loads with bending about the major axis with end loads that have large eccentricities (50 to 100% of the profile depth). Mottram tested different types of eccentric loading and included double curvature (reverse end moments). In both studies, local buckling was the primary failure mode of the eccentrically loaded columns (where testing was not halted due to large deflections). As expected, the failure loads of the eccentrically loaded columns were less than the failure loads under concentric axial loads, since the axial stress in the flanges is larger due to the bending moment.

Since pultruded rigid frames are not possible to build due to the flexibility of the materials[16] and should therefore not be designed, a pultruded beam-column will typically form part of a braced framed structure that will have simple frame connections. End moments on the beams will not exist (theoretically), and sidesway does not need to be considered. As such, a pultruded beam-column will usually be loaded by concentric axial loads and transversely distributed lateral loads along its length as shown in Fig. 14.7.

In what follows, design equations are presented for pultruded beam-columns subjected to combined axial load and major axis bending only. The design equations are based primarily on the test data and results of Mottram et al. (2003b) since the tests were conducted on specimens subjected to major axis bending, which is expected to be the manner in which most pultruded beam-columns will be used in practice. It should be noted, however, that the exterior columns in a simple frame (or columns in frames with unequal beam spans) will be subjected to load eccentricity. This is because the shear force transferred at the beam end to the column through the clip angle connections used in these types of frames will be eccentric to the column axis (see Chapter 15 for more detail on these types of connections). In this case the column needs to be designed as a beam-column with an eccentric axial load and not a uniformly distributed load. Equations for deflections and maximum mo-

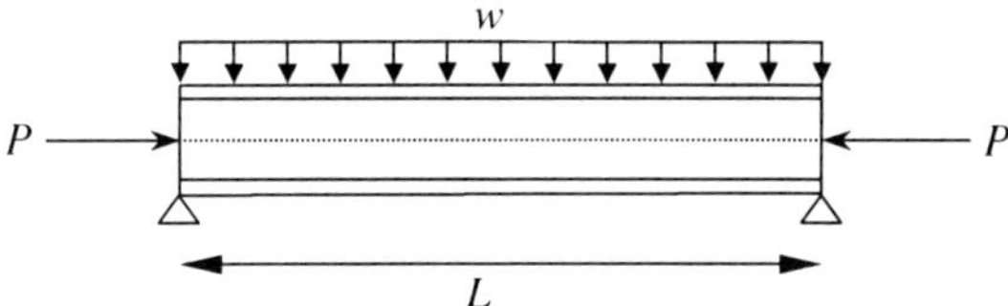

Figure 14.7 Loading for a pultruded beam-column.

[16]This subject is discussed further in Chapter 15.

ments in beam-columns of this type are given in Timoshenko and Gere (1961). Their method of analysis follows that presented for the case shown in Fig. 14.7.

In typical beam-column design equations for steel members, the interaction equation for the case described (uniform lateral load and bending about the major axis with lateral buckling and minor axis buckling restrained) is typically given in the form of the strength or load interaction equation using the moment magnifier form [equation (14.56)]. The beam-column is critical when the strength interaction equation is satisfied and no consideration is typically given to the deflections in beam-columns. In conventional pultruded columns, however, the lateral deflection of the beam-column must be considered as one of the critical design states due to the low *EI* value of conventional pultruded profiles relative to steel profiles. Therefore, the design equations for pultruded beam-columns must include an equation to explicitly calculate the lateral deflection for the beam-column for a given set of loads. The permissible lateral deflection for the beam-column can be taken as the maximum deflection under service loads. Therefore, the design of pultruded beam-columns requires a serviceability calculation and a strength calculation in much the same way that the regular design of beams does.

Serviceability Design of Beam-Columns The exact solution for the maximum midspan deflection of the beam-column shown in Fig. 14.7 is given (Timoshenko and Gere, 1961, p. 9) as

$$\delta_{\text{max}} = \frac{5wL^4}{384EI}\left[\frac{12(2 \sec u - 2 - u^2)}{5u^4}\right] \tag{14.58}$$

where

$$u = \frac{L}{2}\sqrt{\frac{P}{EI}} \tag{14.59}$$

It can be shown that that as P approaches P_E, u approaches $\pi/2$, and the term in parentheses in Eq. (14.58) approaches infinity. As P approaches zero, u approaches zero, and the term in the square brackets approaches 1. The usual expression for the midspan deflection of a simply supported uniformly loaded beam, not including the effects of shear deformation, is then recovered. Given service loads P and w, the midspan deflection for the beam-column is calculated and compared with allowable service load deflections. Deflection functions for other common loading cases, if required, are given in Timoshenko and Gere (1961).

Ultimate Design of Pultruded Beam-Columns The ultimate strength or resistance of the pultruded beam-column considered depends on the failure mode of the profile. Testing by Barbero and Turk (1994) and Mottram et al.

(2003b) have shown that the likely failure mode of pultruded beam-columns, if not by excessive deformation, is one of local buckling of compression flanges. Mottram et al. (2003) have shown that the linear summation of stresses due to the axial load alone and due to the bending moment alone (including the effects of magnification due to the axial load) can be used to obtain the total normal stress in the compression flange. This total normal compressive stress at local buckling failure is close to the local buckling load of the flange under concentric loading. Barbero and Turk (1994) showed similar results. Therefore, a linear interaction between the axial stresses and the bending stresses can be used to predict the strength of pultruded beam-columns.

The exact expression[17] for the maximum bending moment for a pultruded beam-column (not including shear deformation effects) is (Timoshenko and Gere, 1961, p. 10)

$$M_{\max} = \frac{wL^2}{8}\left[\frac{2(1 - \cos u)}{u^2 \cos u}\right] \tag{14.60}$$

where once again the term in the square brackets approaches infinity as P approaches P_E. Moment functions for other common loading cases, if required, are given in Timoshenko and Gere (1961).

The maximum normal stress due to bending in a homogeneous member is

$$\sigma_{z,\max}^{\text{flex}} = \frac{M_{\max}}{S_x} \tag{14.61}$$

The maximum axial load in a homogeneous member is equal to the applied axial load, P, and the axial stress is

$$\sigma_{z,\max}^{\text{axial}} = \frac{P}{A_z} \tag{14.62}$$

The interaction equation for the nominal[18] strength of the pultruded beam-column considered for ASD design is given as

$$\frac{\sigma_{z,\max}^{\text{axial}}}{\sigma_{z,\text{cr}}^{\text{axial}}} + \frac{\sigma_{z,\max}^{\text{flex}}}{\sigma_{z,\text{cr}}^{\text{flex}}} \leq 1.0 \tag{14.63}$$

where $\sigma_{z,\text{cr}}^{\text{axial}}$ is the lesser of:

[17]This can also be written in an approximate form using a moment magnifier.
[18]Not including any safety factors.

1. σ_{cr}^{local}, the local buckling stress of the extreme compression flange of the profile when loaded by pure axial load
2. σ_{cr}^{comp}, the compressive strength of the pultruded material in the extreme compression fiber of the profile
3. $\sigma_E = \pi^2 E/(L/r)^2$, the Euler buckling stress of the profile about the axis of bending.

and $\sigma_{z,cr}^{flex}$ is the lesser of:

1. σ_{cr}^{local}, the local buckling stress of the extreme compression flange of the profile when loaded by pure bending moment
2. σ_{cr}^{comp}, the compressive strength of the pultruded material in the extreme compression fiber of the profile

The interaction equation for the nominal[19] resistance of the pultruded beam-column considered for LRFD design is given as

$$\frac{P}{P_{cr}} + \frac{M_{max}}{M_{cr}} \leq 1.0 \tag{14.64}$$

where P_{cr} is the lesser of:

1. P_{cr}^{local}, the axial load that causes local buckling failure of the extreme compression flange of the profile
2. P_{cr}^{comp}, the axial load that causes compressive failure of the pultruded material in the extreme compression fiber of the profile
3. $P_E = \pi^2 EI/L^2$, the Euler buckling load for the profile about the axis of bending.

and M_{cr} is the lesser of:

1. M_{cr}^{local}, the critical bending moment that causes local buckling of the extreme compression flange of the profile
2. M_{cr}^{comp}, the critical bending moment that causes compressive failure of the pultruded material in the extreme compression fiber of the profile

To use the foregoing interaction equations for design appropriate safety factors for ASD and resistance factors for LRFD are used to modify the nominal strength and capacities. For axial contributions, compressive factors should be used, whereas for bending contributions, flexural factors should be used. The interaction equations above can also be written in the from of Eq.

[19]Not including any resistance factors.

(14.56), where the bending moment term is written with a moment magnifier term in parentheses.

The equations presented above do not include the effects of shear deformation. The effects of shear deformation of the deflection equations used for the serviceability design can be included in adding to Eq. (14.58) the approximate maximum shear deflection term. Neither is shear deformation included in the critical global buckling load. This can be included by modifying the expression for the Euler buckling load to include shear deformation. As discussed in prior sections, shear deformation effects will be quite small (5 to 10%) for conventional profiles, with anisotropy ratios from 4 to 6; however, they should be considered for beams with higher anisotropy ratios and when their spans are short relative to their depths ($l/d < 10$).

The effects of interaction between global and local buckling for intermediate-length beam-columns are also not included in the equations above. If this is desired, the critical axial load capacity for the intermediate column can be found using the interaction equation proposed by Barbero and Tomblin described previously.

14.13.2 Members Subjected to Combined Flexure and Tension

Pultruded profiles subjected to combined flexure and tension occur in braced frames and in bottom chords of simply supported trusses. In general, the magnitude of the axial tensile force is such that no part of the member cross section ever experiences a compressive normal stress. Since instability due to buckling does not occur in such members, the combined member is designed to resist the maximum normal stresses that may develop at either the gross or net cross section. Since the normal stress due to the axial load and the bending moment are additive, a linear interaction equation can be used.

The maximum normal tensile stress in a homogeneous member due to bending on the gross cross section is

$$\sigma_{z,\max}^{\text{flex}} = \frac{M_{\max}}{S_x} \tag{14.65}$$

The maximum normal tensile stress in a homogeneous member due to bending on the net cross section is

$$\sigma_{z,\max}^{\text{net-flex}} = \frac{M_{\max}}{I_{x,\text{net}}} c_{\text{tens}} \tag{14.66}$$

where $I_{x,\text{net}}$ is the second moment of the net cross section and c_{tens} is the distance from the neutral axis to the extreme tensile fiber. The maximum axial tensile stress in a homogeneous member on the gross cross section is

$$\sigma_{z,\max}^{\text{axial}} = \frac{P}{A_z} \tag{14.67}$$

The maximum axial tensile stress in a homogeneous member on the net cross section is

$$\sigma_{z,\max}^{\text{net-axial}} = \frac{P}{A_{z,\text{net}}} \tag{14.68}$$

The interaction equation for the nominal strength of the gross cross section of the pultruded combined flexural tension member for ASD design is given as

$$\frac{\sigma_{z,\max}^{\text{axial}}}{\sigma_{z,\text{cr}}^{\text{axial}}} + \frac{\sigma_{z,\max}^{\text{flex}}}{\sigma_{z,\text{cr}}^{\text{flex}}} \leq 1.0 \tag{14.69}$$

where $\sigma_{z,\text{cr}}^{\text{axial}} = \sigma_{z,\text{cr}}^{\text{flex}} = \sigma_{\text{cr}}^{\text{tens}}$, the tensile strength of the pultruded material in the extreme tension fiber of the gross cross section of the profile.

The interaction equation for the nominal strength of the net cross section of the pultruded combined flexural tension member for ASD design is given as

$$\frac{\sigma_{z,\max}^{\text{net-axial}}}{\sigma_{z,\text{cr}}^{\text{net-axial}}} + \frac{\sigma_{z,\max}^{\text{net-flex}}}{\sigma_{z,\text{cr}}^{\text{net-flex}}} \leq 1.0 \tag{14.70}$$

where $\sigma_{z,\text{cr}}^{\text{net-axial}} = \sigma_{z,\text{cr}}^{\text{net-flex}} = \sigma_{\text{cr}}^{\text{net-tens}}$, the net tensile strength of the pultruded material in the extreme tension fiber of the net cross section of the profile.

The interaction equation for the nominal resistance of the gross cross section of the pultruded combined flexural tension member for LRFD design is given as

$$\frac{P}{P_{\text{cr}}} + \frac{M_{\max}}{M_{\text{cr}}} \leq 1.0 \tag{14.71}$$

where $P_{\text{cr}} = P_{\text{cr}}^{\text{tens}}$, the axial load that causes tensile failure of the pultruded material in the extreme tension fiber of the gross cross section of the profile, and $M_{\text{cr}} = M_{\text{cr}}^{\text{tens}}$, the critical bending moment that causes tensile failure of the pultruded material in the extreme tension fiber of the gross cross section of the profile.

The interaction equation for the nominal resistance of the net cross section of the pultruded combined flexural tension member for LRFD design is given as

$$\frac{P}{P_{cr}^{net}} + \frac{M_{max}}{M_{cr}^{net}} \leq 1.0 \tag{14.72}$$

where $P_{cr}^{net} = P_{cr}^{net\text{-}tens}$, the axial load that causes tensile failure of the pultruded material in the extreme tension fiber of the net cross section of the profile, and $M_{cr}^{net} = M_{cr}^{net\text{-}tens}$, the critical bending moment that causes tensile failure of the pultruded material in the extreme tension fiber of the net cross section of the profile.

To use the interaction equations above for design, appropriate safety factors for ASD and resistance factors for LRFD are used to modify the nominal strength and capacities. For axial contributions, tensile factors should be used, whereas for bending contributions, flexural factors should be used.

PROBLEMS

14.1 For profiles *a* to *e* from U.S. manufacturers (Strongwell, Creative Pultrusions, and Bedford Reinforced Plastics) and profiles *f* to *h* from European manufacturers (Fiberline or TopGlass) listed in Table P14.1,[20] determine the axial rigidity, *EA*, the flexural rigidity, *EI*, and the shear rigidity, *KAG*, about both the strong and weak axes[21] of the profile. The profile is to be used as a column under concentric compressive loading. Calculate the shear rigidity using the $A_{web}G_{LT(web)}$ method for both the strong and weak axis buckling. For nonhomogeneous profiles, use the composite mechanics method to determine *EA*, and either the transformed section method or the composite mechanics method to determine *EI*.

TABLE P14.1 Pultruded Profiles

a	Glass–polyester $8 \times 8 \times \frac{3}{8}$ $(d \times b \times t)$ in. wide-flange I-shaped section
b	Glass–vinylester $12 \times 12 \times \frac{1}{2}$ $(d \times b \times t)$ in. wide-flange I-shaped section
c	Glass–polyester $8 \times 4 \times \frac{3}{8}$ $(d \times b \times t)$ in. narrow-flange I-shaped section
d	Glass–polyester $3 \times \frac{3}{8}$ $(b \times t)$ in. square tube
e	Glass–polyester $2 \times \frac{1}{4}$ $(b \times t)$ in. square tube
f	Glass–vinylester $200 \times 100 \times 10$ $(d \times b \times t)$ mm narrow-flange I-shaped section
g	Glass–vinylester 120×8 $(b \times t)$ mm square tube
h	Glass–vinylester 50×5 $(b \times t)$ mm square tube

[20] Instructors may assign only selected profiles or may choose other doubly symmetric sections from manufacturers' design guides.

[21] Also called the major and minor axes.

14.2 A $12 \times 12 \times \frac{1}{2}$ $(d \times b \times t)$ in. glass–vinylester profile from Strongwell is used as a 12-ft-high concentrically loaded column in a building frame. It is loaded by a sustained dead load of 6000 lb and a live load of 8000 lb. Determine **(a)** the short-term (instantaneous) axial shortening of the column, and **(b)** the long-term axial shortening after 20 years of service. Give your answers in absolute values and also in height ratios (e.g., $L/1500$). You are not required to check the ultimate limit state in this problem.

14.3 For I-shaped profiles *a* to *c* from U.S. manufacturers (Strongwell, Creative Pultrusions, and Bedford Reinforced Plastics) and *f* from European manufacturers (Fiberline or TopGlass) listed in Table P14.1,[22] determine the global flexural buckling load (account for the effects of shear deformation) for the following conditions:

(a) L = 10 ft, fixed base, pinned top, restrained about the weak axis.

(b) L = 12 ft, pinned at the base and the top, no restraints.

(c) L = 18 ft, pinned base and top about the strong axis, pinned top and bottom, and braced midheight about the weak axis.

(d) L = 20 ft, fixed base and free top (strong axis), pinned base and top (weak axis).

14.4 Consider the $12 \times 12 \times \frac{1}{2}$ $(d \times b \times t)$ in. pultruded column of Problem 14.2. Determine **(a)** the critical short-term global flexural buckling load, and **(b)** the critical global flexural buckling load after 20 years of service. Assume the same ratio of dead to live load on the column as given in Problem 14.2. The column has pinned ends about both axes. Determine the percent reduction in the buckling load after 20 years of sustained dead load on the column.

14.5 For the small tube (box)-shaped profiles *d* and *e* from U.S. manufacturers (Strongwell, Creative Pultrusions, and Bedford Reinforced Plastics) and *h* from European manufacturers (Fiberline or TopGlass) listed in Table P14.1, determine the global flexural buckling load, accounting for the effects of shear deformation, for the following conditions:

(a) L = 3 ft, pinned at the base and the top.

(b) L = 5 ft, pinned at the base and the top (see Figs. 1.19, 15.1, and 15.2 for this type of member in a truss).

(c) L = 30 ft, pinned at 6-ft intervals by cross-bracing members (see Figs. 1.17, 15.3, and 15.4 for this type of member in a stick-built light frame).

[22] Instructors may assign only selected profiles or may choose other doubly symmetric sections from manufacturer design guides.

14.6 For I-shaped profiles *a* to *c* from U.S. manufacturers (Strongwell, Creative Pultrusions, and Bedford Reinforced Plastics) and *f* from European manufacturers (Fiberline and TopGlass) listed in Table P14.1, determine the global torsional buckling load. Assume that the columns have no torsional end restraints. Check J/I_p ratios and account for shear deformations effects if appropriate. Consider column heights of 10, 15, and 20 ft.

14.7 For I-shaped profiles *a* to *c* from U.S. manufacturers (Strongwell, Creative Pultrusions and Bedford Reinforced Plastics) and *f* from European manufacturers (Fiberline and TopGlass) listed in Table P14.1, determine the critical local in-plane buckling stress, the critical concentric axial load, and the buckle half-wavelength, *a*, when the profile is subjected to uniform concentric compression. For profiles *a* and *f*, compare your results with those obtained for the local in-plane buckling stress when the profile was used as a beam in flexure (see Problem 13.10).

14.8 For the tube (box)-shaped profiles *d* and *e* from U.S. manufacturers (Strongwell, Creative Pultrusions and Bedford Reinforced Plastics) and *g* and *h* from European manufacturers (Fiberline or TopGlass) listed in Table P14.1, determine the critical local in-plane buckling stress, the critical concentric axial load, and the buckle half-wavelength when the profile is subjected to uniform concentric compression.

14.9 Redesign the column in Design Example 14.1 using a glass–vinylester wide-flange I-shaped homogeneous profile from Strongwell or Bedford Reinforced Plastics. Design the column using **(a)** the ASD basis and **(b)** the LRFD basis. Compare the section selected to those published in the manufacturer's design guides in their column load tables. Compare the profile selected with Design Example 14.1 and with the profiles suggested in the load tables from Strongwell, Bedford Reinforced Plastics, and Creative Pultrusions.

14.10 Redesign the column in Design Example 14.1 using the same series of pultruded beams as used in the design example, but assume that the column is fully restrained from global buckling about its weak axis by an interior wall system. Design the column using **(a)** the ASD basis and **(b)** the LRFD basis. Compare your results to those of the design example.

14.11 Redesign the column in Design Example 14.1 using the same series of pultruded beams as used in the design example, but assume that the column has a height of 15 ft. Design the column using **(a)** the ASD basis and **(b)** the LRFD basis.

14.12 Redesign the column in Design Example 14.1 using a glass–vinylester square tube section from a U.S. or European manufacturer. Design the column using **(a)** the ASD basis and **(b)** the LRFD basis.

14.13 A $3 \times \frac{1}{4}$ in. glass-vinylester square tube with U.S. manufacturers' typical properties (Table 13.7, for example) is to be used as a concentrically loaded column in a stick-built cooling tower. The column is 40 ft high and is braced at 8-ft intervals by transverse beams (see Fig. 1.17 for an example of this type of construction). Using the ASD basis and the LRFD basis, determine the maximum nominal axial load, P_n, and the maximum design load, P_u, that can be carried by the column. Assume a dead load/live load ratio of 1 : 1 for the LRFD basis.

14.14 A 7-ft-long diagonal compression strut is used in a cross-braced simply supported pultruded pedestrian bridge that spans 45 ft. See Figs. 1.19, 15.1, and 15.2 for examples of this construction type. The maximum factored compression load on the diagonal closest to the support is found by structural analysis to be 9 kips. Using typical U.S. manufacturer properties (Table 13.7) and the LRFD basis, design a glass–polyester tubular section to carry the factored load. Assume that the compression strut is unbraced along its full 7-ft length and has pinned ends.

14.15 A $3 \times 3 \times \frac{1}{4}$ glass–vinylester pultruded equal-leg angle from a U.S. manufacturer is used as a tension brace in a pultruded frame structure. (This type of bracing member is shown in the connection detail in Fig. 15.7.) The angle is attached to a gusset plate using a $\frac{5}{8}$-in. (diameter) stainless steel bolt connected through one leg of the angle as shown in Fig. 15.7.

Determine the nominal concentric (i.e., assume that the load is applied at the member centroid) tensile load, P_n, and the ASD design load, P_u, that the brace can carry. Assume a strength reduction factor, k_F, of 0.8 at the hole and that the hole is oversized by $\frac{1}{16}$ in.

14.16 A 7-ft-long diagonal tension tie is used in a cross-braced simply supported pultruded pedestrian bridge that spans 45 ft. (See Figs. 1.19, 15.1, and 15.2 for examples of this construction type.) The maximum factored tensile load on the diagonal closest to the support is found by structural analysis to be $P = 9$ kips. Using typical U.S. manufacturer properties (Table 13.7) and the LRFD basis, design a glass–polyester tubular section to carry the factored load. Assume that the tube is connected to the top and bottom chords of the truss using single $\frac{3}{4}$-in. (diameter) stainless-steel bolts (see Fig. 15.2 for an example). Determine the nominal concentric tensile load, P_n, that the brace can carry (i.e., assume that the load is applied at the member centroid). Assume a strength reduction factor, k_F, of 0.8 at the hole and that the hole is oversized by $\frac{1}{16}$ in.

14.17 A 8 × 8 × $\frac{1}{2}$ in. glass–vinylester pultruded I-shaped profile with typical U.S. manufacturer properties (Table 13.7) forms part of the an exterior wall system of a framed structure. It is subjected to distributed wind load and axial load simultaneously and behaves as a beam-column. It is pinned at its ends over a 10-ft floor height. The distributed wind load produces a uniformly distributed line load on the member of 120 lb/ft. The beam-column is subjected to bending about its major axis and is restrained from buckling out of plane about its minor axis. Determine the maximum nominal axial load, P_n, that can be applied to the beam-column. A maximum deflection of $L/400$ is permitted.

15 Pultruded Connections

15.1 OVERVIEW

In this chapter, design procedures are presented for connecting pultruded FRP profiles with mechanical fasteners. The term *pultruded connection* is used to describe the pultruded profiles being connected, the connection parts that are used to connect, or join, pultruded profiles, and the mechanical fasteners that are used to join the individual parts and profiles. These connection parts may include pieces of pultruded profiles themselves (such short lengths of single-leg angles and channels) or pultruded plate materials (for use as gusset plates). The mechanical fasteners may include nuts, bolts, threaded rods, screws, or rivets that are used to draw together and physically join the profiles and the parts. Both FRP and metallic fasteners are discussed.

The design of adhesively bonded pultruded connections is not covered in this book. Although adhesives play an important and ever-increasing role in the construction of composite material structures of all kinds, their use in the construction of pultruded structures is limited. Where adhesives are used in pultruded structures in load-carrying members, they are typically used in conjunction with mechanical fasteners, either as an aid to construction or to improve the serviceability of the structure. They are very seldom relied on to transfer all the design loads between pultruded profiles in truss and frame structures.

Only an allowable stress design (ASD) procedure is presented in this chapter for the design of pultruded connections. At the present time, a load and resistance factor design (LRFD) procedure cannot be presented with confidence. Material partial factors taken from the Eurocomp design code presented previously for member design are not suitable for the simplified design procedure presented in what follows. The state of the practice of pultruded connection design is not as developed as that for pultruded member design, and safety factors are larger than those for pultruded member design. All safety factors used in connection design are taken as 4.0 for the ASD procedure. This applies to the fasteners, connected members, and connection parts.

Theoretical mechanics-based equations are presented in this chapter for analyzing the strength limit states of pultruded connections. It is important to note at the outset that the design of pultruded connections for strength capacity recognizes that inelastic deformation will occur in the connection parts at the ultimate state and that multiaxial stress states will exist. Since the profiles being connected and the connection parts are orthotropic, great care must be taken to identify the load paths in the pultruded connection and to orient the connection parts to carry these loads. Simplified mechanics procedures are presented to determine the load paths and the loads on fasteners and connection parts in a manner analogous to the design of structural steel connections.

15.2 INTRODUCTION

15.2.1 Conventional Pultruded Connections

Pultruded connections are used in two primary types of pultruded structural systems: light truss or stick-frame systems or heavy braced frame[1] systems. A light truss can be seen in the FRP pedestrian bridge in Fig. 15.1. A close-up of the connections in this type of structure is shown in Fig. 15.2.

The pultruded connections in this structure are typical pinned truss connections that have the line of action of the axial force in the truss members meeting at a point. The connections consist of mechanical fasteners that join tubular or channel sections in single or double shear planar configurations. Additional pultruded connection parts are generally not used in such connections, and the pultruded members are connected directly to each other using

Figure 15.1 FRP bridge at Point Bonita, California.

[1] The terms *heavy* and *light* are qualitative and are used to define the appearance of the structure, not to relate to its actual weight. Relative to steel members, pultruded members are about one-fourth to one-fifth of the weight.

Figure 15.2 Connection details on the Point Bonita bridge.

a fastener that passes though all the members in the connection region. Small spacer elements may be used to improve the geometric fit-up and to allow members to pass through each other to create symmetric double shear connections. Small plate and tube inserts may be used in the connection region to improve the local bearing strength of individual highly loaded members.

A stick-frame system consists of small pultruded profiles assembled in a vertical closely spaced gridlike pattern as shown in Fig. 15.3. The pultruded members in the wall system carry wind load and form part of the lateral load-carrying system. In a stick-built frame the small tubular pultruded members are typically constructed from continuous horizontal and vertical pultruded tubular profiles that overlap at their intersections. They are typically connected to one another at intersections with single galvanized or stainless steel through bolts. These lattice–wall systems are usually braced with diagonal continuous members that are not necessarily connected at all the horizontal and vertical connections. Details of the connections can be seen in Fig. 15.4.

The lattice–wall is then covered with a reinforced or unreinforced plastic panel system (see Fig. 1.16 for example). The panel may or may not have lateral load-carrying functions. The system is similar to that in light wooden frames used in residential construction, consisting of closely spaced members (e.g., 2 by 4's) and plywood sheathing. This is referred to as *balloon framing.*

An example of pultruded connections in a heavy braced frame system can be seen in Fig. 15.5. These types of connections resemble typical steel frame connections. The pultruded connections in these structures are the typical beam-to-column connections seen in most building frame systems. In this type of construction, known as *simple framing,* the beams are connected to the column using shear connections, and it is assumed that no moment is transferred between the members by the connection. The lateral force-resisting system usually consists of diagonal bracing members (X, knee, or K bracing)

Figure 15.3 Stick-built light-frame pultruded structure. (Courtesy of Strongwell.)

or may involve a secondary and independent system (e.g., a shear wall). In these types of connections, the members being connected carry (unintended) bending moments, shear forces, and axial forces, and therefore the connections are subjected to multiaxial stresses. The connections generally use additional parts such as single-leg angles and channels (used as clip angles and seats) and plate stock (used for gusset plates, fillers, and local strengthening). Tubular parts may also be used for local stiffeners in the connection region. The forces at the connection are transferred through these connection parts and not directly through the mechanical fasteners as in the case of light-truss or stick-built frame connections. The members do not overlap or telescope, and the line of action of the forces does not pass through a single point in space. Therefore, the connection parts and fasteners are subjected to load eccentricity.

Connections of this type mimic steel simple framing connections (known as *type 2* in AISC ASD and as *simple connections* in AISC LRFD). They have been used in numerous pultruded frame structures to date and are the most common type of framing used for current pultruded structures. Most

Figure 15.4 Connection in a stick-built structure. (Courtesy of Strongwell.)

large U.S. pultrusion manufacturers' design guides provide drawings and details for typical simple framing connections. They also provide nominal load capacities for selected commonly used connections such as the beam shear connections shown in Fig. 15.6. Different capacities are given for bolted-only and bolted-and-bonded connections.

Details for typical bracing connections are also provided by most manufacturers. An example is shown in Fig. 15.7. However, load capacities are not provided for these connections.

Manufacturers also typically provide guidance on minimum hole spacing and on end and side clearances for bolts for use with pultruded profiles. They also provide tables for the allowable bearing loads for different diameter fasteners and pultruded material thicknesses based on the bearing strength of the pultruded material. However, no specific design procedures are provided for these connections in manufacturers' guides. Beam shear-clip bolted connections are assumed to fail due either to bearing in the clip angle material at the fastener locations or to longitudinal shear failure of the clip angle at the heel of the angle. Published connection capacity data are believed to be based on full-scale tests conducted on different-sized connections or on simple calculations. As noted in Chapter 12, manufacturers use a safety factor of 4.0 when reporting connection capacities.

15.2.2 Custom Pultruded Connections

In addition to the conventional light-truss or heavy-frame connections described above, many pultruders produce custom pultruded connections for

Figure 15.5 Heavy braced frame pultruded connection (Courtesy of Strongwell.)

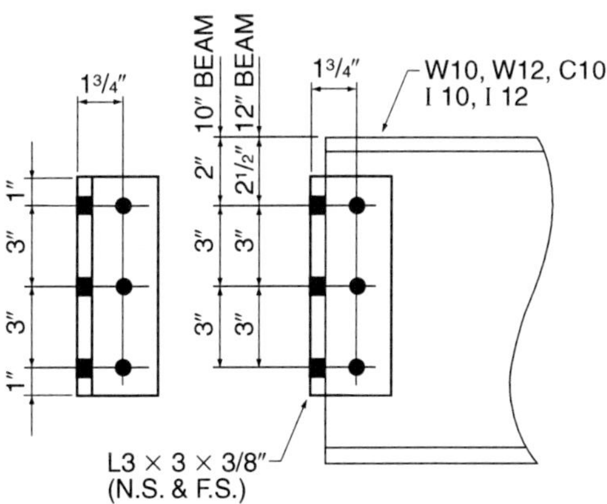

Figure 15.6 Typical details for a heavy pultruded beam simple shear connection. (Courtesy of Strongwell.)

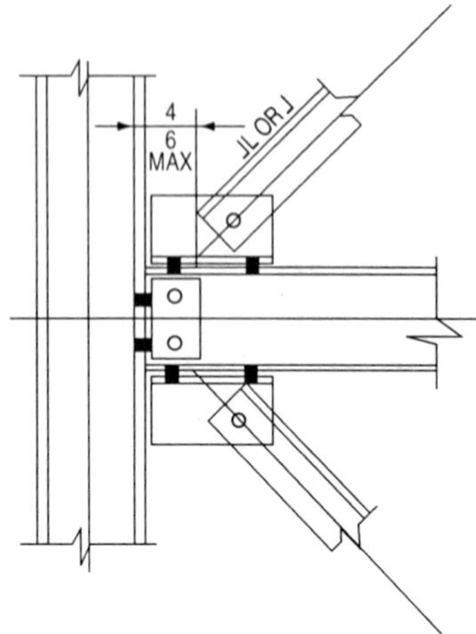

Figure 15.7 Typical details for a heavy pultruded bracing connection. (Courtesy of Strongwell.)

their custom pultruded parts. Such connections can only be used with the specific pultruded parts and are not generally "designed" by the structural engineer. The load-carrying capacity is usually specified for the entire pultruded structural system in this case. Pultruded connections of this type may be fastened mechanically, bonded adhesively, or have unique clipping and locking devices. Examples of such pultruded systems include panelized cooling tower systems (Green et al., 1994), transmission towers (Goldsworthy and Hiel, 1998; Yeh and Yang, 1997), and panelized building systems (Raasch, 1998). Connection details for a proprietary cooling tower system are shown in Fig. 15.8.

15.3 MECHANICAL FASTENERS AND CONNECTION PARTS

Commercially produced pultruded structures (light truss or heavy frame) generally use galvanized or stainless steel mechanical fasteners and may also use steel connection parts (such as clip angles), even though all large pultrusion manufacturers in the United States produce a line of FRP nuts, bolts, and threaded rods (which are used in place of bolts). FRP nuts are compression molded and the threads on FRP bolts and rods are machined into a mat surface layer of a smooth pultruded rod, leading to products that have limited strength and serviceability. FRP mechanical fasteners are typically more expensive than their stainless steel counterparts, and FRP nuts are usually larger than steel nuts, leading to larger required bolt hole clearances. However, in highly corrosive environments where metallic by-products may create a problem (e.g., food, chemical processing) or where electromagnetic transparency is

Figure 15.8 Connection details for a custom pultruded system. (Courtesy of Creative Pultrusions, Inc.)

required, FRP mechanical fasteners are used. Steel or stainless steel connection parts (angles and gusset plates) are less frequently used in pultruded connections but are occasionally required where connection geometries are tight. Steel side gusset plates together with rivets and screws are often used on light-frame and light-truss pultruded structures to connect members that cannot overlap to maintain the concurrent line of action of the forces (see Fig. 15.4 for example).

The use of adhesive bonding in addition to bolting in pultruded connections is typically recommended by pultrusion manufacturers. Although bonding does not generally increase the ultimate capacity of the connection, it does improve the connection stiffness, especially when oversized holes and non-tensioned bolts are used, as is generally the case in conventional pultruded construction. This improves the serviceability of the connection with respect to rotations under service loads and with respect to durability of the connection. However, it has been shown in numerous studies (Lopez-Anido et al., 1999; Mottram and Zheng, 1999a,b) that this increased initial stiffness comes at the expense of the connection ductility, and bonded-and-bolted connections tend to fail in a brittle fashion, whereas bolted-only connections tend to fail in a much more ductile fashion. The ultimate moment capacity of the connection is, however, not typically increased by the use of the adhesive. The adhesive typically fails in tension or shear at a low strain, and when this happens, all the load is suddenly transferred to the mechanical fasteners, which leads to brittle failure of the connection.

Manufacturers recommend that low or high torques be used (corresponding to 37.5 or 75% of the bolt proof load) when using standard steel bolts. This produces a clamping pressure on the connection, which produces good fit-up and helps where bonded and bolted connections are used to achieve a good bond line. (Recommended maximum torques for FRP nuts and bolts are discussed below.) However, the recommendation to use steel bolt torques is not supported by research, which recommends tightening nuts to only finger-tight condition [typically about 2.5 ft-lb (3 N · m)] (Cooper and Turvey, 1995). This is because pultruded connections cannot be relied on to develop any frictional capacity (or slip-critical capacity), due to the properties of the base pultruded material. Higher bolt torques may create a misleading impression of the strength of the connection, which realistically can only be designed as a "bearing" connection.

Adhesively bonded-only connections are rarely used in pultruded structures. Besides the reasons given above, it is recognized that most civil structures are intended for 30 years or more of service life. The long-term properties of pultruded connections under sustained loads using currently available adhesives in the corrosive environments in which pultruded structures are typically used is still a subject of research (Keller et al., 1999; Cadei and Stratford, 2002).

15.3.1 FRP Nuts and Bolts

If FRP nuts are used with FRP bolts, to avoid stripping of the threads in the nuts or bolts, they should not be tightened above the manufacturer's recommended torque. When using FRP nuts and bolts, the permissible clamping torque depends on the nominal diameter of the FRP bolt. Under no circumstances should steel nuts be used with FRP bolts or FRP nuts with steel bolts. FRP nuts should be lubricated with motor oil to aid in constructability. FRP nuts may be square or hexagonal, depending on the manufacturer. Clearance for square nuts is generally larger than that needed for hexagonal nuts since socket heads for wrenches for square nuts are large. The designer should keep this in mind when detailing pultruded connections. Figures 15.9 and 15.10 show pultruded connections made with $\frac{3}{4}$-in. threaded rods and square nuts.

Typical properties of FRP bolt (or threaded rods) and FRP nuts are given in Table 15.1. Current manufacturer specifications must be consulted for specific design values. In the event that steel bolts are used, care must be taken not to overtighten the nuts so as to cause out-of-plane crushing of the base material. Finger-tight tightening is recommended.

15.4 RESEARCH ON HEAVY BEAM-TO-COLUMN PULTRUDED CONNECTIONS

Research studies on the behavior of heavy pultruded beam-to-column connections have been conducted since the late 1980s. Details of the performance of a variety of beam-to-column connections can be found in the literature (e.g., Bank et al., 1990, 1994a, 1996b; Bass and Mottram, 1994; Mottram

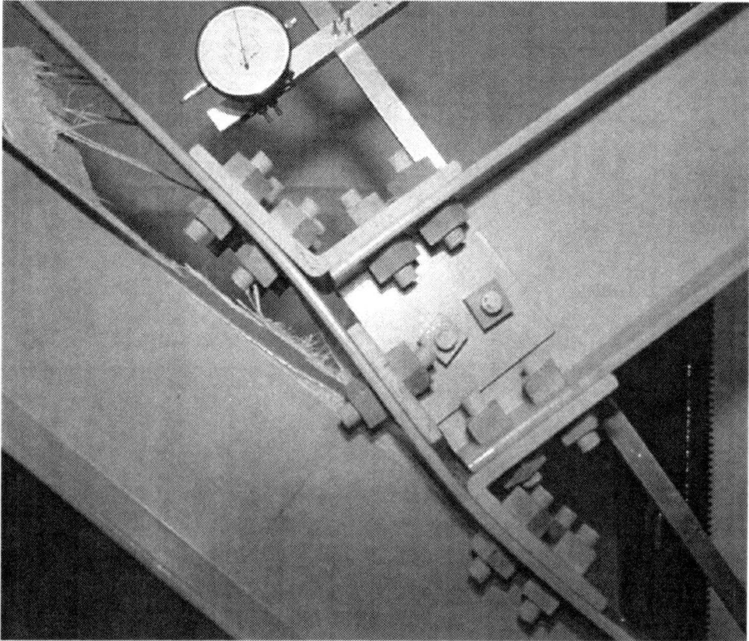

Figure 15.9 Local failure of the web–flange junction in a pultruded member at the connection.

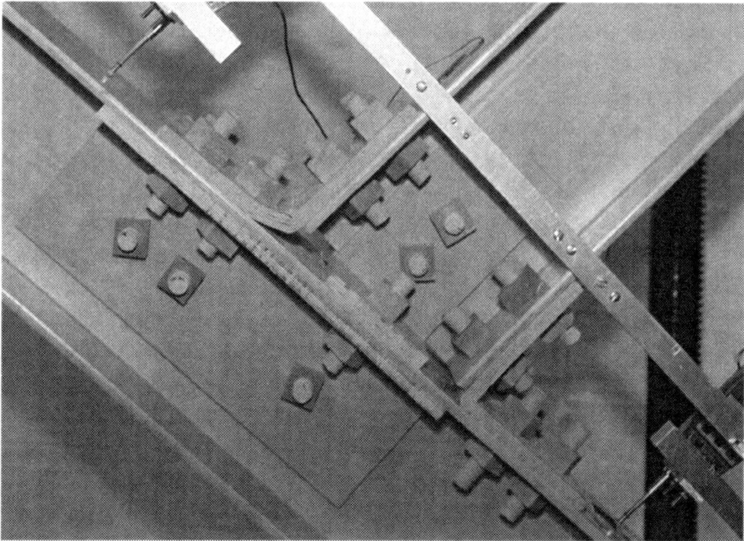

Figure 15.10 Local failure of the pultruded top clip angle in a pultruded connection.

TABLE 15.1 Typical Properties of FRP Nuts and Bolts

	Bolt Diameter				
Property	$\frac{3}{8}$ in.	$\frac{1}{2}$ in.	$\frac{5}{8}$ in.	$\frac{3}{4}$ in.	1 in.
Max installation torque, ft-lb (N · m)	4 (5.4)	8 (10.8)	16 (21.6)	24 (32.5)	50 (67.7)
Max. bolt tensile load based on thread shear, lb (N)	1,000 (4,445)	2,000 (8,890)	3,100 (13,780)	4,500 (20,003)	6,200 (27,560)
Max. transverse shear capacity in double shear,[a] lb (N)	3,000 (13,335)	5,000 (22,225)	7,500 (33,338)	12,000 (53,340)	22,000 (97,790)

[a] In single shear, divide the value given by $\frac{1}{2}$.

and Zheng, 1996, 1999a,b; Turvey and Cooper, 1996; Mottram and Turvey, 1998; Smith et al., 1999; Lopez-Anido et al., 1999; Turvey, 2000). Most available test data and test results in the literature on pultruded beam-to-column connections have been assembled and reviewed by Turvey and Cooper (2004). The purpose of these research investigations has been to develop data on the failure modes, ultimate strength, and stiffness of heavy pultruded beam-to-column connections with the objective of providing designers with more guidance on the connection details that are most appropriate for pultruded structures.

Most researchers feel that "steel-like" connection details are not appropriate for pultruded structures, due to the orthotropic properties of the profiles and the parts being connected and the properties of the fasteners being used. These research studies have brought to light many unique failure modes in pultruded connections that do not occur in steel connections. One such example is the failure of the web–flange junction in the connected member in a simple clip angle and seated shear connector, which is a direct function of the transverse tensile strength of the web of the pultruded material and the properties of the web–flange junction (Mosallam and Bank, 1992; Bank and Yin, 1999; Turvey and Zhang, 2005). An example of a failure of this type is shown in Fig. 15.9. If the web–flange junction is stiffened, failure can occur in the clip angle itself, as shown in Fig. 15.10. Neither of these types of failure is usually encountered in steel bolted or welded connections.

An important objective of the research on pultruded beam-to-column connections has been to attempt to develop recommendations on design details for pultruded connections to enable designers to use flexible (or semirigid) connections in pultruded structures. The stiffness of a beam-to-column connection is typically characterized by a moment–rotation curve of the connection itself (called the *rotational stiffness*) and the unit stiffness (EI/L) of the connecting beam member. The moment–rotation curve, which is typically nonlinear, is defined as

$$K = \frac{M}{\theta} \tag{15.1}$$

where K is the rotational stiffness (kip-in./rad or kNm · rad), M the moment transferred between the beam and the column at the connection, and θ is the relative rotation between the beam and column members (preferably measured at the member centerlines). If K is infinitely large, the connection is termed *rigid;* if K is zero, the connection is termed *simple.* Between these two theoretical extremes lie the semirigid connections. For actual construction, numerical values of KL/EI are used to define if the connection behaves as a rigid, semirigid, or simple connection. In steel structures the connection is termed simple if $K_sL/EI \leq 2$ and rigid if $K_sL/EI > 20$. For steel structures, $K = K_s$, which is the secant stiffness of the connection at service loads (AISC, 2005). For pultruded connections, Mottram and Zheng (1996, 1999a,b) have suggested that the initial rotation, which allows a beam span deflection ratio of $L/250$ (service load range), be used as the criterion for "adequate" rotation of the nominally simple connection. They recommend a value of $\theta = 0.025$ rad at the service load moment for simple web clip and top and seat clip flange connections. This translates to a joint rotational stiffness of $K_{\text{service}} = 40\ M_{\text{service}}$, where M_{service} is the service load moment. At this time, however, anlaytical methods are not available to predict the connection rotational stiffness, and test data should be used for this purpose.

It is well known that designing (and detailing) rigid frame connections in frame structures serves the purpose of enabling the frame to carry lateral loads by frame action and of reducing the deflections in the beam members under gravity loads. In pultruded structures the latter purpose is especially important, since as discussed in Chapter 13, beam design is often controlled by serviceability (i.e., deflections). If rotational restraint can be provided by the beam-to-column connection, the pultruded frame design can be more efficient. Due to the inherent lower stiffness of the pultruded profiles that are connected in a pultruded structure, it appears not to be realistic to attempt to develop a fully rigid connection in pultruded structures but rather, to attempt to develop semirigid connections (Turvey, 2001). Turvey and Cooper (2004) have cataloged the rotational stiffnesses of the key beam-to-column connections (for pultruded connection in I-shaped sections larger than 6 in. and primarily in the 8-in. series) and have divided these connections into three categories: (1) web clip angles only ($K = 30 - 80$ kN · m/rad, $M_{\text{ult}} = 1.7 - 2$ kN · m); (2) top and bottom flange clip angles ($K = 500 - 1100$ kN · m/rad, $M_{\text{ult}} = 2.0 - 6.0$ kN · m; and (3) customized "stiff" connections ($K = 500 - 3000$ kN · m/rad, $M_{\text{ult}} = 8.5 - 30$ kN · m).

At the present time, notwithstanding the research that has been conducted into the behavior of pultruded beam-to-column connections, semirigid frame design using pultruded connections cannot be performed with sufficient confidence based solely on the published data and current analytical models. Only simple framing should be used, and lateral resistance developed using bracing members or supplementary lateral load–resisting systems. In the future, the data tabulated by Turvey and Cooper (2004) may be used to provide design guidance to designers wishing to perform linear (service load range) and nonlinear (ultimate load range) semirigid frame analysis. More test data are needed on the service load rotations of pultruded beam-to-column connections (Turvey and Cooper, 2004). Therefore, in what follows, design guidance is given for simple bolted beam-to-column connections for pultruded structures, on the assumption that the connection will not allow any moment to be transferred. In addition, design guidance is only provided for determining the ultimate strength of the connection and not the deformations of the connection under service load, such that at this time no serviceability design can be performed for pultruded connections.

15.5 BOLTED PULTRUDED CONNECTIONS

The vast majority of light-truss and heavy-frame pultruded connections in load-bearing pultruded structures that are made with mechanical fasteners use bolted connections with either steel (typically, galvanized or stainless) or FRP fastener hardware. As noted previously, adhesive bonding may be used in addition to the bolting but is very rarely used as the sole means of making a pultruded truss or frame connection.

There are many reasons for using bolted connections in pultruded structures[2]:

1. The fabrication of the individual connection parts and the profiles to be connected is relatively easy and is generally familiar to construction workers skilled in steel and wood frame construction.
2. The connection is easy to assemble in the field or in the shop, and no surface preparation of the base materials is required (as in the case of adhesive bonding).
3. The connections are easy to inspect after assembly.
4. The connection can be assembled quickly and achieve its full strength and stiffness immediately (as opposed to an adhesively bonded connection).
5. If no adhesive is used in addition to the bolting, the connection can be disassembled easily if necessary.
6. Bolted connections can be economical when the cost of both shop and field labor work is taken into account. In the United States, fabrication of drilled holes in pultruded materials can be quite costly and is generally in the order of $1 per hole.

[2] Many of these reasons are similar to those given for the use of steel-bolted connections.

7. Specified manufacturing tolerances (out-of-straightness and twist) in pultruded profiles can be accommodated since the connection parts can usually be worked into place, due to their low stiffness.
8. Minor misfits due to bolt hole sizes or locations can usually be fixed in the field using simple hand tools.

However, there are also a number of issues to consider when using bolted connections in pultruded structures:

1. The bolt holes cause stress concentrations and reduce the net section in the pultruded material and thereby reduce the efficiency of the connection.[3] In addition, the load on fastener groups does not typically distribute evenly to multiple rows of bolts as in steel-bolted connections.
2. The pultruded parts used in the connection are made of orthotropic materials, and therefore the orientation of the individual parts in the connection is critical, unlike in steel-bolted connections, where the base material is isotropic and not orientation dependent. This is particularly relevant for the pultruded angles used in beam-to-column connections, where the major fiber orientation (in the longitudinal pultrusion direction) is often placed perpendicular to the primary load direction (see Fig. 15.10, for example, where the top seat angle is in the transverse direction).
3. Conventional glass pultruded materials have low through-the-thickness stiffness and strength properties and can therefore be crushed if high bolt torques are used. Since the base material can creep, the bolt tension can decrease over time, due to strain relaxation. If FRP bolts are used, the bolts themselves will lose tension due to strain relaxation. In addition, if small nuts (without washers) are used, the entire fastener can punch through or crush the base pultruded material when the connection is loaded. For this reason, only bearing connections are used in pultruded bolted connections, and slip-critical or friction connections are not possible in pultruded connections.
4. There is limited availability of FRP nuts and bolts or threaded rods sizes, and no FRP parts are available that have unthreaded shanks such that material will bear on the threaded portion of the fastener, which is undesirable. Since the threads are machined into FRP bolts and threaded rods, they have a tendency to strip-off under bolt shear loads, and thus the tensile capacity of the FRP bolt is quite limited.
5. There is limited availability of pultruded angle thicknesses ($\frac{1}{4}$, $\frac{3}{8}$, $\frac{1}{2}$ in.) and sizes (2 to 6 in leg equal angles) used in typical beam-to-column

[3]The *efficiency* of a connection is defined as the maximum load that can be transferred through the connection, divided by the maximum load that can be carried away from the connection location by the connected member.

connections. This limits the designer in the types of connection that can be made.

6. The properties of pultruded plate material used for connection gusset plates and miscellaneous filler parts are typically less than the properties of the pultruded profiles in the connection.
7. Holes drilled in the pultruded material for a bolted connection provide a pathway for ingress of moisture and other chemical agents that can degrade the pultruded material. Therefore, all drilled holes in pultruded materials must be sealed with a thin epoxy resin coating. This is typically done in the shop after the parts are drilled.
8. Holes in pultruded material must be drilled, preferably using special diamond-tipped bits. Holes cannot be punched like steel parts, due to the properties of the pultruded materials.
9. Bolted pultruded connections can fail in significantly different ways from steel connections even though the connection geometries may look similar. Pultruded connections can fail due to failure in the pultruded material of the members being connected, in the pultruded material of the pultruded connection parts (angles and gussets), or in the mechanical fasteners themselves (the bolts, nuts, rods, and washers). FRP bolts or threaded rods can fail in transverse shear, in longitudinal shear due to thread stripping, and in longitudinal tension or compression. FRP nuts can fail due to longitudinal thread shear. The pultruded material in the connected members or in the parts can fail due to in-plane or out-of-plane loads or a combination of the two, depending on the type of pultruded connection.[4]

Bolted connections are used for both light-truss and heavy-frame pultruded connections, as noted previously; however, different considerations apply in their design, which are due primarily to the types of forces and eccentricities of the fasteners relative to the centerline of the members being connected.

15.6 LIGHT-TRUSS PULTRUDED CONNECTIONS

In light-truss pultruded connections the forces being transferred by the pultruded connection are axial forces and the fasteners are typically positioned on (or symmetrically with) the lines along the centroids of the members, which intersect at a point, as shown in Fig. 15.2. Therefore, the load transfer in the connection is by in-plane forces parallel to the member axes, through either single or double overlapping members (known as lap joints).

[4] The discrete failure modes of pultruded materials are discussed further below.

15.6.1 Lap Joint Connections

The transfer of forces in double lap joints[5] in pultruded plate materials has been studied in some detail by a number of researchers since the early 1990s (Abd-el-Naby and Hollaway, 1993a,b; Cooper and Turvey, 1995; Erki, 1995; Rosner and Rizkalla, 1995a,b; Pabhakaran et al., 1996; Hassan et al., 1997a,b; Turvey, 1998; Wang, 2002). In light-truss connections telescoping pultruded square tubular members are often used for both aesthetic and geometric reasons. Test data for these types of connections can be found in Merkes and Bank (1999) and Johansen et al (1999).

In the studies referred to above, the researchers have investigated the influences of pultruded material orientation, fastener size, bolt torque, pultruded plate thickness, bolt spacing, washer presence and size, and single- and multibolt configurations on connection failure modes and capacities. Most of these data, together with more recent test results, has been assembled by Mottram and Turvey (2003) in a comprehensive review to that date. Recommendations for lap joint connection geometries and equations for checking different elementary failure modes, based on the research cited above, are presented in what follows.

The failure modes of single-bolt lap joints are usually identified as net-tension failure, shear-out failure, splitting (also called *cleavage*) failure, cleavage (also called *block shear*) failure, and bearing failure. These are shown graphically in Fig. 15.11. It is important to recognize that these failure modes are expected to occur in FRP plates with unidirectional, bidirectional, or quasi-isotropic layups and are not necessarily applicable to conventional pultruded materials unless they are loaded at either 0° or 90° (i.e., parallel to transverse to the pultrusion direction). If a pultruded material is loaded in an off-axis orientation (such as the chord of the truss shown in Fig. 15.2) the

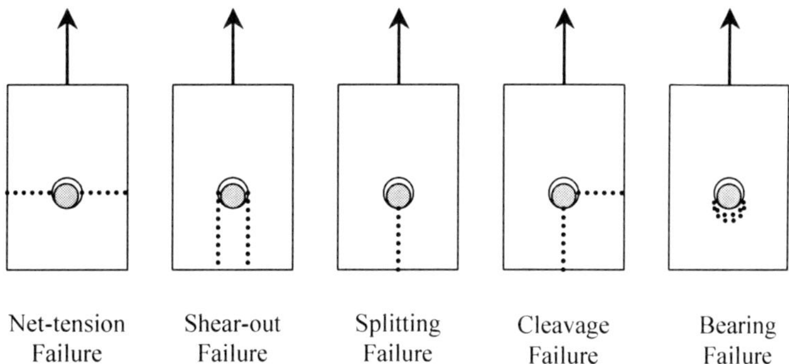

Figure 15.11 Failure modes of single-bolt lap joint in-plane connections.

[5] And to a limited extent in single lap joints.

failure will typically occur in combined net tension and shear along a plane that is parallel to the roving orientation (Cooper and Turvey, 1995). In this case, detailing recommendations based on assuming the failure modes shown above may not be appropriate. Therefore, a designer should attempt to place the longitudinal direction of the pultruded part used in the connection either parallel (preferably) or perpendicular to the load, or conduct tests of the pultruded material (or profile) loaded in the direction of interest.

A review of the extensive research data obtained for pultruded plate materials has led to recommendations, shown in Table 15.2, for spacing and edge distances for lap joints in pultruded connections made of conventional glass FRP pultruded plate loaded by in-plane loads in tension (Mottram, 2001; Mottram and Turvey, 2003). These recommendations apply to structures in operating environments of normal room temperature and humidity. The pertinent geometric parameters for single- and multibolt joints are shown in Fig. 15.12.

TABLE 15.2 Recommended Geometric Parameters for Lap Joint Connections

	Research Data		Manufacturer[a]	
	Recommended	Minimum	Recommended	Minimum
End[b] distance to bolt diameter, e/d_b	≥ 3	2	≥ 3	2
Plate width to bolt diameter, w/d_b	≥ 5	3	≥ 4	3
Side distance to bolt diameter, s/d_b	≥ 2	1.5	≥ 2	1.5
Longitudinal spacing (pitch) to bolt diameter, p/d_b	≥ 4	3	≥ 5	4
Transverse spacing (gage) to bolt diameter, g/d_b	≥ 4	3	≥ 5	4
Bolt diameter to plate thickness, d_b/t_{pl}	≥ 1	0.5	2	1
Washer diameter to bolt diameter, d_w/d_b	≥ 2	2	NR	NR
Hole size clearance, $d_h - d_b$	tight fit ($0.05d_b$)	$\frac{1}{16}$ in.[c]	$\frac{1}{16}$ in.	NA

[a] NR, no recommendation; NA, Not applicable.
[b] Also called edge distance.
[c] Maximum clearance.

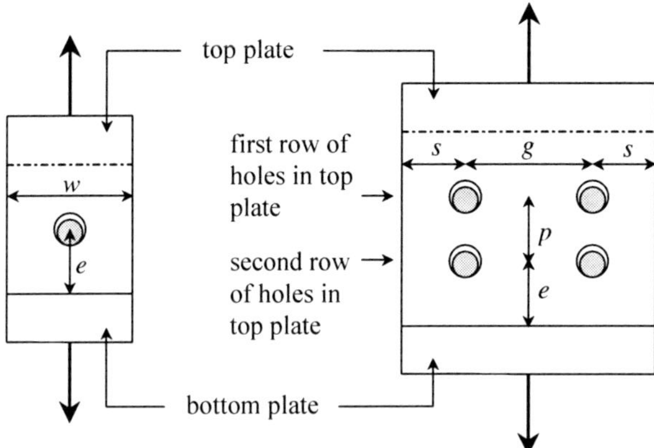

Figure 15.12 Geometric parameters for single- and multibolt lap joints.

It is important to note that these recommendations were obtained from testing pultruded plate (flat sheet), which has properties that differ from those of the pultruded material in conventional profiles (Table 1.4). It can be seen that the pultruded material in profiles typically has a higher degree of orthotropy than plate material and has higher longitudinal strength and stiffness but transverse properties similar to those of plate material. The geometric design recommendations for plates are therefore expected to be conservative when used for profiles loaded in their longitudinal directions, but applicable to profiles when loaded in their transverse directions. Based on the data presented in the research literature, the design geometric parameters given in Table 15.2 are recommended for in-plane lap-joints in pultruded materials.

In addition to recommendations obtained from test data, pultrusion manufacturers provide their own recommendations for the geometric detailing of lap joints. They typically provide ranges of values based on data published originally in the *Structural Plastics Design Manual* (ASCE, 1984). Manufacturers' recommendations are intended to be used with pultruded materials from plates and profiles. These recommendations are also shown in Table 15.2.

It is important to note that the geometric design recommendations presented above are for lap joints loaded in tension. The geometric recommendations are intended to cause bearing failure in the pultruded base material at the locations of the fasteners when the material is loaded in the longitudinal direction, and to avoid net-tension or cleavage failure modes, which are brittle failures modes. However, this is generally true only for single-bolt joints.

Bearing failure is regarded as a ductile failure mode and consists of local crushing and delamination of the pultruded material in direct contact with the bolt. As the lap joint continues to be displaced longitudinally, the bearing failure can become a shear-out failure. Shear-out failure can occur either between the bolt and the part end, or between fasteners in a column of bolts parallel to the load direction. If the end distance (or the longitudinal spacing) is sufficiently long (e.g., $e/d_b > 5$), this failure can also be a progressive pseudoductile failure mode in pultruded materials (Merkes and Bank, 1999; Lamanna et al., 2001).

When more than one bolt is placed in a row (perpendicular to the load direction), net-tension failures have been found to occur even with the spacing recommended. In multirow joints, the bolt (or row of bolts) closest to the edge of the material, where the tensile load is applied, typically carries more load than do those farther back. This is unlike in a steel connection, where it is assumed that due to local yielding, the load is distributed evenly among all the bolts in all the rows and columns at the ultimate limit state.

The distribution of the load between bolt rows is the subject of current research (Mottram and Turvey, 2003). Table 15.3 lists the Eurocomp-recommended load distributions between rows in fastener groups with multiple rows of fasteners. Recommendations are given for both glass FRP profiles to glass FRP parts and for glass FRP profiles to steel parts (when steel plates are used with pultruded profiles). It is recommended that bolted pultruded connections have at least two bolts in a row and not more than four bolts in a row.

For constructability, bolt holes are generally oversized by $\frac{1}{16}$ in., as is the practice in bolted steel connections. However, if possible, holes should not be oversized, and a clearance of 5% of the bolt diameter is recommended (Eurocomp, 1996), but not more than $\frac{1}{16}$ in. As noted previously, pultruded connections are bearing connections, and oversizing holes allows the connection to slip when loaded before the fasteners bear on the base material. This causes undesirable rotations and deflections in the pultruded structure as it "settles in" when loaded. It is possible not to oversize holes and still fit the connection together (provided that holes are drilled accurately) since the pultruded material is relatively compliant, and bolts (especially steel bolts) can be inserted with light tapping using a rubber mallet.

TABLE 15.3 Distribution of Load in Rows of Multirow Lap Joints[a]

Number of Rows	Material	Row 1	Row 2	Row 3	Row 4
1	FRP to FRP	1.0			
2	FRP to FRP	1.0	1.0		
3	FRP to FRP	1.1	0.8	1.1	
4	FRP to FRP	1.2	0.8	0.8	1.2
1	FRP to steel	1.0			
2	FRP to steel	1.15	0.85		
3	FRP to steel	1.50	0.85	0.65	
4	FRP to steel	1.7	1.0	0.7	0.6

[a] The value given is the factor by which the average load on the fasteners is multiplied.

15.7 HEAVY-FRAME PULTRUDED CONNECTIONS

In heavy-frame connections, the connection parts used to connect the members, which are usually at right angles, are eccentrically located with respect to the centroid of the members being connected. As a result, the connection parts are subjected to loads that are offset from, or eccentric to, the line of the fasteners. This causes a localized bending moment to develop in the plane of the web, in addition to the in-plane loads that occur in lap joint connections seen in light trusses. For the in-plane component of the forces, the same discussion and the guidance cited previously for lap joints is applicable to the parts of the heavy pultruded connection that carry in-plane loads.

The moment causes out-of-plane forces to develop at the face of the connected members which generate prying forces at the top and compressive forces at the bottom of the connection. For example, a web clip subjected to a bending moment causes the top of the clip to pull away, or pry away, from the column flange, and the bottom part of the clip angle to be compressed into the flange of the column (Fig. 15.13). This eventually leads to the failure of the clip angle, due to delamination due to tensile and flexural stresses at the top of the angle, as shown in Fig. 15.14.

Failure may also occur at a web–flange junction in the column behind the top set of fasteners on the column flange. If the beam shown in Fig. 15.14 were allowed to continue to rotate, the column could eventually fail in the mode shown in Fig. 15.9.

Figure 15.13 Out-of-plane forces in clip angle simple shear beam-to-column connections. (Courtesy of J. T. Mottram.)

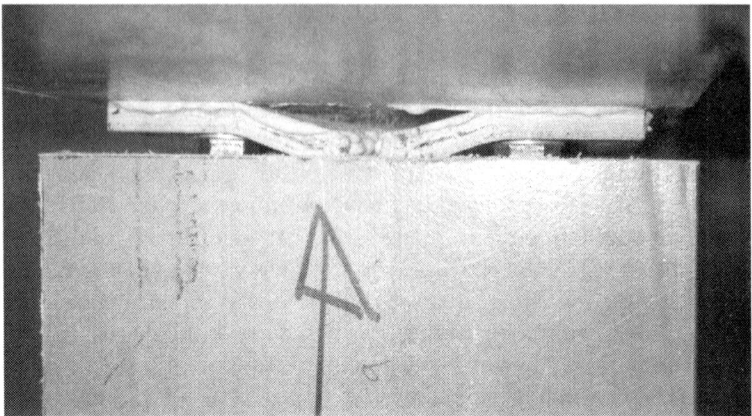

Figure 15.14 Delamination of a web clip angle due to prying action. (Courtesy of J. T. Mottram.)

15.8 DESIGN OF BOLTED PULTRUDED CONNECTIONS

Guidance is given in this section for determining the critical stresses in pultruded connections. The guidance is based on simplified mechanics considerations and assumes that only in-plane tension or out-of-plane tension or compression loads are applied to an individual fastener or connection part (as in a shear clip connection). Combined in-plane tension and shear on bolt groups is not considered. Equations, however, are provided in the Eurocomp design code for calculating stresses at the circumference of a hole in an orthotropic material under multiaxial stress states. It is not generally accepted that the level of complexity involved in performing such calculations is appropriate at this time, given the limited data available on pultruded materials in bolted connections, especially beam-to-column connections. In what follows, guidance is given only for the following connections:

1. In-plane lap-joint connections for in-plane tensile or compressive loads found in pultruded connections in light-frame overlapping joints and in gusset plates for bracing members (see Fig. 15.7) and in elements of heavy-frame beam-to-column connections.
2. Out-of-plane beam-to-column connections for shear forces (and secondary local moments) found in simple regular beam and column frames. At this time this includes only (a) web clip angle connections (see Fig. 15.13) and (b) flexible seated connections. Guidance provided for shear clips can also be used to design simple column base plates. These connections may be designed to account for the eccentricity of the shear load according to traditional elastic vector analysis that is used for steel bolted connection (Salmon and Johnson, 1996, p. 139). The

resulting load on each of the bolts due to a combination of direct shear and torsion on the fastener group (due to the eccentric shear load) is obtained using this traditional procedure. Using the resulting load, the critical stresses at the individual bolt locations are determined.

In the absence of specific guidance for either in-plane or out-of-plane pultruded connections, the basic principles used to design steel-bolted simple shear connections can be followed for the design of bolted pultruded connections (Salmon and Johnson, 1996). However, the orthotropic properties of the pultruded based materials and FRP fasteners (if used) and geometric spacing recommendations for pultruded connections should be followed. In particular, the orientation of the pultruded material in the connection should always be considered (and indicated on detailing drawings of the connection). In the event that the connection capacity cannot be realized with an existing profile wall thickness and pultruded plate thicknesses, the thickness of the pultruded material should be increased by adhesively bonding (during shop fabrication) a piece of pultruded plate material where required to strengthen the existing part. In these situations it is recommended that a piece be bonded with its major fiber orientation at 90° to the fiber orientation in the existing part to create a material with more isotropic parts at the critical location.

15.9 DETERMINATION OF STRESSES IN IN-PLANE LAP JOINTS

In what follows, elementary one-dimensional mechanics-based equations are presented for determining stresses in in-plane single- and multibolt lap joints. The equations for calculating the stresses on the connection parts are based on assumptions of linear elastic material behavior and small deformations and give design stresses that can be compared with coupon-level material test data. However, these equations are used only as a guide to the strength of the connection, since at the ultimate load the connection behavior is not linear, and large deformations can significantly alter the stress distributions and change the failure modes.

15.9.1 Bearing Stress in the Base Pultruded Material

The average bearing stress at the hole in the base pultruded material is given as

$$\sigma_{\mathrm{br}} = \frac{P_b}{d_b t_{\mathrm{pl}}} \tag{15.2}$$

where P_b is the load transferred at an individual bolt location, d_b the bolt diameter, and t_{pl} the thickness of the base pultruded material.

15.9.2 Net-Tension Stress in the Base Pultruded Material

The net-tension design stress at the location of the hole in the base pultruded material is given as

$$\sigma_{\text{net}} = \frac{P_t}{A_{\text{net}}} \tag{15.3}$$

where P_t is the tensile load transferred by the entire lap joint, consisting of a number of columns of bolts, and the net area, A_{net}, is taken as

$$A_{\text{net}} = t_{\text{pl}}(W - nd_h) \tag{15.4}$$

where n is the number of bolts in the row, and W the width of the plate at the critical section, d_h the hole diameter, and t_{pl} the thickness of the base pultruded material. This assumes that the critical section will be through the row of holes and perpendicular to the longitudinal direction of the material. As in steel connections, the critical section may be staggered though multiple rows when holes are not in rows perpendicular to the load direction (Salmon and Johnson, 1996, p. 74). In orthotropic pultruded materials, the failure may occur in a staggered fashion even when the holes are in regular rows (Prabhakaran et al., 1996). This may be considered to be a type of block shear failure (discussed in what follows) in this case. At this time, the empirical formula used for staggered rows for joints in steel members is recommended as a first approximation for orthotropic pultruded materials:

$$A_{\text{net}} = t\left(W - nd + \sum \frac{p^2}{4g}\right) \tag{15.5}$$

where the summation is over all diagonal distances in an assumed failure "path." This equation assumes that the strength of the material is the same along all paths, which is not actually the case for orthotropic materials. Where elements of profile sections are connected in tension (such as flanges of angles and I-sections), it is recommended that only the width of the outstanding element be considered to be effective (not the width of the entire section). The effective net area used in steel design should not be used for pultruded materials. For multiple rows of bolts, the load distributions in Table 15.3 should be used.

15.9.3 Shear-Out Stress in Base Pultruded Material

The shear-out stress at the bolt location at the material end edge in the direction of the load is given as

$$\tau_{\text{shear-out}} = \frac{P_b}{2t_{\text{pl}}\, e} \tag{15.6}$$

where P_b is the load transferred at an individual bolt location, e the end distance, and t_{pl} the thickness of the base pultruded material.

The shear-out stress at the bolt location in the material between two bolts in a column in the direction of the load is given as

$$\tau_{\text{shear-out}} = \frac{P_b}{2t_{\text{pl}}\, p} \tag{15.7}$$

where P_b is the load transferred at an individual bolt location, p the pitch distance, and t_{pl} the thickness of the base pultruded material.

15.9.4 Shear Stress on a Bolt

The shear stress on a bolt is given as

$$\tau_b = \frac{V_b}{A_b} \tag{15.8}$$

where V_b is the shear force on the bolt (accounting for single or double shear configurations) and A_b is the cross-sectional area of the bolt shank.

15.10 STRESSES IN OUT-OF-PLANE SHEAR CONNECTIONS

In simple shear beam-to-column connections, in addition to the stress state related to in-plane loading of lap-jointed parts noted above, additional stresses are determined in the profiles themselves and the connected parts. These stresses are developed due to the eccentricity of the shear load on the connection, which causes bending and shear stresses to develop in the connection parts. For simple shear double clip angles connections, the following states are considered.

15.10.1 Longitudinal Shear Stress at the Heel of an Angle

Shear stress in the heel of the connection angles is calculated as

$$\tau_{\text{pl}} = \frac{V}{2A_{\text{heel}}} \tag{15.9}$$

where V is the total design shear force (one-half to each clip angle) at the beam end and $A_{\text{heel}} = t_{\text{pl}}\, l_{\text{angle}}$ is the shear area of the heel, equal to the thickness of the angle multiplied by the length of the angle parallel to the shear force. If this stress exceeds the ultimate shear strength of the pultruded material, the connection will fail, due to shear failure of the clip angles.

15.10.2 Flexural Stress in the Leg of an Angle Bolted to a Column Member

Due to the prying action of an eccentric connection, the upper portion of the clip angle is subjected to a tensile force which causes the connected leg of the angle to bend and shear as shown in Fig. 15.15. Assuming that the double angle acts as a fixed beam loaded with a concentrated load (equal for the prying force), the bending and shear stress in the leg of the angle can be calculated. Assuming that the prying force, R, on a bolt a distance d from the center of rotation, determined using elastic vector analysis, is

$$R = \frac{Md}{\sum d^2} \tag{15.10}$$

the moment at the connection, $M = Ve_v$, is equal to the shear force multiplied by the eccentricity, e_v, from the column face to the centroid of the bolts. The local moment applied to the leg of each angle based on two back-to-back angles (assuming a fixed–fixed beam with a central load equal to R) is

$$m_{\text{leg}} = \frac{R(b_{\text{leg}} - t_{\text{pl}} - s)}{8} \tag{15.11}$$

where b_{leg} is the width of the angle leg and s is the side distance from the center of the bolt hole to the edge of the leg. The local shear force is, $v_{\text{leg}} = R/2$.

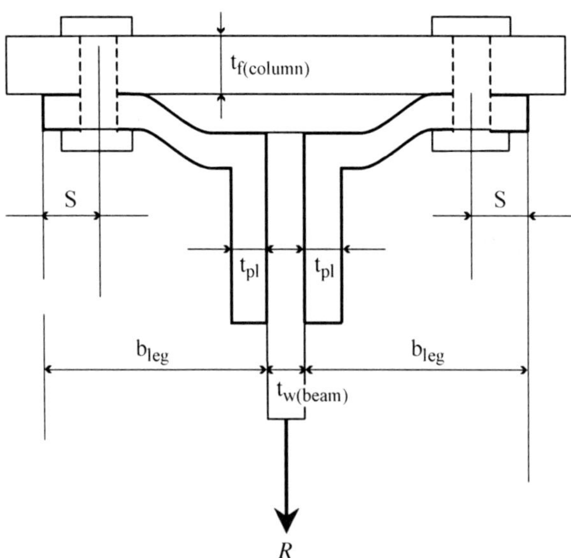

Figure 15.15 Deformation and forces in back-to-back angles.

The flexural stress in the leg of the angle is then given as

$$\sigma_{\text{flex}} = \frac{m_{\text{leg}}}{S} = \frac{m_{\text{leg}}}{(2e)t_{\text{pl}}^2/6} \tag{15.12}$$

where the effective length over which the angle bends is assumed to be twice the distance from the top of the angle to the bolt (i.e., the edge distance, e).

The shear stress in the leg is given as

$$\tau_{\text{leg}} = \frac{V}{A} = \frac{R/2}{t_{\text{pl}}\,(2e)} \tag{15.13}$$

where it is assumed that the shear stress is distributed over a length equal to twice the edge distance. This shear stress is the major cause of the delamination failure shown in Fig. 15.14. Failure will occur when this shear stress is equal to the interlaminar shear stress of the pultruded material.

15.10.3 Transverse Tensile Stress in a Web–Flange Junction of a Column

When the beam is connected to the flange of a column, the prying force exerted by the top half of the back-to-back clip angles will subject the web–flange junction of the column to a transverse tensile force that will tend to pull the flange away from the web of the profile. This transverse tensile force is assumed to act over an area equal to the thickness of the column web, t_{web}, multiplied by a length equal to half the angle height, $l_{\text{angle}}/2$. The transverse tensile stress of the web–flange junction is therefore given as

$$\sigma_{\text{trans}} = \frac{R}{t_{\text{web}}(l_{\text{angle}}/2)} \tag{15.14}$$

15.10.4 Block Shear in a Beam Web

Block shear failure, or cleavage failure (in Fig. 15.11) occurs in the base material over an area that fails due to a combination of net tension failure and shear failure. This type of failure can occur in the web of the beam when the top flange of the beam is coped for construction reasons. The failure of the pultruded material in the web depends on both the net tensile strength of the material and the shear strength of the material assuming that the axes of orthotropy of the pultruded material are aligned either parallel or perpendicular to the load direction. The tensile and shear stresses depend on the area of the through-the-thickness surface that is subjected to tension and shear, and they cannot be calculated separately. An interaction equation, given below, is used to determine the capacity of the connection.

15.10.5 Flexural and Shear Stresses in Flexible Seated Connections

In beam-to-column connections in which a bottom seat is used to support the beam at the column, critical shear and flexural stresses develop at the face of the outstanding leg, which acts as a cantilever, as shown in Fig. 15.16. The local bending moment for the flexural stress calculation depends on the location of the shear force reaction on the seat. Generally, this is taken as half the distance between the beam end and the end of the angle leg. In seated connections a top clip must also be provided to stabilize the top flange of the beam. Seated clip angles using pultruded angles are generally not used in typical designs, due to the low flexural stiffness of the pultruded angle in the "closing" mode.

When both top and bottom clip angles (seats) and web clip angles are used, a pultruded connection (shown in Fig. 15.9) can develop a sufficient rotational stiffness to allow semirigid frame analysis and design. However, as discussed previously, the rotational stiffness can not be determined analytically at this time and must be obtained by full-size experiments, if required.

15.11 CRITICAL CONNECTION LIMIT STATES

The critical strength when the pultruded base material fails due to bearing is taken as either the longitudinal or the transverse bearing strength of the pultruded material. This is regarded as a basic material property and is obtained from testing, which is generally reported by pultrusion manufacturers. If the bearing strength is not available from test data, it can be approximated as the material compressive strength for approximate calculations. Either the longitudinal bearing strength or the transverse bearing strength is used, depending on the direction of the member or element relative to the load:

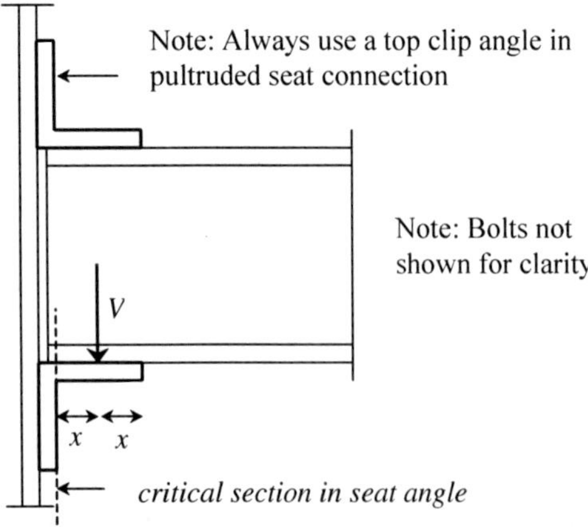

Figure 15.16 Seated angle connection.

$$\sigma_{cr}^{bear} = \sigma_{L,br} \quad \text{or} \quad \sigma_{T,br} \tag{15.15}$$

Pultruded connections are currently designed only for the ultimate strength limit state. The bearing strength used for pultruded materials is the ultimate bearing strength (i.e., the load required to cause a shear-out failure behind the hole), not the serviceability bearing strength (often defined as the load causing an axial deformation equal to 4% of the hole diameter according to ASTM 953).

The critical strength for tensile failure of the pultruded material in the net section[6] either in the longitudinal or transverse direction is given as

$$\sigma_{cr}^{net\text{-}tens} = 0.9\sigma_{L,t} \quad \text{or} \quad 0.9\sigma_{T,t} \tag{15.16}$$

The critical strength for shear failure (and shear-out failure) of the parts of the connection is the in-plane shear strength of the pultruded material, given as

$$\tau_{cr} = \tau_{LT} \tag{15.17}$$

When block shear is a possible failure mode, the net tension and the shear strength of the pultruded material are used together in a combined stress interaction equation. Since both net tension and shear failure in pultruded materials are considered brittle failure modes, a linear interaction is used. The nominal tensile capacity, P_n, of the connection is given as

$$P_n = \sigma_{cr}^{net\text{-}tens} A_{tens} + \tau_{LT} A_{shear} \tag{15.18}$$

where, A_{tens} is the area along the tensile failure path and A_{shear} is the area along the shear failure path.

The critical flexural stress for flexural failure of the leg of the web clip angle due to prying action is

$$\sigma_{cr}^{flex} = \sigma_{L,flex} \quad \text{or} \quad \sigma_{T,flex} \tag{15.19}$$

where $\sigma_{L,flex}$ and $\sigma_{T,flex}$ are the longitudinal and transverse flexural strength of the pultruded material. Flexural strengths are typically reported by manufacturers separately from tensile and compressive strengths. It is important to note that in the web clip angle configuration, the transverse flexural stress will control, due to the orientation of the angle relative to the bending, as shown in Fig. 15.15.

The critical shear stress for the interlaminar shear failure of the leg of the web clip angle due to prying action is

[6]See the important discussion in Chapter 14 related to the strength reduction factor used in the net-section equations.

$$\tau_{cr} = \tau_{TT} \tag{15.20}$$

where τ_{TT} is the interlaminar shear strength of the pultruded material in the leg of the angle. This value is typically reported by manufacturers.

The critical shear strength of a bolt is the ultimate shear strength of the bolt (either steel or FRP),

$$\tau_{cr}^{b} = \tau_{ult}^{b} \tag{5.21}$$

The critical tensile strength of an FRP bolt (or threaded rod) is determined by the maximum (bolt) tensile load (see Table 15.1, line 2) based on thread shear failure, T_{max}, given as

$$\sigma_{cr}^{b} = \frac{T_{max}}{A_b} \tag{15.22}$$

15.12 DESIGN PROCEDURE FOR A PULTRUDED CONNECTION

The design procedure presented in what follows for pultruded connections permits design only by the allowable stress design (ASD) basis, as discussed in Chapter 12. Connection design at this time consists of dimensioning the connection using the recommended geometric parameters and then checking the stresses in the connected members, fasteners, and connection parts using simple calculations described previously in order to determine gusset plate or angle thicknesses and bolt sizes. For beam simple shear connections, the load tables presented in manufacturers' design guides can be used to estimate the size or the connection, or its load-carrying capacity, if necessary, for the purposes of a preliminary design.

Step 1. Determine the design loads and ASD factors. The tensile load for an in-plane connection (or parts of an out-of-plane connection) or the shear load for an out-of-plane connection is determined from the structural geometry and loading. A safety factor of 4 is used for all strengths in the connection parts.

Step 2. Select the connection parts and fasteners and a trial geometry. Estimate the number of bolts based on the total tensile or shear force to be transferred. Determine the connection part thicknesses based on the bearing strength of the pultruded material. Using the geometric parameters, dimension the connection.

Step 3. Determine the maximum design stresses. Determine the design stresses on the connected members, fasteners, and connection parts using the trial geometric parameters.

Step 4. Determine critical stresses. Determine the critical stresses for the connected members, fasteners, and connection parts. It is important to identify the load direction in the connection and use the appropriate strength values based on the orientation of the material relative to the load direction.

Step 5. Determine the factored critical stresses. Divide the critical stresses determined in step 4 by the appropriate safety factor.

Step 6. Check the ultimate strength of the trial connection. Check that the design stresses are less than the allowable stresses. Return to step 2 if the trail connection does not work.

Step 7. Dimension the connection and call out all the parts. Dimension (or design if quantitative procedures are available) all the connection parts and geometries. Provide a sketch of the connection showing all orientations of pultruded materials used for gusset plates or doubler plates.

Design Example 15.1: Pultruded Beam-to-Column Clip Angle Connection Design a simple shear web double clip-angle connection for the beam-to-column connection for the floor system presented previously in the design examples for the beam in Chapter 13 and the column in Chapter 14. The floor plan is shown in Fig. 13.13. The beam size was determined previously to be a wide-flange W $10 \times 10 \times \frac{1}{2}$ profile with material properties given in Table 13.7. The column size was determined previously to be a W $8 \times 8 \times \frac{3}{8}$ profile with material properties given in Table 14.1. Pultruded material from the same manufacturer as the beam section and shown in Table 15.4 is to be used for the clip angles. Note that flexural and bearing strength values have been added to the properties provided for the beam in Chapter 13. Design the connection using FRP bolts and nuts with the properties shown in Table 15.1.

SOLUTION

Step 1. Determine the design loads and ASD factors. The beam is simply supported and the maximum shear force at the end of the beam is

$$V_{\max} = \frac{wl}{2} = \frac{372(15)}{2} = 2790 \text{ lb}$$

Since only ASD design is to be performed, no load factors are used. A safety factor of 4 is used for all strengths in the connection parts.

Step 2. Select the connection parts and fasteners and a trial geometry. For preliminary sizing, consider the bearing capacity of the beam web, clip angles, and column flanges. Since the beam web carries the same end shear with half the number of holes as the clip angles and the column flanges, the beam web

TABLE 15.4 Pultruded Material Properties in the Clip Angles

Symbol	Description	Value[a]
E_L^c	Longitudinal compressive modulus	2.6×10^6 psi
E_L^t	Longitudinal tensile modulus	2.6×10^6 psi
E_T^c	Transverse compressive modulus	1.0×10^6 psi
E_T^t	Transverse tensile modulus	0.8×10^6 psi
G_{LT}	In-plane shear modulus	NR
v_L	Major (longitudinal) Poisson ratio	0.33
$\sigma_{L,c}$	Longitudinal compressive strength	30,000 psi
$\sigma_{L,t}$	Longitudinal tensile strength	30,000 psi
$\sigma_{L,\text{flex}}$	Longitudinal flexural strength	30,000 psi
$\sigma_{L,\text{br}}$	Longitudinal bearing strength	30,000 psi
$\sigma_{T,c}$	Transverse compressive strength	16,000 psi
$\sigma_{T,t}$	Transverse tensile strength	7,000 psi
$\sigma_{T,\text{flex}}$	Transverse flexural strength	10,000 psi
$\sigma_{L,\text{br}}$	Transverse bearing strength	NR
τ_{TT}	Interlaminar shear strength	4,500 psi
τ_{LT}	In-plane shear strength	NR
E_b	Full-section flexural modulus	2.5×10^6 psi
G_b	Full-section shear modulus	0.425×10^6 psi

[a]NR, not reported.

will be the controlling part in the connection (assuming that the clip angle and the column flange thicknesses are at least half the thickness of the beam web). In addition, the beam web is loaded transverse to its primary fiber orientation, and its transverse bearing strength properties will be key. The column flanges and the clip angles will be loaded in bearing in their longitudinal directions and will therefore have higher bearing strengths.

Since the beam web is loaded transverse to its primary fiber direction, the transverse bearing strength of the web is needed for the calculation. Since the manufacturer does not report this value, the transverse compressive strength (16,000 psi) of the pultruded material is used in its place as a conservative estimate. With an ASD safety factor of 4.0, the allowable bearing stress in the web is taken as 4000 psi. For the $\frac{1}{2}$-in.-thick web material, the required number of bolts, n, is determined based on an assumed bolt diameter of $\frac{3}{4}$ in.

$$n = \frac{V}{d_b t_{\text{pl}} \sigma_{\text{br,allow}}} = \frac{2790}{0.75(0.50)(4{,}000)} = 1.9 \qquad \therefore \text{try two } \tfrac{3}{4}\text{-in. bolts}$$

Detail the connection using the recommended geometric parameters presented in Table 15.2, assuming that equal leg angles of the $4 \times 4 \times t_{\text{pl}}$ series are used (where t_{pl} is to be determined in what follows). The recommended and minimum values for the $\frac{3}{4}$-in.-diameter bolt are shown in Table 15.1.

A $\frac{1}{2}$-in. clearance between the end of the beam and the column face is typically used in construction. Therefore, a $\frac{1}{2}$-in. thickness of the clip angle is preferred and 8-in.-long $4 \times 4 \times \frac{1}{2}$ equal leg angles are chosen for this connection. The connection is detailed using the recommended spacing shown in Table 15.5 for the $\frac{3}{4}$-in. bolts. In addition, the actual spacings chosen for the connection are also shown. It can be seen that these fall within the geometric ranges recommended. Note that the minimum values are not violated. Side and plan views of the selected connection geometry are shown in Figs. 15.17 and 15.18.

Step 3. Determine the maximum design stresses.

Bearing Stresses Bearing stresses are calculated in the beam web, clip angles, and column flange. To calculate bearing stresses at the top bolt hole on the beam web, the additional bearing force at the hole due to the eccen-

TABLE 15.5 Geometric Parameters for Example Connection

	Spacings for ¾-in. Bolt			¾-in. Bolt
		Recommended	Minimum	Actual Selected (Controlling Element)
End[a] distance to bolt diameter, e/d_b	e	≥ 2.25 in.	1.50 in.	2.5 in. (angle leg)
Plate width to bolt diameter, w/d_b	w	≥ 3.75 in.	2.25 in.	4.0 in. (angle leg)
Side distance to bolt diameter, s/d_b	s	≥ 1.5 in.	1.13 in.	1.5 in. (column flange)
Longitudinal spacing (pitch) to bolt diameter, p/d_b	p	≥ 3.0 in.	2.25 in.	3.0 in. (angle leg)
Transverse spacing (gauge) to bolt diameter, g/d_b	g	≥ 3.0 in.	2.25 in.	NA
Bolt diameter to plate thickness, d_b/t_{pl}	d_b	≥ 0.5 in.	0.25 in.	2.0 (column flange)
Washer diameter to bolt diameter, d_w/d_b	d_w	≥ 1.5 in.	1.5 in.	use 1.5 in.
Hole size clearance, $d_h - d_b$	tight fit ($0.05d_b$)		1/16 in.[b]	use 1/16 in.

[a] Also called edge distance.

[b] Maximum clearance.

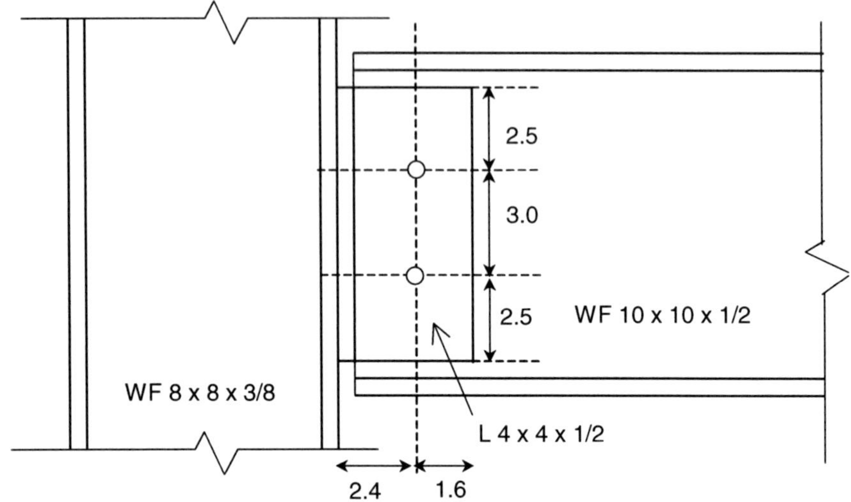

Figure 15.17 Beam-to-column connection with clip angles: side view.

tricity of the shear force is calculated. The moment due to the total shear force eccentricity relative to the face of the column is

$$M = Ve_v = 2790(2.4) = 6696 \text{ in.-lb}$$

The horizontal force (in the plane of the web) at the top bolt due to the moment is

$$R_x = \frac{My}{\sum d^2} = \frac{6696(1.5)}{2(1.5)^2} = 2232 \text{ lb}$$

The vertical force at the bolt due to the vertical shear force is

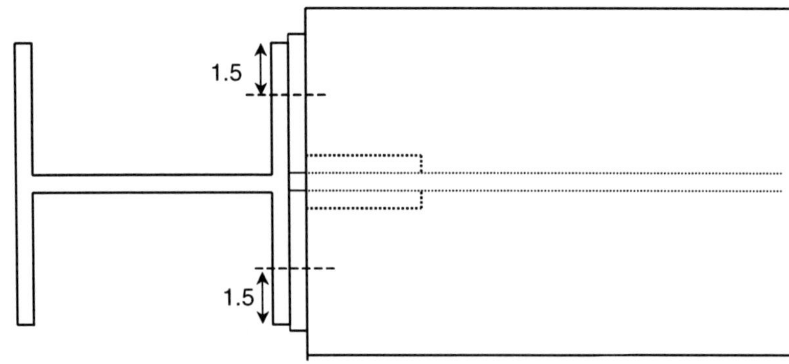

Figure 15.18 Beam-to-column connection with clip angles: top view.

$$R_v = \frac{V}{2} = \frac{2790}{2} = 1395 \text{ lb}$$

The resulting force on the bolt is

$$R_b = \sqrt{R_x^2 + R_v^2} = \sqrt{(2232)^2 + (1395)^2} = 2632 \text{ lb}$$

and the bearing stress on the bolt hole in the beam web is

$$\sigma_{\text{br}} = \frac{R_b}{d_b t_{\text{pl}}} = \frac{2632}{0.75(0.5)} = 7018 \text{ psi}$$

If only the vertical shear is considered (i.e., eccentricity is not considered), the bearing stress at the bolt hole is

$$\sigma_{\text{br}} = \frac{V_b}{d_b t_{\text{pl}}} = \frac{2790/2}{0.75(0.5)} = 3720 \text{ psi}$$

Note that the resultant force acts at an angle to the horizontal of

$$\theta = \arctan \frac{1395}{2232} = 32^\circ$$

Therefore, the resultant force does not act either parallel or perpendicular to the major fiber (roving) direction in the web.

The bearing stresses in the angle legs will be half of those in the beam web since the two angles share the load. The maximum bearing stress, including the effects of the shear force eccentricity, is

$$\sigma_{\text{br}} = \frac{R_b/2}{d_b t_{\text{pl}}} = \frac{2632/2}{0.75(0.5)} = 3590 \text{ psi}$$

The bearing stress on the top bolts of the angles assuming only the horizontal component due to the eccentricity is

$$\sigma_{\text{br}} = \frac{R_x/2}{d_b t_{\text{pl}}} = \frac{2232/2}{0.75(0.5)} = 2976 \text{ psi}$$

Note that in this case the bearing forces are applied at an orthogonal orientation to the pultruded material axes of orthotropy as compared with the beam web (i.e., the horizontal component produces bearing stress in the weak direction of the pultruded material in the clip angle).

The bearing stresses in the column flanges do not include effects of eccentricity and are due only to the vertical shear transferred to the column flanges by the clip angles. The bearing in this case is aligned with the longitudinal material axis of the flange and is

$$\sigma_{\text{br}} = \frac{V_b}{d_b t_{\text{pl}}} = \frac{2790/4}{0.75(0.375)} = 2480 \text{ psi}$$

Shear stresses in the clip angles Shear in the angles can result in as shear-out at the bolt holes and longitudinal shear failure at the heel of the angle:

$$\tau_{\text{shear-out}} = \frac{V_b}{2t_{\text{pl}}\, e} = \frac{2790/2}{2(0.5)(2.5)} = 558 \text{ psi}$$

$$\tau_{\text{pl}} = \frac{V}{2A_{\text{heel}}} = \frac{2790}{2(0.5)(8)} = 349 \text{ psi}$$

Flexural and shear stresses in the angles due to prying action It is assumed that the prying force is equal to the horizontal component of the bolt force, $R_x = 2232$ lb. The flexural stress and shear stress due to prying are calculated as

$$m_{\text{leg}} = \frac{R(b_{\text{leg}} - t_{\text{pl}} - s)}{8} = \frac{2232(4.0 - 0.5 - 1.6)}{8} = 530 \text{ in.-lb}$$

$$v_{\text{leg}} = \frac{R}{2} = \frac{2232}{2} = 1116 \text{ lb}$$

$$\sigma_{\text{flex}} = \frac{m_{\text{leg}}}{S} = \frac{m_{\text{leg}}}{2et_{\text{pl}}^2/6} = \frac{530}{2(2.5)(0.5)^2/6} = 2544 \text{ psi}$$

$$\tau_{\text{leg}} = \frac{v_{\text{leg}}}{A} = \frac{v_{\text{leg}}}{t_{\text{pl}} \cdot 2e} = \frac{1116}{0.5(2)(2.5)} = 446 \text{ psi}$$

It can be seen that the end distance, e, plays a part in decreasing the local bending and shear stresses due to prying action. Decreasing the end distance will increase these stresses. As noted in the text and shown in Fig. 15.14, the failure of clip angle connection usually initiates due to delamination at the top of the clip angle due to these local stresses.

Transverse tensile stress at the web–flange junction in the column The transverse tensile stress at the web–flange junction is calculated as

$$\sigma_{\text{trans}} = \frac{R}{t_{\text{web}}(l_{\text{angle}}/2)} = \frac{2232}{0.375(8/2)} = 1488 \text{ psi}$$

Note that in the case of a top and bottom seated connection, the tensile prying force on the column flange will be exerted on the column at the location of the top bolt row in the top seat. If the connection develops moment resistance, this prying force can be large and the web–flange junction is susceptible to transverse tensile failure, as noted in the text. The calculation presented above is highly approximate, and full-scale testing should be conducted if the connection is intended to carry moment. This is required to determine the failure modes of the connection in addition to its rotational stiffness for semirigid frame analysis.

Shear and tensile stresses on the bolts The bolts are subjected to single shear at the column flanges and double shear at the column web,

$$\tau_b = \frac{V_y}{A_b} = \frac{2790/4}{\pi(0.75)^2/4} = \frac{698}{0.44}$$

$$= 1586 \text{ psi} \qquad \text{(column flange: single shear, no eccentricity)}$$

$$\tau_b = \frac{R}{A_b} = \frac{2632/2}{\pi(0.75)^2/4} = \frac{1316}{0.44}$$

$$= 2991 \text{ psi} \qquad \text{(beam web: double shear, with eccentricity)}$$

Tensile stress on bolt due to prying action Assume that the two top bolts at the column flange carry all the prying force.

$$\sigma_b = \frac{R_x}{2A_b} = \frac{2230}{2(0.44)} = 2534 \text{ psi}$$

Steps 4–6. Determine the critical stresses and factored critical stresses, and check the ultimate strength of the trial connection. Critical limit states, critical stresses, safety factors, allowable stresses, and stresses calculated for the actual connection geometry and loads are shown in Table 15.6. Comments to explain the results are provided in the last column of the table.

Based on the data provided in Table 15.6, the connection design is accepted but with a recommendation to strengthen the web of the section in the connection region by bonding (in the shop) two-sided $\frac{1}{8}$ or $\frac{1}{4}$-in.-thick plates of pultruded material with their longitudinal directions perpendicular to the web. Plates measuring the size of the leg of the web clip angle should be used. This is a highly conservative approach but is recommended since the transverse bearing strength of the pultruded material is not reported and the bearing load is at an inclined angle to the hole.

Note that the most critical elements of the design are the bearing stresses in the web and the local bending and shear stresses in the angles. Also, note that the safety factor of 4.0 that is used for all pultruded connection designs at this time reflects the uncertainty in the calculations and the material properties when multiaxial complex stress states exist in pultruded parts.

TABLE 15.6 Limit States, Critical Stresses, and Actual Stress in Connection

Limit State	σ_{cr} (psi)	SF	σ_{allow} (psi)	σ_{actual} (psi)	Comments
Longitudinal bearing on web	30,000 or 16,000	4	7,500 or 4,000	7,081	NG; the resultant force is at a 32° angle and the full longitudinal bearing strength will not be effective.
Transverse bearing on web[a]	16,000	4	4,000	3,590	OK due to vertical component only.
Longitudinal bearing on angle	30,000	4	7,500	3,720	OK but resultant force is at a 32° angle.
Transverse bearing on angle	16,000	4	4,000	2,976	OK due to horizontal component only.
Longitudinal bearing on column[b]	38,000	4	9,500	2,480	OK.
Shear stress on angle at end (shear out)	4,500[c]	4	1,125	558	OK.
Shear stress on angle at heel	4,500[c]	4	1,125	349	OK.
Transverse flexural stress on angle due to prying	10,000	4	2,500	2,544	NG; however, flexural stress on the angle is found by a simple method.
Interlaminar shear stress on angle due to prying	4,500	4	1,125	446	OK.
Shear stress on $\frac{3}{4}$-in. FRP bolts	13,636	4	3,409	2,991	OK.
Tensile stress on $\frac{3}{4}$-in. FRP bolts (thread shear)	10,277	4	2,659	2,534	NG; however, calculation of the prying force is approximate.

[a] Transverse bearing stress not provided; transverse compressive strength substituted.
[b] Superstructural column longitudinal bearing strength per manufacturer's specifications.
[c] In-plane shear strength not provided; interlaminar shear strength substituted.

Pultrusion manufacturers provide load tables for clip angle connections (like the one discussed in this design example) which are based on only three possible failure modes of the connection: bearing at the fasteners (not accounting for eccentricity or orthotropic bearing properties of the pultruded material parts), longitudinal shear failure at the heel of the angle, and bolt shear failure. Testing by researchers has demonstrated that many of these connections can safely carry the design loads published (Lopez-Anido et al., 1999; Mottram and Zheng, 1999a,b), although the failure modes do not agree with the modes described above. In actual applications of beam-to-column clip angles, manufacturers always recommend using adhesive bonding together with bolting. This is primarily for serviceability of the structure as $\frac{1}{16}$-in. hole clearances are also recommended. Full-scale testing of pultruded connections is recommended for all nonstandard connections to understand failure modes and connection behavior in the service load range. Full-scale tests are recommended for efficient connection design of pultruded structures.

PROBLEMS

15.1 For $\frac{1}{2}$- and 1-in.-diameter FRP and steel bolts (or studs) listed in Table P15.1, obtain the following properties from the manufacturer or standards organizations specifications: **(a)** ultimate tensile capacity (and yield strength for steel); **(b)** ultimate shear capacity; **(c)** maximum bolt torque; **(d)** dimensions of the standard nut; **(e)** cost (assume a 2.5-in.-long bolt.)

15.2 Pultruded plate (also called pultruded flat sheet or pultruded sheet) is produced in thicknesses from $\frac{1}{8}$ to 1 in. Pultruded plates are often used as gusset plates, bearing plates, and stiffeners in pultruded connections. Obtain the mechanical properties listed in Tables 13.7 and 14.1 for $\frac{1}{4}$- and $\frac{3}{4}$-in.-thick glass–polyester and glass–vinylester pultruded plates produced by **(a)** Strongwell, **(b)** Creative Pultrusions, and **(c)** Bedford Reinforced Plastics. List the percentage difference

TABLE P15.1 Fasteners for Bolted Pultruded Connections

FRP	a. Strongwell Fibrebolt Stud
	b. Creative Pultrusions Superstud
Hot-dipped galvanized steel	a. SAE grade 5
	b. SAE grade 8
	c. A307 grade B
	d. A325 type 1
Stainless steel	a. Type 316 SS
	b. Type 18-8
	c. Type 304 SS

between the property of the pultruded material in a typical profile (Table 13.7 or Table 14.1) and the property of the pultruded material in the plate.

15.3 For the recommended spacing requirements for lap joints shown in Table 15.2, determine the maximum number of $\frac{1}{2}$-in.-diameter bolts that can be used in one row in the elements of the pultruded profiles listed below. Assume that the bolt row is perpendicular to the load direction, which is along the axis of the profile and in the longitudinal direction of the part. Sketch, to scale, the element of the part and show the geometry of the bolt holes.

(a) The flange of a $8 \times 8 \times \frac{3}{8}$-in. I-shaped section

(b) The flange of a $12 \times 12 \times \frac{1}{2}$-in. I-shaped section

(c) The web of a $8 \times 8 \times \frac{3}{8}$-in. I-shaped section

(d) The web of a $12 \times 12 \times \frac{1}{2}$-in. I-shaped section

(e) One leg of a $6 \times \frac{3}{8}$ pultruded angle

(f) A wall of a $6 \times \frac{1}{4}$-in. square tube

(g) A 18-in.-wide pultruded plate

(h) The web of a $24 \times 3 \times \frac{1}{4}$-in. channel section

15.4 For the lap-joint connections listed in Problem 15.3, determine the nominal, P_n, and the design, P_u, in-plane tensile load that can be carried out by the element under consideration [assume that only the element (e.g., flange, web) is effective in carrying the tensile load]. Consider the failure modes shown in Fig. 15.11. [Do not consider splitting failure; splitting failure can occur in highly orthotropic (typically, unidirectional) composite materials where the transverse tensile strength is much lower than longitudinal strength, shear strength, and bearing strength and is unlikely to occur in conventional pultruded composites.]

15.5 Redesign the web clip angle simple shear connection in Design Example 15.1 assuming that the beam is a $12 \times 6 \times \frac{1}{2}$-in. narrow-flange I-shaped section. Consider the typical conventional pultruded material properties given in Table 15.4. Design the connection using $\frac{1}{2}$-in.-diameter type 306 stainless steel bolts. Size the clip angle such that three bolts can be used to attach the angles to the beam web and column flange. Use the same loading as in Design Example 15.1.

15.6 Redesign the web clip angle simple shear connection in Design Example 15.1 assuming that the beam is a W $10 \times 5 \times \frac{1}{2}$-in. beam and that it is connected to the web of the column instead of the flange (see Mottram and Zheng, 1999a for examples of these "minor-axis web-cleated" simple shear connections). Consider the typical conventional pultruded material properties given in Table 15.4. Design the connection using galvanized steel grade 5 bolts. Use the same loading as in Design Example 15.1.

15.7 U.S. manufacturers (Strongwell and Creative Pultrusions) publish load tables for beam clip angle simple beam-to-column connections. List (from manufacturer's design guides) the clip angle connection details (bolt size, angle length, angle thickness) that are recommended by manufacturers for the connection loads in Design Example 15.1.

15.8 Two $4 \times 4 \times \frac{1}{4}$ in. $8\frac{1}{4}$ in. long pultruded angles (glass–vinylester material properties from **(a)** Strongwell and **(b)** Creative Pultrusions) are used in a clip angle connection to connect a beam to a column. Determine the maximum shear force that can be transferred by this clip angle connection assuming three $\frac{1}{2}$-in. steel bolts are used in each leg of the connection. Assume that the connection is designed for only (1) shear through the heel of the angles and (2) bearing at the bolt holes. Do not consider eccentricity. Assume a bolted-only connection and that the geometric spacing parameters (Table 15.2) are satisfied for this connection.

15.9 A stick-built frame is constructed using $3 \times 3 \times \frac{1}{4}$ in. pultruded tubes as the vertical members and $4 \times 1\frac{1}{8} \times \frac{1}{4}$ in. channel sections as the horizontal members (see a schematic of this type of construction in Fig. 15.4). Both vertical members (columns) and horizontal members (girts) are continuous. Two channels are attached on the opposite sides of each tube, creating a lattice-type frame that is stabilized with diagonal braces. A single $\frac{5}{8}$-in. stainless steel through-bolt is used at each overlap location. Determine the vertical force that can be transferred from the horizontal member to the vertical member at each overlapping joint. If the columns are 6 ft apart, what transverse load can one of the channels carry, assuming a concentrated load at its midspan (based on the connection capacity only)?

15.10 Consider the braced frame connection detail shown in Fig. 15.7. Assume that the beam and the column are both $8 \times 8 \times \frac{1}{2}$ in. pultruded sections. Dimension the remaining parts of this connection using standard pultruded parts (angles, channels, I-shaped sections) and locate all holes. Use Table 15.2 to dimension the hole locations in the pultruded parts. Assume that the braces can carry both tension and compression. Provide a sketch, to scale, showing all the connection parts. Draw an exploded sketch of the connection and show the forces to be transferred between the various parts in the connection. List and discuss all the possible failure modes of this connection, especially considering the fiber orientations in the various pultruded parts in the connection. No calculations are required for this problem; only a qualitative analysis is expected.

15.11 Design a double-clip-angle coped-beam shear connection for a $6 \times 6 \times \frac{3}{8}$ in. stringer that is connected to a $10 \times 10\frac{1}{2}$ in. girder such that the top surfaces of the top flanges are at the same elevation. Design the connection for a 1200-lb shear force at the stringer end. Make sure to consider the block-shear failure mode in the web of the stringer. Use profiles with homogeneous glass–vinylester pultruded properties (Table 13.7).

15.12 Consider the pultruded heavy-frame structure used in the design examples in Chapters 13 and 14 with the floor plan shown in Fig. 13.13. In the direction of the 6-ft-wide bays, the lateral load in the three-story pultruded structure is carried by braced frames (9-ft-high story height) 6 ft wide by 27 ft high as shown in Fig. P15.12. At each floor level $6 \times 6 \times \frac{3}{8}$ in. stringers are used to frame between the columns. (Note that the beams frame into the weak axis direction of the columns, as shown in Fig. 13.13.) Assume that columns are pinned at their bases and that the vertical braced frame is subjected to a lateral load at each floor of 2000 lb. The lateral drift is limited to $L/500$. Design the braces assuming a lateral load from either direction. Use double angles for the braces (see Fig. 15.7 for a strong axis beam–column connection for this type of connection, and consult manufacturer design guides for other possible configurations). Design the bolted connection between the bracing member and the column–beam connection area (connection A). Use recommended geometric connection limits to detail the brace connections. Account for the gravity loads on the columns, but do not redesign the columns for the additional axial load that develops due to the lateral loads (however, this

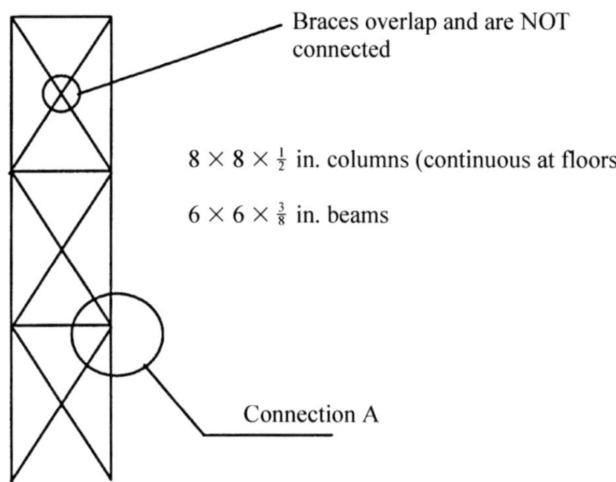

Figure P15.12

would need to be done in an actual design). Use the design basis of your choice.

SUGGESTED PULTRUDED STRUCTURE DESIGN PROJECTS

The following pultruded design projects are suggested for the students in a composites for construction design class. The design project should preferably be done in groups of two or three students. Students should be given 4 to 6 weeks to compete the design project (which therefore needs to be assigned early in the semester). The final deliverable should be a design proposal[7] that includes the problem statement, scope, codes and specifications, loads, materials, design calculations, design drawings, and a cost analysis. A presentation of the design proposal should be given in class. Invite owners and designers to form part of the project jury to obtain feedback from industry on the designs presented.

15.1 *Pultruded light-truss pedestrian bridge.* Design a truss bridge to replace an existing pedestrian bridge using FRP pultruded materials. Students should scout the local area (campus, downtown, etc.) to find a pedestrian bridge with a 30- to 70-ft span and use this bridge as the basis for the design. Pay attention to details of connections and attachments to the foundation or abutment. Use the AASHTO *Guide Specification for Design of Pedestrian Bridges* (1997) for loads and dynamic requirements.

15.2 *Pultruded heavy-girder pedestrain bridge.* Design a girder and stringer bridge to replace an existing pedestrian bridge using FRP pultruded materials. Students should scout the local area (campus, downtown, etc.) to find a pedestrian bridge with a 20- to 40-ft span and use this bridge as the basis for the design. Pay attention to details of connections and attachments to the foundation or abutment. Consider using built-up beam members if conventional shapes do not work. Use the AASHTO *Guide Specification for Design of Pedestrian Bridges* (1997) for loads and dynamic requirements.

15.3 *Pultruded highway sign bridge.* Design a cantilever or a frame highway sign bridge to replace an existing sign bridge using FRP pultruded materials. Students should scout the local area (campus, downtown, etc.) and find either a cantilever sign bridge (typical arm of 20 ft) or frame sign bridge (typical span of 60 to 100 ft) and use this actual sign bridge as the basis for the design. Pay attention to details of connections and attachments to the foundation. Consider using built-

[7] Items listed are at the discretion of the instructor.

up or latticed members if conventional shapes do not work. Consider a truss bridge. Use the AASHTO *Standard Specifications for Structural Supports for Highway Signs, Luminaires and Traffic Signals* (2001 and interims) for loads and other design requirements.

15.4 *Pultruded light-frame structure.* Do a schematic design of a pultruded light frame structure to house a cooling tower system consisting of two 18-ft-diameter fan units (see Fig. 1.16 for an example of this type of structure and visit the Cooling Tower Institute website (www.cti.org) for more information and links to company literature with more details on these types of structures). The external dimensions of the structure are: 30 ft wide (5 bays at 6 ft), 48 ft long (8 bays at 6 ft), and the fan deck is 32 ft above the concrete mat foundation. The operating weight of each fan unit is 80,000 lb. Assume that the tower is to be constructed in an industrial park in rural Kansas. Design for a wind pressure of 30 psf.

References

AASHTO (1997), *Guide Specification for Design of Pedestrian Bridges,* American Association of State Highway and Transportation Officials, Washington, DC.

——— (2001), *Standard Specifications for Structural Supports for Highway Signs, Luminaires and Traffic Signals,* 4th ed., American Association of State Highway and Transportation Officials, Washington, DC.

——— (2002), *Standard Specification for Highway Bridges,* 17th ed., American Association of State Highway and Transportation Officials, Washington, DC.

——— (2005), *LRFD Bridge Design Specifications,* 3rd ed., American Association of State Highway and Transportation Officials, Washington, DC.

Abd-el-Naby, S. F. M., and Hollaway, L. (1993a), "The experimental behavior of bolted joints in pultruded glass/polyester material, 1: single bolt joints, *Composites,* Vol. 24, pp. 531–538.

——— (1993b), The experimental behavior of bolted joints in pultruded glass/polyester material, 2: two-bolt joints, *Composites,* Vol. 24, pp. 539–546.

AC 125 (1997), *Acceptance Criteria for Concrete and Reinforced and Unreinforced Masonry Strengthening Using Fiber-Reinforced Polymer (FFP) Composite Systems,* ICC Evaluation Service, Whittier, CA.

AC 187 (2001), *Acceptance Criteria for Inspection and Verification of Concrete and Reinforced and Unreinforced Masonry Strengthening Using Fiber-Reinforced Polymer (FFP) Composite Systems,* ICC Evaluation Service, Whittier, CA.

ACI (1995), *Building Code Requirements for Structural Concrete and Commentary,* ACI 318-95, American Concrete Institute, Farmington Hills, MI.

——— (1996), *State-of-the-Art Report on Fiber Reinforced Plastic (FRP) Reinforcement for Concrete Structures,* ACI 440R-96, American Concrete Institute, Farmington Hills, MI.

——— (1999), *Building Code Requirements for Structural Concrete and Commentary,* ACI 318-99, American Concrete Institute, Farmington Hills, MI.

——— (2001), *Guide for the Design and Construction of Concrete Reinforced with FRP Bars,* ACI 440.1R-01, American Concrete Institute, Farmington Hills, MI.

——— (2002), *Guide for the Design and Construction of Externally Bonded FRP Systems for Strengthening Concrete Structures,* ACI 440.2R-02, American Concrete Institute, Farmington Hills, MI.

——— (2003), *Guide for the Design and Construction of Concrete Reinforced with FRP Bars,* ACI 440.1R-03, American Concrete Institute, Farmington Hills, MI.

——— (2004a), *Guide Test Methods for Fiber Reinforced Polymers (FRP) for Reinforcing or Strengthening Concrete Structures,* ACI 440.3R-04, American Concrete Institute, Farmington Hills, MI.

——— (2004b), *Prestressing Concrete Structures with FRP Tendons,* ACI 440.4R-04, American Concrete Institute, Farmington Hills, MI.

——— (2005), *Building Code Requirements for Structural Concrete and Commentary,* ACI 318-05, American Concrete Institute, Farmington Hills, MI.

——— (2006), *Guide for the Design and Construction of Structural Concrete Reinforced with FRP Bars,* ACI 440.1R-06, American Concrete Institute, Farmington Hills, MI.

Adams, D. F., Carlsson, L. A., and Pipes, R. B. (2003), *Experimental Characterization of Advanced Composite Materials,* 3rd ed., CRC Press, Boca Raton, FL.

Agarwal, B. D., and Broutman, L. J., (1990), *Analysis and Performance of Fiber Composites,* 2nd ed., Wiley, New York.

AISC (2005), *Specification for Structural Steel Buildings and Commentary,* ANSI/AISC 360-05, American Institute of Steel Construction, Chicago.

Alias, M. N., and Brown, R. (1992), Damage to composites from electrochemical processes, *Corrosion,* Vol. 48, pp. 373–378.

Alkhrdaji, T., Nanni, A., and Mayo, R. (2000), *Upgrading Missouri transportation infrastructure: Solid reinforced-concrete decks strengthened with fiber-reinforced polymer systems,* Transportation Research Record 1740, National Research Council, pp. 157–163.

Alqam, M., Bennett, R. M., and Zureick, A. H. (2002), Three-parameter vs. two-parameter Weibull distribution for pultruded composite material properties, *Composite Structures,* Vol. 58, pp. 497–503.

An, W., Saadatmanesh, H., and Ehsani, M. R. (1991), RC beams strengthened with FRP plates, II: analysis and parametric study, *Journal of Structural Engineering,* Vol. 117, No. 11, pp. 3434–3455.

ANSI (2000), *American National Standard for Ladders—Portable Reinforced Plastic—Safety Requirements,* ANSI A14.5-2000, American National Standards Institute, Chicago.

Antonopoulos, C. P., and Triantafillou, T. C. (2003), Experimental investigation of FRP-strengthened RC beam–column joints, *Journal of Composites for Construction,* Vol. 7, No. 1, pp. 39–49.

Arduini, M., and Nanni, A. (1997), Behavior of precracked RC beams strengthened with carbon FRP sheets, *Journal of Composites for Construction,* Vol. 1, No. 2, pp. 63–70.

ASCE (1984), *Structural Plastics Design Manual,* ASCE Manuals and Reports on Engineering Practice 63, American Society of Civil Engineers, Reston, VA.

——— (2002), *Minimum Design Loads for Buildings and Other Structures,* ASCE 7-02 American Society of Civil Engineers, Reston, VA.

——— (2003), *Recommended Practice for Fiber-Reinforced Polymer Products for Overhead Utility Line Structures,* ASCE Manuals and Reports on Engineering Practice 104, American Society of Civil Engineers, Reston, VA.

ASTM (2006), *Book of Standards,* American Society for Testing and Materials, West Conshohocken, PA.

Bakht, B., et al. (2000), Canadian bridge design code provisions for fiber-reinforced concrete, *Journal of Composites for Construction,* Vol. 4, No. 1, pp. 3–15.

Bakis, C. E., Bank, L. C., Brown, V. L., Cosenza, E., Davalos, J. F., Lesko, J. J., Machida, A., Rizkalla, S. H., and Triantafilliou, T. C. (2002), Fiber-reinforced polymer composites for construction: state-of-the-art review, *Journal of Composites for Construction,* Vol. 6, No. 2, pp. 73–87.

Balsamo, A., Colombo, A., Manfredi, G., Negro, P., and Prota, A. (2005), Seismic behavior of a full-scale RC frame repaired using CFRP laminates, *Engineering Structures,* Vol. 27, pp. 769–780.

Bank, L. C. (1987), Shear coefficients for thin-walled composite beams, *Composite Structures,* Vol. 8, pp. 47–61.

——— (1989a), Flexural and shear moduli of full-section fiber reinforced plastic (FRP) pultruded beams, *Journal of Testing and Evaluation,* Vol. 17, No. 1, pp. 40–45.

——— (1989b), *Properties of Pultruded Fiber Reinforced Plastic (FRP) Structural Members,* Transportation Research Record 1223, National Research Council, Washington, DC, pp. 117–124.

——— (1990), Modifications to beam theory for bending and twisting of open-section composite beams, *Composite Structures,* Vol. 15, pp. 93–114.

——— (1993a), FRP reinforcements for concrete, in *Fiber-Reinforced Plastic (FRP) for Concrete Structures: Properties and Applications* (ed. A. Nanni), Elsevier Science, New York, pp. 59–86.

——— (1993b), Questioning composites, *Civil Engineering,* Vol. 63, No. 1, pp. 64–65.

Bank, L. C., and Melehan, T. P. (1989), Shear coefficients for multicelled thin-walled composite beams, *Composite Structures,* Vol. 11, pp. 259–276.

Bank, L. C., and Mosallam, A. S. (1992), Creep and failure of a full-size fiber reinforced plastic pultruded frame, *Composites Engineering,* Vol. 2, pp. 213–227.

Bank, L. C., and Yin, J. (1996), Buckling of orthotropic plates with free and rotationally restrained unloaded edges, *Thin-Walled Structures,* Vol. 24, pp. 83–96.

——— (1999), Analysis of progressive failure of the web–flange junction in post-buckled pultruded I-beams, *Journal of Composites for Construction,* Vol. 3, pp. 177–184.

Bank, L. C., and Xi, Z. (1993), Pultruded FRP grating reinforced concrete slabs, in *Fiber-Reinforced Plastic for Concrete Structures: International Symposium* (ed. A. Nanni and C. W. Dolan), SP-138, American Concrete Institute, Farmington Hills, MI, pp. 561–583.

Bank, L. C., Barkatt, A., and Gentry, T. R. (1995b), Accelerated test methods to determine the long-term behavior of FRP composite structures: environmental effects, *Journal of Reinforced Plastics and Composites,* Vol. 14, No. 6, pp. 559–587.

Bank, L. C., Gentry, T. R., and Nadipelli, M. (1996a), Local buckling of pultruded FRP beams: analysis and design, *Journal of Reinforced Plastics and Composites,* Vol. 15, pp. 283–294.

Bank, L. C., Mosallam, A. S., and Gonsior, H. E. (1990), Beam-to-column connections for pultruded FRP structures, in *Proceedings of the First Materials Engineering Congress* (ed. B. Suprenant), Denver, CO, August 13–15, ASCE, Reston, VA, pp. 804–813.

Bank, L. C., Mosallam, A. S., and McCoy, G. T. (1994a), Design and performance of connections for pultruded frame structures, *Journal of Reinforced Plastics and Composites,* Vol. 13, pp. 199–212.

Bank, L. C., Nadipelli, M., and Gentry, T. R. (1994b), Local buckling and failure of pultruded fiber-reinforced plastic beams, *Journal of Engineering Materials and Technology,* Vol. 116, pp. 233–237.

Bank, L. C., Puterman, M., and Katz, A. (1998), The effect of material degradation on bond properties of FRP reinforcing bars in concrete, *ACI Materials Journal,* Vol. 95, No. 3, pp. 232–243.

Bank, L. C., Yin, J., and Nadipelli, M. (1995a), Local buckling of pultruded beams: nonlinearity, anisotropy and inhomogeneity, *Construction and Building Materials,* Vol. 9, No. 6, pp. 325–331.

Bank, L. C., Yin, J., Moore, L. E., Evans, D., and Allison, R. (1996b), Experimental and numerical evaluation of beam-to-column connections for pultruded structures, *Journal of Reinforced Plastics and Composites,* Vol. 15, pp. 1052–1067.

Bank, L. C., Gentry, T. R., Thompson B. P., and Russell, J. S. (2003), A model specification for FRP composites for civil engineering structures, *Construction and Building Materials,* Vol. 17, No. 6–7, pp. 405–437.

Bank, L. C., Oliva, M. G., Russell, J. S., Jacobson, D. A., Conachen, M., Nelson, M., and McMonigal, D. (2006), Double layer prefabricated FRP grids for rapid bridge deck construction: case study, *Journal of Composites for Construction,* Vol. 10, No. 3.

Barbero, E. J. (1999), *Introduction to Composite Materials Design,* Taylor & Francis, Philadelphia.

Barbero, E., and Tomblin, J. (1993), Euler buckling of thin-walled composite columns, *Thin-Walled Structures,* Vol. 17, pp. 237–258.

——— (1994) A pheomenological design equation for FRP columns with interaction between local and global buckling, *Thin-Walled Structures,* Vol. 18, pp. 117–131.

Barbero, E. J., Fu, S.-H., and Raftoyiannis, I. (1991), Ultimate bending strength of composite beams, *Journal of Materials in Civil Engineering,* Vol. 3, pp. 292–306.

Barbero, E. J., Dede, E. K., and Jones, S. (2000), Experimental verification of buckling-mode interaction in intermediate-length composite columns, *International Journal of Solids and Structures,* Vol. 37, pp. 3919–3934.

Barbero, E. J. and DeVivo, L. (1999), Beam-column design equations for wide-flange pultruded structural shapes, *Journal of Composites for Construction,* Vol. 3, No. 4, pp. 185–191.

Barbero, E. J., and Turk, M. (2000) Experimental investigation of beam-column behavior of pultruded structural shapes, *Journal of Reinforced Plastics and Composites,* Vol. 19, pp. 249–265.

Bass, A. J., and Mottram, J. T. (1994), Behaviour of connections in frames of fibre-reinforced-polymer section, *Structural Engineer,* Vol. 72, pp. 280–285.

Bedford (2005), *Bedford Reinforced Plastics Design Guide,* Bedford Plastics, Bedford, PA; www.bedfordplastics.com.

Benmokrane, B., Chaallal, O., and Masmoudi, R. (1996a), Flexural response of concrete beams reinforced with FRP reinforcing bars, *ACI Structural Journal,* Vol. 93, No. 1, pp. 46–55.

Benmokrane, B., Tighiouart, B., and Chaallal, O. (1996b), Bond strength and load distribution of composite GFRP reinforcing bars in concrete, *ACI Materials Journal,* Vol. 93, No. 3, pp. 246–253.

Benmokrane, B., El-Salakawy, E., Desgagne, G., and Lackey, T. (2004), FRP bars for bridges, *Concrete International,* Vol. 26, August, pp. 84–90.

Berg, A. C., Bank, L. C., Oliva, M. G., and Russell, J. S. (2006), Construction and cost analysis of an FRP reinforced concrete bridge deck, *Construction and Building Materials,* Vol. 20, No. 8, pp. 515–526.

Bisby, L. A., Kodur, V. K. R., and Green, M. F. (2005), *Fire endurance of fiber-reinforced polymer-confined concrete columns,* ACI Structural Journal, Vol. 102, No. 6, pp. 883–891.

Bizindavyi, L., and Neale, K. W. (1999), Transfer lengths and bond strengths for composites bonded to concrete, *Journal of Composites for Construction,* Vol. 3, No. 4, pp. 153–160.

Blankenship, L. T., White, M. N., and Puckett, P. M. (1989), Vinyl ester resins: versatile resins for composites, *Proceedings of the 34th International SAMPE Symposium,* Vol. 34, pp. 234–245.

Bleich, F. (1952), *Buckling Strength of Metal Structures,* McGraw-Hill, New York, NY.

Bradberry, T. E., and Wallace, S. (2003), FRP reinforced concrete in Texas transportation past, present, future, in *SP215: Field Applications of FRP Reinforcement: Case Studies* (ed. S. Rizkalla and A. Nanni), American Concrete Institute, Farmington Hills, MI, pp. 37–54.

Braestrup, M. W. (1999), Footbridge constructed from glass-fiber-reinforced profiles, Denmark, *Structural Engineering International,* Vol. 9, No. 4, pp. 256–258.

BRI (1995), *Guidelines for Structural Design of FRP Reinforced Concrete Building Structures,* Building Research Institute, Tsukuba, Japan. See also Design guidelines of FRP reinforced concrete building structures, *Journal of Composites for Construction,* Vol. 1, No. 3, pp. 90–115, 1997.

Burgoyne, C., and Head, P. R. (1993), Aberfeldy bridge: an advanced textile reinforced footbridge, presented at the TechTextil Symposium, Frankfurt, Germany, June.

Buyukozturk, O., and Hearing, B. (1998), Failure behavior of precracked concrete beams retrofitted with FRP, *Journal of Composites for Construction,* Vol. 2, No. 3, pp. 138–144.

Buyukozturk, O., Gunes, O., and Karaca, E. (2004), Progress on understanding debonding problems in reinforced concrete and steel members strengthened using FRP composites, *Construction and Building Materials,* Vol. 18, No. 1, pp. 9–19.

Cadei, J., and Stratford, T. (2002), The design, construction and in-service performance of the all-composite Aberfeldy footbridge, in *Advanced Polymer Composites for Structural Applications in Construction* (ed. R. A. Shenoi, S. S. J. Moy, and L. C. Hollaway), pp. 445–455. Thomas Telford, London.

CEN (2002a), *Reinforced Plastics Composites: Specifications for Pultruded Profiles,* Part 1: Designation; Part 2: Method of Test and General Requirements; Part 3: Specific Requirements, EN 13706, Comité Européen de Normalisation, Brussels, Belgium.

——— (2002b), *Eurocode 1: Actions on Structures,* Comité Européen de Normalisation, Brussels, Belgium.

Chaallal, O., and Benmokrane, B. (1993), Pullout and bond of glass-fiber rods embedded in concrete and cement grout, *Materials and Structures,* Vol. 26, pp. 167–175.

Chai, Y. H., Priestley, M. J. N., and Seible, F. (1991), Seismic retrofit of circular bridge columns for enhanced flexural performance, *ACI Structural Journal,* Vol. 88, No. 5, pp. 572–584.

Chajes, M. J., Januszka, T. F., Mertz, D. R., Thomson, T. A., and Finch, W. W. (1995), Shear strengthening of reinforced concrete beams using externally applied concrete fabrics, *ACI Structural Journal,* Vol. 92, No. 3, pp. 295–303.

Chambers, R. E. (1997), ASCE design standard for pultruded fiber-reinforced plastic (FRP) Structures, *Journal of Composites for Construction,* Vol. 1, pp. 26–38.

Chen, J. F., and Teng, J. G. (2003a), Shear capacity of fiber-reinforced polymer-strengthened reinforced concrete beams: fiber reinforced polymer rupture, *Journal of Structural Engineering,* Vol. 129, No. 5, pp. 615–625.

——— (2003b), Shear capacity of FRP strengthened RC beams: FRP debonding, *Construction and Building Materials,* Vol. 17, pp. 27–41.

Cofie, E., and Bank, L. C. (1995), Analysis of thin-walled anisotropic frames by the direct stiffness method, *International Journal of Solids and Structures,* Vol. 32, No. 2, pp. 235–249.

Connolly, M., King, J., Shidaker, T., and Duncan, A. (2005), Pultruding polyurethane composite profiles: practical guidelines for injection box design, component metering equipment and processing, presented at the Composites 2005 Convention and Trade Show, September 28–30, American Composites Manufacturers Association, Columbus, OH, (CD-ROM).

Cooper, C., and Turvey, G. J. (1995), Effects of joint geometry and bolt torque on the structural performance of single bolt tension joints in pultruded GRP sheet material, *Composite Structures,* Vol. 32, pp. 217–226.

Cosenza, E., Manfredi, G., and Realfonzo, R. (1997), Behavior and modeling of bond of FRP rebars to concrete, *Journal of Composites for Construction,* Vol. 1, No. 2, pp. 40–51.

Cowper, G. R. (1966), The shear coefficient in Timoshenko's beam theory, *Journal of Applied Mechanics,* Vol. 33, pp. 121–127.

Creative Pultrusions (2004), *The Pultex Pultrusion Global Design Manual,* 4th ed. Creative Pultrusions, Alum Bank, PA; www.creativepultrusions.com.

CSA (2002), *Design and Construction of Building Components with Fibre-Reinforced Polymers,* CSA-S806-02, Canadian Standards Association, Toronto, Ontario, Canada.

CTI (2003), *Fiberglass Pultruded Structural Products for Use in Cooling Towers,* STD-137, Cooling Technology Institute, Houston, TX.

Czaderski, C., and Motavalli, M. (2004), Fatigue behavior of CFRP L-shaped plates for shear strengthening of RC T-beams, *Composites: Part B,* Vol. 35, pp. 279–290.

Daniali, S. (1990), Bond strength of fiber reinforced plastic bars in concrete, *Proceedings of the First Materials Engineering Congress* (ed. B. Suprenant), Denver, CO, August 13–15, ASCE, Reston, VA, pp. 501–510.

Daniel, I. M., and Ishai, O. (1994), *Engineering Mechanics of Composite Materials,* Oxford University Press, New York.

Davalos, J. F., and Qiao, P. (1997), Analytical and experimental study of lateral and distortional buckling of FRP wide-flange beams, *Journal of Composites for Construction,* Vol. 1, pp. 150–159.

Dietz, D. H., Harik, I. E., and Gesund, H. (2003), Physical properties of glass fiber reinforced polymer rebars in compression, *Journal of Composites for Construction,* Vol. 7, No. 4, pp. 363–366.

Dietz, D. H., Harik, I. E., Gesund, H., and Zatar, W. A. (2004), Barrier wall impact simulation of reinforced concrete decks with steel and glass fiber reinforced polymer bars, *Journal of Composites for Construction,* Vol. 8, No. 4, pp. 369–373.

Eberline, D. K., Klaiber, F. W., and Dunker, K. (1988), *Bridge Strengthening with Epoxy-Bonded Steel Plates,* Transportation Research Record 1180, pp. 7–11.

Ellingwood, B. R. (2003), Toward load and resistance factor design for fiber-reinforced polymer composite structures, *Journal of Structural Engineering,* Vol. 129, pp. 449–458.

El-Mihilmy, M. T., and Tedesco, J. W. (2000), Deflection of reinforced concrete beams strengthened with fiber-reinforced polymer (FRP) plates, *ACI Structural Journal,* Vol. 97, No. 5, pp. 679–688.

El-Salakawy, E., Benmokrane, B., Masmoudi, R., Briere, F., and Breaumier, E. (2003), Concrete bridge barriers reinforced with fiber-reinforced polymer composite bars, *ACI Structural Journal,* Vol. 100, No. 6, pp. 815–824.

El-Sayed, A., El-Salakawy, E., and Benmokrane, B. (2005), Shear strength of one-way concrete slabs reinforced with fiber-reinforced polymer composite bars, *Journal of Composites for Construction,* Vol. 9, No. 2, pp. 147–157.

Erki, M. A. (1995), Bolted glass-fibre-reinforced plastic joints, *Canadian Journal of Civil Engineering,* Vol. 22, pp. 736–744.

Eurocomp (1996), Structural design of polymer composites, *Eurocomp Design Code and Handbook* (ed. J. Clarke), E&FN Spon, London.

Fardis, M. N., and Khalili, H. (1981), Concrete encased in fiber-glass reinforced plastic, *ACI Journal,* Vol. 78, No. 6, pp. 440–446.

Faza, S. S., and GangaRao, H. V. S. (1990), *Bending and Bond Behavior of Concrete Beams Reinforced with Plastic Rebars,* Transportation Research Record 1290, pp. 185–193.

——— (1993), Theoretical and experimental correlation of behavior of concrete beams reinforced with fiber reinforced plastic rebars, in *Fiber-Reinforced-Plastic Reinforcement for Concrete Structures,* SP-138, American Concrete Institute, Farmington Hills, MI, pp. 599–614.

FIB (2001), *Externally Bonded FRP Reinforcement for RC Structures,* International Federation for Structural Concrete, Lausanne, Switzerland.

Fiberline (2003), *Fiberline Design Manual,* Fiberline, Kolding, Denmark; www.fiberline.com.

Foccaci, F., Nanni, A., and Bakis, C. E. (2000), Local bond slip relationship for FRP reinforcement in concrete, *Journal of Composites for Construction,* Vol. 4, pp. 24–31.

Frosch, R. J. (1999), Another look at cracking and crack control in reinforced concrete, *ACI Structural Journal,* Vol. 96, No. 3, pp. 437–442.

Fujisaki, T., Sekijima, K., Matsuzaki, Y., and Okumura, H. (1987), New materials for reinforced concrete in place of reinforced steel bar, presented at the IABSE Symposium, Paris-Versailles, September, pp. 413–418.

Fukuyama, H. (1999), FRP composites in Japan, *Concrete International,* October, pp. 29–32.

Galambos, T. V. (ed.) (1998), *Guide to the Stability Design Criteria for Metal Structures,* 5th ed., Wiley, New York, NY.

Gentry, T. R., and Husain, M. (1999), Thermal compatibility of concrete and composite reinforcements, *Journal of Composites for Construction,* Vol. 3, No. 2, pp. 82–86.

Gere, J. M., and Timoshenko, S. P. (1997), *Mechanics of Materials,* 4th ed., PWS, Boston.

Gilstrap, J. M., Burke, C. R., Dowden, D. M., and Dolan, C. W. (1997), Development of FRP reinforcement guidelines for prestressed concrete structures, *Journal of Composites for Construction,* Vol. 1, pp. 131–139.

Giroux, C., and Shao, Y. (2003), Flexural and shear rigidity of composite sheet piles, *Journal of Composites for Construction,* Vol. 7, No. 4, pp. 348–355.

Gjelsvik, A. (1981), *The Theory of Thin-Walled Bars,* John Wiley & Sons, New York, NY.

Goldsworthy, B. (1954), The continuous extrusion of RP, *Proceedings of the 9th SPI RPD Conference,* Chicago, February 3–5, Section 13.

Goldsworthy, W. B., and Hiel, C. (1998), Composite structures are a snap, *SAMPE Journal,* Vol. 34, No. 1, pp. 24–30.

Goldsworthy, W. B., and Landgraf, F. (1959), Apparatus for producing elongated articles from fiber-reinforced plastic material, U.S. patent 2,871,911, February 3.

Green, A., Bisarnsin, T., and Love, E. A. (1994), Pultruded reinforced plastics for civil engineering structural applications, *Journal of Reinforced Plastics and Composites,* Vol. 13, pp. 942–951.

Hancock, G. J. (1978), Local, distortional and lateral buckling of I-beams, Journal of the Structural Division, Proceedings of the ASCE, Vol. 104, No. ST11, pp. 1787–1798.

Hart-Smith, L. J. (1978), Mechanically fastened joints for advanced composites: phenomenological considerations and simple analyses, *Proceedings of the 4th Conference on Fiberous Composites in Structural Design,* Plenum Press, New York, pp. 543–574.

Hassan, N. H., Mohamedien, M. A., and Rizkalla, S. H. (1997a), Multibolted joints for GFRP structural members, *Journal of Composites for Construction,* Vol. 1, pp. 3–9.

——— (1997b), Rational model for multibolted connections for GFRP members, *Journal of Composites for Construction,* Vol. 1, pp. 71–78.

Head, P., and Templeman, R. B. (1986), The application of limit states design principles to fibre reinforced plastics, *Proceedings of the British Plastics Federation Congress 86,* Nottingham, England, September 17–19, pp. 69–77.

Hollaway, L. (1993), *Polymer Composites for Civil and Structural Engineering,* Chapman & Hall, New York.

——— (2003), The evolution of and the way forward for advanced polymer composites in the civil infrastructure, *Construction and Building Materials,* Vol. 17, pp. 365–378.

Hollaway, L. C., and Garden, H. N. (1998), An experimental study of the influence of plate end anchorage of carbon fibre composite plates used to strengthen reinforced concrete beams, *Composite Structures,* Vol. 42, pp. 175–188.

Hollaway, L. C., and Head, P. R. (2001), *Advanced Polymer Composites and Polymers in the Civil Infrastructure,* Elsevier, London.

IBC (2003), *International Building Code,* International Code Council, Falls Church, VA.

ICC (2003), *Performance Code for Buildings and Facilities,* International Code Council, Falls Church, VA.

Inokuma, A. (2002), Basic study of performance-based design in civil engineering, *Journal of Professional Issues in Engineering Education and Practice,* Vol. 128, No. 1, pp. 30–35.

ISO (2006), *ISO Catalogue 2006,* Geneva; www.iso.org.

Iyer, L., and Sen, R. (1991), *Advanced Composites for Civil Engineering Structures,* ASCE, Reston, VA.

Johansen, G. E., Wilson, R. J., Roll, F., Levin, B., Chung, D., and Poplawski, E. (1999), Testing and evaluation of FRP truss connections, in *Materials and Construction: Exploring the Connection,* (ed. L. C. Bank) ASCE, Reston, VA, pp. 68–75.

JSCE (1997), *Recommendation for Design and Construction of Concrete Structures Using Continuous Fiber Reinforcing Materials,* Concrete Engineering Series 23, Japan Society of Civil Engineers, Tokyo.

——— (2001), *Recommendation for Upgrading of Concrete Structures with Use of Continuous Fiber Sheets,* Concrete Engineering Series 41, Japan Society of Civil Engineers, Tokyo.

Karbhari, V. M. (1998), *Use of Composite Materials in Civil Infrastructure in Japan,* International Technology Research Institute, World Technology (WTEC) Division, Baltimore, MD.

Katsumata, H., Kobatake, Y., and Takeda, T. (1988), A study on strengthening with carbon fiber for earthquake-resistant capacity of existing reinforced concrete columns, *Proceedings of the 9th World Conference on Earthquake Engineering,* Tokyo–Kyoto, August 2–9, Vol. VII, pp. 517–522.

Katz, A. (2000), Bond to concrete of FRP rebars after cyclic loading, *Journal of Composites for Construction,* Vol. 4, No. 3, pp. 137–144.

Katz, A., Berman, N., and Bank, L. C. (1999), Effect of high temperature on the bond strength of FRP rebars, *Journal of Composites for Construction,* Vol. 3, pp. 73–81.

Keller, T. (1999), Towards structural forms for composite fiber materials, *Structural Engineering International,* Vol. 9, No. 4, pp. 297–300.

Keller, T., Kunzle, O., and Wyss, U. (1999), Finding the right bond, *Civil Engineering,* Vol. 69, No. 3, p. 4.

Khalifa, A., Gold, W. J., Nanni, A., and Aziz, A. (1998), Contribution of externally bonded FRP to shear capacity of RC flexural members, *Journal of Composites for Construction,* Vol. 2, No. 3, pp. 195–202.

Kim, D.-H. (1995), *Composite Structures for Civil and Architectural Engineering,* E&FN Spon, London.

Kollár, L. P. (2002), Buckling of unidirectionally loaded composite plates with one free and one rotationally restrained unloaded edge, *Journal of Structural Engineering,* Vol. 128, No. 9, pp. 1202–1211.

——— (2003), Local buckling of fiber reinforced plastic composite structural members with open and closed cross sections, *Journal of Structural Engineering,* Vol. 129, No. 11, pp. 1503–1513.

Kollár, L. P., and Springer, G. S. (2003), *Mechanics of Composite Structures,* Cambridge University Press, Cambridge.

Lackey, E., and Vaughan, J. (2002), Resin fillers and additives, *Composites Fabrication,* March, pp. 12–17, 36.

Lackey, E. Vaughan, J. G., Washabauch, F., Usifer, D., and Ubrich, P. (1999), Effects of fillers on pultrusion processing and pultruded properties, presented at the International Composites Expo '99, Cincinnati, OH, May 10–12, Session 21-E, pp. 1–10, Composites Institute, New York.

Lamanna, A. J., Bank, L. C., and Scott, D. (2001), Flexural strengthening of RC beams using fasteners and FRP strips, *ACI Structural Journal,* Vol. 98, No. 3, pp. 368–376.

——— (2004), Flexural strengthening of RC beams by mechanically attaching FRP strips, *Journal of Composites for Construction,* Vol. 8, No. 3, pp. 203–210.

Lane, A., and Mottram, J. T. (2002), Influence of modal coupling on the buckling of concentrically loaded pultruded fibre-reinforced plastic columns, *Journal of Materials: Design and Applications, Proceedings of the Institute of Mechanical Engineers, Part L,* Vol. 216, pp. 133–144.

Lees, J. M., Winistorfer, A. U., and Meier, U. (2002), External prestressed carbon fiber-reinforced polymer straps for shear enhancement of concrete, *Journal of Composites for Construction,* Vol. 6, No. 4, pp. 249–256.

Lopez-Anido, R., Falker, E., Mittlestadt, B., and Troutman, D. (1999), Shear tests on pultruded beam-to-column connections with clip angles, in *Materials and Construction: Exploring the Connection,* ASCE, Reston, VA, pp. 92–99.

Maeda, T., Asano, Y., Sato, Y., Ueda, T., and Kakuta, Y. (1997), A study on bond mechanism of carbon fiber sheet, *Non-Metallic (FRP) Reinforcement for Concrete Structures,* Proceedings of the 3rd Symposium, Sapporo, Japan, Vol. 1, pp. 279–286.

Mander, J. B., Priestley, M. J. N., and Park, R. (1988), Theoretical stress–strain model for confined concrete, *Journal of Structural Engineering,* Vol. 114, No. 8, pp. 1804–1826.

MBrace (1998), *MBrace Composite Strengthening System: Engineering Design Guidelines,* Master Builders, OH; www.mbrace.com.

McCormick, F. C. (1988), Advancing structural plastics into the future, *Journal of Professional Issues in Engineering,* Vol. 114, No. 3, pp. 335–343.

Meier, U. (1986), Future use of advanced composites in bridge construction engineering, presented at the 2nd International Conference on Fibre Reinforced Composites, Liverpool, Lancashire, England, April 8–10, Institution of Mechanical Engineers, London.

——— (1995), Strengthening of concrete structures using carbon fibre/epoxy composites, *Construction and Building Materials,* Vol. 9, pp. 341–351.

Meier, U., and Kaiser, H. (1991), Strengthening structures with CFRP laminates, in *Advanced Composites in Civil Engineering Structures* (ed. S. Iyer and R. Sen), ASCE, Reston, VA, pp. 224–232.

Meyer, L. S. (1970), Pultrusion, *Proceedings of the 25th Annual Technical Conference,* Society for the Plastics Industry, New York, Section 6-A, pp. 1–8.

Meyer, R. W. (1985), *Handbook of Pultrusion Technology,* Chapman & Hall, London.

Merkes, D., and Bank, L. C. (1999), Longitudinal lap splices for pultruded FRP tubes, in *Materials and Construction: Exploring the connection* (ed. L. C. Bank), ASCE, Reston, VA, pp. 60–67.

Michaluk, C. R., Rizkalla, S., Tadros, G., and Benmokrane, B. (1998), Flexural behavior of one-way concrete slabs reinforced by fiber reinforced plastic reinforcement, *ACI Structural Journal,* Vol. 95, No. 3, pp. 353–364.

Mirmiran, A., and Shahawy, M. (1997), Behavior of concrete columns confined by fiber composites, *Journal of Structural Engineering,* Vol. 123, No. 5, pp. 583–590.

Mosallam, A. S., and Bank, L. C. (1991), Creep and recovery of a pultruded FRP frame, in *Advanced Composite Materials for Civil Engineering Structures,* ASCE, Reston, VA, pp. 24–35.

——— (1992), Short-term behavior of pultruded fiber-reinforced plastic frame, *Journal of Structural Engineering,* Vol. 118, pp. 1937–1954.

Mottram, J. T. (1991), Evaluation of design analysis for pultruded fibre-reinforced polymeric box beams, *Structural Engineer,* Vol. 69, pp. 211–220.

——— (1992), Lateral torsional buckling of a pultruded I beam, *Composites,* Vol. 23, pp. 81–92.

——— (1993), Short- and long-term structural properties of pultruded beam assemblies fabricated using adhesive bonding, *Composite Structures,* Vol. 25, pp. 387–395.

——— (2001), Analysis and design of connections for pultruded FRP structures, in *Composites in Construction: A Reality,* ASCE, Reston, VA, pp. 250–257.

——— (2004), Determination of critical load for flange buckling in concentrically loaded pultruded columns, *Composites, Part B: Engineering,* Vol. 35, No. 1, pp. 35–47.

Mottram, J. T., and Turvey, G. J. (eds.) (1998), *State-of-the-Art Review on Design, Testing, Analysis and Applications of Polymeric Composite Connections,* EUR 18172 EN, European Community, Brussels, Belgium.

——— (2003), Physical test data for the appraisal of the design procedures for bolted connections in pultruded FRP structural shapes and systems, *Progress in Structural Engineering and Materials,* Vol. 5, pp. 195–222.

Mottram, J. T., and Zheng, Y. (1996), State-of-the-art review on the design of beam-to-column connections for pultruded structures, *Composite Structures,* Vol. 35, pp. 387–401.

——— (1999a), Further tests on beam-to-column connections for pultruded frames: web-cleated, *Journal of Composites for Construction,* Vol. 3, pp. 3–11.

——— (1999b), Further tests on beam-to-column connections for pultruded frames: flange-cleated, *Journal of Composites for Construction,* Vol. 3, pp. 108–116.

Mottram, J. T., Brown, N. D., and Anderson, D. (2003a), Physical testing for concentrically loaded columns of pultruded glass fibre reinforced plastic profile, *Proceedings of the Institution of Civil Engineers, Structures, and Buildings,* 156, pp. 205–219.

——— (2003b), Buckling characteristics of pultruded glass fibre reinforced plastic columns under moment gradient, *Thin-walled structures,* Vol. 41, pp. 619–638.

Nagaraj, V., and GangaoRao, H. V. S. (1997), Static behavior of pultruded GFRP beams, *Journal of Composites for Construction,* Vol. 1, pp. 120–129.

Nanni, A. (1993a), Flexural behavior and design of reinforced concrete using FRP rods, *Journal of Structural Engineering,* Vol. 119, No. 11, pp. 3344–3359.

——— (ed.) (1993b), *Fiber-Reinforced Plastic (FRP) for Concrete Structures: Properties and Applications,* Elsevier Science, New York.

——— (1995), Concrete repair with externally bonded FRP reinforcement: examples from Japan, *Concrete International,* June, pp. 22–26.

Nanni, A., and Bradford, N. M. (1995), FRP jacketed concrete under uniaxial compression, *Construction and Building Materials,* Vol. 9, No. 2, pp. 115–124.

Nanni, A., and Norris, M. S. (1995), FRP jacketed concrete under flexure and combined flexure–compression, *Construction and Building Materials,* Vol. 9, No. 5, pp. 273–281.

Nanni, A., Bakis, C. E., and Boothby, T. E. (1995), Test methods for FRP-concrete systems subjected to mechanical loads: state of the art review, *Journal of Reinforced Plastics and Composites,* Vol. 14, No. 6, pp. 524–558.

Nanni, A., and Tumialan, G. (2003), Fiber reinforced composites for the strengthening of masonry structures, *Structural Engineering International,* Vol. 13, No. 4, November, pp. 271–278.

Nawy, E. G. (2003), *Reinforced Concrete: A Fundamental Approach,* 5th ed., Prentice Hall, Upper Saddle River, NJ.

Nawy, E. G., and Neuwerth, G. E. (1977), Fiberglass reinforced concrete slabs and beams, *Journal of Structural Engineering,* Vol. 103, No. ST2, pp. 421–440.

Nawy, E. G., Neuwerth, G. E., and Phillips, C. J. (1971), Behavior of fiberglass reinforced concrete beams, *Journal of Structural Engineering,* Vol. 97, No. ST9, pp. 2203–2215.

NCHRP (2004), *Bonded Repair and Retrofit of Concrete Structures Using FRP Composites,* National Cooperative Highway Research Program, Report 514, Transportation Research Board, Washington, DC.

NFPA (2003), *Building Construction and Safety Code,* NFPA 5000, National Fire Protection Association, Quincy, MA.

Nilson, A. H., Darwin, D., and Dolan, C. W. (2004), *Design of Concrete Structures,* 13th ed., McGraw Hill, New York.

Nowak, A. S., and Collins (2000), *Reliability of Structures,* McGraw-Hill, New York.

Ospina, C. E. (2005), Alternative model for concentric punching capacity evaluation of reinforced concrete two-way slabs, *Concrete International,* September, pp. 55–57.

Ozel, M., Bank, L. C., Arora, D., Gonenc, O., Gremel, D., Nelson, B., and McMonigal, D. (2003), Comparison between FRP rebar, FRP grid and steel rebar reinforced concrete beams, *Proceedings of the 6th International Symposium on FRP Reinforcement for Concrete Structures: FRPRCS6* (ed. K. H. Tan), Singapore, July 8–10, World Scientific, Singapore, pp. 1067–1076.

Pantelides, C. P., Nadauld, J., and Cercone, L. (2003), Repair of cracked aluminum overhead sign structures with glass fiber reinforced polymer composites, *Journal of Composites for Construction,* Vol. 7, No. 2, pp. 118–126.

Park, R. J. T., Priestley, M. J. N., and Walpole, W. R. (1983), The seismic performance of steel encased reinforced concrete bridge piles, *Bulletin of the New Zealand National Society for Earthquake Engineering,* Vol. 16, No. 2, pp. 123–140.

Paterson, J., and Mitchell, D. (2003), Seismic retrofit of shear walls with headed bars and carbon fiber wrap, *Journal of Structural Engineering,* Vol. 129, No. 5, pp. 606–614.

Paulay, T., and Priestley, M. J. N. (1992), *Seismic Design of Reinforced Concrete and Masonry Buildings,* Wiley, New York.

Pecce, M., and Cosenza, E. (2000), Local buckling curves for the design of FRP profiles, *Thin-Walled Structures,* Vol. 37, pp. 207–222.

Pleimann, L. G. (1991), Strength, modulus and bond of deformed FRP rods, in *Advanced Composites for Civil Engineering Structures* (ed. L. Iyer, and R. Sen), ASCE, Reston, VA, pp. 99–110.

Prabhakaran, R., Razzaq, Z., and Devera, S. (1996), Load and resistance factor design (LRFD) approach to bolted joints in pultruded structures, *Composites: Part B,* Vol. 27B, pp. 351–360.

Priestley, M. J. N., and Park, R. (1987), Strength and ductility of concrete bridge columns under seismic loading, *ACI Structural Journal,* Vol. 84, No. 1, pp. 61–76.

Priestley, M. J. N., and Seible, F. (1995), Design of seismic retrofit measures for concrete and masonry structures, *Construction and Building Materials,* Vol. 9, No. 6, pp. 365–377.

Priestley, M. J. N., Seible, F., and Fyfe, E. R. (1992), Column seismic retrofit using fiberglass/epoxy jackets, in *Advanced Composite Materials in Bridges and Structures* (*ACMBS*) (ed. K. W. Neale, and P. Labossiére), Canadian Society of Civil Engineering, Montreal, Quebec, Canada, pp. 287–298.

Priestley, M. J. N., Seible, F., and Calvi, G. M. (1996), *Seismic Design and Retrofit of Bridges,* Wiley, New York.

Qiao, P., Davalos, J. F., and Wang, J. (2001), Local buckling of composite FRP shapes by discrete plate analysis, *Journal of Structural Engineering,* Vol. 127, No. 3, pp. 245–255.

Raasch, J. E. (1998), All-composite construction system provides flexible low-cost shelter, *Composite Technology,* March–April, pp. 56–58.

Razzaq, Z., Prabhakaran, R., and Sirjani, M. M. (1996), Load and resistance factor design (LRFD) approach for reinforced-plastic channel beam buckling, *Composites: Part B,* Vol. 27B, pp. 361–369.

Replark (1999), *Replark System: Technical Manual,* Mitsubishi Chemical Corporation, Sumitomo Corporation of America, New York; www.sumitomocorp.com.

Rizzo, A., Galati, N., Nanni, A., and Dharani, L. R. (2005), Material characterization of FRP pre-cured laminates used in the mechanically fastened FRP strengthening of RC structures, *Proceedings of the 7th International Symposium on FRP Reinforcement for Concrete Structures FRPRCS7,* ACI, SP-230, Farmington Hills, MI, pp. 135–152.

Roberts, T. M. (2002), Influence of shear deformation on buckling of pultruded fiber reinforced plastic profiles, *Journal of Composites for Construction,* Vol. 6, pp. 241–248.

Roberts, T. M., and Al-Ubaidi, H. (2002), Flexural and torsional properties of pultruded fiber reinforced plastic I-profiles, *Journal of Composites for Construction,* Vol. 6, No. 1, pp. 28–34.

Rosner, C. N., and Rizkalla, S. H. (1995a), Bolted connections for fiber-reinforced composite structural members: experimental program, *Journal of Materials in Civil Engineering,* Vol. 7, pp. 223–231.

——— (1995b), Bolted connections for fiber-reinforced composite structural members: analytical model and design recommendations, *Journal of Materials in Civil Engineering,* Vol. 7, pp. 232–238.

S&P (1998), *Clever Reinforcement Company,* Schere & Partners, Brunnen, Switzerland; www.sp-reinforcement.ch.

Saadatmanesh, H. (1994), Fiber composites for new and existing structures, *ACI Structural Journal,* Vol. 91, No. 3, pp. 346–354.

Saadatmanesh, H., and Ehsani, M. R. (1991), RC beams strengthened with GFRP plates, I: experimental study, *Journal of Structural Engineering,* Vol. 117, No. 11, pp. 3417–3433.

Saadatmanesh, H., Ehsani, M. R., and Li, M. W. (1994), Strength and ductility of concrete columns externally reinforced with fiber composite straps, *ACI Structural Journal,* Vol. 91, No. 4, pp. 434–447.

Saha, M., Prabhakaran, R., and Waters, W. A. (2004), Compressive behavior of pultruded composite plates with holes, *Composite Structures,* Vol. 65, pp. 29–36.

Salmon, C. G., and Johnson, J. E. (1996), *Steel Structures: Design and Behavior,* 4th ed., HarperCollins, New York.

Samaan, M., Mirmiran, A., and Shahawy, M. (1998), Model of concrete confined by fiber composites, *Journal of Structural Engineering,* Vol. 124, No. 9, pp. 1025–1031.

Schwartz, M. M. (1997a), *Composite Materials, Vol. I, Properties, Nondestructuve Testing and Repair,* Prentice Hall, Upper Saddle River, NJ.

——— (1997b), *Composite Materials, Vol. II, Processing, Fabrication and Applications,* Prentice Hall, Upper Saddle River, NJ.

Schwegler, G. (1994), Masonry construction strengthened with fiber composites in seismically endangered zones, in *Proceedings of the 10th European Conference on Earthquake Engineering,* Vienna, Austria, pp. 2299–2303.

Scott, D. W., and Zureick, A. (1998), Compression creep of a pultruded E-glass/vinylester composite, *Composites Science and Technology,* Vol. 98, pp. 1361–1369.

Scott, D. W., Zureick, A., and Lai, J. S. (1995), Creep behavior of fiber reinforced polymeric composites, *Journal of Reinforced Plastics and Composites,* Vol. 14, pp. 590–617.

Seible, F., Priestley, J. N., Hegemeier, G. A., and Innamorato, D. (1997), Seismic retrofit of RC columns with continuous carbon fiber jackets, *Journal of Composites for Construction,* Vol. 1, No. 2, pp. 52–62.

Seymour, R. B. (1987), *Polymers for Engineering Applications,* ASM International, Materials Park, OH.

Shao, Y., and Shanmugam, J. (2004), Deflection creep of pultruded composite sheet piling, *Journal of Composites for Construction,* Vol. 8, No. 5, pp. 471–479.

Shield, C. K., Busel, J. P., Walkup, S. L., and Gremel, D., (eds.) (2005), *7th International Symposium on Fiber-Reinforced (FRP) Polymer Reinforcement for Concrete Structures,* Kansas City, SP-230, American Concrete Institute, Farmington Hills, MI.

Sika (1997), *Sika Carbodur: Engineering Guidelines for the Use of Sika Carbodur (CFRP) Laminates for Structural Strengthening of Concrete Structures,* Sika Corporation, Lyndhurst, NJ; www.sikaconstruction.com.

——— (2000), *Sika Carbodur: Carbon Fiber Laminate for Structural Strengthening*, Sika Corporation, Lyndhurst, NJ; www.sikausa.com.

Smallowitz, H. (1985), Reshaping the future of plastic buildings, *Civil Engineering, ASCE,* May, pp. 38–41.

Smith, C. (2002), A comparison of A&E glass mats for pultrusion, *Composites Fabrication,* July, pp. 12–19.

Smith, C., McEwen, S., and Tillson, J. (1998), The effect of different continuous strand mats on pultrusion property performance, *Proceedings of the International Composites Expo,* Nashville, TN, January 19–21, Section 19A, pp. 1–10.

Smith, S. J., Parsons, I. D., and Hjelmstad, K. D. (1999), Experimental comparison of connections for GFRP pultruded frames, *Journal of Composites for Construction,* Vol. 3, pp. 20–26.

Sonobe et al. (1997), Design guidelines of FRP reinforced concrete building structures, *Journal of Composites for Construction,* Vol. 1, pp. 90–115.

Sourcebook, (2006), *Sourcebook Industry Directory,* Ray Publishing, Wheat Ridge, CO.

Spadea, G., Bencardino, F., and Swamy, R. N. (1998), Structural behavior of composite RC beams with externally bonded CFRP, *Journal of Composites for Construction,* Vol. 2, No. 3, pp. 132–137.

Spoelstra, M. R., and Monti, G. (1999), FRP-confined concrete model, *Journal of Composites for Construction,* Vol. 3, No. 3, pp. 143–150.

Starr, T. (ed.), (2000), *Pultrusion for Engineers,* CRC Press, Boca Raton, FL.

Steffen, R., Scott, D., Goodspeed, C., Bowman, M., and Trunfio, J. 2003, Design issues and constructibility of a CFRP grid reinforced bridge deck, *Bridge Materials 2001: High Performance Materials in Bridges* (ed. A. Azizinamini, A.Yakel, and M. Abdelrahman), ASCE, Reston, VA, pp. 106–116.

Strongwell (2002), *Strongwell Design Manual* (CD ROM), Strongwell, Bristol, VA; www.strongwell.com.

Sugita, M. (1993), NEFMAC: grid type reinforcement, in *Fiber-Reinforced-Plastic (FRP) Reinforcement for Concrete Structures: Properties and Applications* (ed. A. Nanni), Elsevier, Amsterdam, pp. 355–385.

Szerszen, M. M., and Nowak, A. S. (2003), Calibration of design code for buildings (ACI 318), Part 2: Reliability analysis and resistance factors, *ACI Structural Journal,* Vol. 100, No. 3, pp. 383–391.

Taerwe, L. R., and Matthys, S. (1999), FRP for concrete construction: activities in Europe, *Concrete International,* October, pp. 33–36.

Täljsten, B. (1997), Strengthening of beams by plate bonding, *Journal of Materials in Civil Engineering,* Vol. 9, No. 4, pp. 206–212.

——— (2004), *FRP Strengthening of Existing Concrete Structures: Design Guideline,* 3rd ed., Lulea University of Technology, Lulea, Sweden.

Tan, K. H. (ed.) (2003), Fibre-reinforced polymer reinforcement for concrete structures, *Proceedings of the 6th International Symposium on FRP Reinforced Concrete Structures: FRPRCS6,* Singapore, World Scientific, Singapore.

Tan, K. H., and Patoary, M. K. H. (2004), Strengthening of masonry walls against out-of-plane loads using fiber-reinforced polymer reinforcement, *Journal of Composites for Construction,* Vol. 8, No. 1, pp. 79–87.

Teng, J. G., Chen, J. F., Smith, S. T., and Lam, L. (2001), *FRP Strengthened RC Structures,* Wiley, New York.

Thériault, M., Neale, K. W., and Claude, S. (2004), Fiber-reinforced polymer-confined circular concrete columns: investigation of size and slenderness effects, *Journal of Composites for Construction,* Vol. 8, No. 4, pp. 323–331.

Timoshenko, S. P. (1921), On the correction for shear of the differential equation for transverse vibration of prismatic bars, *Philosophical Magazine,* Vol. 41, pp. 744–746.

Timoshenko, S., and Gere, J. (1961), *Theory of Elastic Stability,* 2nd ed., McGraw-Hill, New York.

Timoshenko, S., and Woinowsky-Krieger, S. (1959), *Theory of Plates and Shells,* McGraw-Hill, New York.

Tingley, D. A., Gai, C., and Giltner, E. E. (1997), Testing methods to determine properties of fiber reinforced plastic panels used for reinforcing glulams, *Journal of Composites for Construction,* Vol. 1, No. 4, pp. 160–167.

Todeshini, C. E., Bianchini, A. C., and Kesler, C. E. (1964), Behavior of concrete columns with high strength reinforced steels, *Journal of the American Concrete Institute, Proceedings,* Vol. 61, pp. 701–715.

Tomblin, J. and Barbero, E. (1994), Local buckling experiments on FRP columns, *Thin-walled Structures,* Vol. 18, pp. 97–116.

Tonen (1996), *Forca Towsheet Technical Manual,* Rev. 5.0, Tonen Corporation, Tokyo.

Torres-Acosta, A. A. (2002), Galvanic corrosion of steel in contact with carbon–polymer composites, II: Experiments in concrete, *Journal of Composites for Construction,* Vol. 6, No. 2, pp. 116–122.

TR 55 (2004), *Design Guidance for Strengthening Concrete Structures Using Fibre Composite Materials,* The Concrete Society, Camberley, Surrey, England.

TR 57 (2003), *Strengthening Concrete Structures with Fibre Composite Materials: Acceptance, Inspection and Monitoring,* The Concrete Society, Camberley, Surrey, England.

Triantafillou, T. (1998), Shear strengthening of reinforced concrete beams using epoxy-bonded FRP composites, *ACI Structural Journal,* Vol. 95, pp. 107–115.

Triantafillou, T. C., Deskovic, N., and Deuring, M. (1992), Strengthening of concrete structures with prestressed fiber reinforced plastic sheets, *ACI Structural Journal,* Vol. 89, No. 3, pp. 235–244.

Tsai, S. W. (1988), *Think Composites,* 4th ed., Think Composites Publishers, Dayton, OH.

Tsai, S., and Hahn, T. (1980), *Introduction to Composite Materials,* Technomic, Lancaster, PA.

Tureyen, A. K., and Frosch, R. J. (2002), Shear tests of FRP-reinforced concrete beams without stirrups, *ACI Structural Journal,* Vol. 99, No. 4, pp. 427–434.

——— (2003), Concrete shear strength: another perspective, *ACI Structural Journal,* Vol. 10, No. 5, pp. 609–615.

Turvey, G. J. (1996), Effects of load position on the lateral buckling response of pultruded GRP cantilevers: comparisons between theory and experiment, *Composite Structures,* Vol. 35, pp. 33–47.

——— (1998), Single-bolt tension tests on pultruded GRP plate: effects of tension direction relative to pultrusion direction, *Composite Structures,* Vol. 42, pp. 341–351.

——— (2000), Bolted connections in PFRP structures, *Progress in Structural Engineering and Materials,* Vol. 2, pp. 146–156.

——— (2001), Pultruded GRP frames: simple (conservative) approach to design, a rational alternative and research needs for improved design, in *Composites for Construction: A Reality,* ASCE, Reston, VA, pp. 258–266.

Turvey, G. J., and Cooper C. (1996), Characterization of the short-term static moment-rotation response of bolted connections between pultruded GRP beam and column WF-sections, *Proceedings of Advanced Composite Materials in Bridges and Structures 2,* Montreal, Quebec, Canada, pp. 295–300.

——— (2004), Review of tests on bolted joints between pultruded GRP profiles, *Proceedings of the Institute of Civil Engineers, Structures and Buildings,* Vol. 157, No. SB3, pp. 211–233.

Turvey, G. J., and Wang, P. (2003), Open hole tension strength of pultruded GRP plate, *Proceedings of the Institution of Civil Engineers, Structures and Buildings,* Vol. 156, No. 1, pp. 93–101.

Turvey, G. J., and Zhang, Y. (2005), Tearing failure in web–flange junctions in pultruded GRP profiles, *Composites: Part A,* Vol. 36, pp. 309–317.

Tyfo (1998), *Design Manual for the Tyfo Fibrwrap System,* Fyfe Co. LLC, San Diego, CA; www.fyfeco.com.

Van Den Einde, L., Zhao, L., and Seible, F. (2003), Use of FRP composites in civil structural applications, *Construction and Building Materials,* Vol. 17, pp. 389–403.

Wambeke, B. W., and Shield, C. K. (2006), Development length of glass fiber reinforcing bars in concrete, *ACI Structural Journal,* Vol. 103, No. 1, pp. 11–17.

Wang, Y. (2002), Bearing behavior of joints in pultruded composites, *Journal of Composite Materials,* Vol. 36, pp. 2199–2216.

Wang, C.-K., and Salmon, C. G. (2002), *Reinforced Concrete Design,* 6th ed., Wiley, New York.

Wang, Y., and Zureick, A. (1994), Characterization of the longitudinal tensile behavior of pultruded I-shape structural members using coupon specimens, *Composite Structures,* Vol. 29, pp. 463–472.

White, J. R., and Turnbull, A. (1994), Weathering of polymers: mechanisms of degradation and stabilization, testing strategies and modeling, *Journal of Materials Science,* Vol. 29, No. 3, pp. 584–613.

Whitney, C. S. (1937), Design of reinforced concrete members under flexure or combined flexure and direct compression, *Journal of the American Concrete Institute,* Vol. 33, pp. 483–498.

Xiao, Y., and Ma, R. (1997), Seismic retrofit of RC circular columns using prefabricated composite jacketing, *Journal of Structural Engineering,* Vol. 123, No. 10, pp. 1357–1364.

Yeh, H. S., and Yang, S. C. (1997), Building a composite transmission tower, *Journal of Reinforced Plastics and Composites,* Vol. 16, No. 5, pp. 414–424.

Yoon, S. J., Scott, D. W., and Zureick, A. (1992), An experimental investigation of the behavior of concentrically loaded pultruded columns, in *Advanced Composite*

Materials in Bridges and Structures (*ACMBS*) (ed. K. W. Neale and P. Labossiére), Canadian Society of Civil Engineering, Montreal, Quebec, Canada, pp. 309–317.

Yost, J. R., Gross, S. P., and Dinehart, D. W. (2001), Shear strength of normal strength concrete beams reinforced with deformed GFRP bars, *Journal of Composites for Construction,* Vol. 5, No. 4, pp. 268–275.

Zureick, A. (1998), FRP pultruded structural shapes, *Progress in Structural Engineering and Materials,* Vol. 1, No. 2, 143–149.

Zureick, A., and Scott, D. (1997), Short-term behavior and design of fiber-reinforced polymeric slender members under axial compression, *Journal of Composites for Construction,* Vol. 1, pp. 140–149.

Zureick, A., and Steffen, R. (2000), Behavior and design of concentrically loaded pultruded angles struts, *Journal of Structural Engineering,* Vol. 126, pp. 406–416.

INDEX

Lightning Source UK Ltd.
Milton Keynes UK
UKOW03n0928241113

221628UK00002B/145/P

9 780471 681267